Utopian Universities

Utopian Universities

*A Global History of the New Campuses
of the 1960s*

Edited by
Jill Pellew and Miles Taylor

BLOOMSBURY ACADEMIC
LONDON • NEW YORK • OXFORD • NEW DELHI • SYDNEY

BLOOMSBURY ACADEMIC
Bloomsbury Publishing Plc
50 Bedford Square, London, WC1B 3DP, UK
1385 Broadway, New York, NY 10018, USA

BLOOMSBURY, BLOOMSBURY ACADEMIC and the Diana logo
are trademarks of Bloomsbury Publishing Plc

First published in Great Britain 2021

Copyright © Miles Taylor and Jill Pellew, 2021

For legal purposes the Acknowledgements on p. xv constitute an
extension of this copyright page.

Cover design: Terry Woodley
Cover photo © Douglas Miller/Keystone/Hulton
Archive/Getty Images.

Jill Pellew and Miles Taylor have asserted their right under the Copyright, Designs and
Patents Act, 1988, to be identified as Author of this work.

All rights reserved. No part of this publication may be reproduced or
transmitted in any form or by any means, electronic or mechanical, including
photocopying, recording, or any information storage or retrieval system,
without prior permission in writing from the publishers.

Bloomsbury Publishing Plc does not have any control over, or responsibility for,
any third-party websites referred to or in this book. All internet addresses given
in this book were correct at the time of going to press. The author and publisher
regret any inconvenience caused if addresses have changed or sites have ceased
to exist, but can accept no responsibility for any such changes.

A catalogue record for this book is available from the British Library.

Library of Congress Cataloging-in-Publication Data
Names: Taylor, Miles, editor. | Pellew, Jill, editor.
Title: Utopian universities : a global history of the new campuses of the
1960s / edited by Miles Taylor and Jill Pellew.
Description: London ; New York : Bloomsbury Academic, 2020. | Includes
bibliographical references and index.
Identifiers: LCCN 2020027583 (print) | LCCN 2020027584 (ebook) | ISBN
9781350138636 (hardback) | ISBN 9781350138643 (ebook) | ISBN
9781350138650 (epub)
Subjects: LCSH: Public universities and colleges–History–20th century.
Classification: LCC LB2328.6 .U86 2020 (print) | LCC LB2328.6 (ebook) |
DDC 378/.05–dc23
LC record available at https://lccn.loc.gov/2020027583
LC ebook record available at https://lccn.loc.gov/2020027584

ISBN: HB: 978-1-3501-3863-6
       ePDF: 978-1-3501-3864-3
       eBook: 978-1-3501-3865-0

Typeset by Integra Software Services Pvt. Ltd.

To find out more about our authors and books visit www.bloomsbury.com
and sign up for our newsletters.

In memory of two founding fathers
Geoff Taylor (1928–2009; University of Waikato, 1966–70)
Frank Thistlethwaite (1915–2003; University of East Anglia, 1961–80)

Contents

List of Illustrations	ix
List of Contributors	xi
Acknowledgements	xv
Foreword *by Laurie Taylor*	xviii
Introduction *by Jill Pellew and Miles Taylor*	1

Part 1

1. Learning from Redbrick: Utopianism and architectural legacy of the civic universities *by William Whyte* — 21
2. Keele: Post-war pioneer *by Miles Taylor* — 37
3. Sussex: Cold War campus *by Matthew Cragoe* — 55
4. 'A poor man's Oxbridge'?: The founding of the University of York, 1960–73 *by Allen Warren* — 73
5. The University of East Anglia: From mandarins to neo-liberalism *by John Charmley* — 89
6. Great expectations: Sloman's Essex and student protest in the long sixties *by Caroline Hoefferle* — 105
7. Science and the new universities *by Jon Agar* — 121
8. The new and the old: The University of Kent at Canterbury *by Krishan Kumar* — 141
9. Social history comes to Warwick *by Carolyn Steedman* — 157
10. Innovation and evolution: Lancaster learning, 1964–74 *by Marion McClintock* — 175
11. Failed utopia? The University of Stirling from the 1960s to the early 1980s *by Holger Nehring* — 191
12. The New University of Ulster and the Northern Ireland crisis *by Thomas G. Fraser and Leonie Murray* — 207
13. The new British campus universities of the 1960s and their localities: The culture of support and the role of philanthropy *by Jill Pellew* — 223

Part 2

14 California dreaming: Clark Kerr and the University of California
    by Christopher Newfield   251
15 Utopian universities of the British Commonwealth *by Miles Taylor*   269
16 The other 1960s: University leaders as agents of change in Canadian
    higher education *by Paul Axelrod*   287
17 From progressive pedagogy to 'capitalist fodder': The new universities in
    Australia *by Hannah Forsyth*   305
18 Jawaharlal Nehru University: A University for the Nation *by Rajat Datta
    and Shalini Sharma*   323
19 From American dream to nightmare on the left: Student revolts,
    the 'crèche sauvage' and the slums: The University of Nanterre,
    1962–71 *by Victor Collet*   341
20 The reform universities of West Germany: Bochum, Konstanz and
    Bielefeld *by Stefan Paulus*   357

Afterword: The utopian universities in historical perspective
by Peter Mandler   375

Index   380

Illustrations

Figures

1 Hans Werner Rothe's ideal campus (1961) (Rothe, *Über die Gründung einer Universität in Bremen* from Rolf Neuhaus (ed.), *Dokumente zur Gründung neuer Hochschulen. Empfehlungen und Denkschriften auf Veranlassung von Länden in der Bundesrepublik Deutschland in den Jahren 1960–1966* (Wiesbaden: Franz Steiner Verlag, 1968)   240
2 Students astride a walkway over the 'moat' at the University of Sussex, c. 1963 (© Yves Fedida)   240
3 James Birrell's James Cook University Library (James Birrell Archive, James Cook University Library Special Collections)   241
4 Julian Elliott and Anthony Chitty's University of Zambia campus (© Dr Ruth Craggs)   241
5 Arthur Erickson's University Hall, University of Lethbridge, Canada (Courtesy of the University of Lethbridge Archives)   242
6 Sir Keith Murray receiving an honorary degree from the University of Western Australia, 1963 (National Archives of Australia)   242
7 University of East Anglia, Norwich: an exhibition display showing a model of the proposed university being scrutinised by visitors (Lasdun Archive, RIBA Collections)   243
8 'Some people who matter, and some who think they do' by 'Tom', Essex, Winter 1968 (by permission of Tom Brass)   244
9 Pradip Nayak, Founding President of the Students' Representative Council, University of York, 1964 (© *Yorkshire Post*)   245
10 John Fulton receiving an honorary degree from the Chinese University of Hong Kong, 1964 (Courtesy of the Chinese University of Hong Kong)   246
11 Demonstration against Ronald Reagan's sacking of Clark Kerr, University of Irvine, 23 January 1967 (University Communications photographs. AS-061 (Box 37, A67-002) Special Collections and Archives, University of California Irvine Libraries)   246
12 'University for Derry' protest march, 18 February 1965 (© *Derry Journal*)   247

13 Siné (Maurice Sinet), 'Debout les damnés de Nanterre', *Action*, 7 mai 1968 (© Éditions du Crayon) 247
14 Sitaram Yechury, President of the Jawaharlal Nehru University Students' Union, reading a memorandum to Indira Gandhi, demanding her resignation as chancellor of the university 5 September 1977 (© *Hindustan Times*) 248
15 The Queen's visit to the University of Stirling, 12 October 1972 (© PA Images) 248

Tables

13.1 Sources of university income in Britain, before and after the Second World War, 1938/9 and 1950/51 224
13.2 Recurrent grants from public funds (Exchequer and local authority) for three universities over early ten-year period 228
13.3 Appeal campaigns in the new universities 232
13.4 Income from research grants and contracts to Britain's new universities in their early years showing its percentage of total income 233
18.1 Rural-urban distribution of new entrants, 1983–84 to 2015–16 332
18.2 Gender distribution of students across all disciplines 333
18.3 Profile of social reservations for students 333
18.4 Gender and social distribution of faculty 334
18.5 Admissions at JNU, 2008–14 334

Maps

1 New public universities founded worldwide, 1961–70 xvii
2 New public universities founded in Europe, 1961–70 xvii

Contributors

Jon Agar is Professor of Science and Technology Studies at University College London. He researches and writes on the history of recent science and technology, including radio astronomy (*Science and Spectacle*, 1994), mobile phones (*Constant Touch*, 2003), government and technology (*The Government Machine*, 2003), and is also the author of a synthetic overview of the subject in *Science in the Twentieth Century and beyond* (2012).

Paul Axelrod is Professor and former Dean in the Faculty of Education at York University, Canada. His main books include: *Values in Conflict: The University, the Marketplace and the Trials of Liberal Education* (2002) and *Making a Middle Class: Student Life in English Canada during the Thirties* (1990).

John Charmley is Pro Vice-Chancellor for Academic Strategy at St Mary's University, Twickenham. Previously he was at the University of East Anglia, where he taught since 1979, and was latterly Head of the Interdisciplinary Institute for the Humanities. He is the author of a series of influential books on mid-twentieth-century foreign policy, and most recently has co-edited *Documents on Conservative Foreign Policy, 1852–1878* (2012). He is also the author of *A History of Conservative Politics since 1830* (2008).

Victor Collet was awarded a PhD in Political Science from the University of Paris Ouest in 2013, now published as *Nanterre, du bidonville à la cité* (2019). He is currently a research officer for the 'The past in the present' Labex, for which he participates in the collective project 'Exploring the history of the University of Nanterre'.

Matthew Cragoe is Professor of Modern History at the University of Lincoln where he is also Pro Vice-Chancellor (Arts). He has published widely in the social and political history of nineteenth- and twentieth-century Britain, including most recently a study of Asa Briggs' time at the University of Sussex, and a co-edited volume, *The Land Question in Britain, 1750–1950* (2010).

Rajat Datta is Professor of History at the Centre for Historical Studies, Jawaharlal Nehru University, New Delhi. He studied at JNU and King's College London. His books include *Society, Economy and the Market: Commercialization in Rural Bengal, c. 1760–1800* (2000), *Rethinking a Millennium: Perspectives on Indian History from the Eighth to the Eighteenth Century* (2008), and the forthcoming *Market, Subsistence and Transition in Early Modern India: Perspectives from Eighteenth Century Bengal*.

Hannah Forsyth is Senior Lecturer in History at the Australian Catholic University in Sydney. She is the author of *A History of the Modern Australian University* (2014). She is currently working on the political economics of 'survival' and the structures of opportunity in rural Australia, especially for Aboriginal communities.

Thomas G. Fraser is Emeritus Professor of History at Ulster University. He worked at the New University of Ulster/University of Ulster from 1969 to 2006, retiring as Provost of its Magee campus. His most recent books being: edited, with John Hume and Leonie Murray, *Peacemaking in the Twenty-first Century* (2013); *The Arab–Israeli Conflict* (4th edition, revised and expanded, 2015); (ed.) *The First World War and Its Aftermath: The Shaping of the Modern Middle East* (2015); and (co-ed.) *Ethics, Governance and Corporate Social Responsibility in India: Issues and Perspectives* (2016).

Caroline Hoefferle is a Professor of History at Wingate University, North Carolina. She graduated from Central Michigan University and the University of Strathclyde Joint PhD in History Programme in 2000. Her books include *British Student Activism in the Long Sixties* (2012).

Krishan Kumar is University Professor and William R. Kenan, Jr., Professor of Sociology at the University of Virginia, the United States. He was previously Professor of Social and Political Thought at the University of Kent at Canterbury, where he taught for many years. Among his publications are *Utopia and Anti-utopia in Modern Times* (1987), *From Post-industrial to Post-modern Society*, 2nd ed. (2005), and *The Making of English National Identity* (2003). A volume of his essays *The Idea of Englishness: English Culture, National Identity and Social Thought* was published in 2015.

Peter Mandler is Professor of Modern Cultural History at the University of Cambridge and a fellow of Gonville and Caius College. Between 2012 and 2016 he was President of the Royal Historical Society. His most recent book is *The Crisis of the Meritocracy: Britain's Transition to Mass Education since the Second World War* (2020).

Marion McClintock has written two histories of the University of Lancaster: *University of Lancaster: Quest for Innovation* (1974) and *Shaping the Future: A History of the University of Lancaster 1961–2011* (2011). She is currently Honorary University Archivist and Honorary Fellow at Lancaster.

Leonie Murray is Lecturer in International Politics at Ulster University. She works on modern American foreign policy. She is co-editor of *Peacemaking in the Twenty-first Century* (2013).

Holger Nehring is Professor of Contemporary European History at Stirling. He studied at Tübingen, LSE and Oxford, and previously taught at the University of Sheffield. He is the author of *Politics of Security. The British and West German Protests against Nuclear Weapons and the Early Cold War, 1945–1970* (Oxford, 2013) and works on 'Cold War Environments: Infrastructure, Landscape, and Material Culture in the UK's Gun Belt'.

Christopher Newfield is Professor of Literature and American Studies at the University of California at Santa Barbara. His books include *Ivy and Industry: Business and the Making of the American University, 1880–1980* (2003), and *Unmaking the Public University: The Forty Year Assault on the Middle Class* (2008), and, most recently, *The Great Mistake: How We Wrecked Public Universities and How We Can Fix Them* (2016).

Stefan Paulus is currently a Research Fellow at the University of Augsburg. He has published on the history of higher education in the post-1945 federal republic, including a monograph, *Vorbild USA?: 'Amerikanisierung' von Universität und Wissenschaft in Westdeutschland 1945–1976* (2010).

Jill Pellew is a Senior Research Fellow at the Institute of Historical Research. She is a student of British institutions, in particular the Civil Service (Home Office) and universities. She is currently working on the role of benefactors in the founding of British universities. Recently, with Lawrence Goldman, she co-edited *Dethroning Historical Reputations: Universities, Museums and the Commemoration of Benefactors (2018)*.

Shalini Sharma studied at Cambridge, JNU and SOAS. Since 2005 she has been lecturer in colonial and post-colonial history at Keele University. She is the author of *Radical Politics in Colonial Punjab: Governance and Sedition* (2009), and is currently working on a book about America and Indian cultural and intellectual life after 1947.

Carolyn Steedman is Emeritus Professor of History at the University of Warwick. She came to Warwick in 1984 and joined the Centre for Social History as Reader in 1995, before becoming its last Director in 1998–9. Her books include *Landscape for a Good Woman: A Story of Two Lives* (1986), *Margaret McMillan. Childhood, Culture and Class in Britain, 1860–1931* (1990), *Dust: The Archive and Cultural History* (2001), and *An Everyday Life of the English Working Class* (2013). Her most recent book is *Poetry for Historians: or, W. H. Auden and History* (2017).

Laurie Taylor is a broadcaster (BBC Radio 4) and columnist (*Times Higher*). He taught sociology at the University of York from 1965 until retirement in 1993.

Miles Taylor is Professor of Modern History at the University of York. He was Director of the Institute of Historical Research from 2008 to 2014. His recent books include (ed.) *The Age of Asa: Lord Briggs, Public Life and History since 1945* (2015) and *Empress: Queen Victoria and India* (2018).

Allen Warren was educated at Bristol Grammar School and New College, Oxford. After a junior research fellowship at Oxford he joined the History Department at York in 1971, subsequently serving as Head of Department from 1996 to 2003. He was also Provost of Vanbrugh College from 1984 until 2008.

William Whyte is Professor of Social and Architectural History and Fellow of St John's College, Oxford. He is the author of *Oxford Jackson: Architecture, Education, Status, and Style, 1835–1924* (Oxford, 2006) and *Redbrick: A Social and Architectural History of Britain's Civic Universities* (Oxford, 2013). A member of the International Commission for the History of Universities, he is currently writing *The University: A Material History* for publication by Harvard University Press.

Acknowledgements

This volume is derived from a project that ran for two years at the Institute of Historical Research in London and then moved to the University of York. Generous financial support came from a variety of sources, and it is a pleasure to acknowledge them here: the British Society for the History of Science; Shepherd, Epstein & Hunter; the Shepherd Building Group, York; the UK Canada Foundation; and the York Civic Trust. Two anonymous donors made all the difference too. The vice-chancellors of the 'new' universities of East Anglia, Essex, Keele, Kent, Lancaster, Stirling, Sussex, Ulster and Warwick all contributed as well. At York, the then Dean of the Arts & Humanities, Mark Ormrod provided university backing, as did the Centre for Modern Studies. Publication has been supported by an award from the Scouloudi Foundation.

Our book has benefitted immeasurably from the peer-review undertaken of the project by Lawrence Black and Michael Shattock. Peter Mandler has, as ever, been a critical friend of critical importance throughout. Conversations with Sylvie April, Jonathan Beecher, the late Asa Briggs, Heike Jons, John Kingman, John Melvin, Edward Mortimer, the late Colin Seymour-Ure and David Vincent have all enhanced our ideas and arguments.

We used two conferences to shape the project into a book, and are grateful for the stimulating presence of so many, amongst whom we would like to single out the following: Jenny Abramsky, Robert Anderson, Sanjoy Bhattacharya, Peter Buckley, Barnabas Calder, Geoffrey Crossick, Helen Cowie, Carol Dyhouse, Frank Furedi, Mark Goldie, Esmée Hanna, Michael Hebbert, the late Lisa Jardine, Nancy Weiss Malkiel, Joanna Motion, Lucy Robinson, Gervase Rosser, Peter Scott, Bill Sheils, Nick Thomas, Malcolm Tight, Keith Vernon, Chris Warne and Dan Weinbren.

Charlie Berry and Mark Merry at the IHR drew the maps. Over the course of the project the resources of several libraries and archives have been immensely useful: the London Library, the library of the Royal Institute of British Architects and the UCL Institute of Education Library. Sean McLaughlin and Catherine Sinet helped track down key images.

For permission to quote from and cite material, we acknowledge the following: the University Archives University of East Anglia, the estate of Noel Annan, Albert Sloman Library University of Essex, University Archives University of Lancaster, Keele University Special Collections & Archives, University of Kent Special Collections & Archives, the University of Leeds Library, the University of Liverpool Library, LSE Library, the University of Manchester Library, University of Stirling, Ulster University, the Modern Records Centre University of Warwick, the Borthwick Institute for Archives University of York, the University of Durham Library, the Bodleian Library University of Oxford, and The Keep University of Sussex.

Lastly, both editors spent some of their early years on 'utopian' campuses. With fond memories, we dedicate this volume to the two men who made that, and much else, happen in our lives.

Jill Pellew
Miles Taylor
London and York, September 2019

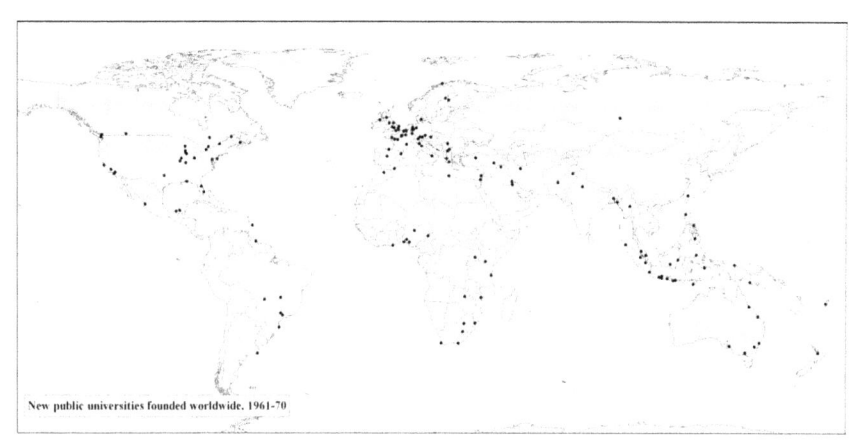

Map 1 New public universities founded worldwide, 1961-70.

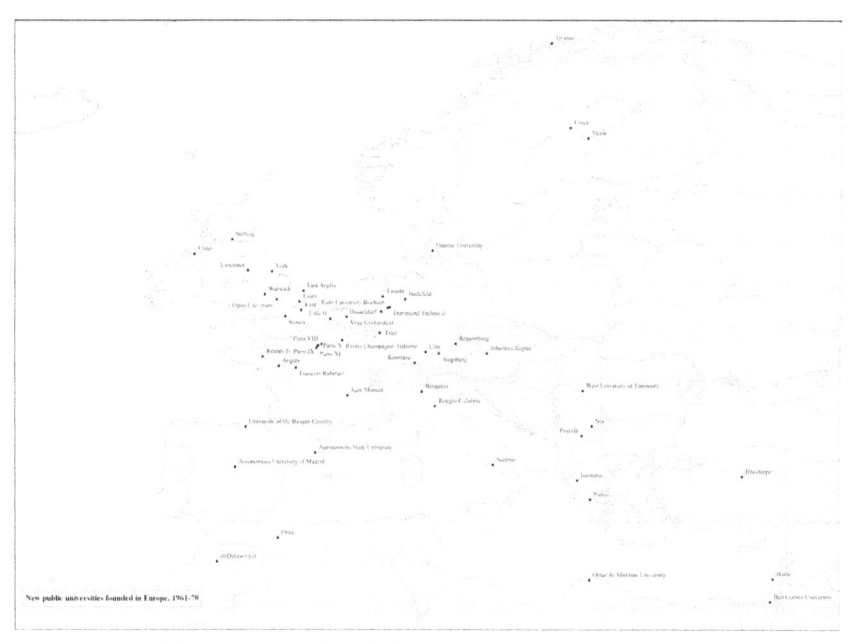

Map 2 New public universities founded in Europe, 1961-70.

Foreword

by Laurie Taylor

The letter from the Registrar of the University of York was dated 15th April 1964. It read: 'Thank you for coming here for interview yesterday. I now have much pleasure in writing on behalf of the University to offer you the post of Assistant Lecturer in Sociology. The salary offered is £1,050 pa.'

It was, I can now readily admit, a surprising appointment. My first degree, gained after four years evening study at Birkbeck College, was in psychology and my only sociological qualification was a borderline MA from the University of Leicester.

At the time I could only suspect that some good advice from a friend at the University of Essex had tipped the scales in my favour. 'You must say "interdisciplinary" whenever you have the opportunity' he'd advised. 'Make a virtue out of your background in two disciplines. Talk about your intention to bring sociology and psychology together in your teaching and research. And then, if there's a space in the conversation, talk about your wish to be associated with the idealism of the new universities, their collegiate system, and their determination to create a community of staff and students on a single out-of-town campus.'

Now that I've had an opportunity to read the many excellent essays in this volume, I realize I might also have benefitted from a form of hiring which, in the words of John Charmley, 'would have had modern human resources departments in despair'. At York, newly installed Vice-Chancellor Lord James, former headmaster of Manchester Grammar School, displayed far more interest in a candidate's ability to teach than in their research record or their commitment to a specific discipline or indeed their university background. As Allen Warren notes in his discerning chapter on the University of York, 'of the ten founding departmental heads, only one was already a tenured professor'.

Although the University of York's 'ever so "umble prefabricated units" (Reyner Banham)' hardly echoed the grandiose architectural constructions at other new universities, its commitment to a new vision of higher education was clear from the first day that I took up my appointment. Nowhere was this more evident than in its determination to break down disciplinary silos. All social science students were required to spend the first five terms of their degree studying sociology, politics, economics, economic history and statistics, before adopting one of these subjects for their final degree – a degree which was described on their certificate as a BA in Social Science, followed by a bracketed reference to their specialism.

But as other writers in this volume make admirably clear, nothing epitomized the manner in which the new universities departed from their original pedagogic idealism than the gradual abandonment of such interdisciplinary commitment. Good intentions slowly succumbed to the traditional power of the single discipline.

At York, despite some spirited protests from students, the five-term interdisciplinary component was first reduced to five terms and then to four and then to three, and was attenuated even further when the Economics Department declared UDI and began teaching single subject Economics from day one.

This pedagogic transformation was paralleled by a gradual subversion of the university's original determination to mix staff from different disciplines. College senior common rooms which had originally provided tea and biscuits and a copy of the *Times Higher Educational Supplement* for lecturers from several disciplines became increasingly known by their specialism: the Sociology SCR and the English SCR.

This gradual reversion to the disciplinary mean was only hastened by the career ambitions of academic staff. Breaking down barriers between disciplines might have been fine as an ideal but it seemed that few serving academics were happy to imply a lack of specialism by adopting such titles as Professor of Social Science or Reader in Natural Science.

This increasing disciplinary confinement was mirrored by other withdrawals from the initial ideals of the new universities. York's requirement that staff lived within a few miles of campus was slowly and quietly abandoned, while the expectation that York in common with the other new universities would abolish the 'nine to five' university by encouraging students to spend as much time as possible on the university sites was visibly contradicted by the line of weekend coaches waiting to carry them to more appealing destinations.

These destinations did not include the city of York. Initial talk about the need for a close 'town and gown' relationship was undermined by the dull fact that its students were increasingly recruited from well outside the York area and were, in many cases, wary of even venturing into town because of local hostility. (As I remember, only one brave academic, historian Professor Gwyn Williams, attempted to break the cycle by giving his lectures in a room above a city centre pub.)

Although other essays in this volume find space to acknowledge more positive elements of the new universities, it is impossible to ignore their dominant theme: the new universities' slow and irreversible retreat from pedagogic idealism and collegiality.

It is perhaps a mark of how much the new universities have come to resemble the higher education institutions which they once set out to reform, that they are not now so much recognized by their distinctive utopian ideals as by their brute position in the latest university league table.

Introduction

by Jill Pellew and Miles Taylor

In a remarkable decade of public investment in higher education, some 200 new university campuses were established across the globe between 1961 and 1970.[1] One of the most famous instances of this expansion came in Britain, where seven new campuses were founded in a four-year spell (Sussex, East Anglia, York, Lancaster, Kent, Essex and Warwick) with two more following towards the end of the decade (Stirling and Ulster). There was also in Britain a new 'university of the air': the Open University based in Milton Keynes, itself a new town, perhaps the most well-known of the new urban areas planned and built in the years either side of the Second World War.

The new university phenomenon was not confined to Britain. In different ways and in a variety of contexts – the post-war 'baby boom', the drive for research and development in science and technology, and the end of empire in Africa, Asia and the Caribbean – university campuses appeared from Adelaide (Flinders, 1966) to Vancouver (Simon Fraser, 1965), from the Western Cape (1970 in South Africa) to Tromso (1968 in Norway). Several were architectural innovations: Niemeyer's Brasilia campus (1961), or the epitome of Danish functionalism at Odense (1966), or the Central Florida campus of concentric rings at Orlando (1963 inspired by Walt Disney). Others experimented with radical new curricula, such as the postgraduate Jawaharlal Nehru University (JNU) in Delhi (1969) and Simon Fraser. Many were conceived as experiments in social and national engineering: the Canadian French-speaking campuses of Quebec (1969–70) and Moncton, New Brunswick (1963), the Dutch-speaking breakaway campus of Vrije University in Brussels (1970) and the new Serbian universities of Nis (1965) and Pristina (1969) in Yugoslavia. In France the monolith that was the University of Paris was broken up, and several new outposts were created, including Nanterre (1964, Paris X), designed to accommodate, amongst others, students originating from North Africa. In the United States, the state university system became increasingly co-educational. In Spain, it began to go bi-lingual: in the new autonomous universities of Bilbao (Basque-speaking, 1968) and Catalonia (Barcelona, 1968).

Of course, there were precursor and pioneer campuses – the Australian National University in Canberra (1946), the Freie Universität in Berlin (1948) and Keele in England (1949, known as the University College of North Staffordshire until 1962). There were parts of the world, notably the Soviet Union and China where expansion came later or not at all (after all, during the Chinese 'cultural revolution' of 1966–70, the universities were shut down completely). Elsewhere, such as in Argentina, Brazil and South Korea, the mood of innovation did not last, with a crackdown on campus reform setting in by the middle of the decade. As critics pointed out at the time, the contribution of the new universities to the overall growth of higher education in Britain in the period, for example, was relatively small.[2] But they were harbingers of change and they were a novelty. The new universities of the 1960s belong to a unique historical moment, one that demands closer inspection and analysis of a comparative and trans-national kind. They are part of that 'moment of modernity' that is associated with the 1960s, as different national cultures around the world shook off the legacies of the past.[3] Seldom has there been so much experimentation in what a university should look like physically; how, what and whom it might teach; and how it should be governed. Never before, nor since, has higher education been such a priority of public spending. Only in very few parts of the world, for instance, the Philippines and Lebanon, were the new universities largely the result of private enterprise.[4] Rarely – especially in the Anglophone world – has the expansion of higher education been driven to such an extent by central government, and by the individual vision and entrepreneurship of men like Alec Lindsay (Keele) and Clark Kerr (California). And perhaps at no other time than in the 1960s have the universities shaped wider political and social developments, both overtly as in the protest movements that dominated the decade, and more incrementally as laboratories of changing lifestyles and attitudes.

To date, there has been no single volume devoted to the 1960s campuses, although each new university has spawned a small industry of institutional histories.[5] Moreover, whilst most histories of twentieth-century higher education have much to say about the establishment of the new universities in Britain, they seldom join up the dots that link developments at home with what was happening around the world.[6] New ideas about academic spaces, curriculum and governance were shared across borders and overseas. Expertise on the reform of higher education travelled widely, usually centrifugally, from Europe and North America outwards into the developing world. At the same time, student radicalism on the new campuses flowed from the periphery to the centre, with activism and its consequent repression in parts of the global south such as sub-Saharan Africa and Latin America often fuelling politicization back in the Western world. At the heart of this book lies the task of capturing this interdependence of forces that went into the making of the new universities of the 1960s, as well as the aim of making comparisons between different institutions and national systems.

We have chosen the concept of the 'utopian university' as a way of considering a range of interconnecting themes associated with the new campuses of the 1960s. 'Utopias' then as now were in vogue. From Robert Hutchins's 1953 *University of Utopia*, through to Robert Birley's *Universities and Utopia* (1965) and onto Robert Wolff's *The Ideal of the University* (1969, with its concluding chapter on 'Practical Proposals for Utopian Reform'), there was frequent discussion of the university as a form of utopia.[7] More recently, Hanna Holborn Gray, in her *Searching for Utopia: Universities and Their Histories* (2012), has deployed the concept in a lament for the golden age of liberal arts higher education.[8] Even if these invocations are rhetorical tropes, we suggest that the notion of 'utopia' is a useful way to categorize and analyse the new universities of the period. They might be deemed utopian in four main ways. First, these universities were the outcome of bold schemes or blueprints, for example, those of Lindsay at Keele, Kerr in California, Darcy Ribeiro in Brazil, Murray Ross in Canada or Hans Werner Rothe in West Germany.[9] In other words, it is possible to trace the progress of the idealized institution from the imagination to the drawing board and then onto the construction site. Visionary vice-chancellors worked, often single handedly, with architects to design new sites, or dreamed up new and untested curricula. Innovation was in the air, fed both by the wider schemes of social and economic planning that shaped government thinking after the Second World War – the so-called warfare state – and by aspirations prevalent across the 1960s towards a society that was based on 'meritocracy', the 'knowledge economy' and 'post-industrial' realities.[10]

Secondly, like most modern utopias, the new campus universities were the products of careful state planning. Although the 1960s did see the emergence of the 'free university' movement, notably in California, Australia and West Germany, the creation of new institutions of higher education was *dirigiste* and mostly publicly funded, controlled by central and regional governments, to varying degrees of intervention. In Britain, the University Grants Committee, which reported to the Treasury, broke with decades of unsystematic approaches to higher education. 'There is no university system in Britain,' declared Lord Simon of Wythenshawe, chairman of the Council of the University of Manchester in 1947.[11] Within a year that began to change. The University Grants Committee (UGC) was given an enhanced remit in 1948 and, under its dynamic chair, Sir Keith Murray, who served two successive terms of office (1953–63), took up the recommendation of successive government committees for expanding the universities to satisfy the demand for an estimated 124,000 places, and also to keep up with scientific research. Faced with a muted response from existing universities, and given the go-ahead to spend by the Treasury in 1958, Murray spearheaded the search for new campuses, laying down a list of specific criteria they were required to fulfil through their 'Academic Planning Boards'. Two future VCs of the new campuses, Asa Briggs (Sussex) and Eric James (York), joined his team.[12] Elsewhere, even more control from the centre was exerted. Enabling legislation was passed to authorize the setting up of new universities, again with special conditions laid down. In the United States the 'G. I. Bill' of 1944 had already helped to subsidize the provision of new universities, and in 1963, another federal measure, the Higher Education Facilities Act, gave further support. In West Germany, the 'reform universities' of Bochum (1962), Konstanz (1966) and Bielefeld (1969) were signed off by the federal government in Bonn, then taken forward by the regional Länder. Similarly, in France, new universities in Paris and elsewhere – Orleans (1960), Angers (1970), Reims (1967), St Etienne (1969) and Tours (1969) – were created before and after the unrest of 1968 by central government initiative.[13] In some countries, notably in Asia, new universities were expected to embody 'national' values, usually those of a post-colonial nation-building kind, for example, the ten 'flagship' universities of South Korea (established mainly in the 1950s); the two dozen new universities of Sukarno's Indonesia, many of which took their names from figures from the pre-colonial past or heroes of the national revolution Syiah Kuala (1961), Brawijaya (1963), Jendaral Soedirman (1963); and in India, where the full apparatus of the Congress Party state was devoted to the establishment of JNU in Delhi.[14] In this way, the new campuses of the 1960s were planned political experiments, breaking down social and ethnic barriers. In some cases, for example, apartheid South Africa (Rand Afikaans 1967) and Israel (Negev 1969), they actually reinforced separation.[15]

Thirdly, these universities were utopian spaces, insofar as many of them were based on ground-breaking plans for community living, new kinds of curriculum and pedagogy, and different forms of governance from the rest of the higher education sector. Architects were given licence to design with boldness, but also with an eye on speed of construction and affordability. Some sites were repurposed from older uses such as heritage buildings: Mataram in Indonesia (1962), in the former eighteenth-century Water Palace. Others, for example, Illinois-Chicago (1965), went into the inner city to stimulate renewal. By contrast, in Barcelona and Madrid (1968) new campuses were deliberately located far from city centres, in an attempt to isolate students from street politics downtown. Some were experiments in ecology: the new campus of the Middle East Technical University in Turkey (1963) planted its own forest.[16] Mostly, however, universities of the era chose virgin greenfield locations with all the space that entailed, purpose-built for a residential, largely pedestrian population of staff and students. Communal life, it was argued by campus advocates, would nurture and enhance the educational experience. In Britain back in 1957, guided by the radical educationalist, Roy Niblett, the UGC under Keith Murray set up a subcommittee on university halls of residence. New vice-chancellors such as Albert Sloman at Essex argued that 'it does most students good to get away from their families'; he thought that the university should 'provide an experience of living as well as an opportunity for learning'.[17] An integrated, close-knit sequence of spaces and buildings would also make it easier for different subject-disciplines to meet and to interact. A new 'map of learning' – the phrase was coined in 1960 by Asa Briggs, on the eve of his arrival at the University of Sussex (and it spread) – might arise from this unusual spatial order, less specialist, and committed, particularly in the first year of undergraduate studies, to a general syllabus of education, combining the sciences and non-sciences.[18] Liberated from their subject silos, conventional university departments would forge new hybridized programmes of study and research, in areas fit for the late twentieth century: for example, sociology, computer and environmental science, and business economics. And as with all utopias, these would be self-governing communities, devising new rules and regulations for the conduct of everyday life as well as for the constitution of an institution of learning.

Fourthly, the history of the new campuses of the 1960s perhaps exemplifies the precarious nature of all utopian experiments: a tendency to drift into the opposing extremes of anarchy or ideology, and the fine line between utopian experiment and dystopian regime. In this way, the student radicalism and disorder prevalent in the new campuses across the world in the later 1960s and early 1970s might be seen as demonstrating the iron law of utopias that they are in some way always bound to fail, as they rest on idealistic versions of how people would like to live, rather than how they actually behave when placed in an unfamiliar environment. Although the new campuses were not disproportionately in the vanguard of student protest after 1968, there were enough famous examples to coin a cliché: Essex in the UK, Nanterre and Vincennes (1968, Paris VIII) in France, Trento (1972) in Italy (the birthplace of the 'Brigate Rosse'). In other cases, for example, Warwick, or 'Warwick University Ltd', struggles between progressive teaching aspirations and local business sponsors left a pall that hung over the new campus for many years.[19] By the early 1970s, then, the 'utopian' moment had passed. In Western Europe, Australasia and North America, a backlash set in against the baby-boomers and the public taxation that indulged their lifestyles. 'Neo-liberal' voices began to demand more student self-responsibility, that is to say loans to pay for their studies, and less student self-government.[20] Private universities began to be championed as the necessary corrective to the errors of the post-war higher education reforms.

In documenting the rise and fall of the utopian universities, this book seeks to place the British story within a wider comparative framework, both to point up connections and contrasts and also to capture the unique features of the new campuses of the period. Around half the volume is devoted to Britain. We look at the seven new English universities, and at Ulster and Stirling, with two additional chapters looking at the British experience around science in research and the curriculum, and at the role of philanthropy in starting up the new campuses. We have also included Keele, seen as a pioneer, although it was founded originally in 1949. Another chapter takes a sideways view at innovation and expansion in the existing civic universities. Then the book turns overseas, taking in case studies from West Germany, France, Australia, California, Canada and India, including a survey of Britain's involvement in the new universities of the Commonwealth. Inevitably, many more institutions, countries and themes might have been selected. In Britain, the Colleges of Advanced Technology and then the burgeoning polytechnic sector ran parallel to some of the developments described in this book. Similarly, the ancient universities of Oxford and Cambridge joined in this moment of modernity with two new colleges – St Catherine's (1962) and Churchill (1960), respectively – both important statements of inventive design and ethos.[21] Moreover, there are other countries that might have furnished examples of the phenomenon of the utopian university: notably Brazil (Brasilia), the Netherlands (Twente 1961) and Denmark (Odense).[22] That would have made for an even longer book, and the gain in breadth could also have meant a loss of depth, as a study framed around key research questions became too like an encyclopaedia, each chapter an individual entry. University histories are in any event notorious for their narrow vision, their inability to look outwards to understand what is happening within.[23] This book attempts to break with the genre. Although many of its contributors are insiders, and some of the chapters take the form of memoir, they have all been encouraged to sever umbilical cords and put the early years of their institution into a broader context. As a result, six main themes run across the chapters that follow.

Firstly, how distinct is the chosen decade 1961–70? From a British perspective, these years constitute a discrete moment, spanning the establishment of the University of Sussex in 1961 through to the start of the final year of the first intake at the New University of Ulster (founded in 1968). This decade is dominated by a sea change in higher education policy, its centrepiece being the Report of the Committee on Higher Education chaired by Lionel Robbins, published in October 1963, which advocated expansion of undergraduate numbers (envisaging over half a million by 1980).[24] The publicity surrounding the Robbins report together with the media spotlight shone on the first of the new campuses – Sussex, East Anglia and York – gave unprecedented coverage to the universities. There were new journals and forums of comment, for example, *Minerva* began publication in the autumn of 1962, the Society for Research into Higher Education started up in 1965. The leaders of the new universities captured the airwaves and issued influential manifestoes, notably Albert Sloman's Reith lectures of 1963, David Daiches's *The Idea of a New University* (1964), featuring Sussex, and Eric James's *The Start of a New University* (1966), about York.[25] The new campuses inspired satire, for example, James Dundonald's *Letters to a Vice-Chancellor* (1962), a novel (Malcolm Bradbury), and made good copy for the tabloid press.[26] At the height of the student militancy in 1968, Michael Beloff produced a survey of the new campuses: *The Plateglass Universities*, that emphatically identified Sussex, UEA, York, etc. as products of the 1960s, with all that this entailed.[27] But is it legitimate to separate out the 1960s campuses in this way? Whilst many of the new universities were physically established in the 1960s, the thinking and planning behind them came in the late 1940s and 1950s. The Robbins report anticipated future planning scenarios of the late 1960s and 1970s, not earlier. The new campuses in Britain emerged from post-Second World War concerns about the 'crisis in the university' (Walter Moberly's phrase of 1949), which in turn reflected the recommendations of a series of government white papers. Two committees appointed by Parliament – on Higher Technological Education (1945, chaired by Lord Percy) and on Scientific Manpower (1946, chaired by Sir Alan Barlow) – recommended improving the quality and quantity of scientists, summed up in the call for a British MIT (Massachusetts Institute of Technology). It was no coincidence that two prominent movers-and-shakers within the new universities in Britain were Solly Zuckerman and Noel Annan, both of whom had seen distinguished war service and then moved into advisory roles in the 1950s.[28] If we look overseas, particularly in West Germany, in India and in Brazil, we can see a similar pattern whereby the new campus universities of the 1960s were the result of plans laid down in the 1950s. Perhaps the greatest experiment of all in publicly funded higher education of the century, Clark Kerr's California plan, began in the 1950s, as Chris Newfield's chapter shows. And although these developments were concerned with expanding capacity to absorb the expected demographic bulge, they also stemmed from a critique of the existing universities and their inability to modernize. So, an account of the 1960s campuses needs to pay as much attention to continuity as well as change across the 1950s and 1960s. This caveat also applies when we look at the enduring influence of British and American higher education models on the new campus universities in the developing world. As Miles Taylor's chapter reminds us, new universities of the 1960s were established in Commonwealth countries along the lines recommended by a series of reports and commissions dating back to the late 1940s, led by British experts. New campuses in Australia, as Hannah Forsyth shows, and in New Zealand (Waikato), still looked to British advice for guidance, whilst Australian planners were behind the University of Papua New Guinea (1965).[29] In South America, Iran and in the Philippines, American advisors continued to play an influential role in the shaping of new universities from the early 1950s down to the late 1960s.[30] In an age of post-colonial nationalism, the old imperial powers were still at work in the business of higher education. At the same time, there was an acceleration of change after 1960. In Britain the traditional mandarin instincts and networks were in play, as the chapters by

John Charmley, Krishan Kumar and Allen Warren all suggest, but the new university leaders were younger men. Of the seven new English VCs, five were in their mid- to late forties. Likewise, Paul Axelrod and Hannah Forsyth demonstrate in their chapters in this volume how in Canada and Australia although change was driven from above, there was much that was radical and rule-breaking about the university principals who delivered the new campuses.

Our second theme concerns campus architecture. As Beloff's 1968 *Plateglass Universities* book indicated, the new campuses were most striking in their physical appearance. The design and construction of new universities proved an attractive and lucrative lure for some of the best modernist architects of the period, not just in the UK, but overseas as well. Basil Spence (Sussex), Denys Lasdun (East Anglia) and William Holford (Kent) stole the headlines at home, for better or worse, although two other firms, Shepheard and Epstein, and Robert Matthew-Johnson Marshall ('Romjam') were responsible for most of the work beyond the famous boutique projects. Overseas, renowned names were involved: Walter Gropius at Baghdad, James Cubitt in Libya, Oscar Niemeyer at Brasilia, Jane Drew and Maxwell Fry in Ibadan, Nigeria. The novelty of the new buildings – Spence's sublime shapes at Sussex (the first buildings of the 1960s to be Grade 1 listed), the clasp method at York, the ziggurat waves at East Anglia – and the relative speed with which they were erected, drew plenty of admirers and critics. There were two symposia on the architecture of the new universities hosted by the Architectural Association and one by the Royal Institute of British Architects.[31] Not surprisingly, this aspect of the history of the 1960s campus universities has been well-covered in the secondary literature.[32] However, we go further than some of these studies currently take us. In particular, our contributors look at function as well as form, to assess how well new ideas about the university curriculum fitted with the layout of buildings and their internal design. After all, the out-of-town 'campus' was not a new model. It had been pioneered in the United States during the eighteenth and nineteenth centuries, as the states expanded, setting aside space for universities through the land-grant scheme.[33] Were the new campuses simply an American import, as in the case of Twente in the Netherlands or the Aryamehr University of Technology in Tehran (1966)? To what extent did they seek to reproduce the style and substance of the existing 'red-brick' and ancient universities? For example, the new campus at San Diego, California (1960), for example, was modelled on an Oxford college. In their architecture, did these new designs merit the label 'utopian' – from ideals of communal living to practical issues around traffic containment and landscaping? Or were they simply part of the same revolution in post-war public building design and delivery, that produced the high-rise housing estates ('streets-in-the-sky'), transport hubs, shopping centres and schools, all loosely inspired by a new wave of architects, but not specifically focused on the new campuses?[34] As William Whyte argues, the older civic universities were themselves no strangers to architectural rejuvenation, on their main sites, and also in their halls of residence.[35] However, some of the new campuses – such as Essex with its accommodation towers – came up with housing solutions that were not tried elsewhere and which were the focus of much discussion, both positive and negative, as Caroline Hoefferle's chapter argues. The new campuses were unlike the old in other ways too. They were planned as secular spaces. In Britain, only Lancaster included a chapel or faith area at the outset, although one came later at Kent too.[36]

Thirdly, our contributors look in detail at curriculum design. A powerful vision – a new 'map of learning' – drove the planning agenda for many of the new universities in the UK and elsewhere. They sought to break with over-specialization and with learning by rote. They placed interdisciplinarity and teaching by seminar at the heart of their curricula as they sought to move away from older conventions of lecturing and tutorials. In a bold step, the whole academic structure of some of the new campus universities was given over to 'schools', rather than faculties and departments, in the conviction that these would facilitate combined studies. What drove such thinking? Was it the contemporary debate over 'two cultures', fired up by C. P. Snow's 1959 Rede lecture, an extended criticism of how the education system privileged the humanities (especially the classics) over the teaching of science?[37] Or was it a throwback to older aspirations of a liberal arts curriculum, again imported from the United States, and given new life after the Second World War by the methods coming out of Robert Hutchins' University of Chicago and James Conant's Harvard? Many of the early visionaries and new academic staff in Britain had studied in American universities, for example, Keith Murray, chair of the UGC, Jack Butterworth and Frank Thistlethwaite amongst the VCs, and Lionel Robbins (York), Charles Wilson (East Anglia) and Derman Christopherson (Kent), who led Academic Planning Boards. Robbins's committee members spent over three weeks in the spring of 1962 gathering information in the United States, homing in on the California 'master-plan', far more time than they spent anywhere else on their foreign fact-finding missions.[38] Curriculum design was also affected by the government requirements of the Cold War in the space-race era, especially in research and development. The 'Sputnik' moment triggered new demands for better provision for science and technology in the universities: where better to deliver that than on the new campuses? The merger of the ministries of science and education in Britain in 1964 was a telling sign of the times.[39] But how did this 'Cold War' priority sit with the commitment in the new universities to teaching across specialisms, blurring the distinction between science and non-science? Jon Agar's chapter surveys the rhetoric and the reality behind these imperatives, and, in the case of Sussex, Matthew Cragoe describes the Cold War mind at work in other ways. Other chapters in the book examine how the new universities pioneered new subject disciplines. The expansion of the social sciences in this decade is often linked to the new campuses, particularly in sociology, social history, American studies, applied economics and management and business studies, although social sciences enjoyed an ascendancy in the 1960s in old as well as in new universities. Miles Taylor's opening chapter on Keele shows how the fortunes of British sociology fared at Keele, whilst Marion McClintock discusses the repertoire of distinctive new departments developed at Lancaster. In West Germany Stefan Paulus describes the full-frontal assault made by the 'reform' universities on the old faculty structures, a sea-change that came to France too with the 'loi Faure' of 1968.[40] The 1960s campuses also honed new ways of teaching established subjects, notably music at York, geography at Sussex, and also fast-tracked new trends, such as sociology and American studies.[41] Again, attempts at curriculum reform were not limited to the new sector. In Oxford for example, two reports (by William Kneale in

1965 and by Oliver Franks in 1966) were published in the 1960s, proposing substantial alterations, notably the introduction of four-year courses and mixed subject degrees.[42] Most excitement around reform, however, was generated on the new campuses, as the epithets 'Keele experiment' and Sussex 'the pacesetter' implied.

A fourth theme covered in the book is the role of the state in the establishment of the new universities. In Britain the Robbins report confirmed the pattern of planning and the governance of the new institutions worked out earlier with the UGC. Three years before Robbins, a committee chaired by Sir Robert Anderson had recommended an increase in student grants, and that duly happened in 1962.[43] The British experience of state-funded higher education was considerably less interventionist than other countries, and especially less so than the new federal or state universities established in Australia, Brazil, India and the United States. Arguments were also made for new campuses as vehicles of economic regeneration and regional integration, for example, in Italy's *mezzogiorno* (Salerno 1968, Reggio 1968).[44] For these reasons, in Britain as elsewhere, the management of the universities became more of a political question than ever before, as general taxation funded what remained the higher education of a small elite. How was the status of the new universities as utopian communities of learning presented for political consumption? More generally, to what extent were the new universities across the globe seen as agents of social mobility and national integration? And conversely, how were they viewed within the regions where they were built? The 'white heat' revolution of technology in the 1960s brought heightened expectations of new synergies between the universities (including the new campuses), industry and business, the Confederation of British Industry amongst those heralding a fresh start.[45] As Jill Pellew shows in her chapter, local business and philanthropists contributed significantly to the founding of the new campuses. Strings were not attached, but clearly as Carolyn Steedman shows in the case of Warwick, mixing economic enterprise with radical pedagogy brought tension.

Fifthly, there is of course the student and staff experience of the new campuses. Here there has been a great opportunity to draw from the rich resource of oral history testimony from the initial cohorts of undergraduates at the new campuses. Many of the new universities made the collection of alumni reminiscences a focal point of their jubilee anniversaries, notably Essex, Kent and Warwick. From these sources we can begin to find out more about the first cohorts of undergraduates and postgraduates in this brave new world. Conceived as residential universities and breaking with the idea of local 'catchment' and '9-to-5' attendance, new kinds of communities were created. The balance between the sexes was skewed in many new campuses towards men. Not until the 1990s was parity reached in the numbers of male and female school-leavers going to university (in the United States it was reached in the 1970s).[46] Given this, what patterns of gendered behaviour were regulated, adopted and tolerated on the new campuses? Additionally, how might we relate the student experience to the myth and reality of the 'swinging '60s'? Did 'town' and 'gown' boundaries evolve as they did in other college environments?[47] Especially relevant here is the question of whether the new campus universities were utopian organizations in their forms of governance. In Britain, for the most part, the traditional constitutional structure of the older university sector, that is to say, an executive led by a vice-chancellor, overseen by an academic Senate and lay Council, was taken up in most of the new universities. As the decade went on there were widespread calls for greater better student representation, academics such as Graeme Moodie at York, and the leadership of National Union of Students contributing to the campaign.[48] Similarly, overseas, new campuses actually became centres of alternative forms of university governance, especially in California, in France and in India. Universities, it was claimed, should be run by their members – staff and students – as self-governing republics. So onto student radicalism in the 1968–72 period: why did a higher proportion of the new universities participate in the student unrest of these years, or is that itself a myth? And what were the contributory factors? Did students on the new campuses emerge 'out of apathy', radicalized by the Vietnam War, apartheid in South Africa and national liberation struggles elsewhere, or were they fighting smaller campus battles about housing, co-ed status, student representation and extra-curricular activities? Or a mixture of both? Whilst many student politicians belonged to the militant left, many did not. New campuses such as Essex and York also spawned headline-grabbing Young Conservative branches.[49] Two chapters in this collection, Victor Collet on Nanterre in Paris, and Rajat Datta and Shalini Sharma on JNU in Delhi, bring out how student demands shaped institutions from the moment of their inception, and many other contributors to the book reflect on the impact of 1968 and after.

Finally, the book looks at what happened next. How did the new universities fare after their first decade, and what can we learn now from their experience? In Britain in 1970, the UGC recommended that 'several more universities should be built'.[50] There have been none, only the foundation in 1976 of a private university, the University of Buckingham, situated on a loosely connected campus, a mixture of repurposed existing buildings and some new-build. Even as the new campuses were going up, political enthusiasm was waning. Although committed to developing the 'white-heat of technology', the Labour government of Harold Wilson pressed for university expansion outside the traditional sector: in the distance learning of the Open University and also through the Colleges of Advanced Technology. Critical of meritocracy and the elitism of the universities, Wilson's Secretary of State for Education, Anthony Crosland, introduced a new dual system of funding: state support via the UGC for the universities, and local authority financing for the polytechnics, their degrees validated by the National Council for Academic Awards.[51] Public coffers were running low too. By the time projections were being made for the 1967–72 quinquennium of spending on higher education, there was less money on the table, as the pangs of slow growth suffered by later newcomers such as Ulster and Stirling, examined in this book by Tom Fraser and Leonie Murray, and by Holger Nehring, showed up only too well. Add in the lingering unrest on campuses up and down the country, especially over secret student files held by university authorities, and criticism from within the universities that unbridled expansion was a bad idea – 'more means worse' as Kingsley Amis put it in 1960 – and all the factors were in play for a retreat from the optimum levels of funding that came in with the Robbins and Anderson reports.[52]

Thus in Britain since 1970, further expansion of higher education has come within the established universities. In 1992, the polytechnics were upgraded to enjoy university status, and in 2013 restrictions on admissions numbers were lifted entirely.

Around the rest of the world, however, new campuses have flourished, particularly in the BRIC economies, and in the oil-rich states of the Middle East. Many seem far removed from the utopian communities of the 1960s, more like vast 'univer-cities' as they have been described, some dependent on non-resident 'flying faculty' to deliver their teaching, hardly the face-to-face learning environments championed by the visionaries of the 1950s and 1960s.[53] Indeed, it is now hard to imagine another wave of utopian universities. There is less public appetite for subsidized tertiary education, and universities have had to evolve different business models. Private or for-profit universities now lead the way in the higher education sector in many high-growth economies.[54] Student finance has switched from grants to loans and other forms of payment, as the central state in the UK has retreated from directly funding the universities. Student accommodation no longer features as an integral part of the campus landscape, linking living and learning spaces. Instead, universities have increasingly devolved the management, and even the construction, of student housing to private companies, whose 'campus villages' on brownfield sites resemble budget hotels more than university residences. Towns adjacent to greenfield campuses are now experiencing 'studentification' as the demand for short-term rented housing spirals.[55] New technologies in teaching: moocs, VLE, social media, and digital information systems more generally have lessened reliance on the static learning environment of the out-of-town campuses.

Idealism has declined. Where are the blueprints for our universities today? In the UK, veterans of the new campuses look back now with a world-weary nostalgia, whilst the language of higher education reform has become dominated by voices that denounce Robbins and the baby-boomers, and place their trust in market forces.[56] Similar ideological shifts towards neo-liberal 'academic capitalism' have been observed in the USA, India, Germany and France.[57] Utopian visions are few and far between, restricted to laudable plans for environmentally sustainable campuses or new ecosystems of learning. There are some exceptions. In Nalanda in northern India, a new postgraduate campus has been established: green and car-free, on the site of a twelfth century monastery to which scholars originally came from afar. With its connected walkways and information bazaars, it evokes the spirit of the 1960s. 'Starchitecture' is still evident in university construction where it can be afforded, and some innovations speak to changing pedagogy: 'wall-less labs' and 'learning commons', for example.[58] Generally, however, little is left of the pioneering design of the holistic campus. Nature not nurture is all that is left of the utopian ideal.

So where might our conclusions about the legacy of the utopian universities take us? Whilst some commentators might say it was the 1960s campuses that gave universities a bad name, we can point also to their positive role: how they rescued the liberal arts, nurtured new subjects, freed up intellectual enquiry and revitalized sleepy and declining parts of the economy. Many of the perceived ills of the 1960s universities in fact stemmed from short-term thinking: building materials that would not last; a level of student financial assistance that was unsustainable (Robbins for one advocated moving to a mix of loans and grants); new technologies that took decades not years to develop. However the period is judged, the utopianism of the moment shines through. As we look back after fifty years from the very different situation in which publicly-funded higher education now sits, there is surely much to admire and plenty to ponder in the achievements of the new universities of the era.

Notes

1. For expansion 1945–70, see Heike Jöns, 'Modern School and University', in *A Companion to the History of Science*, ed. Bernard Lightman (Oxford: Wiley Blackwell, 2016), 324. For an influential series of surveys of the phenomenon in America and Europe since the Second World War, see Martin Trow, *Twentieth-century Higher Education: Elite to Mass to Universal*, ed. Michael Burrage (Baltimore: Johns Hopkins University Press, 2010), chs 2, 16, 17.
2. John Vaizey, 'The New Universities', *Times Literary Supplement*, 13 February 1969, 160. By 1970, the new English universities comprised just 8.5 per cent of the total student population: calculated from 'Returns from universities and university colleges in receipt of Treasury grant', 1970–1, TNA, UGC 3/42.
3. Arthur Marwick, *The Sixties. Cultural Revolution in Britain, France, Italy and the United States, c.1958–c.1974* (Oxford: Oxford University Press, 1998); *Moments of Modernity: Reconstructing Britain, 1945–1964*, ed. Becky Conekin et al. (London: Rivers Oram Press, 1999); *Preserving the Sixties: Britain and the 'Decade of Protest'*, ed. Trevor Harris and Monia Carla O'Brien Castro (Basingstoke: Palgrave Macmillan, 2014); *The Global 1960s: Convention, Contest, and Counterculture*, ed. Tamara Chaplin and Jadwiga E. Pieper Mooney (London: Routledge, 2018); Christopher B. Strain, *The Long Sixties: America, 1955–1973* (Chichester: John Wiley, 2017); Tom Hayden, *The Long Sixties: From 1960 to Barack Obama* (Boulder: Paradigm Publishers, 2009).
4. By the end of the 1960s of the thirty-one universities in the Philippines, twenty-five were private: Epifania R. Castro Resposo, *The Role of Universities in the Developing Philippines* (New York: Asia Publishing House, 1971), 106; cf. Samir A. Jarrar, 'Lebanon', in *International Higher Education. An Encyclopedia*, ed. Philip G. Altbach, 2 vols. (London: Routledge, 1991), ii, 1055–63.
5. *The New University*, ed. John Lawlor (London: Routledge & Kegan Paul, 1968); *New Universities in the Modern World*, ed. Murray Ross (London: Macmillan, 1966); Harold Perkin, *New Universities in the United Kingdom* (Paris: OECD, 1969). For a special issue devoted to the new universities in the UK, see *Higher Education Quarterly*, 45 (October 1991), 285–381; cf. Tony Rich, 'The 1960s New Universities', in *The State of UK Higher Education*, ed. David Warner and David Palfreyman (Buckingham: Open University Press, 2001), 49–56.
6. W. A. C. Stewart, *Higher Education in Postwar Britain* (Basingstoke: Macmillan, 1989); Robert Anderson, *British Universities Past and Present* (London: Continuum, 2006); Malcolm Tight, *The Development of Higher Education in the United Kingdom since 1945* (Maidenhead: Society for Research into Higher Education and Open University Press, 2009); Michael Shattock, *Making Policy in British Higher Education, 1945–2011* (Maidenhead: Open University Press, 2012).
7. Robert Hutchins, *The University of Utopia* (Chicago: University of Chicago Press, 1953); Eric Ashby, 'The British Universities: Utopias', *Listener*, 24 March 1955, 513–14; Robert Birley, *Universities and Utopia* (Johannesburg: Witwatersrand University Press, 1965); Robert Wolff, *The Ideal of the University* (Boston: Beacon Press, 1969).

8   Hanna Holborn Gray, *Searching for Utopia: Universities and Their Histories* (Berkeley: University of California Press, 2012).
9   For Lindsay, see Tom Steele and Richard Taylor, *British Labour and Higher Education, 1945 to 2000: Ideologies, Policies and Practice* (London: Continuum, 2011), ch. 4. The literature on Kerr is extensive. See, inter alia, Cristina González, *Clark Kerr's University of California: Leadership, Diversity, and Planning in Higher Education* (New Brunswick: Transaction, 2011); *Clark Kerr's World of Higher Education Reaches the 21st Century: Chapters in a Special History*, ed. Sheldon Rothblatt (London: Springer, 2012); Simon Marginson, *The Dream Is Over: The Crisis of Clark Kerr's California Idea of Higher Education* (Oakland: University of California Press, 2016). For Ribeiro, see Rodrigo Reis et al. (ed.), *Darcy Ribeiro* (Rio de Janeiro: Beco do Azougue, 2007), 14–35. For Ross, see his *The University: The Anatomy of Academe* (New York: McGraw-Hill, 1976), ch. 10. For Rothe, see Wilfried Rudloff, 'Die Gründerjahre des Bundesdeutschen Hochschulwesens: Leitbilder Neuer Hochschulen Zwischen Wissenschaftspolitik, Studienreform und Gesellschaftspolitik', in *Zwischen Idee und Zweckorientierung: Vorbilder und Motive von Hochschulreformen Seit 1945*, ed. Barbara Wolbring and Andreas Franzmann (Berlin: Akademie Verlag, 2007), 97.
10 Alec Cairncross, *Years of Recovery: British Economic Policy, 1945–51* (London: Methuen, 1985), ch. 11; *Scientific Governance in Britain, 1914–79*, ed. Charlotte Sleigh and Don Leggett (Manchester: Manchester University Press, 2016). For the 'warfare state', see David Edgerton, *The Rise and Fall of the British Nation. A Twentieth Century History* (London: Allen Lane, 2018), 341–2. For these keywords, see Michael Young, *The Rise of the Meritocracy 1870–2033: An Essay on Education and Equality* (London: Thames and Hudson, 1958); Peter Drucker, *The Age of Discontinuity: Guidelines to Our Changing Society* (London: Heinemann, 1969), ch. 12 (for the 'knowledge economy'); Alain Touraine, *La Société Post-Industrielle* (Paris: Editions Denoël, 1969; English trans. 1974).
11 Ernest Simon, 'The Universities and the Government', *Universities Quarterly*, 1 (1946–7), 79.
12 For UK higher education policy in the period, see John Carswell, *Government and the Universities in Britain: Programme and Performance, 1960–1980* (Cambridge: Cambridge University Press, 1985); Shattock, *Making Higher Education Policy*, ch. 2; Peter Mandler, 'Educating the Nation, II: Universities', *Transactions of the Royal Historical Society*, 25 (2015), 1–26; Ourania Filippakou and Ted Tapper, 'Policymaking and the Politics of Change in Higher Education: The New 1960s Universities in the UK, Then and Now', *London Review of Education*, 14 (2016), 11–22.
13 For the USA: Suzanne Mettler, *Soldiers to Citizens: The G.I. Bill and the Making of the Greatest Generation* (New York: Oxford University Press, 2005); Marvin Lazerson, *Higher Education and the American Dream: Success and Its Discontent* (New York: Central European University Press, 2010), ch. 1. For Germany, see Rudloff, 'Die Gründerjahre des bundesdeutschen Hochschulwesens'; Antoine Prost, *Éducation, Société et Politiques: Une Histoire de l'Enseignement de 1945 à Nos Jours* (Paris: Editions du Seuil, 1992), ch 6.
14 For some of the background in South Korea, see Sung Chul Yang, 'Student Political Activism: The Case of the 1960 April Revolution in South Korea', *Youth and Society*, 5 (1973), 47–60. For Indonesia, see R. Murray Thomas, *A Chronicle of Indonesian Higher Education: The First Half Century* (Singapore: Chopmen Enterprises, 1973), chs 5, 10. For India, see *Report of the Education Commission (1964–66): Education*

and National Development (New Delhi: Ministry of Education, 1966), 275–81. This part of the report was influenced by, possibly written by, the Chicago sociologist, Edward Shils.

15   Mary Beale, *Apartheid's Goals in the 1960s: The Creation of the University of Port Elizabeth and the Randse Afrikaanse Universiteit* (Johannesburg: Institute for Advanced Social Research, University of the Witwatersrand, 1995); John Davies, 'The State and the South African University System Under Apartheid', *Comparative Education*, 32 (1996), 319–32; Yehuda Gradus Fred Lazin, 'The Role of a University in a Peripheral Region: The Case of Ben Gurion University of the Negev', *Policy Studies Journal*, 14 (1986), 487–94.

16   Howard A. Reed, 'Haceteppe and Middle East Technical Universities: New Universities in Turkey', *Minerva*, 13 (1975), 200–35; Fred W. Beuttler, 'Envisioning an Urban University: President David Henry and the Chicago Circle Campus of the University of Illinois, 1955–75', *History of Higher Education Annual*, 23 (2003–4), 107–41.

17   Albert Sloman, 'The Fulfilment of Lives', *A University in the Making*, Reith Lecture 4 (1 December 1963), https://www.bbc.co.uk/sounds/play/p00hbdmq (accessed 4 September 2018); University Grants Committee, *Report of the Sub-Committee on Halls of Residence* (London: HMSO, 1957). For Niblett, see his *Life, Education, Discovery: A Memoir and Essays* (Bristol: Pomegranate Books, 2001).

18   Matthew Cragoe, 'Asa Briggs and the University of Sussex', in *The Age of Asa: Lord Briggs, Public Life and History in Britain since 1945*, ed. Miles Taylor (Basingstoke: Palgrave, 2015), 225–47; *Times*, 15 November 1969, 7.

19   *Warwick University Ltd*, ed. E. P. Thompson (Harmondsworth: Penguin, 1970; repr. Nottingham: Spokesman Books, 2014); Guy Berger and Wiltrud Ulrike Drechsel, 'Conflits et Expériences Dans Deux Universités "De Gauche": Bremen, Paris VIII, Vincennes', *Paedagogica Europaea*, 13 (1978), 77–90; Gianni Statera, 'Student Politics in Italy: From Utopia to Terrorism', *Higher Education*, 8 (1979), 657–67.

20   Catherine Coron, 'Reshaping the Model: Higher Education in the UK and the Anglo-Saxon Neo-Liberal Model of Capitalism since 1970', *Revue LISA*, 14 (2016). http://lisa.reviews.org (accessed 4 September 2018). For a summary and a critique of neo-liberalism in American higher education since the 1970s, see David R. Shumway, 'The University, Neoliberalism, and the Humanities: A History', *Humanities*, 83 (2017), www.mdpi.com/journal/humanities (accessed 4 September 2018).

21   John Pratt, *The Polytechnic Experiment, 1965–1992* (Buckingham: Open University Press and the Society for Research into Higher Education, 1997); M. Davies and D. Davies, *Creating St. Catherine's College* (Oxford: St Catherine's College, 1997); Mark Goldie, *Corbusier Comes to Cambridge: Post-War Architecture and the Competition to Build Churchill College* (Cambridge: Churchill College, 2012).

22   For Brazil, see Jadiel Vieira da Silva, *Higher Education and University Reform in Brazil* (East Lansing: Latin American Studies Center, Michigan State University, 1977); J. De Boer and J., Drukker, *High Tech, Human Touch 1961–2011: A Concise History of the University of Twente* (Eindhoven: Lecturis, 2012); Per Boje et al., *Pastures New: Odense University, the First 25 Years* (Odense: Odense University Press, 1991).

23   For some recent assessments of the genre, see Sheldon Rothblatt, 'The Writing of University History at the End of Another Century', *Oxford Review of Education*, 23 (1997), 151–67; W. Bruce Leslie, 'Writing Postwar Institutional Histories', *History of Higher Education Annual*, 20 (2000), 83–91; David Hayes, 'The Hard Task of Writing a University History', *University Affairs/Affaires universitaires*, 11 February

2015, https://www.universityaffairs.ca/features/feature-article/hard-task-writing-university-history/ (accessed 4 September 2018); *University Jubilees and University History Writing: A Challenging Relationship*, ed. Pieter Dhondt (Leiden: Brill, 2015); Robert Anderson, 'Writing University History in Great Britain, from the 1960s to the Present', *CIAN-Revista de Historia de las Universidades*, 20 (2017), 17–40.

24 Susan Howson, *Lionel Robbins: A Life in Economics, Government and the Arts* (Cambridge: Cambridge University Press, 2011), ch. 23; *Shaping Higher Education. 50 Years after Robbins*, ed. Nicholas Barr (London: LSE, 2014); 'Remembering Robbins', Special Issue of *Higher Education Quarterly*, 68 (April 2014), 105–240; Mandler, 'Educating the Nation, II: Universities'.

25 Albert Sloman, *A University in the Making* (London: BBC, 1963); *The Idea of a New University: An Experiment in Sussex*, ed. David Daiches (London: A. Deutsch, 1964); Eric James, *The Start of a New University* (Manchester: Manchester Statistical Society, 1966).

26 James Dundonald, *Letters to a Vice-Chancellor* (London: Edward Arnold, 1962); Malcolm Bradbury, *The History Man* (London: Secker and Warburg, 1975). The British tabloids headed for Sussex; see Cragoe, 'Asa Briggs', 234.

27 Michael Beloff, *The Plateglass Universities* (London: Secker & Warburg, 1968).

28 For the committees, see Shattock, *Making Policy*, 19–22; Walter Moberly, *The Crisis in the University* (London: SCM Press, 1949); John Peyton, *Solly Zuckerman: A Scientist Out of the Ordinary* (London: John Murray, 2001); Noel Annan, *Our Age: Portrait of a Generation* (London: Weidenfeld and Nicolson, 1990). See also Harold Silver, 'The Making of a Missionary: Eric Ashby and Technology', *History of Education*, 31 (2002), 557–70.

29 David Hughes Parry, former vice-chancellor of the University of London, chaired the 1959 enquiry that led eventually to the University of Waikato (1964): Noeline Alcorn, *Ko Te Tangata: A History of the University of Waikato: The First Fifty Years* (Wellington: Steele Roberts, 2014), 13. For the University of Papua New Guinea, see Ian Howie-Willis, *A Thousand Graduates: Conflict in University Development in Papua New Guinea, 1961–1976* (Canberra: Australian National University 1980), chs 3–4.

30 I. C. M. Maxwell, *Universities in Partnership: The Inter-University Council and the Growth of Higher Education in Developing Countries 1946–1970* (Edinburgh: Scottish Academic Press, 1980); Ted Goertzel, 'American Imperialism and the Brazilian Student Movement', *Youth and Society*, 6 (1974), 123–50; W. A. Copeland, 'American Influence on the Development of Higher Education in Iran' (unpublished PhD thesis, University of Pennsylvania, 1977); Daniel C. Levy, *To Export Progress: The Golden Age of University Assistance in the Americas* (Bloomington: Indiana University Press, 2005).

31 'The Design of New Universities', *Royal Institute of British Architects Journal*, 71 (1964), 301–10; *University Planning and Design*, ed. Michael Brawne (London: Lund & Humphries, 1967); Tony Birks, *Building the New Universities* (Newton Abbot: David & Charles, 1972).

32 Stefan Muthesius, *The Postwar University: Utopian Campus and College* (London & New Haven: Yale University Press, 2000); John McKean, 'The English University of the 1960s: Built Community, Model Universe', in *Man-Made Future: Planning, Education and Design in Mid-20th Century Britain*, ed. Iain Boyd Whyte (London: Routledge, 2006), 205–22; Anthony Ossa-Richardson, 'The Idea of a University in its Concrete Form', in *The Physical University: Contours of Space and Place in Higher*

Education, ed. Paul Temple (Abingdon: Routledge, 2014), 131–58; Elain Harwood, *Space, Hope and Brutalism: English Architecture, 1945–75* (New Haven and London: Yale University Press, 2015), 256–64.

33   Richard Guy Wilson, 'The University of Virginia and the Creation of the American Campus', *CIAN. Revista de Historia de las Universidades*, 17 (2014), 59–79; *The Modern Land-Grant University*, ed. Robert J. Sternberg (West Lafayette, Indiana: Purdue University Press, 2014).

34   Janina Gosseye, 'Milton Keynes' Centre: The Apotheosis of the British Post-War Consensus or the Apostle of Neo-Liberalism?', *History of Retailing and Consumption*, 1 (2015), 209–29; Guy Ortolano, 'Planning the Urban Future in 1960s Britain', *Historical Journal*, 54 (2011), 477–507; Ben Highmore, 'Streets in the Air: Alison and Peter Smithson's Doorstep Philosophy', in *Neo-Avant-Garde and Postmodern: Postwar Architecture in Britain and Beyond*, ed. Mark Crinson and Claire Zimmerman (New Haven and London: Yale University Press, 2010), 79–100; Alistair Black, *Libraries of Light: British Public Library Design in the Long 1960s* (Abingdon: Routledge, 2017).

35   See also William Whyte, 'Halls of Residence at Britain's Civic Universities, 1870–1970', in *Residential Institutions in Britain, 1725–1970: Inmates and Environments*, ed. Jane Hamlett et al. (London: Pickering & Chatto, 2013), 155–66.

36   For Lancaster, see Robert Proctor, *Building the Modern Church: Roman Catholic Church Architecture in Britain, 1955–75* (Abingdon: Ashgate, 2014), 309–10. Eliot College (1965) at the University of Kent includes a chapel.

37   Guy Ortolano, *The Two Cultures Controversy: Science, Literature and Cultural Politics in Postwar Britain* (Cambridge: Cambridge University Press, 2008).

38   'Appendix Five to the Report of the Committee Appointed by the Prime Minister Under the Chairmanship of Lord Robbins, 1961-63. Higher Education in Other Countries', *Parl. Papers* (1963) 13, Cd 2154-V, 3–4, 167–88. For Hutchins see Mary Ann Dzuback, *Robert M. Hutchins: Portrait of an Educator* (Chicago: University of Chicago Press, 1991), ch. 6. For Conant, see *General Education in a Free Society: Report of the Harvard Committee; With an Introduction by James Bryant Conant* (Cambridge, MA: Harvard University Press, 1945).

39   *Universities and Empire: Money and Politics in the Social Sciences during the Cold War*, ed. Christopher Simpson (London: I.B. Tauris, 1999).

40   For the 'loi Faure', see John Walsh, 'France: Universities Face Era of Reform by Decree', *Science*, 3 May 1968, 517–20; Christine Musselin, *The Long March of French Universities* (New York: Routledge, 2004), ch. 3; Prost, *Éducation, Société et Politiques*, ch 7. For a critique of the stranglehold of faculties on French higher education, and a call to emulate the interdisciplinarity of Keele and the German reform universities, see Georges Gusdorf, *L'Université en Question* (Paris: Payot, 1964), 164.

41   *Times*, 16 October 1964, 16 (music at York); T. H. Elkins, 'Geography in a New University', *Geographical Magazine*, 38 (1966), 812–15 (Sussex); [Malcolm Bradbury], 'Sociology and Literature', *Times Literary Supplement*, 4 April 1968, 345; idem., 'The Arrival of Am. Lit.', ibid., 25 July 1968, 789.

42   *Times*, 13 March 1965, 9 (on the Kneale report); A. H. Halsey, 'The Franks Commission', in *The History of the University of Oxford, vol. 8: The Twentieth Century*, ed. Brian Harrison (Oxford: Oxford University Press, 1994), 721–36.

43   Nicholas Hillman, 'From Grants for All to Loans for All: Undergraduate Finance from the Implementation of the Anderson Report (1962) to the Implementation of the Browne Report (2012)', *Contemporary British History*, 27 (2013), 249–70.

44   Gino Martinoli, *L'Università Nello Sviluppo Economico Italiano* (Rome: Giuffrè, 1962).

45 *Aspects of Interdependence between Industry and Universities. Report of a Conference Convened by the Committee of Vice-Chancellors and Principals of the Universities of the United Kingdom and the Confederation of British Industry on 10th and 11th December 1965* (London: Confederation of British Industry, 1966). Along with Frederick Dainton (the recently appointed vice-chancellor of Nottingham), Jack Butterworth, vice-chancellor of Warwick University, represented the Committee of Vice-Chancellors and Principals in liaison meetings with the CBI: 'UIJC. Minutes of Meetings', 10 December 1965, CVCP papers, Modern Records Centre, University of Warwick, MSS 399/3/UIJC/1. For criticism of the Robbins report for not doing enough, see Richard Lynn, *The Universities and the Business Community* (London: Industrial and Educational Research Foundation (1969), 10–13.
46 Carol Dyhouse, 'Troubled Identities: Gender and Status in the History of the Mixed College in English Universities since 1945', *Women's History Review*, 12 (2003), 169–93.
47 Keith Vernon, 'Engagement, Estrangement or Divorce?: The New Universities and Their Communities in the 1960s', *Contemporary British History*, 31 (2017), 501–23.
48 Graeme Moodie and Rowland Eustace, *Power and Authority in British Universities* (London: George. Allen & Unwin, 1974); Jack Straw 'Student Participation in Higher Education', in *Universities: Boundaries of Change*, Granada Guildhall Lectures (London: Panther, 1970), 7–27.
49 For the Young Conservatives in the 1960s, see Lawrence Black, 'The Lost World of Young Conservatism', *Historical Journal*, 51 (2008), 991–1024. On the radical 1960s, see Richard Vinen, *The Long '68: Radical Protest and Its Enemies* (London: Allen Lane, 2018), ch. 3; Caroline Hoefferle, *British Student Activism in the Long Sixties* (London: Routledge, 2013); Nick Thomas, 'Challenging Myths of the 1960s: The Case of Student Protest in Britain', *20th Century British History*, 13 (2002), 277–97.
50 *Times*, 23 May 1970, 3.
51 For the 'high noon' under Crosland, see Carswell, *Government and the Universities*, ch. 8; Shattock, *Making Policy*, 58–61; C. D. Godwin, 'The Origin of the Binary System', *History of Education*, 27 (1998), 171–91. For the Open University, see Peter Dorey, '"Well, Harold Insists on Having It!" – The Political Struggle to Establish the Open University, 1965–67', *Contemporary British History*, 29 (2015), 241–72; Daniel Weinbren, *The Open University: A History* (Manchester: Manchester University Press, 2015).
52 Peter Scott, 'British Universities, 1968–1978', *Paedagogica Europaea*, 13 (1978), 29–43; Shattock, *Making Policy*, 122–4. On the student files protests, see Hoefferle, *British Student Activism*, 136–40. For Amis, see Dalton B. Curtis, Jr., 'From Left to Right: The Paradox of Kingsley Amis' Views of University Education in Post-World War II Britain', *Journal of Philosophy and History of Education*, 66 (2016), 27–42.
53 Anthony S. C. Teo (ed.), *Univer-Cities: Strategic View of the Future: From Berkeley and Cambridge to Singapore and Rising Asia* (New Jersey: World Scientific, 2015). *The Global Future of Higher Education and the Academic Profession: The BRICs and the United States*, ed. Philip Altbach et al. (Basingstoke: Macmillan, 2013).
54 Pawan Agarwal, *Indian Higher Education: Envisioning the Future* (New Delhi: Sage, 2009), ch. 3; *Private Prometheus: Private Higher Education and Development in the 21st Century*, ed. Philip Altbach (Westport: Greenwood, 1999); *Private Higher Education: A Global Revolution*, ed. Philip Altbach and Daniel Levy (Rotterdam: Sense Publishers, 2005). For the latest trends towards the privatization of tertiary education across the OECD, see *Economist*, 15 September 2018, 93.

55   Nicholas Hillman, 'From Grants for All to Loans for All'; Malcolm Tight, 'Student Accommodation in Higher Education in the United Kingdom: Changing Post-War Attitudes', *Oxford Review of Education*, 37 (2011), 109–22; Darren P. Smith, Joanna Sage and Stacey Balsdon, 'The Geographies of Studentification: "Here, There and Everywhere" ?', *Geography*, 99 (2014), 116–27. For current trends in the financing of UK universities, see Andrew McGettigan, *The Great University Gamble: Money, Markets and the Future of Higher Education* (London: Pluto Press, 2013); Thomas Hale and Gonzalo Viña, 'University Challenge: The Race for Money, Students and Status', *Financial Times Magazine*, 23 June 2016, https://www.ft.com/content/c662168a-38c5-11e6-a780-b48ed7b6126f (accessed 4 September 2018); Thomas Hale, 'The Financing of Student Accommodation', *Financial Times*, 7 February 2018, https://ftalphaville.ft.com/2018/02/05/1517828056000000/The-financing-of-student-accommodation/ (accessed 4 September 2018).

56   Marina Warner, 'Why I Quit', *London Review of Books*, 11 September 2014, 42–3; Stefan Collini, *Speaking of Universities* (London: Verso, 2017); Frank Furedi, *What's Happened to the University: A Sociological Exploration of Its Infantilisation* (London: Routledge, 2017); cf. David Willetts, *A University Education* (Oxford: Oxford University Press, 2017).

57   Richard Münch, *Academic Capitalism: Universities in the Global Struggle for Excellence* (New York: Routledge, 2014); Marginson, *The Dream Is Over*; Christopher Newfield, *The Great Mistake: How We Wrecked Public Universities and How We Can Fix Them* (Baltimore: Johns Hopkins University Press, 2016); André Béteille, *Universities at the Crossroads* (New Delhi: Oxford University Press, 2010); *Higher Education in India: In Search of Equality, Quality and Quantity*, ed. Jandhyala B. G. Tilak (New Delhi: Orient Blackswan, 2013); Rosalind M. O. Pritchard, *Neoliberal Developments in Higher Education: The United Kingdom and Germany* (Oxford: Peter Lang, 2011); Jack Grove, 'Macron Vows to Push Ahead with Admissions Reform Despite Protests', *Times Higher*, 18 April 2018, https://www.timeshighereducation.com/news/macron-vows-push-ahead-admissions-reform-despite-protests (accessed 4 September 2018).

58   Jonathan Coulson et al., *University Trends: Contemporary Campus Design* (Abingdon: Routledge, 2014); Ronald Barnett, *The Ecological University: A Feasible Utopia* (Abingdon: Routledge, 2018). For Nalanda, see Amartya Sen, 'India: The Stormy Revival of an International University', *New York Review of Books*, 62 (13 August 2015), 69–71; and for the Nalanda campus design for 2020: https://www.nalandauniv.edu.in/campus-life/upcoming-campus/ (accessed 4 September 2018).

Part One

1

Learning from Redbrick: Utopianism and architectural legacy of the civic universities

by William Whyte

> The new universities of the '60s created breathing space amidst the dusty, churchy culture of Oxbridge, Durham and the like. The Victorian red bricks had their own imposing buildings and values – I'd already had a bewildering false start in one of those. But the new universities sat geographically and culturally apart – a fresh start.
>
> Julia Condon, 'Warwick in 1970'.

It is an encouraging thought, especially for a volume devoted to the new universities of the 1960s. Recalling her student days at Warwick, Julia Condon pictured the place as distinctively different: 'a fresh start'.[1] For her, at any rate, the feel, the culture, even the location of the university marked it out as new and different, neither as reactionary as 'dusty, churchy' Oxbridge nor as daunting as the old and 'imposing' civics. As a leading figure in Warwick's student politics, she is someone whose experience is worth attending to. It matters that she found Warwick freeing, fresh and new. It matters, too, that she campaigned – with many other students there – to preserve and enhance this sense of freedom; that she was willing to join in with the occupations and to contribute to the publications which made Warwick such a *cause célèbre*, such an icon of the new universities more generally. That such a first-hand, first-rate witness identified Warwick as a fresh start and also associated this feeling with the physical environment of the university is likewise very telling.

The architecture of the new universities was indeed intended to speak of this 'fresh start'. It was understood as such by visitors, commentators and many of those living and working in Warwick and elsewhere. For contemporaries, these were the 'plateglass' universities: distinct not only from Oxbridge and redbrick, but also from the smaller 'whitetiles' – Exeter, Reading, Nottingham, Southampton, Leicester and Hull – which had been created and chartered in the first half of the twentieth century.[2] For subsequent historians, too, the architecture of the new universities still speaks of a 'utopianist' moment: a radical break with older patterns of British higher education, a caesura which was embodied and expressed in assertive, distinctively different buildings and plans.[3] In that sense, our witness from Warwick does not just speak for herself, she voices a view which is widely held.

It is, however, worth pausing to consider whether these claims of uniqueness and utopianism – this sense of a 'fresh start' both architectural and educational – actually stack up. In the most minimal sense, after all, every university is utopian. Their foundation, their construction and their operation all rest on the assumption that these academic communities will somehow transform their inhabitants and the world around them.

Nor were the new universities the only institutions which built – and often built big – both to express this idealism and to help foster the utopian communities which universities were intended to be. Examples of this can be found all across the globe. Bold, demonstrative architecture has proved especially attractive to British universities in the last 200 years, with reforming Oxbridge and the ambitious, insecure civics each embracing distinctive styles and modes of building.[4] Small wonder, for as Sir Walter Moberly, the chairman of the University Grants Committee (UGC), observed in 1946, 'From Plato downwards, educators have recognized the powerful indirect influence of physical surroundings.'[5]

As this suggests, the new universities of the 1960s had a pre-history – a pre-history which is architectural as well as educational. Indeed, the chief burden of this chapter is precisely to stress the extent to which these institutions shared a similar debt to the older universities. And by this, I don't mean, as some historians have suggested and some contemporaries implied, that the plateglass universities were little more than an updating of Oxbridge. Keele was no more Balliol in the Potteries than Sussex was, as its detractors put it, 'Balliol-on-sea'.[6] Rather, what I want to insist on is the importance of the other, civic, redbrick tradition in shaping these universities and the architecture they embraced. The 'utopian universities' represent, I shall argue, not a radical departure from English precedent, but rather a series of responses to an ongoing debate about the nature of the university and the architectural expression and planning that this implied.

This is not, of course, to downplay the excitement that the new universities engendered. Nor do I wish to dismiss their remarkable architecture, which has recently and rightly become the focus for serious research. We are not wrong to see them as a series of experiments. We are also right to think that of these campuses as aspects of a broader movement, places which share a similar ethos and achieve similar effects on those who visit them. That Keith Joseph, the cabinet minister responsible for universities, marred his visit to UEA in the early 1980s by speaking throughout as though he was visiting Essex does not just speak of his eccentricity.[7] It also reveals the way in which all these plateglass universities participated in a joint architectural enterprise.

Yet this utopianist moment was absolutely not confined to the new universities. Indeed, it is worth noting that in a more global sense, the seven new universities and their forerunner Keele made very little difference to the educational landscape of Britain. Real expansion and far more profound experimentation – both educational and architectural – went on elsewhere. It was the civic universities – Manchester, Liverpool, Leeds and the rest – which took the largest share of student numbers.

Leeds grew from just over 3,000 to just under 10,000 students in the decade from 1963 to 1973.[8] Manchester went from around 5,000 to almost 15,000 at exactly the same time.[9] And, of course, the sixties witnessed the foundation of a bewildering range of new institutions of higher education – from teacher-training colleges (many of them future universities) to technological universities like Loughborough or Aston. Most important – in conception if not, ultimately, in realization – were the polytechnics, who pioneered not just a new map of learning, but wholly new ideas about the nature of higher education. And they were big. Put it this way, of the 400,000 members of the National Union of Students in 1969, only 42 per cent were at university.[10] And of these, only about one in ten attended the universities this book is about.

Architecturally, what this means is that the great projects of Essex, Sussex and elsewhere were more than matched by developments in other British universities. Manchester, indeed, became known as 'an empire on which the concrete never set', whilst Leeds found itself with a mega-structure that may very well have possessed the longest corridor in Europe.[11] Little wonder that the vice-chancellor of Nottingham should observe in 1964 that his life was now 'largely made up of bricks, mortar and money'.[12] Little wonder too that Malcolm Bradbury should diagnose the mental illness suffered by universities in this period in *The History Man*, published in 1975. It was, he said, 'building mania'. Nothing less than 'an Edifice Complex'.[13] It was true; and the architecture of both the new universities and their older, civic rivals – though not, significantly, as we shall see, that of the polytechnics – bore witness to this general truth.

That the seven new universities of the 1960s should have been so successful in asserting their distinctiveness and unique importance owes much to a persistent problem in the analysis of British higher education. To this day, journalists, educationalists, even novelists have focused overwhelming attention on the two ancient English universities.[14] Their history, their wealth and their social – though not necessarily their educational – prestige have made Oxford and Cambridge the datum against which all other universities have been judged. This was a problem in 1910, when the Vice-Chancellor of Birmingham, Oliver Lodge, complained that 'British people when they think of a University always think of those two'. It remains a significant obstacle to a proper understanding of university history. And it has been especially problematic when assessing the new universities of the 1960s. Our Oxbridge obsession has meant that the implicit – and, indeed, often the explicit – point of comparison has almost always been two institutions, which as Lodge pointed out over a century ago 'are quite exceptional in the world. As universities they are unique'.[15]

This preoccupation with Oxbridge has meant that the very real structural and functional similarities between plateglass and redbrick have tended to be ignored or downplayed. Even at the time, there tended to be a lazy assumption – especially amongst journalists – that Oxford and Cambridge were the models on which the new universities were based. Yet, in fact, their development echoed and was strongly shaped by earlier foundations.

The civic universities of the nineteenth and early twentieth centuries had been founded by urban elites and by local government.[16] Business leaders, local aristocrats and often the bishop took a commanding role in overseeing the institution: a relationship which could produce conflict, but which also yielded financial dividends. The civic universities were thus, in origin, literally civic: local foundations, funded by local people, and overwhelmingly – at least until the 1960s – educating local students.[17] This did not mean, however, that they were insular or narrowly parochial. The fact that, until it became a university in its own right, each institution prepared students for University of London degrees ensured that, from earliest days onwards, it was undeniably part of a larger, national system.

Moreover – and more significantly still – it would be a very real mistake to assume that redbrick grew up isolated or insulated from the state. All universities owed their existence to a charter granted by the crown and were often underwritten by an act of parliament.[18] Just as importantly, from the 1880s onwards, government funding became an ever-increasing source of support. Research (especially scientific research); student scholarships (especially the bursaries paid to prospective teachers, which funded most arts undergraduates until the 1960s); capital grants to erect new buildings or supply new equipment: all this meant that by the mid-twentieth century, the civic universities had become dependent on central government.[19] Their students, too, were now increasingly in receipt of grants – a situation that was finally codified in the 1962 Education Act and its provision of universal financial support.

This more generous student support eroded the link between civic universities and their locality. In 1953, it was noted that fewer than half of students now came from within thirty miles of their university.[20] Ten years later, the statistics were stark: at Leicester 97 per cent of students came from more than thirty miles away; at Liverpool, fewer than one in ten came from within the city.[21] At Leeds, too, the trend was undeniable, with two-thirds of students coming from within a thirty-mile radius in 1938 and two-thirds coming from outside it twenty years later.[22] Small wonder that, in 1963, the well-placed commentator Noel Annan felt able to declare that 'civic universities are no longer provincial'.[23]

In almost every respect, the evolution of the new plateglass universities can be seen as a version – albeit a speeded-up version – of this redbrick story. True enough, these were institutions which, from the very first, depended almost wholly on the state. Founded away – often a very long way away – from any city, they also, of course, relied upon their ability to attract students from far beyond their locality. It is important, too, to note that they were established with degree-awarding powers of their own and were thus not in thrall to the University of London in the way that the early civics had been.[24] In those ways their experience did undeniably differ from the nineteenth-century struggles of earlier foundations. It is nonetheless quite clear, however, that these aspects of new university development merely reproduced the changes which had already occurred to the civic university model as it had developed in the twentieth century. State support, student mobility, academic autonomy: these were all experiments that had been tried and tested at redbrick before they were implemented by plateglass.

The process of imposing a model derived from the experience of the civics can be seen not least in the decisions made about where to found these new universities. Most writers have emphasized the fact that the choice of location was one taken centrally and have thus tended to see it as further proof of how different the new universities were from their locally inspired, locally driven civic predecessors.[25] On further investigation, however, this sharp dichotomy soon breaks down, not least – as we have already noted – because the state had always taken an interest in universities and had, in the first half of the twentieth century, begun to take an increasingly strategic interest in where universities should be situated. After all, one of the strongest arguments for the foundation of Keele in 1949 was precisely that it was located in the largest conurbation in England with no permanent provision of higher education.[26]

It is still more noteworthy that the final judgement on sites for the new universities was determined as much by the enthusiasm of local communities as it was by any wider issues. As Asa Briggs observed in 1963, they 'depended on local initiative. They have not simply been located in particular places as part of national educational planning'.[27] To that end, the places which were successful in the government-run bidding war to acquire a university tended to mobilize exactly the same range of civic worthies – councillors, peers and the local bishop – who had occupied a similar role at redbrick.[28]

These structural similarities had architectural effects. For one thing, as Diane Chablo has argued, 'the established and especially "red-brick" institutions became testbeds for planning ideas which would later strongly influence the new foundations'.[29] At a basic – indeed, at a fundamental level – the very terms in which architectural decisions were made owed much to the experience of redbrick. Birmingham offered the first example of a campus university, built outside the centre of town some five decades before the greenfield sites of the plateglass institutions were even contemplated.[30] Hull seems to have been the first university to have been described as a 'campus', as it laid out plans for its site on what were then the outskirts of the town in 1927.[31] And in 1946, the distinguished town planner Lord Holford was employed to produce a report for the University of Liverpool which was to prove seminal for all subsequent universities – civic and 'utopian' alike.[32] Although he recommended remaining within the city, his conclusion that the campus should be 200 acres in size was to become normative for the sector as a whole, not least because it was established as the standard by which the University Grants Committee assessed all possible locations for new universities.[33] It is why every one of the new foundations required a 200-acre greenfield site and why each of the successful bids to obtain a university was supported by a local authority gift of at least that much land.[34]

The redbrick universities also had more direct contributions to make. A competition held in 1953 to re-plan Sheffield University, for instance, proved to be highly influential, not least because it revealed the radical thinking of the architects Alison and Peter Smithson.[35] As Elain Harwood observes, their unplaced entry, together with another losing plan by James Stirling and Alan Cordingly, 'became models for the continuous teaching blocks of the 1960s'.[36] Their decks and high-level walkways would reappear at Lancaster, Essex and elsewhere. The 1960 Leeds *Development Plan*, by Chamberlin, Powell and Bon, was even more important, and, with its organizational diagrams, flowcharts, and time and motion studies, it proved formative for every other subsequent proposal.[37] It mattered not least because it offered a methodology for analysing the architectural needs of universities – and it was one that eagerly seized upon in the rush to build. 'We are always getting requests for this Development Plan,' noted G. B. Oddie of the University Grants Committee in 1962,[38] and its influence can indeed be felt in the approach and even the format taken by the development plans of the new universities.[39]

The Leeds plan, argued Tony Birks in his key text of 1972, *Building the New Universities*, was noteworthy because it 'was the first of its kind to tackle the study of a whole university environment'. It mattered, he went on to argue, because it aimed to create a place 'which would give a sense of completeness from the earliest stage'. It was also, Birks emphasized, 'the origin of the concept of the "ten-minute university."'[40] In all these respects, it was influential on other redbrick and on the newer universities. Indeed, like the 'bulge' and the 'trend',[41] and like the idea of the 'nine to five university', the 'ten-minute university' is a term of art that was ever present in the 1960s even if it now needs some explaining.[42] The 'bulge' was the post-war baby boom. The 'trend' was the increasing tendency of teenagers to acquire the qualifications necessary for entry to higher education. It was these two forces which drove university expansion.[43] The 'ten-minute' university was to be a solution – the solution – to the problem of how this growing group of students was to be integrated into a cohesive community. Good design, it was hoped, would place every part of the university within a ten-minute walk. Or, as an Essex student put it, 'everything will be within walking distance – just like the Queen Mary'.[44] The 'ten-minute university' would bring teaching and socializing together. In that way, it was intended to replace that other sixties institution: the 'nine-to-five university' – a place empty and barren in the evenings and at weekends because everyone had gone home. This, it was believed, would be 'the greatest single development which must be expected to influence the character of the University'.[45]

As this makes plain, the architectural experimentation of the civic universities was not just a pragmatic response to rising student numbers. It was also the product of a more profound attempt to transform and reform university life: one that would help shape the new universities and one that predated their foundation by decades. For the rhetoric of the nine-to-five university was not new in the 1960s; nor was the anxiety that universities were failing to create vibrant academic communities. Indeed, it is clear that the development of higher education and the design of university buildings in this period were both critically shaped by concerns which had grown up about the civic universities in the preceding forty years. Books like Walter Moberly's 1948 *Crisis in the University*, H. G. G. Herklots's *The New Universities* of 1928 and – above all – 'Bruce Truscot's' *Red Brick University* of 1943 had set out a clear critique of the civic universities.[46] Again and again, they were criticized as 'nine-to-five universities'. They were attacked for their lack of residential accommodation, for the separation between staff and students, for the over-specialization of their studies; above all they were damned for their failure to function as real academic communities. This academic critique always had an architectural aspect – indeed, new university buildings were seen as a crucial way of solving the perceived problems of redbrick. 'Only in attractive surroundings,' wrote Herklots, 'will Dental Jones be encouraged to discuss his own problems and the problems of metaphysics with Anthropological Robinson and Architectural Smith.'[47]

In the years after the war, the civic universities sought to remedy their faults. They built halls of residence.[48] Several – like Reading – moved out of town.[49] And even those that remained within their home cities built big in order to reform, seeking to remove the terraced houses that surrounded them, plunging those departments at Leeds which occupied these dwellings into what was described as a 'competition to get oneself demolished'.[50] The rush to plan these expanded campuses; the determination to build halls of residence; the desire to escape 'the nine-to-five university': it was these factors which explain what happened at redbrick in the years after the Second World War. These factors, too, help account for what would happen at the new universities as they sought to overcome the problems their older rivals were seeking to solve. As Michael Beloff acutely noted in his survey of the *Plateglass universities*: 'More significant than any mystical attachment to Oxbridge ideals', amongst the founding vice chancellors, 'was a reaction against the pattern of redbrick life, which most of them knew equally well.'[51] It was a reaction shared with those who remained at redbrick and who had been rebuilding their universities in response.

That this was true for even the most ostensibly radical new institutions can be seen at Essex, which, because of its founding Vice-Chancellor Albert Sloman's Reith lectures in 1963, became – for a while – a particularly important symbol of utopianism.[52] Sloman claimed that the model was American and that Essex would become the British MIT. Yet, it was far from incidental that Sloman had succeeded 'Bruce Truscot' – the redbricks' most acute critic – as professor of Spanish at Liverpool. It was certainly this experience in a civic university, where he served on the committee charged with carrying out Holford's masterplan, rather more than his time as a student at Oxford or even his brief exposure to life in America, that shaped his plans at Essex.[53] In that way, even this apparently radical break with the established order owed much to the debates which also helped to rebuild older, ostensibly less utopian institutions. Indeed, with its insistence on residence, its escape from the city and its attempts to create a coherent, modern 200-acre campus, Essex can be seen as a product of the same reforming impulse that had been reshaping the whole of British higher education for several decades.

The same experience of learning from redbrick – learning, that is from both its perceived failures and from the ways in which the civic universities were seeking to address them – can be found in the institution with which we began. Warwick – yet another university intended to be the British MIT – was far from being a radical break with the redbrick tradition. It was, in many ways, an archetypal civic foundation, with the case for the new university 'championed', in Michael Shattock's words, 'by a mixed group of industrialists, trade unionists, civic authorities and a local intelligentsia of school teachers and college lecturers'.[54] Indeed, as he goes on to point out, unlike several of the other new universities, Warwick initially lacked the kind of county support – from the gentry and political leaders from the surrounding countryside – which would prove so important at Kent, York and elsewhere. It was, in fact, as a means of enticing precisely this sort of backing that the proposed University of Coventry was renamed Warwick. It was a successful move, ensuring that Warwickshire County Council helped provide a site and local district councils were willing to stump up some cash. 'Nuneaton will not be found wanting,' declared at least one local authority.[55] And it was not.

A university for Coventry was a project first proposed in 1943 and then first seriously considered at a meeting in 1954, whose agenda began with a welcome from the Lord

Mayor, proceeded to words from the Lord Bishop and went on to assert 'that in view of the high status in industry now attained by the City', a centre for higher education had now become necessary.[56] 'Civic pride', believed the local Director of Education, 'could sway the city council' – and, indeed, it did, with a newly elected Labour administration coming to embrace the idea with enthusiasm.[57] 'Labour leaders', as Elain Harwood puts it, 'wanted a seat of wider learning appropriate to an ambitious working-class city that had just opened a new theatre, art gallery and a school of music'. They wanted, in other words, a civic university. The decision of the University Grants Committee to support this application for a new Midlands university likewise owed much to the argument that Coventry was now the largest English city without its own place of higher education.[58] In that way, both local and national governments were consciously engaged in the process of creating a modern redbrick university. Of course, the local sponsors were always liable to be overruled by their national paymasters. But that had also been true for the civic foundations as they likewise sought university status.[59]

Architecturally, too, the development of the university followed surprisingly familiar tracks. Like so many redbrick institutions, the embryonic university turned to a leading local architect as it sought to rally support for the enterprise. Arthur Ling, the city architect, consequently drew up plans for an urban campus, part of a civic and cultural precinct which included the art gallery and museum, the swimming baths, the local technical college and the rebuilt Coventry cathedral.[60] Inspired in its scale by Keele and in its form by the desire to 'establish the University as a self-contained community complete in every respect', it was, in other words, an idealized civic university: rooted in its urban environment yet capable of sustaining the academic and social life which redbrick was often thought to lack.[61] Like many of its older rivals, as the ambitions for the university grew greater, so this constricted city-centre site came to seem too small. Instead – inspired by 'the setting of the University of Exeter, whose buildings cluster … on a hill' – the scheme's backers began to look outside Coventry for a new location.[62]

Thus, it was that they settled on Gibbet Hill: a far from romantic name to be sure, but a place helpfully offering the 200 acres of land which were regarded as the bare minimum for university purposes.[63] Ling subsequently produced a new set of proposals for a much larger institution. It would, enthused the *Birmingham Post*, ensure that 'most of the academic areas will be sited within ten minutes' walking distance of each other'. It would, the newspaper went on, guarantee 'that the whole university is a lively place at all times and not merely a "nine-to-five" institution'.[64] The journalist was, of course, just quoting the university's Development Plan, but this – with its 'ten minute' university and its desire to avoid becoming 'a "nine-to-five" institution' – was simply reprising the usual rhetoric about redbrick.[65] There was very little new and much less that was truly distinctive here.

For Ling, the move out of the city was a terrible mistake.[66] Like Holford at Liverpool, the architect could not really accept that a civic university could be built beyond the civic centre.[67] Losing faith in the project, he was eventually succeeded by Yorke, Rosenberg and Mardall, who produced a new plan.[68] Yet even this apparent change of tack concealed some fundamental continuities, not least of which was an ongoing emphasis on the need for residential accommodation and a continued desire to use buildings to foster what the vice-chancellor termed 'a lively university society'.[69] To that end, the methodology used to draw up and explain this new set of proposals was beholden to the Leeds Development Plan, and the debates about redbrick which had preceded it.[70]

Looking at the responses to this architecture, we can also see that the subsequent complaints about the new university have a ring of familiarity. Even the gripe from one undergraduate that it now took fifteen minutes to cross the campus spoke of the old, civic-inspired desire for a 'ten minute university'.[71] The student demand for a union building – a central feature at so many redbrick universities – particularly drew on some ubiquitous tropes. This is especially noteworthy not least because it was the 'struggle' for a 'social building' which sparked the student radicalism for which Warwick became famous.[72] Without a union building, argued one campaigner, 'the habit of only being on campus between the hours of nine and five will be encouraged to grow'. It was an argument that had been used at redbrick for decades; and indeed, redbrick proved to the model to which the radical students of Warwick hoped to cleave. 'A very similar problem was encountered at Bristol University,' observed the student newspaper; 'there the problem of isolation was only overcome by the provision of a union building'.[73]

Warwick was, of course, not Essex, any more than Essex was Lancaster, or Lancaster was Sussex or York. Precisely because each campus was experimental and each institution driven by different leaders and different locations, each university was distinctive. Warwick's history made it particularly close to the civic ideal. Its origins in Coventry; its success in mobilizing local business leaders – not least the motor magnate Lord Rootes; its early incarnation as a campus at the heart of a new cultural precinct; and the continued attempts made to merge it with the local technical college: these – and other –influences made Warwick look very much like a redbrick foundation.[74] In many respects, its evolution also makes it look more a part of that other sixties university movement – the technological universities of Aston, Salford, Loughborough and elsewhere – than just another of the 'utopian foundations' this book deals with. Its migration from town to countryside and its development of a large, low-rise campus, one might be tempted to remark, are especially reminiscent of Bath.[75]

Yet, by the same token, similar points of comparison can be observed in all the other new universities. Essex, for instance, was equally intended to be a local university: a powerhouse of scientific research, closely linked to the electronics giant Marconi and the Post Office Research laboratories which were moving to nearby Suffolk. 'This scientific bias', proclaimed the *Sunday Telegraph*, 'will favour local industry. Town and gown will be linked by hand and research'.[76] Lancaster was likewise hailed as an institution which would abolish the 'nine-to-five university'.[77] 'All seven universities', observed the *Financial Times*, 'have the same end in view. The idea is to encourage the students, including those living in lodgings, to spend as much time as possible actually on the university site'.[78] This was exactly what the civic universities had been attempting to do for the best part of half a century. It was very far from being a wholly 'fresh start'. In their architecture, in their ethos, and in their belief that ethos could be shaped by architecture, all the new universities were learning some very old lessons.

Tracing this genealogy, and observing these family resemblances, does not just call into question the novelty of the new universities. It should also force us to look again at whether they – and their architecture – were distinctively utopian. For the assumptions which justified the massive plans and enormous expenditure on buildings turn out, in this analysis, to be typical of all the universities founded in Britain from the late-nineteenth century onwards and hardly confined to the seven new universities opened in the 1960s. This was true of the ideas that drove design. It was true of the means regularly trotted out to solve problems. And, from the publication of the Leeds Development Plan, onwards, it was also true of the methodologies used to investigate the problems, both at the civic universities and their newer rivals.

In two very specific respects, indeed, we might conclude that the debt which plateglass owed to redbrick produced architectural outcomes which were highly traditional, no matter how 'utopian' the form that it took. The first was the insistence on residence. The desire to create community by providing on-site accommodation was one which was first felt at the civic universities in the interwar period.[79] It was a desire which prompted the founders of Keele to insist on 'total residence', with all students and staff alike housed on the new university campus.[80] And it was this desire which prompted a variety of experiments in the new universities – from the tower blocks of Essex to the Ziggurats of UEA to what were called colleges, but were in fact really halls of residence, at Kent, Lancaster and York. At Warwick, too, all the different plans were predicated on the need to provide accommodation of some sort for undergraduates. In 1938 National Union of Students summed up a generation's discussion of the subject: 'A University without hostels is not a University.'[81] The same idea self-evidently still informed universities being built thirty years later.

The halls, however, are just one aspect of a more profound failure on the part of the new universities really to reimagine university architecture or, for that matter, university life. Their emphasis on accommodation speaks of a profound belief in the power of buildings to create an idealized university. From the nineteenth century onwards, the civic universities had built – and built big. They had built to solve a problem. In 1856, seeking to explain the apparent failure of Owens College Manchester, one academic had pointed unflinchingly at the 'absence of *external symbolism* of Collegiate life …. the *want of a Collegiate building*'.[82] Alfred Waterhouse's magnificent new home on Manchester's Oxford Road had been the solution to this problem – and one that was soon aped elsewhere. It was for this reason, after all, that Bruce Truscott had seized on the term 'red brick' to describe these institutions. At Leeds in the 1920s, Waterhouse's Gothic had come to seem small and cramped and – above all – old-fashioned, and so a new and grander neo-classical building was erected, intended, in the words of its architect, to 'dominate the neighbourhood and be seen from all directions and symbolise the University'. The rebuilt Leeds of the 1960s – *Development Plan* and all – was just another part of this ongoing story, as the university continued to put its faith in grandiose architecture as a way of asserting its importance and resolving its anxieties.[83] It was a process that can be traced at all the redbrick universities – and then at plateglass, too. Moreover, by embracing the rationale of the Leeds *Development Plan*, the new universities were confirming their commitment to this tradition: a tradition which would yield a remarkable range of modernist campuses; all formally very different and new; all functionally, philosophically, very similar and rather traditional.

This belief in magnificent, monumental architecture had justified the original redbrick buildings and continued to legitimate their massive extensions. Even when the new universities built low-rise – as at Lancaster or Warwick, say – they integrated this into precisely sort of monumental masterplan that had been evolved at Liverpool and then updated at Leeds. The belief in residential accommodation, and in the need to house students as well as educate them, likewise led to extraordinary expenditure on halls – a legacy still felt today, with British students still more likely to live away from home than almost any other nationality in Europe.[84] Critics at the time condemned such apparent conventionalism, the perpetuation of old-established tradition. In a remarkable, radical critique, the architect Cedric Price produced a truly utopian plan for a mobile university: the 'Potteries Thinkbelt', carried on the railway lines that snaked through Staffordshire. This would be, he argued, 'a major industrial undertaking, not a service run by gentleman for the few'.[85] It was, without doubt, a visionary proposal. Yet it came to nothing and had no impact on the development of higher education in Britain. Tellingly, the only purchaser of the full documentation was the Ministry of Housing rather than the Department of Education.[86] The monumental tradition established by redbrick was simply too strong to resist.

Price's attack was directed primarily against Keele, the least monumental, but most single-mindedly fixated on student residence of all the post-war universities. Other critics were more wide-ranging in their condemnations.[87] Writing in 1969, Ian Brown deplored the 'sad' and 'ludicrous' Sussex; the 'wedding cake architecture' of UEA; the terrible temptation in the new universities to imitate 'the monumental symbolic stasis of … redbrick'.[88] Writing three years earlier, Reyner Banham was even more assertive. 'Where is the vice-chancellor with sufficient self-confidence to say, "We won't commit ourselves to permanent buildings until we see how it works,"' he asked in 1966. And 'in the absence of such paragons of academic modesty, where is the architect who has the guts to say, "Don't do it. Call me back in three quinquennia, when you begin to know what you're doing"'.[89] It was a rhetorical question and Banham knew it. One of the reasons that the architecture of these institutions was so remarkable is precisely because it was so conventional: so conventionally grand; a legacy of redbrick and its tradition of large-scale university building.

The same was not true – at least at first – in the other institutions founded fifty years ago. Funded by local authorities, and generally built by teams of local authority architects rather than the big names responsible for places like Sussex, the teacher-training colleges, art schools and – above all – the polytechnics did not embrace this monumental architecture, this obsession with residential accommodation or this belief that a community could be produced if only the architect could draw up the right plans. From the start of the polytechnic experiment, writers celebrated 'the dispersed polytechnic', with its split-site, multifunctional, low-key buildings – well adapted, or so they argued, to the more human scale of a non-elite institution.[90] The result was rarely breathtaking. At Leicester Polytechnic, for instance, the central Students' Union was designed as an experimental prototype by the Department of Education, and known simply as 'the shed'.[91] But it was, in many ways, precisely this reticence – this willingness to experiment and to reject the monumental impulses of the past – which marked it out as something special, something new. Here, if you like, was something genuinely utopian: an architecture built at the human scale, well-integrated into the urban fabric, envisaged not as a self-contained, exclusive community, but what the polytechnics' boosters called a 'people's university'.[92]

In the end, of course, that utopian moment proved even shorter than the one this book is actually about. Within decades the polytechnics were aping the civic universities too – in their names and in their form. They are still doing it. Leicester Polytechnic became De Montfort University and now dominates the centre of the town with its prize-winning buildings, well-appointed halls of residence and assertive monumental forms.[93] The University of Lincoln – once Humberside Polytechnic –now proudly proclaims its identity as 'The New Civic University'.[94] It has moved to a sixteen-hectare campus on a greenfield site, the first entirely new university landscape created since the 1960s.[95] This is intended to be a major intervention in the city: symbolizing its importance and linking the university to the urban community which gave its name.[96] It all sounds rather familiar – and it is just one example among many, for the urge to emulate the civic universities has not gone away. In total, it was estimated that British universities expected to issue contracts for £2 billion of building between 2017 and 2020, in addition to the £1.5 billion already signed off in the first six months of 2016.[97] The pressure to build – to build big, to attract students and assert an institution's importance – has, if anything, only got worse. It is a situation instantly recognizable to anyone who has studied the new universities of the 1960s: the playing out of a paradigm established long before this supposed 'fresh start'. Almost everyone, it turns out – even the most apparently new, innovative, and utopian – was then and is now learning from redbrick.

Notes

1 Julia Condon, 'Warwick in 1970', in *Warwick University Ltd*, ed. E. P. Thompson, 2nd edn. (Nottingham: Spokesman Books, 2014), xii–xiii.
2 Michael Beloff, *The Plateglass Universities* (London: Secker and Warburg, 1968), 11–13.
3 Stefan Muthesius, *The Postwar University: Utopianist Campus and College* (New Haven and London: Yale University Press, 2000).
4 William Whyte, *Redbrick: A Social and Architectural History of Britain's Civic Universities* (Oxford: Oxford University Press, 2015).
5 Walter Moberly, *The Crisis in the University* (London: SCM Press, 1949), 218.
6 John McKean, 'Built Community, Model Universe: the English University of the 1960s', in *Man-made Future: Planning, Education and Design in Mid-20th century Britain*, ed. Ian Boyd-Whyte (Abingdon: Routledge, 2007), 211.
7 Michael Sanderson, *The History of the University of East Anglia, Norwich* (London: Hambledon, 2003), 321.
8 P. H. J. H. Gosden, 'The Student Body', in *Studies in the History of a University, 1874–1974*, ed. P. H. J. H. Gosden and A. J. Taylor (Leeds: E. J. Arnold, 1975), 48.
9 Brian Pullan and Michele Abendstern, *A History of the University of Manchester, 1951–73* (Manchester: Manchester University Press, 2000), 265.
10 Eric Ashby and Mary Anderson, *The Rise of the Student Estate in Britain* (London: Macmillan, 1970), 91. The actual figure is 407,000.

11 Pullan and Abendstern, *Manchester*, 106; William Whyte, 'The Modernist Moment at the University of Leeds, 1957–1977', *Historical Journal*, 51 (2008), 169-93.
12 *Gongster*, 20 November 1964, 1.
13 Malcolm Bradbury, *The History Man* (1975 London: Picador, 2012), 50.
14 Ian Carter, *Ancient Cultures of Conceit: British University Fiction in the Post-War Years* (London and New York: Routledge, 1990).
15 J. H. Poynting, *Sir Oliver Lodge: A Biographical Sketch* (Birmingham: privately printed, 1910), 13.
16 David R. Jones, *The Origins of Civic Universities: Manchester, Leeds, and Liverpool* (London: Routledge, 1988).
17 Nuffield College (Education Sub-Committee), *The Problem Facing British Universities* (London: Oxford University Press 1948), 54.
18 Keith Vernon, 'British Universities and the State, 1880–1914', *History of Education*, 30 (2001), 251-71.
19 Peter Alter, *The Reluctant Patron: Science and the State in Britain, 1850–1920* (Oxford: Berg, 1987); Robert O. Berdahl, *British Universities and the State* (Berkeley and London: University of California Press, 1959).
20 University Grants Committee, 'University Development. Report on the Years 1947 to 1952', *Parl. Papers* (1952-3), xvii, 24.
21 Brian Burch, *The University of Leicester: A History, 1921–96* (Leicester: University of Leicester, 1996), 37; Thomas Kelly, *For the Advancement of Learning: University College and University of Liverpool, 1881–1981* (Liverpool: Liverpool University Press, 1981), 330.
22 *University of Leeds Development Plan* (Leeds: University of Leeds 1960), 9.
23 Noel Annan, 'The Universities', *Encounter*, 20 (April 1963), 6.
24 The boldest claim for distinctiveness can be found in *The Idea of a New University: An Experiment in Sussex*, ed. David Daiches (London: Andre Deutsch,1964).
25 Michael Shattock, *Making Policy in Higher Education 1945–2011* (Maidenhead: Open University Press, 2012), 43.
26 Whyte, *Redbrick,* 221-2.
27 Asa Briggs: 'Educational and Social Objectives. The Thinking behind Britain's New Universities', *Architectural Review*, 134 (1963), 233.
28 Michael Shattock, 'The Impact of a New University on Its Community: The University of Warwick', in *Universities and the Creation of Wealth*, ed. Harry Grey (Buckingham: Open University Press, 1999), 109.
29 Diane Chablo, 'University Architecture in Britain, 1950–75' (unpublished DPhil thesis, Oxford University, 1987), 21, 46.
30 Eric Ives, 'A New Campus', in *The First Civic University: Birmingham, 1880–1980*, ed. Eric Ives, Diane Drummond and Leonard Schwartz (Birmingham: University of Birmingham Press, 2000), 111–30.
31 A. E. Morgan, 'The Project of a University of Hull', *University Bulletin* 5 (April 1927), 54.
32 William Holford, *Proposals for the Development of a Site for the University of Liverpool* (Liverpool: University Press, 1949).
33 University Grants Committee, 'University Development 1957–1962', *Parl. Papers* (1963-4), xx, 99.
34 Tony Birks, *Building the New Universities* (Newton Abbot: David and Charles, 1972), 17–18. The one exception was York, where the Rowntree Trust provided this support.
35 Alison Smithson, *Urban Structuring: Studies of Alison & Peter Smithson* (London: Studio Vista, 1967), 46–8.

36 Elain Harwood, *Space, Hope, and Brutalism: English Architecture, 1945–1975* (New Haven and London: Yale University Press, 2015), 241.
37 Whyte, 'The Modernist Moment at the University of Leeds, 1957–1977', 182–6.
38 Oddie to G. Wilson (27 August 1962), Buildings, Development Plan (General), University of Leeds Archive.
39 Harwood, *Space, Hope, and Brutalism*, 255.
40 Birks, *Building the New Universities*, 18.
41 University Grants Committee, 'University Development 1957–1962', 65.
42 J. A. Brennan, *Redbrick University: A Guide for Parents, Sixth-Formers and Students* (Oxford: Pergamon Press, 1969), 10.
43 Shattock, *Making Policy in Higher Education*, 47.
44 Quoted in Beloff, *Plateglass Universities*, 113.
45 Peter Chamberlin, quoted in Whyte, 'Modernist Moment', 183.
46 Whyte, *Redbrick*, ch. 7.
47 H. G. G. Herklots, *The New Universities: An External Examination* (London: Benn, 1928), 55.
48 William Whyte, 'Halls of Residence at the British Civic Universities, 1870–1970', in *Residential Institutions in Britain, 1725–1970: Inmates and Environments*, ed. Jane Hamlett et al. (London: Pickering & Chatto, 2013), 155–66.
49 *Turning Points: the Memoirs of Lord Wolfenden* (London: Bodley Head, 1976), 111.
50 Maurice Beresford, *Walks Round Red Brick* (Leeds: Thoresby Society, 1980), 26.
51 Beloff, *Plateglass Universities*, 189–90.
52 Albert Sloman, *A University in the Making* (London: BBC, 1963).
53 Jules Lubbock, 'The Counter-Modernist Sublime: The Campus of the University of Essex', *Twentieth-Century Architecture*, 6 (2002), 109, 111.
54 Shattock, 'Impact of a New University', 109.
55 *Coventry Evening Telegraph*, 22 June 1964.
56 'Proposed University for Centre of Coventry' (2 April 1954), Henry Rees papers, UWA/F/PP/4/4i Modern Records Centre, University of Warwick.
57 Henry Rees, *A University Is Born: The Story of the Foundation of the University of Warwick* (Coventry: Avalon Books, 1999), 11.
58 Michael Shattock, *Making a University: A Celebration of Warwick's First 25 Years* (Warwick: University of Warwick, 1991), 16.
59 See, for instance, the battles between Nottingham council and the UGC described in John Beckett, *Nottingham: A History of Britain's Global University* (Woodbridge: Boydell Press, 2016), 44–5.
60 Gould Jeremy and Caroline Gould, *Coventry: The Making of a Modern City* (Swindon: Historic England, 2016), 49–53.
61 'Proposed University College' (1958), Henry Rees Papers, UWA/F/PP/4/4i, Modern Records Centre, University of Warwick.
62 Rees, *A University is Born*, 9.
63 In fact, Warwick ended up with a 400-acre site, with half given by Coventry Council and the other half bought from Warwickshire County Council.
64 *Birmingham Post*, 10 April 1964.
65 'University of Warwick Development Plan' (1964), UWA/LIB/1/6, Modern Records Centre, University of Warwick.
66 Gould, *Coventry*, 65.
67 'Coventry University College within the City' (1959), UWA/F/PP/4/4iii, Modern Records Centre, University of Warwick.

68 *The Architecture of Yorke Rosenberg Mardall 1944–1972* (London: Lund Humphries, 1972), 54–9.
69 *The Times*, 28 September 1966, 12.
70 'University of Warwick Development Plan' (1966), UWA/PUB/PD/1/2, Modern Records Centre, University of Warwick.
71 Thompson, *Warwick University Ltd*, 43.
72 Ibid., 59.
73 *Campus* 10 (12 January 1968), 4–5.
74 Plans for a merger with Lanchester College were only abandoned when rejected by the secretary of state, Tony Crosland, in 1965. Lanchester would become Coventry University in 1992.
75 Peter Venables, *Higher Education Developments: The Technological Universities, 1956–76* (London and Boston: Faber, 1978).
76 *Sunday Telegraph*, 27 October 1963.
77 *The Times*, 14 April 1964, 18.
78 *Financial Times*, 9 May 1964, 9.
79 Harold Silver, '"Residence" and "Accommodation" in Higher Education: Abandoning a Tradition', *Journal of Educational Administration and History*, 36 (2004), 123–33.
80 W. B. Gallie, *A New University: A. D. Lindsay and the Keele Experiment* (London: Chatto & Windus, 1960), 62.
81 *The Challenge to the University: A Report of the 1938 Congress of the National Union of Students of England and Wales on University Life and Teaching in Relation to the Needs of Modern Society* (London: National Union of Students, 1938), 44.
82 Manchester University Archive, OCA/5/1/2, Owens Trustees Minutes, 1855–61, W. C. Williamson (1856).
83 Whyte, 'Modernist Moment'.
84 William Whyte, 'Somewhere to Live: Why British Students Study Away from Home – and Why It Matters' (Oxford: HEPI, 2019).
85 Cedric Price, 'Potteries Thinkbelt: A Plan for an Advanced Educational Industry in North Staffordshire', *Architectural Design*, 36 (1966), 484.
86 Royston Landau, 'A Philosophy of Enabling', in *The Square Book*, ed. Cedric Price (Chichester: Wiley-Academy, 2003), 15.
87 See also J. K. Page, 'Academic Goals and University Buildings', *Universities Quarterly*, 18 (1963–4), 301–8.
88 Ian Brown, 'The Irrelevance of University Architecture', *Higher Education Review*, 2 (1969–70), 31–55.
89 Reyner Banham, 'The Outhouses of Academe', *New Society* 8 (1966), 547.
90 Ian Brown, 'The Dispersed Polytechnic', *Higher Education Review*, 3 (1971), 25–45.
91 *Leicester Polytechnic Student Handbook, 1975–6* (Leicester: Leicester Polytechnic, 1975), 8.
92 Eric Robinson, *The New Polytechnics: The People's Universities* (Harmondsworth: Penguin, 1968).
93 Arthur Lyons, *The Architecture of the Universities of Leicester* (Syston: Anchorprint, 2010).
94 University of Lincoln, *The New Civic University: A University Working with Its Communities* (Lincoln: University of Lincoln, 2018).
95 Brian Edwards, *University Architecture* (London: Spon Press, 2000), 31–2.
96 Robert Maxwell et al., *Rick Mather Architects* (London: Black Dog Publishing, 2006), 25–7.
97 https://next.ft.com/content/03522a1c-4a9b-11e6-8d68-72e9211e86ab (accessed 25 July 2016).

2

Keele: Post-war pioneer

by Miles Taylor

Introduction

Like pilgrims treading a path to a holy shrine, the new universities came to Keele in the summer of 1964. They arrived in two waves. At the beginning of May there was a festival of the new universities organized by students and sponsored by the *Sun* newspaper, then, lest we forget, a tribune for progressive causes. There was jazz, theatre, poetry and an 'American' rock band (actually, it turned out, from Sussex). There were seminars and debates, including a forum on 'the aims of education' addressed by the conservative Secretary of State, Edward Boyle. And there was an exhibition of the architectural plans of the new campuses. Then it was the turn of the management. In the middle of July, Dr Harold Taylor, vice-chancellor of the University of Keele, hosted a two-day conference attended by senior delegates from all the new English campus universities. Two delegations came from overseas as well: from the University of Port Elizabeth in South Africa, founded controversially in 1964 as a bilingual Afrikaans-English speaking campus. And from the Centre for the Study of Higher Education at Berkeley in California came Thomas McConnell, collecting material for his comparative study on the changing environment of higher education. The gathering of vice-chancellors was a more serious affair than the students' festival. They discussed the problems of development plans, residence, the curriculum and university government. And they looked forward to the future. As the rapporteur to the conference wrote up his report at the end of July, the government announced plans for a further new university campus, this time at Stirling in Scotland.[1]

Although never considered a 'plateglass university', Keele is always included in any discussion about the new universities of the 1960s. As these two events in the summer of 1964 suggest, Keele was viewed as a pioneer in the modernization of higher education in the 1960s. Founded in 1949 as the University College of North

---

In preparing this chapter I was given invaluable assistance by Helen Burton in the Keele Special Collections and Archives. I am grateful as well to former Keele staff who allowed me to take them down memory lane in a series of interviews: Frank Burchill, David Cohen, Gordon Fyfe, Audrey Newsome and John Rogers. And I would like to thank Tony Halmos for providing information about his father.

Staffordshire and given full university status in 1962, Keele paved the way for many of the developments associated with the new campuses of the 1960s. As John Fulton, the first vice-chancellor of Sussex, put it later, the fight for Keele was 'no push-over', and the battles won at Keele cleared the way for Sussex and the rest.[2] Admittedly, to this day no-one is quite sure where Keele is. The coming of the M6 motorway in 1963 helped, not least because a service station was built at Keele. Nonetheless, the university college at Keele was a landmark in terms of higher education. Keele was a residential campus university, not the first in Britain: that accolade goes either to the University College of Wales at Swansea which relocated out of town to a new campus at Singleton Park in 1948 or to the University of Reading, which went suburban at Whiteknights in 1947. But Keele championed the principle of staff and students living together side-by-side on one site with an enthusiasm like nowhere else. In 1956, one reporter likened the atmosphere to an Israeli kibbutz.[3] Keele also broke the mould in awarding full degrees rather than have them certified by the University of London. And Keele was most unique in its experimental undergraduate curriculum: a four-year degree with the first year devoted to a foundation course in which all undergraduates studied science and non-science subjects alike. By 1962 when Keele became a university proper, it boasted many of the trappings of a 1960s campus. There were signature buildings designed by modernist architects, notably the new student union (John Stillman) with its hints of Bauhaus, and the Chancellors' Building (Bridgeman, Shepheard and Epstein). There were vogueish new disciplines too. Keele was the first campus in the UK to run an American studies programme; it became the editorial headquarters of the prestigious *Sociological Review* in 1953 and the home of the periodical *Universities Poetry* (1958–65). A department of communications was added in 1961, funded by Granada Television. And its university chapel was the first campus faith building in Britain built to offer completely ecumenical Christian worship, laid on in the early years by George Youell, who shocked his fellow churchmen with his 'rebel' con-celebration services.[4] On the face of it, Keele seems to belong to the 'sixties' as much as anywhere else, a story retold in the recollections of some of its more famous alumni of the era.[5]

Yet, there are ways in which Keele differed from the famous campus projects of the 1960s. No revolutionary concept of space or stunning architectural design informed in its development plan. Constraints on building in the 1950s slowed construction progress, with frequent reports of staff and students struggling to teach and study in the ex-army huts strewn across the muddy Staffordshire hill (actually, by all accounts, hut-life, with central heating and running hot water, was more cosy than home for many undergraduates in austerity Britain).[6] Set in beautifully landscaped eighteenth-century grounds, studded with lakes and ancient woodland, Keele did not need its Le Corbusier. There were other differences too. Keele's commitment to a generalized programme of undergraduate study rested not on the fashionable idea of inter-disciplinarity, but on the aspiration to communicate knowledge of the humanities to scientists, and vice versa, to passing on the rudiments of experimental science to arts students. Keele's founding ideology was rooted in the 1940s, in arguments about reconstruction and the expansion of the welfare state. Keele was not so much a campus for the baby boomers, as for their teachers and social workers. If there was a zeitgeist at Keele, it was less about the challenge of 'post-industrial society', as post-war rebuilding, and the harnessing of technology to democracy. In its early years, a large part of what the new campus at Keele did was focused on new philosophies of education, so that, when the generation of 1945 grew into childhood and adolescence, there would be a cohort of women and men trained up and able to prepare them for going out into the world.

It is the discontinuity between Keele and the later 1960s campus universities that is the focus of this chapter, which is in three parts. Firstly, I revisit the founding ideas of Keele in the work of its first principal, Alec Dunlop Lindsay – Sandy Lindsay – Lord Lindsay of Birker, philosophy don, Labour party peer, former master of Balliol College and former vice-chancellor of the University of Oxford. Unusually for a head of a university, Lindsay knew much about utopias, having made his mark early in his career as the country's leading interpreter of Plato's *Republic*.[7] His contribution to Keele has been described on several occasions; however, there is something more to be said, particularly about his experience as adviser on reform in the post-war German university. Then, secondly, the chapter turns to look at one of the founding professors at Keele, Campbell Stewart, the professor of education, who, after Lindsay's death in 1952, upheld and carried through the principles of Keele's approach to teaching, not only in his own department, but across the three faculties and in the foundation year.[8] Stewart was also a driving force behind sociology – *the* new discipline of the era – coming to Keele. Finally, I examine Keele in the 1960s, discussing both how Keele looked outwards to the new campus universities of the 1960s and how the 'swinging sixties' came to Keele. By the early 1970s, as we shall see, the utopian vision that the founding fathers had brought to Keele was being turned back against the university, and in that confrontation, something was lost, but something was gained as well.

Lord Lindsay and the 'Keele experiment'

All roads lead back to Sandy Lindsay and the 'Keele experiment'. As with most of the new universities of the 1960s, the story of the foundation of the University College of North Staffordshire has been told many times and the founding myths deepen with each narration. It is a very British story.[9] Originally a product of the University of Glasgow, Lindsay brought to Keele the values and structure of the Scottish honours system, whereby undergraduates moved from general studies through to greater specialization. He believed that both the English schools and the universities emphasized specialization too early and too much. Lindsay also preached the virtues of fellowship and communal living in the university. Tom Steele has shown, for example, how Lindsay was influenced by his fellow Scot and Balliol colleague, John Macmurray, in these ideas.[10] Thus, Lindsay's influence on Keele was also an Oxford story. He had long been involved in the Oxford Delegacy for Extra-Mural Studies, with R. H. Tawney and others, and had experience of workers' adult education in the Staffordshire area.[11] And so the new college was also obviously a Staffordshire story. Part of Lindsay's achievement lay in the partnership he forged with the Vicar of Etruria, Thomas Horwood, leader of the Stoke Borough Council, one of the sponsoring local authorities for the new college.[12] Together, they managed to persuade the University Grants

Committee that the 'Potteries', the populous industrial district of Staffordshire – then the largest manufacturing and mining conurbation in the UK – was the neediest part of the country as far as a new university was concerned. To make the new institution even more appropriate to its habitat, early plans envisaged innovative programmes of teaching in ceramics and in social studies.[13]

However, Lindsay's thinking on higher education was also derived from his experiences outside of the British system as well. In 1948, he served as special adviser to an Allied commission on the reform of the higher education system in Germany. He worked alongside Robert Birley, the former headmaster of Charterhouse.[14] In several respects Lindsay's complaint about specialization in the British curriculum derived from his observations about the Nazification of the German universities. He was critical of the system in which the training of scientists took place in specialist technical institutions, whilst the ancient universities of Germany were given over to educating students into the political ideology of the Third Reich. For Lindsay, this had created a situation in which scientists were blind to the social effects of the new technologies they created, whilst non-scientists knew nothing of the implications of the latest scientific developments. Lindsay's views here were not particularly original. He drew from the work of Adolf Löwe, in particular, his *Universities in Transformation* (1940), and as Tom Steele has described, he was also influenced by Karl Mannheim, the German sociologist, who worked in exile at the LSE until his death in 1947, and whose *Ideology and Utopia* was a scarcely disguised attack on Nazism. In his official role as adviser working in Germany in the months of 1948, Lindsay was well-placed to put his ideas into practice. Stefan Paulus has shown just how important the reform blueprints of the occupying powers were to post-war Germany. And to post-war Britain. For as Lindsay settled in at the new University College of North Staffordshire, his German experience fed into much of his writing and public statements, in which he set out his ambitions for Keele: a curriculum which brought science and non-science studies together in a joint venture of mutual understanding.[15]

Lindsay also brought an American perspective to the new university college. He was an admirer of the general course of education that had been established at Harvard after 1945, based on the pedagogy of James Conant, president of Harvard from 1933 to 1953. Lindsay often singled out the Massachusetts Institute of Technology (MIT), Harvard's neighbour, as the kind of scientific institution of higher education where there was proper inclusion of an introductory programme of undergraduate studies including the arts and humanities. Shortly before his death, Lindsay also made a visit to Swarthmore College in Pennsylvania, where he rated the structure of an undergraduate curriculum in which the first two years of study were devoted to courses drawn from across the curriculum.[16] Lindsay used this occasion to set up an exchange programme with Swarthmore. An early beneficiary of that programme was Saul Sternberg, later a distinguished social psychologist, who whilst at Keele wrote a penetrating account for *Cum Grano* of the new college's innovative foundation course and the extent to which it fostered civic sensibility and political awareness amongst undergraduates.[17]

Nationally and overseas, Lindsay was an effective communicator for the 'Keele experiment', as it quickly became known, with all the connotations of socialist planning that phrase implied. Lindsay enjoyed good relations with Clement Attlee's

Labour government, and it was rumoured that he had pressured Hugh Dalton, the chancellor of the Exchequer, to smooth through the passage of the college's charter, a claim that was subsequently shown to be unfounded.[18] The new college's reputation as a 'Kremlin on the hill' stuck, however. Amongst the first visitors to the new campus were delegations from the People's Republic of China, and from the Soviet Union.[19] Lindsay's tenure, sadly, was short lived. He died in 1952 and his immediate successor John Lennard-Jones succumbed to an early death after only one year in post, although there long enough to signal that he too abided by the Lindsay philosophy, tilting more in the direction of turning out humane scientists.[20] Fortunately, before he died Lindsay brought to Keele a small but smart cohort of scholars – almost exclusively men, mostly in their late thirties and early forties – who proved as committed to his educational philosophy as he was. There was one exception to the male dominance. The social service programme at Keele was run by Mary Glover (1898–1982), recruited from Oxford by Lindsay.[21] As were some of the first thirteen professors, several of them straight from Lindsay's own college, Balliol. They came too from the Scottish universities, principally Glasgow, and from war service, mainly in intelligence. It was this team who gave life to the 'Keele experiment', that is to say, the foundation year of study and the residential campus set-up for staff and students alike. They were a colourful assortment: Sammy Finer, the vivacious political scientist; Alan Gemmell, the genial botanist and stalwart member of the BBC *Gardeners' Question Time*; John Lawlor, the medieval expert and mischievous pseudonymous author of *Letters to a Vice-Chancellor* (1964), a satirical skit on the other new universities; and two men from the ministry: Arthur Vick, the physicist, who had worked in the Ministry of Supply in the Second World War, and later moved onto the Atomic Energy Research Establishment, and Stanley Beaver, geographer, late of naval intelligence and the wartime Land Survey.[22]

The foundation year, at the heart of the curriculum, was delivered through, by modern standards, a very full course of lectures – several per week – followed up by at least two tutorials. At first sight, the course looks familiar: 'Plato to NATO', or 'History 101: the course of western civilisation', in the classic liberal arts mode. Some commentators (e.g. Norman Bentwich) saw nothing new, simply a return to the old mediaeval programme of the *studium generale*.[23] However, closer inspection reveals Lindsay's concerns at the heart of the list of topics and approaches. Particular importance was attached to the kind of lessons that students needed to draw from the foundation course for the contemporary world. Whilst Lindsay was alive, he began the convention of the college principal delivering the introductory lecture of the course, a *tour d'horizon* of current politics and cultural affairs.[24] The content of the foundation course changed over time and also expanded to reflect additional subject disciplines added to the Keele curriculum during the 1950s, for example, on social science subjects such as psychology.[25] And Lindsay's colleagues took on the philosophy of a broader approach to the subject from the foundation course and into the rest of the degree. For example, John Charlton, the professor of classics, dispensed with rote learning of Latin and Greek and moved more in the direction of using translated texts to get students to understand the ancient world, a forerunner of today's broad approach to classical studies.[26] Lindsay's science colleagues were converted to his philosophy of education as well. Arthur Vick used his presidency of the UK Institute of Physics to spread the word about the foundation course as a new way to teach science, bringing together experimental and non-experimental subjects.[27] The foundation year was complemented by the honours programme that succeeded it whereby undergraduates took a subsidiary subject alongside their main, and many students switched degree programme entirely as a result of exposure to such a range of topics.

By the early 1960s the 'Keele experiment' was well known and highly regarded. It found favour in the first curriculum of the new University of Sussex, and from there John Fulton the Sussex vice-chancellor (and former colleague of Lindsay at Balliol) recommended it to the new Chinese University being set up in Hong Kong.[28] In 1962 the Nuffield Foundation funded a two-year research programme to test the advantages and claims of the Keele foundation year. Run by Alan Iliffe, psychology lecturer at Keele, the project published its findings in 1968. The study compared the experiences and the attainments of first year students at Keele with comparable counterparts at some of the civic universities. Keele students scored highly on awareness of the world around them.[29] And the 'Keele experiment' began to travel more as its pioneers moved on from the college. Bruce Rodda Williams, founding professor of Economics, who left Keele in 1959 to go to Manchester and later became vice-chancellor of the University of Sydney, was very involved in the mid-1960s with the foundation of the University of Lancaster.[30] Similarly, John Blake, first professor of History at Keele, went out in 1963 to become the first chancellor of the new University of Swaziland and took the Keele foundation year with him: that is, the programme, not the students.[31] They stayed behind. Let us find out a little more about what they studied.

A college for the welfare state?

As well as turning out worldly citizens the core business of the University College of North Staffordshire was for many years the training of teachers and social workers. A key feature of the Keele curriculum was the vocational diploma that ran alongside the four-year undergraduate programme, rather than being taken as a one-year course at the end of the BA, or as a one-year course entirely independent of the BA. The rationale behind this fully integrated diploma was that students on the course would benefit from the wider general curriculum being offered to everybody else. Moreover, they were permitted to take a subsidiary course drawn from areas of study most appropriate to their eventual vocation. Two diplomas were offered: one in education and the other in social work, and in both Keele proved pioneering. Well into the 1960s, the teaching profession took the largest proportion of Keele graduates: 17 per cent in 1954 rising to just under one-third in 1963.[32]

Education and social work were, in fact, originally intended to lie at the heart of the 'Keele experiment'. As noted earlier, social studies alongside ceramics were proposed as a new core field for the college. In early discussions around the new college, an Institute of Education was also planned, but postponed as the Universities Grant Committee felt it would distract from the other important work that the new college was to undertake.[33] However, education became a very important part of what Keele did in the 1950s. This aspect of the new college was masterminded by Campbell Stewart, another Scot, appointed in 1950 from University College of Wales at Cardiff. Stewart had studied with Karl Mannheim at the LSE, taught in secondary schools as well as at university, and was the author of a study of punishment in Quaker schools.[34] Stewart took up the Keele approach to the undergraduate curriculum more or less where Lindsay had left off. Like Lindsay he was against specialization coming too early in undergraduate study. He valued teaching over research. And he followed Conant's Harvard model of general education.[35] Indeed, America supplied Stewart with some influential contacts, such as Nevitt Sanford, the psychologist and educationist who was thrown out of Berkeley in 1950 during the purges of McCarthyism, and also Hans Froelicher, headmaster of a progressive school in Baltimore. Stewart spent the summer of 1958 in the United States, working alongside Sanford and Froelicher, a budget trip that must have delighted the Keele finance department, as Stewart took two and a half days to reach New York, travelling via Glasgow, Iceland, Greenland and Canada.[36]

However, in terms of the philosophy of education at Keele, it is Stewart's relationship with the late work of Karl Mannheim that is most interesting and revealing. Campbell Stewart became one of Mannheim's most dogged editors and interpreters. He wrote surveys of the main principles of Mannheim's work, defended him from his many critics and developed ideas about how Mannheim's approach might be transferred into a programme of undergraduate instruction. With the Hungarian, John Eros, a lecturer in psychology at the college, appointed in 1958 (his name was naturalized from the original Janos Istvan Stark: Eros, the psychologist, was surely tongue-in-cheek), Stewart produced an annotated edition of Mannheim's lectures from the 1930s as an undergraduate primer.[37] Stewart laid emphasis in Mannheim's late work on the positive future he envisaged for democracy and state socialism, or the planning of the economy and of society. The task of sociology, stated Stewart, glossing Mannheim was to prepare for 'social reconstruction and planning for freedom'. Teachers were crucial in this role; following Jean Floud, Stewart called them 'the midwives of social mobility'.[38] In this way, Stewart took the education department at Keele in a different direction from the older adult education ethos originally brought to the Potteries by Tawney and others. Here perhaps was more of a 'post-industrial' vision, a university focused on producing the servants of the state, albeit one rooted in post-war ideas of reconstruction. Adult education did return properly in 1967, when Roy Shaw, Stewart's colleague in the Department of Education, and later head of the Arts Council, established the Extra-Mural Studies department at Keele, but to an extent it only existed on the margins of the 'Keele experiment'.[39]

Campbell Stewart not only aligned the training of teachers and social workers with Lindsay's philosophy of education at Keele, he also oversaw the arrival of the *Sociological Review*, at the college in 1953. On the face of it, this was a strange development. Keele did not have a department of sociology on its books or even in its development plans. In fact, although a Statistical Research Unit in Sociology, again funded by the Nuffield Foundation, was set up in 1963, it was not until 1965 that there was an appointment in sociology. And only in 1969 did a founding professor arrive in the shape of Ronnie Frankenberg.[40] Keele did not hothouse new disciplines. Rather, from its conventional departmental structure it forged hybrid approaches, such as psychology and education, that fed into the undergraduate curriculum. In a way this is precisely what happened with the journal. The *Sociological Review* had been published by the Institute of Sociology, based at Le Play House, one of the founding institutions of sociology in this country, since the first decade of the century. By the 1940s, the owners of both Le Play House and the journal, the Farquharsons, were running out of money. They faced competition from the LSE, which was setting up its own journal in sociology, the *British Journal of Sociology*, and also from a foundling organization, the British Sociological Association. Le Play House needed to find a new home not only for the journal, but also for its very extensive library and for its ongoing series of commissioned town and rural investigative studies, the bedrock of its work. In stepped Lindsay. He knew the Farquharsons and, in a characteristically entrepreneurial move, offered to take over the entirety of the Le Play House operations at the new college, promising American funding – he had in mind the Rockefeller Foundation – to provide recurrent support to secure Le Play House's future at the college.[41]

In the event Lindsay died before any new finance could be raised. The library came to Keele, indeed with a large library collection from Glasgow, the Le Play House collection filled the shelves of the new college stacks. But the rest of Le Play House's activities were left to wither on the vine. In 1955 the charity was wound up.[42] The college did however see through its promise to take over the journal, bringing it under the joint editorship of seven of the Keele professors whose disciplines were most associated with its subject matter. Almost seamlessly the journal continued publication. True, Keele academics did seem to feature disproportionately in the issues of the first two or three years. Sammy Finer (on science fiction) and Bruce Williams (on economic science and public policy) both published articles only tangentially concerned with sociology.[43] And many of the book reviews through the 1950s were written by college staff. At the same time, this 'Ministry of the Talents' approach to editing the journal did not miss out on a representative range of all that was trending in British sociology. Early issues of the journal edited out of Keele included Edward Shils and Michael Young's influential article on the British and their monarchy, Michael Banton on immigrant communities, and the early work of A. H. Halsey and Peter Townsend.[44] Over time, the Keele editorial team gave the journal the flavour of applied sociology flavour, distancing it from the more theoretical work coming out of the LSE. In surveys of the sociological scene, attention was drawn to the pioneering departments of investigative sociology around the country.[45] And the journal developed a staple diet of articles based on field-work studies in the British regions.

Another innovation introduced by the Keele editorial board of the *Sociological Review* was a monograph series which commenced in 1958, and was edited for the next fifteen years by Paul Halmos, another central European émigré at Keele, appointed in 1957 as a joint lecturer in psychology and education. The expertise of Halmos lay in social psychology: his LSE thesis on social isolation was published in 1952, and he later went on to publish on counselling and therapy in social work. Lacking any clinical expertise, Halmos's writings now seem very dated, but he developed an interesting, shall we say 'post-industrial', thesis about how the 'personal service society', by which he meant teachers, social workers and, of course, counsellors, had become as important as the 'professional service society' (lawyers, accountants, civil servants) in the post-war world.[46] Halmos missed out on the first chair in psychology at Keele – that went to Ian Hunter, an expert on memory, and so Halmos left for Cardiff in 1965. Before he left he nurtured the monograph series through its first years. Overseen by Campbell Stewart, Halmos focused on research devoted to the interface between educational studies, sociology and psychology, particularly looking at the issue of teaching teacher to deal with personality development in children and adolescents. At its outset, the monograph series derived its content from an annual conference of teachers and social workers.[47]

So the *Sociological Review*, both as a journal and a monograph series, made an effective contribution to a formative moment in the history of the discipline of sociology in Britain, one that is seldom noted in the mainstream history of British sociology which tends to focus on London and Leicester.[48] Under Keele's management, an important native tradition in British sociology – field-work and social inquiry – was kept alive. Moreover, the journal incubated innovative areas of applied sociology such as the psychology of education. Only in Keele's distinct set-up could the fusion of educational studies, psychology and sociology, take place and feed into the coverage and contents of the journal. Moreover, only in Keele, undergoing its own on-campus experiment in social engineering, did the theory and practice of British sociology have such immediate relevance. Another pioneering first at Keele was the UK's first campus counselling service, started up in 1962, with the arrival of Audrey Newsome. Mindful of his own children's negative experience as students at Oxford, Harold Taylor, the Keele vice-chancellor, wanted better provision for the mental welfare of students on campus and turned to Newsome, who had trained in the United States, to develop the programme. Counselling for university students and staff was a common enough element in the United States, but a novelty in Britain. Again, Keele led the way, recognizing the challenge that came with communal living, prefiguring some of the concerns around *anomie* and alienation that developed on other new campuses such as Essex. Campbell Stewart later reflected on how residence posed the 'deeper question of living together', of 'living with and suffering stress through the indifferences or cruelty of other people' and of experiencing 'depression or loneliness yourself'.[49] Lindsay's mantra of the campus community as fellowship, its good and bad sides, was alive and well.

Legacy

Committed to upholding its distinctive undergraduate teaching curriculum and focused as much on its vocational diplomas as its BA honours courses, Keele entered the new decade as a rather different kind of campus university from the new cohort established across the UK in the years after 1961. Keele's principal in this period, Dr Harold Taylor, who became its first vice-chancellor in 1962, was happy to be swept up in the wave of publicity surrounding the launch of Sussex then York and East Anglia. Keele the 'pioneer' sat neatly alongside Sussex the 'pacesetter' in the accolades of the time, and Taylor did his level best to attract media coverage.[50] Taylor's successor as vice-chancellor was more sceptical. In 1967 Taylor retired; the search for a big name from outside to take over Keele was unsuccessful. Overtures were made, for example, to lure Asa Briggs from Sussex.[51] And so for the second time in his career Campbell Stewart was placed in charge of the university, on this occasion not as acting principal (as he had been in 1960), but as fully fledged vice-chancellor, his tenure of the office lasting through to his retirement in 1979.

In March 1968, a year into his vice-chancellorship, Stewart reviewed where Keele was at in relation to recent developments in both the UK and the United States.[52] In these inauguration lectures Stewart reiterated the founding principles of Keele as laid down by Lindsay, noting that some of these had indeed been taken up by the new universities. But he also claimed that Keele remained distinctive: in its foundation year, in its four-year degree, in its insistence on science and non-science subjects be taken by everyone, and in its provision of residence for everyone. There were aspects of the new English campuses that Stewart did not welcome. He argued that there was too much research going on Warwick; he was alarmed that research staff apparently outnumbered teaching staff there. He believed that Sussex was expanding its undergraduate numbers far too quickly, without giving its own teaching experiments in redrawing the 'map of learning' time to bed down properly. Stewart proved no fan of interdisciplinarity either. In fact, Stewart stated that interdisciplinarity was not for Keele, it might consider it in the future, but he did not see how 'schools' grouping together related subjects, such as was happening at Sussex and East Anglia, could achieve interdisciplinarity, for that was something that can only come through the general studies format of teaching. Finally, and somewhat curiously given Keele's location in the heart of the 'Potteries', Stewart was critical of the much-vaunted regional multiplier effect of the new universities in their local economies. 'Any region fathering and giving houseroom to an university', he observed somewhat negatively in 1972, was producing a 'non-local product'. The only example he could come up with of Keele's links with its hinterland were that some of its graduates had become local councillors.[53] Now Stewart was no dyed-in-the-wool conservative. He wrote widely on innovation in teaching: his history of the progressive schools is still worth reading, and he capped his career in retirement with a generous account of changes in higher education across the twentieth century.[54] Rather, he remained committed to a different set of principles derived from the hopes and aspirations around higher education in the late 1940s and early 1950s, centred on nurturing the civic mind and socially responsible young adult, who in one way or another would serve the welfare state. Campbell Stewart's vision remained focused around the ideals of small liberal arts college. But how well were those ideals standing up by the late 1960s, as campuses around the world shook with student rage and radicalism?

Keele's own 1968 started peacefully enough. The university covered itself in glory, as its team progressed through the rounds of *University Challenge*, eventually winning the trophy in July. But in June protests over the lack of student participation in the governance of the university began, quietly at first, with students taking the seats of staff at the high table in the refectory. Then some members of the Keele University Students' Union occupied the Registry, their demands fanning out to include not just student membership of university committees, but also co-residence of men and women in university accommodation, and in student lodging off-campus.[55] For the next three years, 1968 through 1971, the Keele campus experienced episodic student militancy, much of it chronicled in the national newspapers. During the academic year 1969–70, echoing developments in France, Germany and Australia, an 'Action for a Free University' movement was launched, and 'F.U.K' (i.e. the 'Free University of Keele') slogans adorned the campus. There were arson attacks and sun-bathing in the nude in the summer of 1970, ahead of a visit by Princess Margaret. In the autumn of that same year, seeking to emulate similar student activism at the Pentagon in Washington, there was a 'levitation' of the vice-chancellor's residence at the Clockhouse, led by Pete Sykes, editor of *Cygnet*. In the new year of 1971, controversy flared again, this time over the introduction by the university of student ID cards (actually meal vouchers).[56] In the *Guardian*, the journalist, Simon Hoggart wrote shortly afterwards that 'dream Keele is over', and a Granada TV documentary, 'The dream on the hill', made the following year, pursued the same theme.[57] Local opinion turned against Keele: Staffordshire County Council held back its annual funding following the naked frolics on the lawn, and from London, the UGC looked on aghast, vowing never to fund another four-year degree programme ever again.[58]

Throughout all of this, Campbell Stewart, as vice-chancellor, stood up reasonably firm. Some of his responses seem petty in retrospect, especially rusticating and fining the sun-bathers.[59] To his credit, however, in the summer of 1968, as the first wave of protests broke, he set up a wide-ranging review of university governance at Keele. Known as the 'Exploratory Committee', the same title given to the body that scoped the original foundation of the University College of North Staffordshire some twenty years previously, this review looked into how student representation in university business might be increased. Some participants in the review, for example, the 'Critical Education Group' attached to Keele's 'Revolutionary Socialist Students' Society', quoted Lord Lindsay's 'sociology of knowledge' back at the university authorities, decrying the way in which Keele had lost touch with its founding ideals of communal living, become isolated from its local community down the hill in the Potteries, and become too preoccupied with turning out fodder for the big public and private employers.[60] Perhaps stung by this criticism, Stewart went back to basics. In October 1968, he travelled to the United States, ten years after his first trip there. Plenty had changed. At Harvard, he noted how James Conant's general curriculum was no longer held in high favour: it was now regarded as too Eurocentric and triumphalist. Elsewhere, Stewart was worried by evidence of the growing gulf between faculty and students. And he listened as the heads of small colleges told of their fears for their survival in competition with the larger players. Finally, Campbell Stewart found a university where the 'Keele experiment' was alive and well: Reed College, in Portland Oregon, where staff–student relations on a small residential campus were harmonious, and the curriculum was based around a general studies foundation year.[61] Revived and reassured, he returned to the UK, and oversaw a series of changes to Keele's constitution that had the effect of making the student voice heard through improved consultation and representation, if not always actioned.

As it turned out, the 'Keele experiment' did not have much longer to run. Throughout the 1960s, encouraged by the UGC, Keele committed itself to expanding its student numbers from the mid-hundreds into the low thousands. Against the backdrop of the recession of the early 1970s, unlike elsewhere, growth was still encouraged, but only, insisted the UGC, if Keele ran a new three-year degree structure alongside its famous four-year programme, and only if Keele dropped its requirement that all undergraduates be resident on the campus across the programme of study. As vice-chancellor, Stewart accepted these conditions, seeing them very much as exceptions or special arrangements that did not undermine the founding principles of the college dating back to 1950. A three-year degree was agreed to in 1971, and during 1973–4 the whole foundation year programme was made the subject of a major review.[62] Inevitably, conceding the ground to a three-year programme, and giving up 100 per cent residence, did change the character of Keele. Blue-sky thinking continued: literally, for example, the Space Observatory was expanded in 1977. But by the 1980s and beyond it was hard to distinguish between Keele and all the other competitor institutions in the sector. Now, in 2019, foundation courses still exist at Keele, but are run separately in the arts, sciences, medical and social science schools, and students taking them are not eligible for student loans. Residence on campus no longer dominates, although it remains popular, not least amongst the staff, both current and retired.

Little today remains of the legendary vision of the founding fathers of Keele: Lindsay, Stewart and the others. The 'Keele experiment' was quite unlike what happened elsewhere in the 1960s universities, despite some obvious similarities. Keele serviced the generation that came of age in the years immediately after 'Rab' Butler's 1944 Education Act, and not the baby boomers who came in the years after the Robbins report of 1963. Keele turned out teachers, social workers, civil servants, less so scientists or the liberal professions, although that was to change by the end of the 1960s. Keele supplied the personnel of the welfare state, and it did so by delivering a radical curriculum, mostly imported from the United States, but inflected by central European émigré sociology and psychology. No other UK university had quite such a utopian vision of a scholarly residential community, grounded in a distinct philosophy of education. Keele stands out then as Britain's first and finest liberal arts campus. And perhaps its last.

Notes

1 Brochure of the 'New Universities Festival' (1–3 May 1964), Keele University Special Collections & Archives (KUSCA); 'The universities conference, 14–16 July 1964' (transcript), Harold Taylor Papers, KUSCA. For a summary of the conference, see C I. C. Bosanquet and A. S. Hall, *The Creation of New Universities* (Keele: University of Keele, 1964). For McConnell, see his later co-authored *From Elite to Mass to Universal Higher Education: the British and American Transformations* (Berkeley: University of California, Centre for Research and Development in Higher Education, 1973).

2 'Address by Baron Fulton of Falmer', in *University of Keele: Twenty-First Anniversary: Three Addresses* (University of Keele, 1972), 5.

3 *Manchester Guardian*, 21 December 1956, 6.

4   *Times*, 29 July 1965, 6. George Youell (1910–95) was Archdeacon of Stoke-on-Trent from 1956 to 1970.
5   Francis Beckett, *What Did the Baby Boomers Ever Do for Us? Why the Children of the Sixties Lived the Dream and Failed the Future* (London: Biteback, 2010); cf. Michael Mansfield, *Memoirs of a Radical Lawyer* (London: Bloomsbury, 2010), 10–13.
6   See the testimony from alumni memories collected by the 'Keele Oral History Project': www.keele.ac.uk/thekeeleoralhistoryproject/hutlife/ (accessed 1 March 2017).
7   Drusilla Scott, *A. D. Lindsay: A Biography* (Oxford: Basil Blackwell, 1971); Gary McCulloch, 'Lindsay, Alexander Dunlop, First Baron Lindsay of Birker (1879–1952)', *Oxford Dictionary of National Biography* (Oxford: Oxford University Press, 2004), https://www-oxforddnb-com.libproxy.york.ac.uk/view/10.1093/ref:odnb/9780198614128.001.0001/odnb-9780198614128-e-34537 (accessed 1 March 2017).
8   A. H. Halsey, 'Stewart, (William Alexander) Campbell (1915–1997)', *ODNB* https://www-oxforddnb-com.libproxy.york.ac.uk/view/10.1093/ref:odnb/9780198614128.001.0001/odnb-9780198614128-e-66197 (accessed 1 March 2017).
9   W. B. Gallie, *A New University: A. D. Lindsay and the Keele Experiment* (London: Chatto and Windus, 1960); A. E. Teale, 'The Origin of the Keele Experiment', *North Staffordshire Journal of Field Studies*, n. s. 1 (1961), 101–14; J. Mountford, *Keele, An Historical Critique* (London: Routledge & Kegan Paul, 1972); Michael Paffard, *Keele. An Introduction to the Parish and the University* (privately published, 1998), chs 11–16; J. M. Kolbert, *Keele: The First Fifty Years, a Portrait of the University, 1950–2000* (Keele: Melndrium Books, 2000).
10 Richard Taylor and Tom Steele, *British Labour and Higher Education, 1945 to 2000: Ideologies, Policies and Practice* (London: Continuum, 2011), ch. 4.
11 R. A. Lowe, 'Determinants of a University's Curriculum', *British Journal of Educational Studies*, 17 (1969), 41–53; Tom Steele and Richard Taylor, 'R. H. Tawney and the Reform of the Universities', *History of Education*, 37 (2008), 1–22.
12 See Horwood's case for the Potteries in his 'A New University: A Proposal for North Staffordshire', *Higher Education Quarterly*, 2 (1947), 77–81.
13 'Memorandum to the University Grants Committee on the Proposal for a University in North Staffordshire' (11 November 1946), TNA, T227/73, 1, 7–8.
14 Details of the mission to occupied Germany are in the Lindsay Papers, KUSCA, L221, and in Scott, *A. D. Lindsay*, 296–304. Cf. David Phillips, *Investigating Education in Germany: Historical Studies from a British Perspective* (Abingdon: Routledge, 2016), ch. 8.
15 Transcript of 'The Idea of a University' (BBC Third Programme, 6 March 1950), Lindsay Papers, KUSCA, L217; Lindsay, 'The Function of the Universities', *Nature*, 166 (16 December 1950), 1009–10. For Lindsay's friendship with Mannheim, see Scott, *A. D. Lindsay*, 266–8; Steele and Taylor, *British Labour and Higher Education*, 47. On the universities in post-war Germany, see Stefan Paulus, 'The Americanization of Europe after 1945? The Case of the German Universities', *European Review of History*, 9 (2002), 241–53.
16 There a typescript account of Lindsay's visit to Swarthmore in the Lindsay Papers, KUSCA, L217. For references to Harvard and MIT, see Lindsay, 'The University College of North Staffordshire', *Highway*, 42 (February 1951), 100–2; 'The Modern University', *University: A Journal of Enquiry*, 1 (1951), 80–2.

17 Sternberg, 'Student Responsibility and General Education', *Cum Grano*, 2 (March 1953), 18–22.
18 Robert O. Berdahl, *British Universities and the State* (Berkeley: University of California Press, 1959), 142.
19 'Clock House Visitors' Book', KUSCA. The Clock House was the Principal's residence.
20 John Lennard-Jones, *Trends in University* Education. *An Inaugural Address Delivered on May 20th 1953* (Keele: University College of North Staffordshire, 1953), 3–4; John Lennard-Jones, 'A New University College', *Cambridge Review*, 7 March 1953, 374–6. For his life, see Nevill Mott, 'Jones, Sir John Edward Lennard– (1894–1954)', rev. John Shorter, *ODNB* https://www-oxforddnb-com.libproxy.york.ac.uk/view/10.1093/ref:odnb/9780198614128.001.0001/odnb-9780198614128-e-34496 (accessed 1 March 2017).
21 'Mary Reaveley Glover', *St Hugh's College Chronicle*, 55 (1982–3), 46–7.
22 For Finer, see the special issue of *Government and Opposition* 29 (1994). For Gemmell (1913–86), see the appreciation in the *Year Book of the Royal Society of Edinburgh*, 93 (1987), 53–5. For Lawlor (1918–99) there are some biographical details (mainly recollections of Oxford in the mid-1930s) in Lawlor, C. S. *Lewis: Memories and Reflections* (Dallas: Spence, 1998), ch. 2. For Beaver, see 'Obituary: Stanley Henry Beaver, 1907–1984', *Transactions of the Institute of British Geographers*, 10 (1985), 504–6. For Vick, see Michael Thompson, 'Vick, Sir (Francis) Arthur (1911–1998)', *ODNB*, https://www-oxforddnb-com.libproxy.york.ac.uk/view/10.1093/ref:odnb/9780198614128.001.0001/odnb-9780198614128-e-71011 (accessed 1 March 2017).
23 Norman Bentwich, 'Two British Universities: Old and Young', *Journal of Education*, 83 (1951), 474–6.
24 Lindsay, 'Introductory Lecture' (19 October 1950), Lindsay Papers, L209; cf. Lennard-Jones, 'Introductory lecture. Foundation Year Course' (Ms, 8 October 1953), Lennard-Jones Papers, Churchill Archives Centre, Cambridge, LEJO68; George Barnes, 'Foundation Year Lecture' (typescript, 13 October 1958), George Barnes Papers, KUSCA.
25 *University College of North Staffordshire Prospectus* (Session 1958–9), 56–8.
26 John Charlton, 'An Experiment in Classical Education at an English University', *Classical Weekly*, 45 (4 February 1952), 148–9.
27 F. Arthur Vick, 'Education Through Physics', *Bulletin of the Institute of Physics*, 6 (July 1955), 143–50; F. Arthur Vick, 'Science for the Non-Scientist', *Universities Quarterly*, 11 (1957), 245–53; cf. I. N. Sneddon, 'Science Courses at Keele', *Research*, 7 (September 1954), 335–41.
28 *Report of the Fulton Commission … to Advise on the Creation of a Federal-Type Chinese University in Hong Kong* (Hong Kong: Government of Hong Kong, 1963), 24.
29 A. H. Iliffe, *The Foundation Year in the University of Keele*, Sociological Review Monograph no. 12 (Keele: University of Keele, 1968).
30 See his autobiography: *Making and Breaking Universities: Memoirs of Academic Life in Australia and Britain 1936–2004* (Sydney, Australia: Macleay Press, 2005).
31 David Ambrose, *The History of Education in Lesotho: Six Brief Subsectoral Studies* (Roma: House 9 Publications, 2007), 61. Blake later joined another new campus, the New University of Ulster.
32 Derived from data in 'Appointments and Counselling, 1954–73', Registry files, Box 4, KUSCA.

33 See the correspondence between Lindsay, Stewart and the Ministry of Education between November 1950 and April 1951 in ED110, TNA.
34 Halsey, 'Stewart, (William Alexander) Campbell', *ODNB*; Stewart, *Quakers and Education: As Seen in Their Schools in England* (London: Epworth Press, 1953).
35 Stewart, 'University Education in Breadth and Depth', *Listener*, 56 (6 December 1956), 919–20; Stewart, 'The Adaptation of a University to the Changing Needs of the Community', in *Naar een verantwoorde opvoedin: album Prof. J.E. Verheyen: de gehuldigde aangeboden ter gelegenheid van zijn emeritaat door kollega's vakgenote*, ed. Robert Léon René Plancke et al. (Ghent: Rijksuniversiteit, Hoger Instituut voor Opvoedkundige Wetenschappen, 1959), 168–79; 'Higher Education and the Social Context', in *The American College. A Psychological and Social Interpretation of the Higher Learning*, ed. Nevitt Sanford (New York: Wiley, 1961), 894–939.
36 'Diary of a Visit to the USA, 5 July–3 August [1958]', W. A. C. Stewart Papers, KUSCA; For Nevitt Sanford (1909–95), see *New York Times*, 11 July 1995, https://www.nytimes.com/1995/07/11/obituaries/nevitt-sanford-86-psychologist-who-traced-roots-of-prejudice.html (accessed 1 September 2019). For Hans Froelicher (1891–1976), see K. A. Stakem, 'Hans Froelicher, Jr.: Civic Educator', *Maryland Historical Magazine*, 77 (1982), 193–201.
37 Stewart, 'Philosophy, Psychology and the Training of Teachers', *Sociological Review*, 42 (1950), 1–35; 'Karl Mannheim and the Sociology of Education', *British Journal of Educational Studies*, 1 (1953), 99–113; John Eros and W. A. C. Stewart (ed.), *Systematic Sociology: An Introduction to the Study of Society* (London: Routledge, 1957). For the role of Stewart and Mannheim's widow, Julia, in the publication of these lectures, see E. S. Hank, Woldring, *Karl Mannheim. The Development of his Thought: Philosophy, Sociology and Social Ethics, with a Detailed Biography* (Assen: Van Gorcum, 1986), 64–5.
38 Stewart, 'The Role of the Teacher in Advanced Societies', *Proceedings of the Twentieth Symposium of the Colston Research Society held in the University of Bristol, 1–5 April 1969* (London: Butterworths, 1968), 21.
39 Roy Shaw (1918–2012). Interview with Frank Burchill (16 March 2017). For some of the background, see M. Cruickshank (ed.), *30 Years of Education at Keele, 1951–81* (Keele: University of Keele, 1981).
40 For the Statistical Research Unit, see *Nature*, 200 (14 December 1963), 1050. The evolution of sociology at Keele can be reconstructed from two files: 'Sociology, Aug. 1965–Sept. 1976' and 'Centre for Social Science Research Oct. 1966–Sept. 1987', Registry files, Box 15, KUSCA. For Ronnie Frankenberg (1929–2015), see *Guardian*, 4 January 2016, https://www.theguardian.com/science/2016/jan/04/ronald-frankenberg (accessed 1 September 2019); Interview with Gordon Fyfe (12 October 2016).
41 'Le Play House. Memorandum for Lord Lindsay' (29 June 1951), KUSCA, GB172 LP/6/1/3/9.
42 Details of the demise of Le Play House, in which Campbell Stewart represented the interests of the University College, are contained in KUSCA, LP6/1/2/19-24.
43 B. R. Williams, 'Economic Science and Public Policy', *Sociological Review*, new series, 1:1 (1953), 87–109; S. E. Finer, 'A Profile of Science Fiction', *Sociological Review*, 2:2 (1954), 239–55; S. E. Finer, 'The Political Power of Private Capital, Part 1', *Sociological Review*, 3:2 (1955), 279–92, 'The Political Power of Private Capital, Part 2', *Sociological Review*, 4:1 (1956), 1–31.

44 Edward Shils and Michael Young, 'The Meaning of the Coronation', *Sociological Review*, 1:2 (1953), 63–82; Michael Banton, 'The Economic and Social Position of Negro Immigrants in Britain', *Sociological Review*, 1:2 (1953), 43–62; A. H. Halsey, 'Social Mobility in Britain. A Review', *Sociological Review*, 2:2 (1954), 169–79; P. Townsend, 'The Family Life of Old People: An Investigation in East London', *Sociological Review*, 2:2 (1954), 175–97.

45 See the 'research reports' on the *Sociological Review* for Liverpool, 8:1 (1960), 109–19; Edinburgh, 8:2 (1960), 255–66; Institute of Community Studies, 9:2 (1961), 203–13: LSE, 10:3 (1962), 329–42: Nottingham, 11:3 (1963), 355–61.

46 Paul Halmos (1911–77). His major work includes: *Solitude and Privacy. A Study of Social Isolation, Its Causes and Therapy* (London: Routledge and Kegan Paul, 1952); *Towards a Measure of Man. The Frontiers of Normal Adjustment* (London: Routledge and Kegan Paul, 1957); *The Faith of the Counsellors* (London: Constable, 1965); *The Personal Service Society: An Inaugural Lecture* (Cardiff: University of Wales Press, 1966); Halmos and his wife Edith (1917–2000) originally came to Britain in January 1939, joining Halmos's sister, who, coincidentally, rented a house in London from Karl Mannheim. Interview with Tony Halmos (28 April 2017).

47 The first four volumes in the series, all edited by Halmos, were as follows: *The Problems Arising from the Teaching of Personality Development*, 2 vols. (1958–9); *Moral Issues and the Teaching of Teachers and Social Workers* (1960); *The Teaching of Sociology to Students of Education and Social Work* (1961).

48 Keele and the *Sociological Review* receive only a few passing mentions in recent histories of the discipline in the UK, e.g. Baudry Rocquin, 'British Sociology in the Inter-war Years', in *The Palgrave Handbook of Sociology in Britain*, ed. John Holmwood and John Scott (Basingstoke: Palgrave Macmillan, 2014), 189–210; Martin Bulmer, 'The Development of Sociology and Empirical Social Research in Britain' and A. H. Halsey, 'Provincials and Professionals: the British Post-War Sociologists', in *Essays on the History of British Sociological Research*, ed. Martin Bulmer (Cambridge: Cambridge University Press, 1985), 3–36, 151–64.

49 Stewart, 'Recollections and Anticipations by the Vice-Chancellor', in *University of Keele: Twenty-first Anniversary: Three Addresses* (Keele: University of Keele, 1972), 36. Interview with Audrey Newsome (6 February 2017). Findings from the Keele residential experience were later reported in Iliffe, *The Foundation Year in the University of Keele*, 90–1 and Audrey Newsome et al., *Student Counselling in Practice* (London: University of London, 1973), 21–3.

50 *Manchester Guardian*, 3 March 1962, 6; *Times*, 9 October 1963, ii; *Sunday Telegraph*, 20 October 1963, 4.

51 Samuel Finer to Asa Briggs, 14 December 1965, Asa Briggs Papers, Howard Gottlieb Archival Research Center, Boston University, Box 194/F2.

52 *British Universities: Dilemmas and Opportunities. Two Lectures by the Vice-Chancellor … Given in the University of Keele, 4th and 5th March 1968* (Keele: University of Keele, 1968), 30–6.

53 Stewart, 'Recollections and Anticipations', 38–9.

54 Stewart, *The Educational Innovators*, 2 vols. (London: Macmillan, 1967–8); Stewart, *Higher Education in Postwar Britain* (Basingstoke: Macmillan, 1989).

55 *Guardian*, 4 June 1968, 16; *Times*, 19 June 1968, 2 (sit-in at the Registry); *News of the World* (21 June 1970), 1 (nude sunbathing), cf. www.keele.ac.uk/thekeeleoralhistoryproject/theincidentonkeelehalllawn/.

56 'Action for a free university' (2 June 1968) in 'Revolution Keele June 1968' file, Margaret Whiteley Papers, KUSCA; *Newcastle-under-Lyme Times*, 26 June 1968, 1 (slogans); *Guardian*, 4 May 1970, 26 (arson); *Guardian*, 25 January 1971, 4 (ID cards). For the 'levitation', see www.keele.ac.uk/thekeeleoralhistoryproject/levitationsanddemons/
57 *Guardian*, 13 February 1971, 11. 'Dream on the hill' was directed by Carol Wilks.
58 *Guardian*, 28 July 1970, 5.
59 See the Keele Student Union response on the illegality of Stewart's move: 'Nudes versus prudes', *Concourse*, 9 October 1970, 1–2.
60 'A New University' (3 February 1969), 'Revolution Keele June 1968' file, Margaret Whiteley Papers, KUSCA.
61 'Report on American Visit by Professor W. A. C. Stewart 4 September–2 October 1968', Campbell Stewart Papers, KUSCA.
62 *Guardian*, 4 March 1971, 6 (decision to introduce a three-year degree); Senate Minutes, 12 December 1973, Registry/USO records, Box 5, KUSCA; *University of Keele, Report of the Vice-Chancellor to Council* (1973–4), 11.

3

Sussex: Cold War campus

by Matthew Cragoe

In November 1960, a rousing letter soliciting public support for the new universities then being established appeared in the pages of *The Times*.[1] Lord Mackintosh of Halifax, who chaired the University of East Anglia's appeal Committee, Lord Monckton of Brenchley (for Sussex) and Archbishop Michael Ramsay (for York) came straight to the point. While acknowledging the many demands on people's resources, they hoped that 'a special measure of generosity' might be shown to their fledgling institutions as they needed 'the sympathetic support of the nation in a task undertaken for the nation as a whole'.

Contemporary opinion in Britain was well informed regarding the 'national' task the new universities were designed to address. One was the need to meet the steepling demand for higher education. The post-war baby boom ensured that greater numbers were flowing through the school system; by the mid-1950s it was clear that a greater proportion aspired to attend university.[2] The number of students staying on at school after seventeen almost doubled to 12 per cent by 1962.[3] Another 'national' problem related to the shortage of key skills within the country's workforce.[4] A succession of parliamentary committees and ministerial pronouncements pointed out that shortages of scientists and technologists threatened the country's economic future.[5] These were matched by shortcomings in management, 'a failure to use modern techniques of production management, costing, accounting, industrial relations and related commercial disciplines'.[6]

The challenges facing the British economy were thus widely understood at the time the new universities were founded. There was, however, another 'national' dimension to political discourse at this time, which inevitably fed into the debate surrounding the establishment of Sussex and the rest – a lurking, omni-present anxiety about the Cold War and the threat it posed to the British, and indeed, Western way of life. The Conservative Party, which held office from 1951 to 1964, made hostility to Russian and, to a lesser extent, Chinese communism a cornerstone of anti-socialist rhetoric. As Quintin Hogg, Viscount Hailsham, Conservative Party chairman, put it during the 1959 election:

> Communism is a conspiracy not only against us but against all those who are, as the words of the preamble to the North Atlantic Treaty put it, 'determined to safeguard

the freedom, common heritage and civilization of their peoples, founded on the principles of democracy, individual liberty and the rule of law'.[7]

The great struggle of the age was to resist the conspiracy and protect freedom.

In this chapter it will be suggested that, in addition to helping with the problems associated with skilled manpower and the democratic demand for higher levels of participation in higher education, the new universities of the 1960s were expected to play a key role in safeguarding the values which characterized Western civilization. Here, if anywhere was the utopian vision towards which higher education was expected to contribute in this period. The chapter focuses on the experience of the University of Sussex, the first of the new universities to open its doors. To set the context, Part One explores how the wider international tensions gave rise to the so-called Cold War university in the United States. Part Two then examines Sussex and makes the argument that whilst it was not in any way a 'Cold War university' on the American model, it can be described as a 'university *of* the Cold War'.[8] The university famously set out to 'redraw the map of learning' even as the physical map of the world was being recast along Cold War lines: as will be seen, the founders of the new institution took many of their cues from American models as they sought to develop a vision of higher education which could both meet the practical challenges faced by the British economy and play its part in safeguarding an enlightened vision of Western civilization. Part three then briefly considers the way in which these ideas ran aground in the mid-1960s as student protests against American involvement in Vietnam ushered in a very different perspective on the Cold War.

The Western vision

A considerable literature now exists around the concept of the 'Cold War university'. The term applies to American institutions such as Stanford and MIT which received very large sums of money from industry, charitable foundations, the military and agencies of the federal government to pursue research deemed useful to national security. Rebecca Lowen suggests that by 1960 universities received *c.* $1bn per annum from such sources, with half shared by just six institutions.[9] This arrangement was never short of critics. By the mid-1960s, intellectuals such as Noam Chomsky had begun to complain vociferously about the heavy price institutions paid for this largesse. In key polemics such as 'The responsibility of intellectuals', a lecture at New York University that appeared in *New York Review of Books* in February 1967, he argued that external interests had been allowed too much influence over the running of universities by 'cowardly' boards of trustees and university administrators. Whereas the responsibility of intellectuals was 'to expose the lies of governments, to analyze actions according to their causes and motives and often hidden intentions',[10] too many institutions had abdicated this duty and, in the words of Howard Zinn, become 'openly or subtly' subservient to society's rulers.[11] The assaults of McCarthyism on the universities were evidence of this.[12]

Modern accounts have taken issue with some elements of this critique, placing events in a more secular perspective, beyond the ideological fervour of the Cold War years.[13] Lowen, for example, in her study of Stanford, demonstrates that it was the university's engagement with industry before the Second World War, and the ultimate inadequacy of funding from that source, which paved the way for the acceptance of federal contracts.[14] Similarly, she dismisses the notion of universities allowing government agencies to block the appointment of politically undesirable – that is, left-wing – intellectuals. She suggests instead that in universities where a large proportion of the annual budget was derived from military and federal contracts, 'the ability to attract patronage became inseparable from administrators' evaluations of a faculty member's "excellence" and of an entire department's significance'.[15] Senior administrators were not above blocking the appointment of individuals who had no fund-raising profile in favour of those who did.

Most funding went to the natural sciences and certainly played an important role in the growth of whole academic fields – nuclear physics being one example. Some money, however, as David Engerman argues, found its way into the social sciences, where interdisciplinary 'area studies' programmes became very important.[16] In 1946, Columbia University founded a Russian Institute, developing the government's wartime practice of employing interdisciplinary teams of academics to study 'areas' like Russia as an aid to strategy formation.[17] The initiative was backed by the Rockefeller Foundation, and in 1948, the Social Science Research Council argued for a wide expansion of interdisciplinary area studies in American universities.[18] Accordingly, and with support from the charitable foundations, new programmes began to spring up within American universities.[19] From the start, area studies were envisaged as intrinsically multi- and interdisciplinary. Practitioners recognized, as David Szanton has written, that to effect change in Africa, Asia, Central Europe, Latin America and the Middle East, a simple reliance on older American and Euro-centric perspectives would be inadequate. Instead, 'more culturally and historically contexualized knowledge [...] would be essential if the US was to assist their economic development, modernization, and political stability – as well as to compete effectively for their loyalty with the Soviet Union'.[20] The launch of Sputnik in 1957 turbo-charged the logic of this argument: from the late 1950s, the US government itself began to fund new area studies programmes.

Even those universities not in receipt of 'Cold War university' style funding were affected by the prevailing anxiety. Most American universities sought to play their part in producing well-educated student-citizens who could resist the allure of communist doctrines.[21] The belief that an educated citizenry was essential to the survival of democracy was taken as axiomatic.[22] Humanities was incorporated into the undergraduate curriculum at even the most prestigious technological institutions such as MIT in order to 'promote the growth and development of students as "whole" men and women' who placed a premium on individual rights. The aim was to turn out 'well-rounded, socially responsible, self-governing citizens' rather than the 'narrow technocrats who could be manipulated for dubious political ends' purportedly trained in Soviet Russia.[23] A period of study abroad, especially to Europe, was also recommended for these new student-citizens. As one Princeton professor put it in

1965, it was intolerable that 'only the elite of policy makers understands what is going on internationally'; such knowledge should be spread widely, and the rise of the junior semester (or year) abroad became the vehicle for this.[24] Students from the Soviet Union, of course, enjoyed no such privileges.

This softer, more 'cultural' understanding of the Cold War was what the Americans exported after 1945 when Western Europe came to be regarded as the new front line in the war of attrition with the Soviet Union.[25] In Germany, as Stefan Paulus chronicles, American policy sought to make universities advocates for a distinctly American-led vision of what the West stood for.[26] The existing university establishment was damned for its wartime collusion with the Nazi government and, as Natalia Tsvetkova writes, its failure to accept higher education's responsibility for 'training students for effective citizenship'.[27]

Effective citizenship in the post-war context entailed the promotion of democracy,[28] and considerable efforts were made to reinvigorate the social and political sciences. Chairs and institutes were founded, buttressed by huge sums of money from the US government, and the Rockefeller and Ford Foundations.[29] One of the most important new subject areas developed in this period, explicitly designed to inculcate 'a Western and primarily American-influenced world-view', was American studies.[30] It was, however, just one among a broad range of subjects pioneered by the Americans in Germany that were to become very familiar in British universities from the 1960s onwards: 'social science, general science, education and educational research, political science, American history, international relations, and cultural history'.[31]

It would be misleading to suggest that Americans simply imposed their models on passive national groups in Europe: in many areas, 'Americanization' is better understood as the domestication into European culture of certain American emphases and ideas – a subtle hybridization.[32] One area in which a distinctively American way of thinking was pressed home with energy, however, was management education. American management methods had made very little impact in Europe prior to the Second World War; America's involvement in European reconstruction, and their provision of various forms of technical assistance, however, helped change this.[33] Management studies was an area of the academy undergoing considerable growth within America itself at the time: men like Clark Kerr, president of the University of California, whose book *Industrialism and industrial Man* was a foundational text of the 1960s, would argue in the 1950s that while training engineers and scientists was vital to economic growth, so too was 'business administration'.[34] American methods of management were actively promoted under the Marshall Plan: German companies could apply for loans for 'productivity-increasing projects' but had to demonstrate application of 'American principles and methods'.[35] According to Alfred Kieser, 'management training institutions were founded in Germany with the explicit goal of making managers familiar with American management approaches' and there came to be a strong public acceptance that these methods held the key to business success. Within universities the younger generation of researchers 'began intensively to explore American management sciences'.[36] They did so in part because the methods themselves were gaining widespread acceptance within Germany, but also because the 'science' of management offered an antidote to the ideology-driven Nazi era.

'American management sciences offered two options for the advancement of German management sciences: formalization … and a conceptualization based on general, empirically testable laws (hypotheses).' Historians such as Berghahn (1993) have argued 'that the organizational culture and the leadership styles of German managers changed profoundly as a consequence of the Marshall Plan', even if more authoritarian and bureaucratic traditions survived in some areas.[37]

The Cold War university phenomenon was thus not purely exclusive to the United States. In Germany, an educational system which internalized the logic of the Cold War established itself, while a recent study has claimed that the Australian National University, set up in Canberra in 1946, experienced many of the symptoms seen in America, albeit in milder form.[38] Very little scholarly attention, however, has been paid to the experience of British universities established during the Cold War period. It is to an examination of Sussex that attention now turns.

The foundation of Sussex

When Sussex admitted its first fifty-one students in the autumn of 1961, it was difficult to imagine the impact the new institution would make. Yet, within six years, the university numbered in excess of 3,000 students and there were twenty applicants for every place in arts subjects.[39] Newspapers outdid each other to lavish epithets on the 'infant legend' taking shape on its purpose-built campus in the South Downs. In February 1965, *The Listener* described Sussex as 'the "with-it," twenty-first century university' and 'the fashionable sixth former's first choice'.[40] Located close to London – and thus to the headquarters of both television and the press – it attracted talented staff and glamorous students. Half a decade after its founding, Sussex was, concluded the education correspondent of *The Times*, 'an obvious success'.[41]

Half a century later, the rapid growth of Sussex and its ability to charm the media remain impressive. Yet in other respects, the contemporary evaluation appears flawed, nowhere more so than in the admittedly throwaway characterization of Sussex as a 'twenty-first century' university. In this section, it will be argued that while Sussex certainly embodied cutting-edge trends in everything from the organization of its campus to the shape of its curricula, it was distinctly of its time in its assumptions and its goals. It was a university of the Cold War era. The section accordingly falls into two parts – the first examines how the immediate imperatives of British social and economic life shaped the new institution; the second explores the subtle influence of the Cold War on its academic ambitions.

In 1958 the government announced an extra £60m to support the expansion of higher education and ear-marked £1.5m specifically for a new university in Sussex.[42] This was the culmination of an intensive lobbying campaign that began in 1956 when the shape of post-war demand for higher education became clear.[43] Among the membership of advisory committee appointed by the University Grants Committee (UGC) was Dr Ronald Holroyd, deputy chairman of ICI Ltd – a reminder of the government's priority that the new universities produce more of what Frank Thistlethwaite, vice-chancellor at the University of East Anglia (UEA), described as the 'general purpose student' industrialists complained they were lacking.[44] Warwick's Vice-Chancellor, Jack Butterworth, in similar vein, promised his institution's willingness to 'develop the industrial side on up-to-date American lines'.[45]

How important industry was to the projected new universities is revealed by the published lists of subscribers to the public appeals with which this chapter began. Sussex, for example, successfully raised £500,000 to build student residences, an item not covered by the government's initial grant.[46] Some came from national concerns like ICI – Roger Holroyd's presence on the UGC committee was designed to underscore the emphasis on technological education in the new institutions and British Nylon Spinners Ltd, a joint venture begun in 1940 by ICI and Courtaulds duly subscribed £1,750 to the Sussex appeal.[47] Britain's largest electrical manufacturer, Associated Electrical Industries Ltd, also subscribed £1,750.[48] However, several local industrial concerns also subscribed. Two came from Brighton itself: the Talbot Tool Company, which produced parts for the Ministry of Aircraft Production, and Allen West & Co, which produced radar equipment during the war and employed 3,000 people. Both subscribed £1,000. A sense of local patriotism may have influenced them, but these were also the kind of businesses that would benefit from technically literate graduates.[49] The same is true of Buxted Ltd, a battery chicken enterprise located near Uckfield, which gave £3,500.[50] The company invested heavily in technology to meet the growing demand for frozen chickens from customers and supermarket chains.[51] Their super-chilling plant at Aldershot has been described as being 'the most advanced processing plant in the world' when it opened in the late 1950s, ahead even of American facilities.[52] This was exactly the kind of business that required a technologically adaptable and intellectually flexible workforce.[53]

If the principles of the new University were important, so were its principals. Both John Fulton, the first vice-chancellor, and Asa Briggs, ostensibly dean of Social Sciences but in practice the deputy vice-chancellor, commanded the attention of the media, and sought to make the Sussex experiment seem part of a crusade. Fulton, for example, in a long article for *The Times* on the day the institution's Royal Charter of Incorporation took effect, argued that nothing like the current 'explosion' of interest in higher education had 'happened since the foundation in the Middle Ages of the universities of Western Europe':[54]

> Men of every colour, race and creed see in it the key that will open the doors to self-realization for the individual, to the true independence of the nation or group, to the possibility of co-existence in peace for a divided human race; and, above all, to influence over the intellectual leaders of the generations to come.

This was heady stuff, and Sussex seemed determined to break away from the old, mediaeval university model. Here the influence of Briggs was key. He was an adept 'public intellectual' even before joining Sussex, and it is perhaps natural that it was he who came up with the strapline that would serve Sussex so well in its early years – the plan to 'redraw the map of learning'. The phrase formed the title of a lecture delivered by Briggs at the Australian National University in 1961.[55] The thrust of his argument was that universities needed to rise above about the 'rivalry and occasional friction, boundary disputes and far from splendid isolation' which dogged the modern academy and adopt fresh ways of teaching and researching to deal properly with the challenges of the contemporary world.[56] What was required was a more fluid academic system which dissolved boundaries and allowed more holistic approaches to problem solving.

This became the hallmark of the new curriculum promised by Briggs to those considering a degree at Sussex – themes rehearsed in a 1962 article for the *Listener*.[57] The segregation of disciplines, he argued, was more insidious than implied by C. P. Snow's simple binary between the arts and the sciences in his 'Two cultures' lecture of 1959. Equally impenetrable partitions had been raised within these broad disciplinary areas – 'between literature and the social studies, and between biological and physical sciences'. What the current century required of its graduates, however, was not extreme specialization but 'adaptability'. Students required an education that would equip them 'to fill the ever-widening range of graduate employments, many of which are not in the least looking for single academic specialisms as necessary qualifications'. This education the University of Sussex would provide. It had the 'immeasurable advantage of a fresh start', he said, and added, with the kind of astute market differentiation that was a key part of Sussex's appeal, that what Sussex achieved was bound to have repercussions on the older (for which read old-fashioned) universities. 'The wind of change', he concluded, 'is the breath of life'.

The new map of learning was duly realized at Sussex. Departmental structures were done away with,[58] and in their stead teaching was organized around large multi-, and interdisciplinary, schools. First off the blocks were the humanities and social sciences in the shape of European studies, English and American studies, social studies, African and Asian studies (AFRAS) and educational studies. The schools of physical sciences, molecular sciences, biological sciences and applied sciences followed in the next few years.

How the new system worked in practice can be seen in European studies.[59] It was modelled on the Oxford 'Greats' course and sought to explore European civilization through the lenses of history, philosophy and literature; however, whereas the Oxford course focused on the civilizations of antiquity, Sussex students were explicitly 'to concern themselves with contemporary as well as inherited culture, with history in the making as well as history that is already made'.[60] All first-year students in the arts and social sciences were required to take two papers: *Languages and values* challenged students to study the nature of contemporary society; *An introduction to history* demanded an understanding of how historians worked, the questions they asked and why they disagreed. In years 2 and 3, students selected five papers in their major discipline balanced by four 'contextual' papers from other disciplines within the school. The degree was as much about knowledge creation as knowledge retention, and sought to allow the individual student to become as independent as he or she was prepared to be.

The design of the campus emulated the novelty of the curriculum. On the 'utopianist campus', as Stefan Muthesius remarks, there was a 'belief in the behaviour-shaping power of the buildings'.[61] The Vice-Chancellor of UEA, Frank Thistlethwaite, argued that it was 'necessary to arrange the buildings to give a sense of the coherence of knowledge',[62] and Sussex followed this model.[63] Sir Basil Spence designed the central cluster of buildings including Falmer House, built around a quadrangle and designed to be a gatehouse to the campus. It was designed, as Maurice Howard writes, 'as the fulcrum of the university's social gatherings' and brought together the Refectory, staff common room, facilities for student counselling and a debating chamber – an academic community in one building.[64] It won the RIBA Bronze medal in 1964. Spence also designed the physics and chemistry buildings and Arts A and B, before the UGC money ran out, leaving the Meeting House and the Arts Centre to be paid for by public benefactors.[65]

Here then was the 'twenty-first century' university in full swing: a distinctive academic community promoting a novel emphasis on the connectedness of knowledge and its contemporary relevance. Whilst the Sussex offer was deliberately different from that available to students everywhere else (except at Keele), its approach to the organization of education had important echoes of contemporary American practice. One of these was the University's campus setting, but another was the influence of contemporary anxiety about the Cold War.

During the late 1950s and early 1960s, politicians, policymakers and educational leaders emphasized the connection between the reform of higher education and the prevailing Cold War. A speech by John Fulton at the inaugural meeting of the Sussex court in 1961, for example, explicitly juxtaposed the university's goals with the values of communism. 'We in the west are committed to faith in the worth of the individual,' he declared: 'His freedom is not to be subordinated to the claims of the party line' and ensuring the freedom of the individual was thus the highest 'challenge to western education'.[66] The Prime Minister, Harold Macmillan, made a similar point eighteen months later, when accepting an honorary degree from the university.[67] Universities were important for developing not only the nation's economy, he said, but also 'the character of our people', giving 'each young man and woman a sense of not just being a cog in a huge machine but of an individual living soul'. That individuality was the antithesis of Soviet standardization. Catching a similar mood, Alan Richardson from the University of Nottingham urged that new Universities like Sussex engage deeply with 'the whole intellectual and cultural inheritance of western civilization' when setting up their new interdisciplinary schools. Writing to *The Times* in 1961, he justified his stance with reference to the Soviet Union:

> It is well known that in Russia students of science and technology are thoroughly instructed in the philosophy and ethics of communism. Can we afford to deny the students of our universities the opportunity of learning and discussing the values of our western inheritance, and must we not act in the faith that, when they have understood them, they will love what they know?

'The western tradition,' he concluded, 'cannot be understood or appreciated apart from its history.' The crucial function of a School such as European studies was thus not so much redrawing the map of learning as ensuring that those within the Western half of civilization could properly identify and value its salient landmarks.

These kinds of broad statements reflected the common sense of the period. As Lord Hailsham put it on the eve of the Conservatives' triumphant 1959 election campaign, and just a few months after Fulton's appointment at Sussex, there were only three forces for good in the modern world: the British Commonwealth, western Europe and the United States. 'On their ability to win the hearts and minds of mankind for our own precious ideal of liberty under the law,' he added, 'hang the real hopes and chances of the human race.'[68] It is interesting to note how Hailsham's sketch map of the geopolitical world was realized in the organization of the new Schools at Sussex. The first Schools to be created, as noted above, were AFRAS, European studies, and English and American Studies which corresponded almost exactly to the 'forces for good' listed by Hailsham. Equally striking, perhaps, was the absence of schools devoted to eastern Europe or China.[69] While these areas might be studied within the broader curricula of other Schools, such an arrangement inevitably ensured they would be comprehended within broadly Western contexts rather than on their own terms. The map of Cold War geopolitics and the redrawn 'map of learning' at Sussex enjoyed a distinctive overlap.

The nature of the disciplines taught at Sussex reflected a similar tendency. American Studies provides a good example. The School of English and American Studies naturally had an Anglophone focus, but in American Studies it contained the quintessential Cold War discipline. It was actively promoted by the US government in Germany after 1945, and became significant in other countries as the Cold War took hold.[70] In Britain it complemented a sharp post-war upturn of academic interest in American history, particularly those branches which seemed to presage the 'Cold War' partnership between the two countries – foreign affairs; economic and business history.[71] Many of the British scholars who became involved in this intellectual endeavour were strong advocates for the kinds of Cold War positions encountered earlier in this essay. The historian of foreign affairs, H. C. 'Harry' Allen, for example, echoing Churchill's belief in the 'special relationship', was convinced that 'the very peace of the world, or at least the survival of western civilization … rested on a secure Anglo-American partnership'.[72] Similarly, the economic historian, Brinley Thomas, insisted that Britain's future security rested not on its Commonwealth but on an alliance with the United States. 'The only power bloc which has a chance of deterring a would-be aggressor', he argued, 'is a world coalition of nations led by the United States.'[73]

From the early 1950s, the USAID and the Rockefeller Foundation generously supported American studies in Britain, and in 1955, at the prompting of the US government and the US Embassy in London, a formal British Association of American Studies (BAAS) was established.[74] External support remained crucial – the Rockefeller Foundation provided $150,000 for the period 1957–62. Sussex quickly became a leading centre for the discipline, and in 1965 the American government provided the funds necessary to appoint Marcus Cunliffe, a central figure in the foundation of BAAS, to a chair at the university.[75]

Another distinctive feature in Sussex's intellectual landscape was the Institute for Development Studies (IDS); it, too, had a certain Cold War logic. During the 1950s, as Dudley Seers describes, the newly independent former colonies became seen as 'potential breeding grounds for communism'.[76] IDS was a response to a report by the former head of the Civil Service, Lord Bridges, which argued that the leaders of developing countries urgently needed training in the principles of public administration to help counter this tendency.[77] Although an independent body, primarily funded by the government, located within, but distinct from, the university, IDS operated on the same interdisciplinary principles as the area studies institutions discussed earlier. As Paul Streeten, the distinguished American professor of development economics, remarked at the opening of IDS, the problems of the developing world could be solved 'only by collaboration between economists, statisticians, sociologists, anthropologists, political scientists and students of administration and business management'.[78]

With the benefit of government funding and considerable revenue generated through immersive training courses for overseas officials, IDS grew quickly. The Institute had just eight academic staff and ten others engaged in administration and the library in 1966; by 1971 it had quadrupled in size and by 1976 employed seventy-five academics and fifty-four staff in supporting roles.[79] Whilst independent of the university, its location at Sussex, and its global mission, suited well the general approach fostered by the university's founders.

There was, therefore, a distinctive alignment between the cultural framing of the new provision at Sussex and the map of Cold War geopolitics. Sussex did not, of course, emulate all aspects of the modern American curriculum. Notably absent, for example, was a business school – something that other new universities were quick to establish. In 1964, the Ford Motor Company gave £200,000 to Essex 'to support research and invention' in the field and followed this up with £150,000 for the establishment of Business Schools at LSE and Manchester. Warwick founded a similar school, citing the 'urgent need for top quality management' and the responsibility of universities to provide it. It opened in 1967.[80] Nevertheless, the nature of what was offered at Sussex in the early 1960s internalized many approaches familiar within contemporary American higher education – both at home and abroad – and with similar ends in view.

Disenchantment

If Sussex in the early 1960s provides a British example of the 'subtle hybridization' of indigenous and American ideologies that took place in many regions during the Cold War,[81] the culture of the new institution was not uncritically supportive of America or American actions – especially in the field of international affairs. Indeed, from the mid-1960s, the early quietude of the student population was replaced by a new, powerful current of anti-American feeling – a phenomenon closely associated with US involvement in Vietnam.[82] The founding fathers of the university – men like Asa Briggs – recognized that universities had to allow scope for argument and dissent between the generations, but they were unprepared for the militant tone of student activism that emerged from the mid-1960s.

The new timbre of student politics was closely bound up with hostility to the Vietnam War. At Sussex, students protested noisily when the university decided to bestow an honorary degree on the Prime Minister Harold Wilson in 1966.[83] A year later, an official from the US Embassy was attacked with red paint on a visit to the campus and two students, Michael Klein and Sean Linehan, were disciplined.[84] In 1973, meanwhile, the Frank G. Thomson Professor of Government at Harvard University, Samuel P. Huntington, was prevented from speaking on campus.[85] The students believed an article Huntington contributed to *Foreign Affairs* in 1968 offered the rationale for much of America's strategy in Vietnam.

Such student radicalism disrupted those principles of dialogue upon which the university leadership hoped to base their new institution. Asa Briggs, addressing the university's Court in December 1968, reflected unhappily on the actions of hard-core protestors, 'small groups of people who rely on slogans rather than ideas, who are keener to destroy than they are to create and who are fundamentally totalitarian in their approach to personalities and to issues'.[86] And despite the fact that the student body at Sussex was at pains to distance itself from the paint-throwing incident in 1968,[87] public opinion beyond the university chilled significantly as the decade drew to a close. As Harold Perkin wrote, 'People were scandalized by the sight of well-heeled, middle class students demonstrating for rights which they themselves did not have in their factories and offices, at their expense as tax and rate payers.'[88] Newspapers abounded with angry letters suggesting that militant students were 'not fit to study at the taxpayers' expense', and student radicals were denounced as 'communists' or 'Maoists'.[89] The readiness with which people reached for these terms hints at the continuing salience of Cold War cultural tensions. The new universities in which people had invested not only money but also hope had been part of the antidote to British vulnerability in the new Cold War. Yet, far from working an economic miracle which returned the nation to prosperity and a place of leadership in the free world, they seemed only to produce students who took to the streets in protest, 'playing bolshies'.

At Sussex, the Conservative MP for Arundel and Shoreham, Captain Henry Kerby, resigned his position as a member of the university's Court, citing the 'sinister trend' of 'so-called "student power" manifesting itself in British universities'.[90] In his letter of resignation, he cited the revelation of a Gallup Poll that 40 per cent of 273 Sussex students had taken part in some kind of demonstrations in the previous twelve months, and the 'shocking figures' that revealed the 'absolute majorities' opposed to the Smith regime in South Africa and American involvement in Vietnam. Too much money, he concluded, was being spent on British universities.

In fact, the radicalism of students at Sussex, or anywhere else in Britain, can be exaggerated.[91] As Nick Thomas notes, the Gallup poll cited by Kerby, which polled students at Sussex and Cambridge, highlighted an interesting difference in outlook regarding 'domestic' and 'political' questions. On issues such as the representation of students on university committees, for example, fully two-thirds of students were supportive of taking action; on 'political' questions like Vietnam or nuclear disarmament, the proportions fell away sharply: barely a quarter of students polled claimed to have taken part in any demonstrations.[92] For most students at the end of the 1960s, the underlying logic of Lord Hailsham's map seems to have held good.

Conclusion

During the 1950s and early 1960s, the spectre of growing Soviet power provided a backdrop to British life. As with any other conflict, the Cold War had its home front. Anxieties about the Russian threat permeated politics and culture and naturally formed part of the discursive framework within which the expansion of higher education was imagined.[93] From the late 1940s onwards, the notion that the Soviet system was antithetical to 'the British way of life' and all that it stood for in terms of freedom and individuality formed a central element in Conservative party discourse, and educating the nation's youth in the principles of freedom was seen as essential.[94] As a pamphlet produced by the party's 'One Nation' group put it as early as 1952: 'Education is more than a social service; it is part of Defence.'[95]

Just as American cultural policy, in Robert Hewison's words, 'played an important part in promoting the "values of the free world,"' so too American ideas about the organization and content of higher education were readily absorbed by those committed to reforming the university sector in the 1950s and 1960s.[96] Many of those involved in establishing Britain's new universities in the early 1960s had first-hand experience of working within the American university sector, where they were exposed to new ways of thinking and to the Cold War logic that often pervaded them. Albert Sloman, vice-chancellor of Essex, held an appointment at Berkeley before moving to Liverpool and so would presumably have been exposed to the kind of thinking on issues like business organization popularized by Clark Kerr.[97] Frank Thistlethwaite at UEA was an historian of the United States and the key figure in the creation of the BAAS: in recognition of his diverse talents, he was actually offered the post of cultural attaché at the US Embassy in London.[98] And at Sussex, Asa Briggs had spent time at the University of Chicago in the mid-1950s, one of the centres of behavioural science.[99] Whilst he said he had no intention of making Sussex an American-style university, it remained the case that he was very open to the possibility of assimilating American models into his own university.[100]

Sussex, like the other new universities founded in the early 1960s, was not a 'Cold War university' in the sense implied by the use of this term in America; nonetheless, it, and they, could justifiably be called universities *of* the Cold War. These institutions assimilated many of the preoccupations and preconceptions in circulation at the time of their foundation, and promised an education that would produce people equipped not just to take their place in the modern economy, but to uphold the values of Western civilization in its defining battle with the international 'communist conspiracy'. The experience of Sussex offers another example – in a British rather than a European context – of the subtle hybridization of American ideas and methods in the post-war period, an alloy of what Hugh Wilford has described as 'repulsion, attraction and (heavily mediated) influence'.[101] As a consequence when Sussex offered its students a new map of learning, the part with which it insisted they become familiar was the Western portion: schooling people in the values of the known world was seen as the most effective antidote to the mysteries and dangers of the 'other'. Students, as we have seen, ultimately found their own paths through this new landscape; however, the task on which the founders of universities like Sussex embarked, 'undertaken for the nation as a whole' as the appeal chairs put it, was a crusade for an ideal vision – a very Western utopia.

Notes

1 *Times*, 23 November 1960, 13.
2 Michael Sanderson, 'Higher Education in the Post-War Years', *Contemporary Record*, 5 (1991), 417–31.
3 Harold Perkin, 'University Planning in Britain in the 1960s', *Higher Education*, 1 (1972), 112–13.
4 Sanderson, 'Higher Education in the Post-War Years', 418; Peter Mandler, 'Educating the Nation: II. The Universities', *Transactions of the Royal Historical Society*, 25 (2015), 3–4.
5 Matthew Cragoe, 'Asa Briggs and the University of Sussex, 1961–1976', in *The Age of Asa: Lord Briggs, Public Life and History in Britain since 1945*, ed. Miles Taylor (Basingstoke: Palgrave Macmillan, 2015), 227–8; William Whyte, *Redbrick: A Social and Architectural History of Britain's Civic Universitie*s (Oxford: Oxford University Press, 2015), 229–37; Sanderson, 'Higher Education in the Post-War Years', 417; Jean Bocock et al., 'American Influence on British Higher Education: Science, Technology, and the Problem of University Expansion, 1945–63', *Minerva*, 41 (2003), 327–46.
6 Jean Bocock and Richard Taylor, 'The Labour Party and Higher Education: 1945–51', *Higher Education Quarterly*, 57 (2003), 253.
7 Viscount Hailsham, *The Conservative Case* (Harmondsworth: Penguin, 1959), 161.
8 For the distinction, see Joel Isaac, 'The Human Sciences in Cold War America', *Historical Journal*, 50 (2007), 727.
9 Rebecca Lowen, 'The More Things Change …: Money, Power and the Professoriate', *History of Education Quarterly*, 45 (2005), 439.
10 Noam Chomsky, *American Power and the New Mandarins* (Reprinted New York: New Press, 2003), 323.
11 Ibid. Howard Zinn, 'Foreword', iv.
12 Hanna Holborn Gray, 'Tools of Power or Oases of Freedom?', *Foreign Affairs*, 76 (1997), 147–51.
13 Isaac, 'Human Sciences'; cf. David C. Engerman, 'Social Science in the Cold War', *Isis*, 101 (2010), 393–400.
14 Rebecca S. Lowen, *Creating the Cold War University: The Transformation of Stanford* (Berkeley and London: University of California Press, 1997), 6; Roger L. Geiger, *Research and Relevant Knowledge: American Research Universities since World War II*, 2nd edn. (New Brunswick, NJ: Transaction, 2008), xiii.
15 Lowen, 'The More Things Change', 443.
16 David Engerman, 'Rethinking Cold War Universities: Some Recent Histories', *Journal of Cold War Studies*, 5 (2003), 83–6.
17 Victoria E. Bonnell and George W. Breslauer, 'Soviet and Post-Soviet Area Studies', in *The Politics of Knowledge: Area Studies and the Disciplines*, ed. David L Szanton (Berkeley: University of California Press, 2004), 217–61.
18 Vicente L. Rafael, 'The Cultures of Area Studies in the United States', *Social Text*, 41 (1994), 91–7.

19 Inderjeet Parmar, 'American Foundations and the Development of International Knowledge Networks', *Global Networks*, 2 (2002), 13–30.
20 David L. Szanton, 'Introduction: The Origin, Nature and Challenges of Area Studies in the United States', in *The Politics of Knowledge*, ed. Szanton, 10.
21 Peter Justin Kizilos-Clift, 'Humanizing the Cold War Campus: The Battle for Hearts and Minds at MIT, 1945–1965' (unpublished PhD thesis, University of Minnesota, 2009), 5.
22 Jude van Konkelenberg, 'Australia's Cold War University: The Relationship between the Australian National University's Research School of Pacific Studies and the Federal Government, 1946–1975' (unpublished PhD thesis, University of Adelaide, 2009), 27.
23 Kizilos-Clift, 'Humanizing the Cold War Campus', 19.
24 Quoted in Nicholas B. Dirks and Nils Gilman, 'Berkeley's New Approach to Global Engagement: Early and Current Efforts to Become More International', *Research and Occasional Papers Series* (Centre for Studies in Higher Education, Berkeley, 2015), 3.
25 Stefan Paulus, 'The Americanisation of Europe after 1945? The Case of the German Universities', *European Review of History: Revue Européenne d'Histoire*, 9 (2002), 241–53.
26 James F. Tent, *Mission on the Rhine: 'Re-education' and Denazification in American-Occupied Germany* (Chicago: University of Chicago Press, 1983); Natalia Tsvetkova, 'Making a New and Pliable Professor: American and Soviet Transformations in German Universities, 1945–1990', *Minerva*, 52 (2014), 161–85. Russians did the same: John Connelly, *Captive University: The Sovietization of East German, Czech, and Polish Higher Education, 1945–56* (Chapel Hill & London: University of North Carolina Press, 2000).
27 Tsvetkova, 'Making a New and Pliable Professor', 166–7.
28 Paulus, 'The Americanisation of Europe', 249–50.
29 Tent, *Mission on the Rhine*, 286–8.
30 Paulus, 'The Americanisation of Europe', 249–50.
31 Tsvetkova, 'Making a New and Pliable Professor', 168.
32 Paulus, 'The Americanisation of Europe', 245–6; Udi E. Greenberg, 'Germany's Postwar Re-education and Its Weimar Intellectual Roots', *Journal of Contemporary History*, 46 (2011), 10–32.
33 Behlul Usdiken, 'Americanization of European Management Education in Historical and Comparative Perspective: A Symposium', *Journal of Management Inquiry*, 13 (2004), 87–9.
34 Ethan Schrum, 'To "Administer the Present": Clark Kerr and the Purpose of the Postwar American Research University', *Social Science History*, 36 (2012), 499–523. Kerr's *The Uses of the University* (1963) was widely read in British universities: Michael Shattock, 'Parallel Worlds: The California Master Plan and the Development of British Higher Education', in *Clark Kerr's World of Higher Education Reaches the 21st Century: Chapters in a Special History*, ed. Sheldon Rothblatt (New York & London: Springer, 2012), 107.
35 Alfred Kieser, 'The Americanization of Academic Management Education in Germany', *Journal of Management Inquiry*, 13 (2004), 90–7.
36 Ibid., 94.
37 Ibid., 93.
38 van Konkelenberg, 'Australia's Cold War University', 6.
39 *Times*, 26 June 1967, 8.
40 Stuart Maclure, 'The "With-It" University?', *The Listener*, 25 February 1965, 290–2.

41 *Times*, 26 June 1967, 8.
42 *Times*, 21 February 1958, 8.
43 For a full account, see Cragoe, 'Asa Briggs'.
44 *Times*, 26 April 1958, 8. For the UGC, Michael Shattock and Robert O. Berdahl, 'The British University Grants Committee, 1919–83: Changing Relationships with Government and Universities', *Higher Education Quarterly*, 13 (1984), 471–99. For Holdroyd, see J. D. Rose, 'Ronald Holdroyd, 1904–1973', *Biographical Memoirs of the Fellows of the Royal Society*, 20 (1974), 235–45.
45 *Times*, 9 November 1961, 14 (UEA); *Times*, 10 April 1964, 7 (Warwick).
46 *Times*, 11 November 196, 6.
47 http://www.gracesguide.co.uk/British_Nylon_Spinners (accessed 30 July 2016).
48 http://www.gracesguide.co.uk/AEI (accessed 30 July 2016).
49 http://www.gracesguide.co.uk/Talbot_Tool_Co (accessed 30 July 2016); http://www.gracesguide.co.uk/Allen_West_and_Co (accessed 30 July 2016).
50 http://buxtedvillage.org.uk/village-info/history/anecdotes-and-stories/buxted-chickens/ (accessed 30 July 2016).
51 Andrew Godley and Bridget Williams, *The Chicken, the Factory Farm, and the Supermarket: The Emergence of the Modern Poultry Industry in Britain*, University of Reading Working Papers Series, 50 (2007), 9–16.
52 Ibid., 14.
53 For Fisher, see Christopher Muller, 'The Institute of Economic Affairs: Undermining the Postwar Consensus', *Contemporary British History*, 10 (1996), 88–110.
54 *Times*, 16 August 1961, 9.
55 Asa Briggs, *The Map of Learning: The First Annual Lecture of the Research Students' Association Delivered at Canberra on 2nd November, 1960* (Canberra: Australian National University, 1961).
56 Briggs, 'Map of Learning', 10–11.
57 Asa Briggs, 'A New Approach to University Degrees', *Listener*, 24 May 1962, 899–90.
58 Briggs, 'Map of Learning', 19.
59 Martin Wight, 'European Studies', in *The Idea of a New University: An Experiment in Sussex*, ed. David Daiches (London: André Deutsch, 1964), 100–19.
60 Quoted in Michael Beloff, *The Plateglass Universities* (London: Secker & Warburg, 1968), 89.
61 Stefan Muthesius, *The Postwar University. Utopianist Campus and College* (London and New Haven: Yale University Press, 2000), 126, 186.
62 *Illustrated London News,* 26 April 1963, 5 (UEA); 23 October 1963, 6; ibid. 23 October 1965, 26.
63 Ibid., 15 June 1956, 6; 25 March 1960, 8 and 23 February 1962, 5.
64 Maurice Howard, 'Basil Spence and the University's Architecture', in *Making the Future*, ed. Fred Gray (Brighton: University of Sussex, 2011), 31–41.
65 Ibid., 35–7.
66 *Times*, 11 November 1961, 6.
67 *Times*, 12 June 1963, 6.
68 Hailsham, *Conservative Case*, 159.
69 Soviet Studies did feature at other UK universities: Victor S. Frank, 'Soviet Studies in Western Europe: Britain', *Survey: A Journal of Soviet and East European Studies*, 50 (1964), 90–6; Rory M. Miller, 'Academic Entrepreneurs, Public Policy and the Growth of Latin American Studies in Britain during the Cold war', *Latin American Perspectives*, 45 (2018), 46–68.

70 Ali Fisher and Scott Lucas, 'Master and Servant? The US Government and the Founding of the British Association for American Studies', *British Journal of American Culture*, 21 (2002), 24; Hugh Wilford, 'Britain: In Between', in *The Americanization of Europe: Culture, Diplomacy, and Anti-Americanism after 1945*, ed. Alexander Stephan (New York and Oxford: Berghahn Books, 2006), 23–43.
71 Michael Heale, 'The British Discovery of American History: War, Liberalism and the Atlantic Connection', *Journal of American Studies*, 39 (2005), 357, 360; Michael Heale, 'American History: The View from Britain', *Reviews in American History*, 14 (1986), 504.
72 Heale, 'British Discovery', 368.
73 Heale, 'American History', 507.
74 Fisher and Lucas, 'Master and Servant?', 18.
75 *Court*, 7 December 1964. For Cunliffe, see 'Marcus Cunliffe, 'A Pastmaster', in *American Studies: Essays in Honour of Marcus Cunliffe*, ed. Brian Holden Reid and John White (Basingstoke: Macmillan, 1991), 1–20.
76 Dudley Seers, 'Back to the Ivory Tower?', in *Has Universal Development Come of Age?*, ed. Richard Longhurst, IDS Bulleting 48 (October 2017), 9; Carlos Fortin, 'The Experience of the Institute of Development Studies, UK', in *Report on the Inaugural Seminar of the Zimbabwe Institute of Development Studies* (Harare: Zimbabwe Institute of Development Studies, 1982), 77–91. Cf. Richard Jolly, *A Short History of IDS: A Personal Reflection*, IDS Discussion Paper 388 (Brighton: University of Sussex, 2008), 11–12.
77 John Winnfrith, 'Edward Ettingdean Bridges. Baron Bridges', *Biographical Memoirs of Fellows of the Royal Society*, 16 (1970), 35–56.
78 *Times*, 16 September 1966, 11.
79 Fortin, 'The Experience of the IDS', 83.
80 *Times*, 14 November 1962, 14.
81 Paulus, 'The Americanisation of Europe', 245–6; Greenberg, 'Germany's Postwar Re-education'.
82 Tom Wills, 'Student Radicalism', in *Making the Future*, ed. Gray, 148.
83 *Daily Mail*, 16 July 1966, 5.
84 Wills, 'Student Radicalism', 151–2.
85 Rob Skinner, 'The Radical University', in *Making the Future*, ed. Gray, 146–8.
86 This paragraph is based on *Court*, 6 December 1968, 8–10.
87 *Times*, 8 March 1968.
88 Harold Perkin, 'Dream, Myth and Reality: New Universities in England, 1960–1990', *Higher Education Quarterly*, 45 (1991), 303.
89 *Times*, 3 December 1968, 9; *Times*, 19 March 1969, 7, 11 March 1970, 11.
90 *Times*, 2 July 1968, 3.
91 Scott, 'British Universities, 1968–78', *Paedagogica Europaea*, 13 (1978), 40–1.
92 Nick Thomas, 'Challenging the Myths of the 1960s: The Case of Student Protest in Britain', *Twentieth Century British History*, 13 (2002), 282–3.
93 Tony Shaw, *British Cinema and the Cold War the State, Propaganda and Consensus* (London: I.B. Tauris, 2001); Tony Shaw, 'The Politics of Cold War Culture', *Cold War Studies*, 3 (2001), 59–76.
94 Matthew Cragoe, '"We Like Local Patriotism": The Conservative Party and the Discourse of Decentralisation, 1947–51', *English Historical Review*, 122 (2007), 965–85.

95 Hugh Bochel, 'One Nation Conservatism and Social Policy, 1951–64', *Journal of Poverty and Social Justice*, 18 (2010), 123–34.
96 Robert Hewison, *Culture and Consensus: England, Art and Politics since 1940* (London: Methuen, 1997), 132.
97 *Times*, 5 July 1962, 6.
98 Fisher and Lucas, 'Master and Servant?', 21–4.
99 Schrum, 'To "Administer the Present"', 501.
100 Cragoe, 'Asa Briggs', 234–5.
101 Wilford, 'Britain: In Between', 29, 39–40.

4

'A poor man's Oxbridge'?: The founding of the University of York, 1960–73

by Allen Warren

In December 1959, the University Grants Committee accepted two very different schemes for new campuses at Norwich and York. As part of a larger programme, each attracted its own myths. Norwich was seen as the more innovative in terms of vision, architecture and its curriculum, York the more traditional with its college system and familiar academic disciplines. Moreover, York's Vice-Chancellor, Lord James of Rusholme, had a high public profile, closely identified with the 1944 Education Act, and with the place of grammar schools in particular. Described as 'a poor man's Oxbridge' by student delegates at their national conference in October 1963, this chapter is devoted to examining the truth of that short-hand phrase.[1] Five differing perspectives will be examined: the York-based campaign for a university, the impact of Lord James as its founding vice-chancellor, the creation of the campus on a collegiate model, the appointment of academic staff and the degree courses created, and, finally, the character of the students recruited in its first decade along with their experiences.

The campaign for a university after 1945

It is important, at the outset, not to see the group of York citizens actively supporting the campaign for a university out of context, and as exclusively driven by this ambition. Their aspirations were more focused on civic renewal and modernization of the historic city as a whole in the environment of rapid post-war reconstruction. At their core was the York Civic Trust, founded by Oliver Sheldon in 1946, former director of the Rowntree confectionery company, based in York, and out of which a committee was formed to promote the prospect of a university. Alongside this particular development were others: an expansion of the York Georgian Society, the creation of the York Conservation Trust, the revival of the York Livery Companies along with the preservation of their historic halls, the resurrection of the York cycle of Mystery plays. This was an overlapping group of individuals that provided the energy, principally Oliver Sheldon, J. B. Morrell, another former Rowntrees' director and, on two occasions, Lord Mayor of York, Eric Milner-White, dean of York, and later John Shannon, chair of the York Civic Trust, and Donald Barron, another Rowntree director. None were leaders or officials within the York City Council or men with significant experience of running a university.

Their 1947 bid to the UGC was rejected, but not wholly discouraged. Following advice, the Civic Trust worked to try to create some university-level activity in the city, establishing a separate York Academic Trust within its own more general activities. The Trust was chaired by Eric Milner-White, who had come to York in 1941, having been previously dean of King's College, Cambridge. He certainly saw universities and schools through Cambridge eyes, later in life referring rather preciously to King's as 'this blessed place'.[2] But throughout most of the 1950s, he was principally concerned to find a way of realizing the UGC's advice. As a result of recruiting John West-Taylor, a young Cambridge graduate, the Borthwick Institute for Historical Research and the Institute for Advanced Architectural Studies (IAAS) were created; the first, effectively the record office of the diocese of York and its research centre, the second, a focus for advanced professional development for the architectural industry. At the same time, Milner-White dreamed of the possibility of a single donor founding a college in York, as Lord Nuffield had done in Oxford, perhaps a 'college for Britain', building on the increasing numbers of Commonwealth students attending short courses in the IAAS, interested especially in British history and heritage. But no such donor appeared. There were other opportunities. It looked quite likely in 1957–8 that the King's Manor might become the location of a future Local Authority Staff College on lines already developed by other large governmental agencies such as the Post Office and the National Coal Board. As for the unoccupied Heslington Hall, just outside the city's boundaries, it had been acquired by J. B. Morrell and one of his family's trusts, as a potential future civic amenity, to be used in ways, as yet not specified.[3]

Before 1958 the York Academic Trust had no worked-up scheme for a university in any conventional sense. This was wise as there were no national plans for new universities, and certainly not in places like York. The UGC in the early 1950s saw any rising student numbers as being absorbed in existing institutions or in upgraded local university colleges such as Hull. It was recognized that there were shortages of science applicants and places, a point made repeatedly by UGC member Eric James, then High Master of Manchester Grammar School. Only from about 1955 did the UGC become aware, probably through its secretary, Sir Edward Hale, himself a senior Treasury official, that the numbers of suitably qualified boys and girls wishing to continue their education after the age of eighteen were rising more rapidly than predicted.[4]

Not that this helped the York Academic Trust, because it did not imply new institutions, other than Sussex. Indeed, by the autumn of 1958, Milner-White was gloomy about the prospects for York.[5] All of this was transformed with the change in Treasury and UGC thinking by the end of 1958 about additional wholly new universities, which would be national in orientation and degree awarding from the outset. They were to have a wide range of academic disciplines within arts, science and social science; they were to encourage graduate work and research, but, critically, they should represent a wake-up call in relation to the experience of undergraduates, through innovation in teaching, the creation of multi-disciplinary programmes and wider curriculum content, as well as deepening the students' development, maturity and confidence through living on a campus. Such universities were also to be less hierarchical in their governance, creating opportunities for non-professorial staff to play an active and creative part in the evolution of this new sort of university. These were the priorities since, in the short term, these institutions could make only a marginal difference to the challenge of rising student numbers, each only adding about 3,000 places and then only slowly.

In this context, some particular Oxbridge practices were seen as valuable – tutorial teaching, non-compulsory lectures, individual attention and interest on the part of dons, a collegiate life for undergraduates and their teachers, an academic structure that limited professorial and faculty dominance – all of which were aspirations among the new universities. What, of course, Oxbridge did not have was a spirit of innovation in relation to new subjects or combinations of disciplines. The single subject curriculum was difficult to change. Teaching methods were almost wholly individual and collegiate outside the sciences. And the assessment system was almost exclusively through three-hour closed examinations.

With this change of atmosphere in central government, the York campaign group had to work very quickly. John West-Taylor briefed Milner-White on how a new university might be established in terms of legal and governance structures. Sir Keith Murray, chairman of the UGC, was personally encouraging and paid an informal visit to York in June 1959, staying with Milner-White overnight, and meeting with representatives from the York Academic Trust at Bishopthorpe Palace, under the watchful presence of Archbishop Michael Ramsey, himself a very distinguished academic churchman. Murray also visited the proposed site for the campus at Heslington, viewing it from the high ground of the water tower. This made a significant impression: 'The lawns and trees give a very gracious setting – a very important factor.'[6]

John West-Taylor also wrote all of the papers that were presented to Murray on his visit to York and those used for discussion among the expanded band of supporters attending civic meetings at the city's Mansion House later in the year, and finally in the formal submission to the UGC in London in December 1959. Looking over the documents, it is the pragmatism that strikes the reader. It was already clear that cities like York were likely to be favoured. Murray had encouraged West-Taylor not to be too detailed in the proposals. In terms of subject range and approach, the documents identified a number of conventional disciplines on which the academic foundations of the university would be established – architecture, history, English, philosophy, politics, economics, sociology, mathematics, physics, chemistry and biology. There was some uncertainty about foreign languages. Music and education might come later, while classics, theology, geography, the fine and applied arts were not included. Other social sciences including psychology were left as open questions to be worked out in association with a possible collaboration with the various Rowntree Trusts, whose support was a significant element in the overall case being made. It was acknowledged that applied or heavy engineering, medicine and law would not receive support. Included, almost as a last thought, was the 'aspiration' that the university might be organized on collegiate lines.[7] The document, finally submitted to the UGC in December 1959, was essentially that presented at Bishopthorpe the previous June, described by Murray as a 'masterpiece'.[8] It was accepted the same day, as was the bid from Norwich, although both were not publicly confirmed until the following April.

Once made public the UGC decided to follow the precedent of Sussex and to appoint Academic Planning Boards to oversee the individual development of each new university. Lord Robbins, at that point simply the distinguished and long-serving professor of Economics at the London School of Economics, was invited to chair the Board in York's case. He took to the task with enthusiasm, describing York as 'an extraordinarily exciting centre', and continuing 'something is happening, here and now, which is going to affect this city, and may affect the history of the country for the next thousand years'.[9] The Board's first responsibilities were to start the process of drafting a charter and statutes, to appoint the first vice-chancellor, and begin the process of transforming the bid document into an interim development plan. The new board did not want to be too prescriptive, in order that the incoming vice-chancellor would have a brief window of opportunity to comment on the plan, prior to it being sent to the main committee of the UGC in April 1961. As it turned out, Lord James, the first vice-chancellor, wrote later that he had not wanted to make any significant alteration, endorsing the academic shape of the university and its proposed collegiate structure.[10]

'Lord Jim'

Who then was Lord James of Rusholme, recommended by Robbins and his committee as the first vice-chancellor and how much did he contribute to the shape, style and philosophy of the new university? Eric James was one of thirty-three names submitted for consideration by Robbins and his colleagues as the potential vice-chancellor. His was only one of three names of men (there were no women on the list), who were not already employed within the university system – the others being Dr Robert Birley, headmaster of Eton, Norman Fisher, director of the Staff College of the National Coal Board and Sir Andrew Cohen, UK representative to the United Nations Trusteeship Council. Of the other twenty-nine, thirteen were present full fellows of Oxford or Cambridge colleges, eleven held senior appointments in other British universities and five held or had university appointments in South Africa and Australia. No one was suggested from universities in the United States or from continental Europe.[11] James had been High Master of Manchester Grammar School since 1945, having previously spent thirteen years as an assistant master at Winchester College. Since leaving Oxford with a DPhil and bachelor's degree in chemistry, his whole career had been spent in the two most academically selective schools of their type – a famous independent public boarding school and a very successful academic day school, operating within the maintained structure established under the 1944 Education Act. During his time at Manchester, he had become, by both position and choice, the major public intellectual within English secondary education. A prolific writer, speaker and broadcaster, who had lectured widely on education and society, he had been member of the University Grants Committee from 1949 until 1958 (the only schoolmaster). He had also been elected as the chairman of the Headmasters' Conference for 1953–4, the body that spoke on behalf of the headmasters of leading schools, and he was appointed as a member of the Crowther Committee on the raising of the school leaving age in 1956. Knighted in 1956, he had been created a life peer in 1958.[12] Of all of the thirty-three names submitted to Robbins, James's name would have been the best known to the generally educated public. It is probably then no surprise that the York University Promotion Committee, which is what the York Academic Trust had now become, greeted his nomination as vice-chancellor with 'enthusiasm'.[13]

From this earlier career record, it is easy to characterize James as an establishment man, an image which to a degree he encouraged. However, it is important to remember that he was the son of a respectable commercial traveller with both his parents being devout Congregationalists. By good luck, he had attended a local authority grammar school in Southampton founded originally in the eighteenth century, which, by chance, had two closed scholarships to the Queen's College, Oxford. Much of James's future educational thinking was framed by that personal good fortune of getting into Oxford. In a position of considerable informal influence after 1945, James was then able to advocate the values of a selective secondary grammar school system in extending the opportunities to all able children. Such a system, and especially the 150 direct grant schools within it, would inevitably, he predicted, result in larger numbers remaining in school up to the age of eighteen, qualified and wanting to go to university.[14] James may have been well known, but he had no direct previous professional experience of universities, nor had he ever been a fellow of an Oxbridge college, unlike four of the new university vice-chancellors: Fulton, Thistlethwaite, Butterworth and Carter. Moreover, the skills of a prominent headmaster with almost unchallengeable authority do not inevitably sit easily with university committee structures and over-mighty professorial subjects. In addition, James was also inexperienced in planning and executing major capital projects, as only one new building had been necessary during his time at Manchester. To create one of a series of wholly new universities was a unique and challenging state experiment.

From the announcement of his appointment, James brought to the fledgling university educational principles derived from his years as a headmaster. In early 1940, in a talk entitled 'Education and a scientific humanism' he had set out his vision for what he called the 'new Jerusalem' that would follow the war. At its core was a belief in the scientific necessity for selection and specialization in secondary education as the foundation for future prosperity and moral welfare. Based on conversations with his headmaster, Spencer Leeson, the talk outlined the possible patterns of post-war secondary education.[15] James was first of all an able young chemist, who had published a paper jointly with his supervisor in the *Proceedings of the Royal Society* while still a research student. From this perspective, it was only through a proper application of the social, political and moral potential of science and the work of trained scientists that the 'new Jerusalem' might be built. Not all students or their teachers had the ability to excel in this part of the national educational scheme; specialization and selection among pupils and potential teachers alike were a necessary part of creating that better world. Secondly, James was very widely read in philosophy, history, literature and the arts, and at Oxford he had also been a half-blue at chess, a mountaineer, later becoming a knowledgeable amateur Alpine botanist. He also had taught general classes on divinity and Greek philosophy to older pupils at Winchester, even though a sceptic in religious matters.

In particular, he had been influenced strongly by Victorian debates among liberal Anglicans, like Thomas Arnold, headmaster of Rugby, of the need for refashioned elites for the industrial age. The idea of a 'new clerisy', the phrase used by James, with its echoes of a Samuel Taylor Coleridge's secular writings, informs much of his writing, leading some to see him as a twentieth-century Thomas Arnold.[16] In broader terms of the educational curriculum, he was particularly influenced during the 1930s by the work of distinguished scientists like J. D. Bernal and Lancelot Hogben, who had evangelized about the need to widen the public understanding of science. For James, in the new democratic, scientific and technological post-war world, there was an urgent need to develop these new elites to replace earlier aristocratic and Christian authority. This implied new cadres of able, qualified and educated leaders, both humane and scientific, formed by a system of secondary and tertiary education that was both selective and specialist by definition. For James, a new 'aristocracy of talent' (in R. A. Butler's famous phrase), not held back by the lottery of geography or money, was a worthy national ambition that was to provide secondary education for all for the first time.[17] In 1964, giving evidence to the Franks Commission on the University of Oxford, he was as critical as he had been of Winchester twenty-four years previously, calling for more diversity in teaching through the use of seminars, more faculty scrutiny of the quality of dons' teaching, a shake-up of the admissions process with less reliance on A level performance, but above all the abandonment of all the associations with elite privilege. The scholarship boy of the 1920s remained a dynamic reformer of tradition.[18] In 1963 as the first students arrived, James acquired a nick name, not very imaginative but telling, of 'Lord Jim'.[19]

James's first task was to select the architectural practice to work up a draft master plan for approval by the Academic Planning Board. By his own admission, he did not know much about major architectural projects, especially one very much in the public eye. The commissioning of Sir Basil Spence by the University of Sussex increased still further the interest in what became one of the best-known modernist public sector building programmes after 1945. Of the five new universities approved by the end of 1961, three decided to make dramatic architectural statements, using already internationally renowned architects. Only Warwick and York did not, and in Warwick's case largely through confused decision making. At York, a less flamboyant choice was made of Robert Matthew Johnson-Marshall, a public-sector practice at that time largely known for its schools building programmes led by Stirrat Johnson-Marshall, previously chief architect for schools for London County Council. There certainly was interest expressed by more famous architects and the choice of RMJM could be thought of as overly cautious, although its ultimate realization was positively received by Nikolaus Pevsner amongst others (Pevsner thought the lake a 'stroke of genius'). But York never attracted a 'wow' reaction, as was the response to East Anglia's Ziggarut or Essex's assertive brutalism. Patrick Nuttgens, director of the Institute for Advanced Architectural Studies, simply described the new campus as a 'non-pompous place'.[20]

It was not so much that caution lay behind the construction of the York campus, but more the different philosophy chosen. John West-Taylor, the registrar, had become interested social architecture whilst an undergraduate at Cambridge and in his work in establishing the Institute for Advanced Architectural Studies in York. James obviously knew about schools, and through his friendships with Sir John Newsom (a contemporary at Queen's and now Chief Education Office for Hertfordshire), Harry Rée (headmaster of Watford Grammar School, and later founding head of the Department of Education at York) and Norman Fisher (formerly director of Education for Manchester, 1949–55), he was aware of the RMJM record of school building for London and Hertfordshire County Councils. But the inspiring influence behind all these figures was another director of education, Henry Morris. From the 1920s, Morris had been the driving educational force in Cambridgeshire, the poorest local education authority in England. A charismatic, inspiring and difficult man with a strong aesthetic personality, Morris had pioneered a community-based approach to education within his largely rural county through village community colleges, which had secured him an international reputation by 1945. The colleges like those at Impington, Sawston, Bottisham and Comberton served a broad community purpose as well as being schools; they were not intended to be 'nine to five schools' to modify the earlier description of the civic universities in the 1950s. As a community campus, each was to be built on a human scale in a modern mode influenced by Walter Gropius and others; they were to be set in relatively spacious and well-tended grounds and were to have inspirational and interesting artwork internally.[21] John West-Taylor had come to know Morris well as an undergraduate, lodging in his house in Cambridge and also being his travelling companion – a friendship later extended to John's wife, Catherine. Through Morris, he came to know Stirrat Johnson-Marshall, also profoundly influenced by Morris's philosophy of public architecture, and whose architectural practice was based in Hertfordshire. Fifteen years earlier, Harry Rée had also come to know Morris as a Cambridge undergraduate, later writing a short study of his life. Similarly, Norman Fisher had had early experience of Morris at the beginning of his career as a young education officer in Cambridgeshire. Finally, Morris, James, Rée and Fisher were all members of the All Souls Group, an informal think tank of educationalists founded in 1941 that met regularly in Oxford. It is not surprising therefore that RMJM were favoured architects for the first and later stages of planning the new university.[22]

What the choice of RMJM meant in practice, according to James, was a day-long monthly 'dialectical' interrogation by Stirrat Johnson-Marshall and the project architect Andrew Derbyshire. Unusually for the time, RMJM was a very client-focused practice in which James was expected to be able to answer clearly what he wanted and why in relation to the project as a whole. In this case, the brief was ground breaking, not only to build a new university from scratch, but one whose use of space should be collegiate in nature, so that learning, living and enjoyment should be physically and humanly interconnected. Academics and architects, as James told a conference on the new universities in 1964, were in 'collaboration' in developing the new campuses.[23] Johnson-Marshall and Derbyshire demanded to know how subjects were to be taught, how students were to live, eat and learn, what the staff and potentially their families might expect in terms of facilities, housing, recreation, was there to be a chapel, what pastoral duties of care were to be exercised and many more. James and his colleagues had to have answers to many of these questions before *any* heads of department, academic or administrative, had been appointed.[24] Two decisions turned out to be fundamental, both essentially pragmatic but framed by what might be called the spirit of Henry Morris. First, the poorly drained campus would be built around an artificial lake as the crucial landscape setting within which all the other buildings and walkways would be situated. The second was to construct many of the buildings using a prefabricated method, developed for local authorities, known as 'clasp' ('Consortium of Local Authorities Special Programme'). Few commentators at the time or since have argued that the first, almost certainly suggested by Stirrat Johnson-Marshall, was other than inspired. The second is still debated by both architects and York staff and students to this day. Nevertheless, fifty years after the first stages of the development plan had been completed in 2018, Historic England listed the site and its built environment of the Central Hall and the first two colleges (Derwent and Langwith) as grade II.[25]

Departments

Once the architects had been selected, James could move on to appointing senior administrative staff and the founding heads of academic departments, all of whom had to be approved by Robbins and the Academic Planning Board. John West-Taylor had already been confirmed as Registrar. Appointing the founding heads of department was going to be critical as there was anxiety that established academics, especially in the sciences, would be reluctant to take the risk. Robbins took an active role in the first academic appointment as the head of the department of Economics, recommending his former colleague at the LSE, Professor Alan Peacock, of the University of Edinburgh, the only existing chair holder from Britain to be appointed as the founding heads of department.[26] In the case of economics, and subsequently the social sciences more generally, it was the London School of Economics and the Scottish universities, not Oxbridge, that proved to be the dominant influence in appointing heads of department and academic staff and in the shaping of the undergraduate, postgraduate and research programmes. On the other hand, in the arts, the story was different. In the department of English, the James's family friend, Professor L. C. Knights, close ally in Cambridge in the 1930s of the Leavises, advised on the appointment of Philip Brockbank.[27] Mary, Lady Ogilvie, principal of St Anne's College, Oxford, a member of the Academic Planning Board, and a long-standing friend of Eric and Cordelia James, may well have been the caution to the Board not supporting James's first recommendation of Lawrence Stone from Oxford as the founding head of the Department of History.

Some difficulty was experienced initially in securing a librarian, a critical post, until Harry Fairhurst, then based in the University College of Rhodesia and Nyasaland, was appointed. Other positions were also filled from universities within the sphere of British influence. Anne Riddell, deputy registrar and then the first female registrar in a UK university, worked for the University of Khartoum, and David Allen, the second Bursar, had previously been a District Officer in West Africa.[28] James took a pride in his ability to select staff as a headmaster, arguably his most significant responsibility, and he did the same at York, chairing almost all academic interviewing committees until his retirement. Was this Oxbridge on the cheap? Perhaps, partially. Of the ten founding departmental heads only one was already a tenured British professor, three held or had held more non-established posts at Oxford or Cambridge and a majority had attended those universities as undergraduates, with the others graduating from Edinburgh, Dublin, Bristol and London. By 1973 with all the founding departments at a working size, along with others like music and education that had come later, roughly 40 per cent of academic staff had one or more degrees from either Oxford or Cambridge with the departments of English and related literature, history and chemistry being especially strongly represented in this way.[29]

York was often seen as more traditional than either Sussex or East Anglia, in choosing a relatively small number of large departments on familiar disciplinary lines, but with no faculties, and with undergraduate degrees being either single subject or main/subsidiary joint honours.[30] Once again, the reality was somewhat different. In the first place, York's departmental structures were more nuanced. In the case of the social sciences, undergraduates were taught for the first two years within a 'school' of multi-disciplinary social studies. This pattern, insisted on by both Robbins and Peacock, became increasingly unpopular with students and staff alike. Biology, led by Mark Williamson, was also, in effect, a school of biological sciences, bringing together for undergraduates the separate disciplines of zoology, botany and genetics, following the recommendations of a Royal Society working party.[31] In chemistry, founded by Richard Norman, the traditional separation of organic and inorganic chemistry was dissolved within the department.[32]

In the arts, a department of English and Related Literature was created, while English language and foreign languages, European and non-European, were combined into a single Department of Language, with vocational linguistic training being provided through a separate Language Teaching Centre. There were no individual departments of foreign languages and literature. Each of these units broke with conventions and not simply in their undergraduate programmes. In the Department of English, Philip Brockbank (aided by James) brought in from Cambridge F. R. Leavis as a visiting lecturer. Leavis welcomed the appointment, seeing James as a kindred spirit ('I should feel it an honour to be able to take part in establishing a spirit and a tradition'), and he was clearly star quality, with *Nouse*, the student newspaper, later interviewing him.[33] Similarly, in the Department of Language, Bob Le Page, who joined York from the University College of the West Indies via Malaya, looked outward to innovate, encouraging first-year students to learn non-Western tongues such as Hindi, and establishing collaborations with universities in Hyderabad (India) and Dar es Salaam (Tanzania).[34]

Elsewhere across the new campus this pattern of invention, even iconoclasm, within traditional disciplines was repeated. Ronald Fletcher, whose work on the persistence of the nuclear family annoyed trendier commentators, transferred from the LSE to head up, albeit briefly, sociology. Two free-market economists – Alan Peacock and Jack Wiseman – came in to run economics. Further along the ideological spectrum was Harry Rée, probably the best known of all the founding professors because of his wartime escapades with the French resistance. Rée led the new education department and used his position to advocate passionately comprehensive schools (he was a recent convert, having previously championed grammar schools). From Leavis's *Scrutiny* stable came Wilfrid Mellers, the 'Beatles professor', initially to teach a joint programme run by English and music, but then given charge of music itself. Politics, inevitably, bedded down some of the hottest radicalism of the era, not least because it offered refuge to a group of students and staff thrown out of Rhodesia by Ian Smith's government, who helped establish in the early 1970s a cross-departmental Centre for South African Studies.[35]

A notable exception to these signs of the times was the Department of History. In inviting Gerald Aylmer of the University of Manchester to be the founding head of History at York, Eric James could be seen as making the most traditional type of appointment.[36] A product of Winchester College and Balliol College, Oxford, Aylmer had been taught chemistry by James himself at school, later describing James's general lectures on the minor prophets to a science pupil (as he then was) as a 'revelation'. Following war service in the Navy, famously in the company of the jazz entertainer, George Melly, he had been a junior fellow at Balliol before securing an assistant lectureship at Manchester in 1952. But once again appearances are deceptive. Manchester's was a famous and esteemed history department in the 1950s with Sir Lewis Namier at its head and it had gathered an able set of young lecturers – Gerald Aylmer, Brian Pullan, Penry Williams, Alastair Parker, Gordon Leff and Harold Perkin, who found the institutional atmosphere stifling. Many were on the political left and some were active in the Campaign for Nuclear Disarmament or in writing critical journalism in the *Universities Review*. The opportunities to move in an expanding sector were eagerly grasped either to through a return to Oxford (Williams and Parker) or to the 'new' universities (Aylmer, Leff and Perkin).

As the only member of the future Department of History, Aylmer had to write the prospectus entry for the first cohort of applicants in the autumn of 1962 for admission the following October. He paid a visit to Sussex in March 1963, meeting with Asa Briggs, to discuss the approach being adopted there, writing a typically thorough and robust personal commentary, which he later shared with his own early appointments in the department. The resulting prospectus entry is revealing as to his own thinking.[37] At York, British history was not to be taught as a free-standing element, but within the frame of European history, US history was to feature and prospectively the history of Africa, the past was to be seen not simply through a political lens, thematic courses in comparative history were to be developed, lectures would not be compulsory, teaching would be by tutorial and seminars and assessment methods would be diverse, including the possibility of a research dissertation. Historiography and the theory of history were also being contemplated. By the late 1960s all those elements were in place, resulting from a vigorous departmental debate encouraged by Gerald Aylmer among the growing group of junior colleagues, a majority of whom were Oxbridge graduates under the age of thirty.

Students

But what of the students themselves, what did they think about the choices available to them? Did prospective students think that they were applying to a dilute version of Oxbridge, and did those well-qualified and successful candidates get what they had hoped for? In 1968–9 there were over 10,000 plus applicants to the university, for 324 of whom York was essentially a second choice, among others, after Oxford and Cambridge, and of whom 60 per cent were men. A total of 27 per cent came from the north, 18 per cent from the Midlands, 37 per cent from the southeast and 10.5 per cent from the southwest. Over 70 per cent of entrants had come from maintained institutions, overwhelmingly grammar schools, and 15.9 per cent from independent schools and 7.0 per cent from direct grammar schools. Four years later, in 1972–3, there were approximately 8,500 applicants, and among entrants 53 per cent were men. Within individual departments, some 25 per cent of historians had attended independent or direct grant schools in 1968–9 and the figure was the same in 1972–3, for chemistry the figures were 7.5 per cent and 20 per cent respectively, and for social sciences collectively, 36 per cent and 21 per cent.[38] By 1972–3 the regional distribution had changed significantly with the north and midlands together only slightly exceeding the southeast, with a much greater shift among entrants in the arts towards the southeast. The proportion from independent and direct grant schools had fallen to 13 per cent, perhaps surprisingly given the regional shifts. It was a very different pattern from that of either Oxford or Cambridge at that time or since. The UGC's ambition to create new 'national' institutions had clearly been achieved in the case of York.

Student politics were vigorous in what of course was a small institution, only slightly larger than Manchester Grammar School in 1967. Both major parties had their apprentice future politicians with the spectrum represented on the right by Harvey Proctor and on the left by Tony Banks, both honing later careers at Westminster. There were stand-offs between the university and the students over catering, discipline and secret files. Throughout the decade there were calls for greater student representation on university committees, led early on by Pradip (nicknamed 'Pip') Nayak, a Kenyan Indian studying economics and political science, who was the first president of the students' representative council.[39] On several occasions there were protests against right-wing Powellite MPs such as Patrick Wall from the recently founded Monday Club.[40] Just as the politics department felt the effects of UDI and apartheid, so too did students. They protested against white minority rule in 1966, later escalating into opposition to the apartheid regime in South Africa.[41] But there were no major disruptions as occurred at Essex, Warwick and the LSE and on major campuses in Europe and the United States. Demos and sit-ins that took place were often under the paternal eye of university elders. In 1964 Lady James turned up to give hot drinks to members of an all-night vigil for Nelson Mandela. The largest demonstration of the period was on the day following 'Bloody Sunday' in January 1972, when a march of students and staff across the city led to a number of student arrests. Gerald Aylmer, perhaps with his experience of Campaign for Nuclear Disarmament marches in mind, walked alongside the students to monitor the behaviour of the police.[42]

Elitism under fire

By the early 1970s, Eric James found some trends in the universities difficult to stomach. He broadly accepted the increasing student numbers, implied by the Colleges of Advanced Technology, but believed that universities should remain strongly academically selective. Towards the end of his stewardship of the new campus, he was calling for a cap on further expansion of student numbers, for bringing in partial loans to cover fees, and disparaging the politicization of 'bloody minded students'.[43] But at the height of student radicalism in 1968, in his speeches in the House of Lords, in particular, he showed respect for student protests about apartheid, and over the conduct of the Vietnam War. He told fellow peers they 'must keep their nerve' and be tolerant of the young. And he spoke out when necessary, using a trip to South Africa in 1967 to denounce apartheid.[44] James also firmly acknowledged that students had a legitimate interest in their education and environment on campus, although characteristically wondering aloud whether there were not more interesting things to do, if you were twenty, than attend meetings about catering. He believed in and respected rational argument and responded quickly to general questions of student representation. He supported the efforts of Graeme Moodie, head of the Department of Politics, both nationally and on campus, to improve student representation in departmental and university governance and administration. But James did not disguise his belief in academic elitism or his paternalism in relation to individual students, bearing in mind that the age of majority was not raised to twenty-one until 1969. This he expressed in a head-masterly way in the best sense, always dealing with student transgressions personally and compassionately. For example, in the case of a student expelled for anti-semitism, he wrote an employment reference for the student in order that he might have the experience to acquire greater maturity.[45]

By 1973, and as James prepared for retirement, most of what had been outlined in the development plan had been accomplished in terms of the living, learning and recreational environment, and the University was near to its original target of 3,000 students. The campus had been articulated and built on broad collegiate principles. The hope had been that the functions and purposes of a new university would be supported and enhanced by the mix of activities within and between the buildings, moulded by the shape and appearance of the campus as a whole. By that time the term college system had narrowed in its regular usage to that of how it had worked in personal terms for student and academic staff alike, and on that question, James commented that he thought that the 'jury was still out'. In relation to undergraduates, it had probably worked well, despite the dramatic changes in student behaviour and life style over the years and impossible to predict in 1961. As James noted 'college traditions' are difficult to nurture on a campus, in which college buildings appeared to be architecturally similar.[46] Having said that colleges, regiments, cathedral chapters and scout groups are 'imagined communities' as well as utilitarian ways of ordering human affairs. As such, the idea of a college system has an elasticity which enables not too dramatic change to be accommodated and modified, which is perhaps why York's collegiate nature has remained. One of Murray's ambitions for new campus universities in 1960 had been that they should not be 'nine to five' and in that respect York's collegiate approach had succeeded. Noel Annan, *eminence grise* behind so much of the new university thinking of the era, complimented James on giving York 'a reputation of commitment to teaching such as the best Oxford and Cambridge colleges used to have'.[47] College Junior Common Rooms, sports teams, volunteering projects all increased the range of individual student participation outside the curriculum as did university choirs, drama and the activities stimulated by individual departmental staff members, who, in the case of York, overwhelmingly lived in the city just a mile or so away. There was no student union building, no faculty offices and no chapel nor any plans for such.

For the academic staff, the idea of a collegiate campus had to change from the late 1960s and particularly after James's retirement, as their individual personal and professional lives altered in some ways more dramatically than those of their students. Within the wider developments taking place at universities across the globe as well as in research fields, a dispersed departmental structure became increasingly unhelpful. Similarly, the hope that intellectual and interdisciplinary dialogue would happen through college senior common rooms was always probably rather innocent on James's part, perhaps because his most influential collegiate experience in professional life had been Winchester College. Finally, the idea that colleges would provide a recreational sphere for academic families was probably a case of trying to over-plan, especially as very little permanent staff housing could be provided, and as patterns of academic family lives and careers were again on the verge of significant change, also impossible to predict in 1961. A campus planned on collegiate lines was more than just a return to tradition by way of responding to the request for innovation and imagination in planning. York was no 'poor man's Oxbridge', but an individual variant of experimentations in learning and living going on, as this volume shows, across the world.

Notes

1 *Eboracum: the Journal of the University of York*, 1 (December 1965), 1. Cf. *Observer*, 14 June 1964, 12; Stuart McClure, 'Britain's New Universities III: University of York', *The Listener*, 6 May 1965, 663–5. This chapter is derived partially from my *Eric James and the Founding of the University of York*, Borthwick Paper no. 26 (University of York: Borthwick Institute for Archives, 2016).
2 Private information. For Milner-White, see N. Watson, 'White, Eric Milner – (1884–1963), Dean of York', *Oxford Dictionary of National Biography*, http://www.oxforddnb.com.libproxy.york.ac.uk/view/10.1093/ref:odnb/9780198614128.001.0001/odnb-9780198614128-e-47849 (accessed 23 October 2018).
3 For the campaign see Katherine A. Webb, *Oliver Sheldon and the Foundation of the University of York* (York: Borthwick Institute, University of York, 2009); Pat Hill and Katherine Webb, 'John Bowes Morrell (1873–1963)', http://yorkcivictrust.co.uk/heritage/civic-trust-plaques/john-bowes-morrell-1873-1963/ (accessed 23 October 2018).
4 TNA, UGC/1/6 (26 November 1958).
5 For Milner-White's pessimistic report on the correspondence with Murray see BIA, UOY/YAT/MIN/1-7 (7 October 1958).
6 Murray, 'Memo' (25 June 1959), TNA UGC7/169; West-Taylor notes following Murray's visit: BIA, UoY/F/YUPC/1/3 (24/6/1959).
7 For York's submitted bid to the UGC in December 1959, see Warren, *Eric James and the Founding of the University of York*, Appdx 6.

8   For Murray's informal comment, see BIA, UoY/F/YUPC/1/3 (Notes of Bishopthorpe Discussions, 24 June 1959).
9   Robbins (March 1961), quoted in Susan Howson, *Lionel Robbins: A Life in Economics, Government and the Arts* (Cambridge: Cambridge University Press, 2011), 864.
10 For accounts of the founding of the University of York, see Lord James of Rusholme, 'The Starting of a New University', *Transactions of the Manchester Statistical Society* (1965-6), 1–21; idem., 'The University of York', in *New Universities in the Modern World*, ed. Murray G. Ross (London: Macmillan, 1966), 32–52; Christopher Storm-Clark, 'Newman, Palladio and Mrs. Beeton: The Foundation of the University of York', in *York 1831–1981: 150 Years of Scientific Endeavour and Social Change*, ed. Charles Feinstein (York: Ebor Press, 1981), 285–310; David Smith, 'Eric James and the "Utopianist" Campus: Biography, Policy and the Building of a New University During the 1960s', *History of Education*, 37 (2008), 23–42.
11 TNA, UGC 1/8 (1960).
12 For James's career before his appointment at York, see Warren, *Eric James and the Founding of the University of York*, 36–57. For his various public pronouncements, see James, *Education and the Moral Basis of Citizenship: A Lecture Given at the Annual General Meeting of the Association for Education in Citizenship at Burlington House, London, 17th March 1954* (London: Heinemann, 1955); *The Content of Education: Oration Delivered at the London School of Economics and Political Science on Friday, 5th December, 1958* (London: LSE, 1959); *Education and Democratic Leadership ... Delivered before the University of St. Andrews 26 October 1960* (London: Oxford University Press, 1961).
13 Minutes, 23 January 1961, BIA, UoY/F/YUPC/1/4.
14 'High Master's Annual Report' (16 July 1946), Manchester Grammar School Archives, Minutes of the Governors.
15 A. O. J. Cockshutt in conversation with the author, 3 May 2013; *Spencer Leeson: Shepherd, Teacher, Friend. A Memoir by Some of His Friends* (London: SPCK, 1958), 60–88. 'Education and a Scientific Humanism' is reprinted in Warren, *Eric James and the Founding of the University of York*, Appdx 1.
16 HL Debs, 123 (11 May 1960), 691.
17 James, speech to the Northern Heads Conference, Charlotte Mason College, Ambleside, 9–10 October 1954, reprinted in Warren, *Eric James and the Founding of the University of York*, Appdx 3.
18 'Written Evidence Given to the Franks Commission of the University of Oxford', BIA, UoY/JAM/2/15.
19 *Observer*, 14 June 1964, 12.
20 Nikolaus Pevsner, *Yorkshire: York and the East Riding* (Harmondsworth: Penguin, 1972), 251. For Nuttgens's comment, see British Library, Archive of Recorded Sound, C447/34/01.
21 For Morris, see Tony Jeffs, *Henry Morris: Village Colleges, Community Education and the Ideal Order* (Nottingham: Educational Heretics Press, 1998); David Rooney, *Henry Morris: The Cambridgeshire Village Colleges and Community Education* (Haslingfield: Henry Morris Memorial Trust, 2013); cf. P. Cunningham, 'Morris, Henry (1889–1961), Educational Administrator', *Oxford Dictionary of National Biography*, http://www.oxforddnb.com.libproxy.york.ac.uk/view/10.1093/ref:odnb/9780198614128.001.0001/odnb-9780198614128-e-63830 (accessed 23 October 2018).
22 Catherine West-Taylor, in conversation with the author, 5 June 2014 and 1 September 2015. For Johnson-Marshall, see Andrew Derbyshire, 'Marshall, Sir Stirrat Andrew

William Johnson – (1912–1981), Architect', *ODNB*, http://www.oxforddnb.com.libproxy.york.ac.uk/view/10.1093/ref:odnb/9780198614128.001.0001/odnb-9780198614128-e-31292 (accessed 23 October 2018). For the All Souls group, see UCL Institute of Education Archives, GB/366/DC/ASG; Harry Rée, *Educator Extraordinary: The Life and Achievement of Henry Morris* (Harlow: Longman, 1973).

23 James, contribution to 'Shaping University Development Plans', Session II of 'New Universities Conference' (Keele, 15 July 1964), fol. 36, Keele University Special Collections and Archives.

24 For oral testimony, see Stuart Sutcliffe, 'Interview with Lord James', BIA, UoY/HIS/3/1/9. Stuart Sutcliffe also interviewed Andrew Derbyshire at that time. See also the interviews carried out by Andrew Saint in preparation for *Towards a Social Architecture: The Role of School Building in Post-War Britain* (1987) British Library, Archive of Recorded Sound, C447/15/01 (Patrick Nuttgens, Lord James of Rusholme), C447/34/01 (John West-Taylor), C447/09/01 (Andrew Derbyshire), and C447/31/01 (Harry Rée).

25 For the debate surrounding 'CLASP', see Andrew Saint, 'Reflections on Clasp' (unpublished typescript, January 1985), BIA, UoY/JAM/2/4/3. For the 2018 listing, see https://historicengland.org.uk/listing/the-list/list-entry/1457040 (accessed 6 August 2019).

26 For Peacock, see his *Anxious to Do Good: Learning to Be an Economist the Hard Way* (Exeter: Imprint Academic, 2010); G. C. Peden, 'Alan T. Peacock', *Biographical Memoirs of Fellows of the British Academy*, 14 (2015), 495–516; Ilde Rizzo and Ruth Towse, 'In Memoriam Alan Peacock: A Pioneer in Cultural Economics', *Journal of Cultural Economics*, 39 (2015), 225–38. Neither Peacock nor Rizzo and Towse make much of his time at York.

27 For Philip Brockbank, see his obituary: *Guardian* (25 July 1989), 35; Judith Woolf, 'English at York: "Anything but Traditional,"' https://yumagazine.co.uk/english-at-york-anything-but-traditional/ (accessed 23 October 2018).

28 For the academic background of early senior appointments at the University, see BIA, UoY/UP/3/1/1, UoY/F/APB/2/1, 3.

29 For the degree backgrounds of the academic staff appointed from 1963 to 1973, see BIA, UoY/UP/3/1/1-10.

30 For the early discussions on academic, collegiate and departmental matters, see notes on academic staff meetings 1962–3, BIA, UoY/F/ST.

31 Mark Williamson and David White, *A History of the First Fifty Years of Biology at York* (University of York: Department of Biology, 2013), 9.

32 For Norman see D. Hughes, 'Norman, Sir Richard Oswald Chandler (1932–1993), Chemist', *Oxford Dictionary of National Biography*, http://www.oxforddnb.com/view/10.1093/ref:odnb/9780198614128.001.0001/odnb-9780198614128-e-52420 (accessed 23 October 2018); *50 Years of Chemistry at York, 1965–2015* (University of York: Department of Chemistry, 2016).

33 Leavis to Brockbank, 18 November 1964, BIA UOY/PP5/1; *Nouse*, 15 May 1969, 5. For Leavis at York, see Steven Cranfield, *F. R. Leavis: The Creative University* (London: Springer, 2016), 58–60.

34 For Le Page, see his obituary, *Guardian*, 16 February 2006, https://www.theguardian.com/news/2006/feb/16/guardianobituaries.mainsection1 (accessed 1 September 2019). Page, *Ivory Towers: The Memoirs of a Pidgin Fancier; a Personal Memoir of Fifty Years in Universities around the World* (Cave Hill: Society for Caribbean Linguistics,

1988). For undergraduate degree courses, see University of York Prospectuses from 1963, BIA, UoY/UP/3/1/1-10.

35 For Fletcher, see his *Britain in the Sixties: The Family and Marriage, an Analysis and Moral Assessment* (Harmondsworth: Penguin, 1962); Irving Horowitz, *Tributes: Personal Reflections on a Century of Social Research* ch. 13. For Peacock, see Peacock, *Anxious to Be Good*. For Wiseman, see Keith Hartley, 'Jack Wiseman, 1919-91', *Economic Journal*, 110 (2000), 445-54. Peacock and Wiseman were early advocates of student loans: *Education for Democrats: A Study of the Financing of Education in a Free Society* (London: Institute of Economic Affairs, 1964), and by the end of the decade were involved in the movement that led to the establishment of a private university at Buckingham. For Rée, see Jonathan M. Daube, *Educator Most Extraordinary: The Life and Achievements of Harry Rée* (London: Institute of Education Press, 2017). For Mellers, see John Paynter, 'Introduction', to *Between Old Worlds and New: Occasional Writings on Music*, ed. Paynter (London: Golden Cockerel Press, 1997), 8-9. For the Centre for South African Studies: 'Interview with Christopher Hill', 3 July 2019.

36 For Gerald Aylmer, see Gordon Leff, 'Gerald Aylmer in Manchester and York', in *Public Duty and Private Conscience in Seventeenth-Century England: Essays Presented to G. E. Aylmer*, ed. John Morrill et al. (Oxford: Oxford University Press, 1993); Keith Thomas, 'Gerald Edward Aylmer, 1926-2000', *Proceedings of the British Academy*, 124 (2004) 3-21.

37 See University Prospectus 1963, BIA, UoY/UP/3/1/1.

38 See BIA, UoY/Admissions Reports, 1968-9, 1972-3.

39 Minutes of the Staff-Student Committee (24 May 1965), 65/47, BIA UoY/M/SSC/1963-6. For Nayak, see *East Africa Standard* (Nairobi), 19 January 1966, in 'Cuttings', BIA UoY/PUB/PC/1/2.

40 Extensively covered in *Nouse*, the student newspaper, and in the local press. See *Nouse*, 4 December 1968, 4, 12 March 1970, 1; *Northern Echo*, 9 November 1968; *York Evening Press*, 17 January 1969, in 'Cuttings', BIA UoY/PUB/PC/1/18.

41 *York Evening Press*, 28 November 1966, in 'Cuttings', BIA UOY/PUB/PC/1/13. For recollections of student protest on campus over UDI and apartheid, see Julian Friedmann, 'The Original York Student', https://yumagazine.co.uk/the-original-york-student/ (accessed 6 August 2019).

42 *Yorkshire Post*, 3 June 1964, in 'Cuttings', BIA UoY/PUB/PC/1/9 (Mandela vigil) UoY/PUB/PC/22 (Bloody Sunday March).

43 *Birmingham Post*, 27 February 1970, BIA UoY/PUB/PC/1/18 ('Bloody Minded'); HL Debs, 311 (15 July 1970), 640 (loans).

44 HL Debs, 293 (19 June 1968), 55-6. For the trip to South Africa, see *Guardian*, 15 August, 1967, 1; 18 August 1967, 9.

45 BIA, UoY/VC/1/3.

46 Sutcliffe, 'Interview with Lord James'.

47 Annan to James, 16 July 1973, Annan Papers, Archives Centre, King's College, Cambridge, NGA5/508.

5

The University of East Anglia: From mandarins to neo-liberalism

by John Charmley

The 'new men'

The idea of founding a University in Norwich went back before the great war of 1914–18, received an airing after the Second World War, but only looked like becoming a reality in the late 1950s when the British state showed an interest in major investment in higher education.[1] This was inspired more by *raison d'état* than by anything in John Henry Newman's classic *Idea of a university*. Clement Attlee's consensus, as historians have labelled the post-war settlement which lasted until the late 1970s, was based upon a paternalistic version of state socialism, most notably embodied in the welfare state. The drive for new universities was a product of that paternalism and so were the men who helped found and run what became the University of East Anglia. C. P. Snow, the Cambridge scientist, novelist and government administrator, called these meritocrats 'the new men'.[2] Educated, for the most part, at the ancient universities of Oxford and Cambridge, the men behind the new universities saw them in three ways: in the first place as providing a larger, better-educated work force, so that Britain could keep pace with the United States and USSR; secondly, as offering higher educational opportunities for a considerably expanded cohort of able sixth formers as a result of the post-war 'baby bulge'; and thirdly, as places where their own innovative ideas on higher education could be put into practice. In the authentic language of the Attlee consensus, the prospectus soliciting money for the new University of East Anglia (UEA) emphasized the national context, along with the comment (made with true socialist seriousness) that even its modest programme proposed 'calls for great efforts', adding that 'more university places are essential if Britain is to survive as a major power'.[3]

These 'new men' developed their ideas and had them institutionally realized in familiar working practices: through a network of committees closely linked to existing universities (particularly the ancient universities) and to Whitehall. Part of this network were the powerful academic planning boards (APBs) – one for each new university, composed of distinguished academics, civil servants and the occasional corporate leader. As well as advising, in the very early stages, they provided the legitimacy for the thinking and direction of the new vice-chancellors as they set out to create their institutions. Not only did the APBs select the founding vice-chancellors, but they also participated in early discussion about the overall structure of curricula and methods of teaching, and in the selection of the initial consulting architect. UEA's planning board, appointed in 1960, was composed of one vice-chancellor, two heads of Cambridge colleges, the Warden and Secretary of Rhodes House, Oxford, the chief scientific adviser to the Ministry of Defence, a senior chemist from University College, London, a senior industrialist and the chief education officer of Norfolk.[4] But if the need for the new universities was cast in the language of C. P. Snow's 'new men' with their slide rules, their plans and their faith in state activity, then the character of them owed more to the ideas of their founders. In the case of the fledgling University of East Anglia (UEA) that meant one man – Frank Thistlethwaite.

The founding vice-chancellor

It is impossible to over-emphasize the influence that Thistlethwaite exercised on the character and structures of UEA. Thistlethwaite belonged to a generation of graduates steeped in the English idealist tradition founded by T. H. Green, which allowed its adherents to accept the moral teachings of Christianity without having to accept its dogmatic elements. Green, like his admirers, believed in the redemptive power of education and the utility of the state as the vehicle for its promotion; this was a core part of Thistlethwaite's thinking. A Lancastrian, whose own Christian background was Congregationalist and Quaker (he had been at Bootham School in York), Thistlethwaite embodied the best aspects of the idealist tradition: a high seriousness combined with an appetite for hard work. He was formidably well-networked, an excellent administrator and a man whose experience of Whitehall and of education in the United States equipped him well for the task which fell to him in 1959.[5]

Thistlethwaite, like many other British academics of his generation, found himself irked by the 'mental and institutional barriers between disciplines in British universities'.[6] A fellow of St John's, Cambridge, he took a keen interest in higher education, and he responded to the state's decision to found new universities, and the chance to 'redraw the maps of learning'. In a scene straight out of a C. P. Snow novel, he later recalled discussing the possibilities of breaking down the old barriers with his slightly younger contemporary, Asa Briggs, on a vaparetto on Lake Maggiore in Italy, as the two men skipped a session of the International Sociological Association at Stresa in 1959.[7] Within two years both men would find themselves in a position to put their ideas into effect.

Thistlethwaite was firmly in favour of an interdisciplinary approach to curriculum planning. He was partly influenced by the precedents of Keele and Sussex and partly by his own somewhat unorthodox pattern of study at Cambridge, moving between history, English and economics. 'Many of the great breakthroughs,' he argued, 'have recently been made at the interstices between established disciplines.'[8] The initial UEA prospectus may have talked reassuringly about the new university restricting itself to 'the study of those basic subjects in arts and sciences which are recognised as being central to civilization as we know it',[9] but thanks to Thistlethwaite, UEA took an avowedly interdisciplinary approach to these subjects, a fact evidenced by its very structure. Instead of the traditional departments, UEA had schools of study, where subjects which were taught in separate silos in traditional universities were brought together in the firm faith that placing cognate disciplines together without the traditional barriers would produce 'new knowledge'. While the tendency of the government was to direct universities to be more serviceable to the economic needs of the country, the 'new men' charged with the task had a wider vision. For all the criticism it later incurred (not least from within the new universities) the establishment connection worked well in terms of academic freedom. Trust in the professional judgement of the new vice-chancellors allowed them to deliver what they felt was wanted without asking too many questions – or demanding anything much by way of scrutiny. Later generations of vice-chancellors, trammelled by 'key performance indicators' and 'intended learning outcomes', can only look with envy at the freedom enjoyed by Thistlethwaite's generation. UEA became an interdisciplinary university with structures to facilitate that because Thistlethwaite believed that was where the universities of the future would make their mark. Against any avowed utilitarianism of state design, he shared Newman's view that, since it is impossible for the state to know precisely what skills would be needed in the future, what was required were generally educated men and women who knew how to think, and who would be able to approach different kinds of problems with intelligence, coherence and precision. Thistlethwaite thought his interdisciplinary approach the best method of doing this for the modern age. As he would later put it: 'We enjoyed a remarkable freedom to design our universities in our own individual ways. It was a unique experience and I suspect it could never happen again.'[10] Thistlethwaite had no idea what the result of putting cognate disciplines together in the same school of studies would be, but that was just the point – if new knowledge was to be generated, you could not predict in advance what it would be. He had faith in the generative power of what intellectuals could produce when put in close proximity to each other and to young, enquiring minds.

Again, influenced by positive American experience, Thistlethwaite insisted that the major vehicle of pedagogy would be the seminar – not only cheaper than the Oxford and Cambridge system of tutorials (or supervisions), but also facilitating group discussion in a way that the Oxbridge system did not. He was also adamant that course work would count towards the final degree classification. At the start of their course at UEA, undergraduates took a preliminary course for two terms in order to consider their options for their main honours courses. This was less appreciated by the scientists and economists whose subjects, they felt, demanded sequential block-building at the start. Others felt that the provision of such choice was confusing for English sixth form graduates with their focused 'A' level exams.[11] As far as interdisciplinary studies were concerned, students in the early years could choose flexibly within their schools of study, majoring in one subject and minoring in one or more others. Thus in the School of English and American Studies, one student might – after the first two terms – go on and read mainly history courses with an 'English minor', while another – ostensibly doing the same degree – might read mainly literature courses with a 'history minor'. It was compulsory for all students to do an avowedly interdisciplinary seminar in their final year. Seminars (performance which contributed to a student's final degree in the continuous assessment system) received a mixed reception. The system had the virtue of insuring that students looked at their seminar topics within a wider vision of the period they were studying – which was the job of the lecturers to provide. No new lecturer could retreat into the safety of their own 'subject', and the ability to lecture widely on one's period (and outside it) was an essential prerequisite for new appointments. At their best, seminars were true voyages of discovery with conventional boundaries between subjects being fluid, if not irrelevant, while at worst, they created seminar rooms of Trappist students and nervily loquacious lecturers. A student survey of 1968 identified a marginally unfavourable view of seminars overall.[12] Yet it is a mark of the success of Thistlethwaite's founding enthusiasms that within half a decade, these innovations were becoming commonplace in university teaching.

The early days of the university were later described by one of the early leading academics as 'the golden days of UEA, Garden of Eden before the Fall'.[13] Before and at the start of major building work on the main campus (known as 'the Plain'), the offices and early teaching facilities were located in Earlham Hall, an eighteenth-century house (and former home of members of the Gurney family) on the premises, and a 'village' of temporary, prefabricated buildings. Here the first students were enrolled, taught and granted their degrees with appropriate ceremony; plays, parties, even balls took place. The intention was not so much to recreate Oxford or Cambridge on the River Wensum, as to create a modern version of the same – doing for the 'atomic age' what those ancient universities had done for the Renaissance. The desire to be self-consciously modern did not, of course, preclude a touch of the Oxbridge culture, and so UEA managed to get £10,000 from the City Council to lay down a good wine cellar and a suitably dignified butler to go with it.[14] Thus Thistlethwaite, the erstwhile Cambridge don (who had been wine steward of his college), built up a 'senior common room' with regular dining in style among early senior academics. When the time came for the gradual transfer of schools of study, library and common rooms to the new, permanent buildings on the main site, 'there was a sense of dislocation and regret and, for a time, a real loss of morale'.[15] The development of what was becoming a major institution, with its own bureaucracy and internal hierarchies, inevitably cast aside – or absorbed – the cosiness of the founding group of pioneers. This seems to have been a common phenomenon among the new universities.

Part of a sense of community is a feeling of identity and, in trying to create this, the new universities followed their predecessors with coats of arms, mottos and degree-giving ceremonies enhanced with formal academic dress. At UEA the motto 'Do different' was taken from an old Norfolk saying. The pageantry of its degree ceremonies, which were exploited as major civic occasions and, in early days, held in the centre of Norwich, were enhanced with gowns and Tudor-style 'bonnets' designed by Cecil Beaton for the university leadership. The decision of the APB that this University should not be collegiate (others were) but should be highly residential provided a challenge to the vice-chancellor and the consulting architect for the university's initial development plan, Denys Lasdun, in creating community feeling among the students. Thistlethwaite's idea was that students should identify with their schools of study on the one hand and with their neighbours in their living quarters on the other. Lasdun responded to the challenge in the design of his 'ziggurats' which catered for 'staircases' of groups of about twelve students' study bedrooms having access to a communal breakfast-kitchen area, and also in the overall design of the campus in which a 'walkway' system above vehicular service traffic linked much of the campus.[16] Thistlethwaite liked to describe the effect they were after as akin to an Italian Renaissance hill town – something that became somewhat belied by the poor-weathering effect of the chosen medium of construction, concrete.[17]

Creating schools of study

Between 1963 and 1967 four new schools of study were established at UEA in the humanities and four in the sciences; in true UEA style the social sciences were embedded within the interdisciplinary Schools. In the humanities, the founding school was the School of English Studies (that, within four years, became the School of English and American Studies), followed by the School of European Studies, the School of Social Studies (which was to spawn the School of Development Studies) and the School of Fine Arts and Music. The first of the science schools was the School of Biological Sciences, followed by the School of Environmental Sciences, the School of Chemistry and the School of Maths and Physics. Each school was to be coordinated by its founding dean whose selection was given careful consideration by the vice-chancellor whose firm belief in the potential and virtues of interdisciplinarity created problems for him from the start. Naturally enough, most prominent scholars in their fields believed in the virtues of the current way of organizing higher education, and this led UEA to look towards younger scholars, when failing to attract more established ones. For his interdisciplinary School of English Studies, Thistlethwaite originally wanted the Cambridge literary maverick, Raymond Williams. He chose a literary scholar rather than an historian because his own experience of historians had shown him that the latter were likely to be 'less responsive to an intimate collaboration' than the more 'plastic' discipline of literary studies.[18] As it turned out this was harder to achieve than he had imagined, even in that area, whilst in others, such as the planned School of Environmental Sciences, it led to several failures to appoint a suitable dean. In that case, it was the flexibility of the geographer, which provided the answer when biologists, chemists and physicists turned their noses up at the post.

Having failed to lure Raymond Williams to become dean of the School of English Studies, Thistlethwaite found a congenial collaborator in the form of his old college friend, Ian Watt – then a literary scholar at the University of California, Berkeley – whose 'historical cast of mind towards literary criticism ... reputation and his sheer ability ... made him an obvious choice'.[19] As was so often the case, he was hired in a way which would have modern human resources departments in despair (not to mention the university in court). On a visit to America, Thistlethwaite stopped off in Berkeley to talk to Watt and 'had little difficulty in persuading him to become the founding Dean of UEA's new and pioneering School of English Studies'.[20] Watt himself lasted but a year – he and his family missed the Californian lifestyle – and sunshine – but his influence proved more enduring.

Watt's replacement was an historian, whose manner of appointment had been even more unorthodox (though typical of the Thistlethwaite style). Robert (Bob) Ashton, an economic historian of seventeenth-century England, happened to be on a visiting professorship in Berkeley, and Watt had thought someone with his interdisciplinary bent would make an ideal founding history professor. So, Jack Plumb, the eminent Cambridge historian, interviewed him in New York and recommended his appointment.[21] Thistlethwaite did not wait for the new professor to make the junior appointments, but went ahead and began appointing. One of the first Ashton interviewed was, his curriculum vitae stated, born in Fiji, and he and the vice-chancellor were both taken aback when the very non-Fijian looking Dick Shannon turned up; he was duly appointed, and the story would improve with the telling down the years. Under Ashton the school began to develop a distinctive offering in American Studies which led to debate about forming a separated School of American Studies, and then to the later merger. The relationship between Thistlethwaite and Ashton would never be as close as the one he had had with Watt: for one thing Ashton had his own very clear ideas on the types of history appropriate to a university placed on the edge of the second largest city in medieval England, and insisted that medieval history had to be part of the offering; for another, the two men were, perhaps, too similar in temperament for either to defer easily to the other. The new school kept the seminar style which Watt had wanted, and its curriculum, if that was not too old-fashioned a term for it, was the one hammered out by Watt with Ashton's help.

It would be Ashton's proud boast that undergraduates should not expect anything covered in seminars to come up in finals papers; they were, after all, supposed to be 'reading for a degree'. Brave words, which sent a shiver of fear through generations of finalists. It was left to his juniors to banish by pointing out that if students looked at past finals papers, they would see topics they recognized from seminars. But it did have the virtue of ensuring that students looked at their seminar topics within a wider vision of the period they were studying – which was the job of the lecturers to provide. No new lecturer could retreat into the safety of their own 'subject', and the ability to lecture widely on one's period (and outside it) was an essential perquisite for new appointments. I recall being told, casually, that I would be expected to lecture on political and diplomatic topics in the period after 1760 – without anyone quite telling me precisely what that involved. But that was part of the ethos of the 'pioneering' School – it was as much a learning experience for the lecturers as it was for the students. There was a real sense of everyone being in it together, and of conventional boundaries between subjects being fluid, if not irrelevant.

The School of European Studies began in 1964. There had been much discussion about the direction of this school in the APB many of whose ideas (such as Noel Annan's general opposition to European studies) did not necessarily take hold. Again, Thistlethwaite's initial approach for the post of dean – to the historian, Geoffrey Barraclough, whose interests in fact took him to the University of California – was unsuccessful.[22] The founding dean was James (Mac) McFarlane, a Germanist and Scandinavian specialist, who would go on to shape an innovative School of European Studies with an outstanding Scandinavian section.[23] What was intended was a northern European thrust. But a correction had soon to be made when student demand for French language within a year or two led to the offering of a wide range of studies in various European studies and languages, with provision for a year's study and work abroad in students' chosen specialisms. The School of Social Studies, the third humanities school, had C. R. 'Dick' Ross, fellow of Hertford College, Oxford as its founding dean.[24] The school was set up to teach economics, sociology, economic history and philosophy. The omission of politics was contrary to Thistlethwaite's strong inclination because both Ross and his colleague Marcus Dick had joined the school from the Oxford programme of philosophy, politics and economics whose teaching of politics they abominated and did not wish to see replicated at UEA.[25] A successful offshoot of the school, which emerged in 1967, was the Overseas Development Group (set up by Athole Mackintosh) which both the APB and the vice-chancellor saw as being important and relevant in its contemporary focus in extending the university's vision and activity beyond Europe.

Thistlethwaite was imaginative at utilizing East Anglian resources to add lustre to the institution. One of the most prominent, indeed it might be argued, feature of English and American Studies, a programme in creative writing, happened almost as an afterthought. Someone pointed out to Ian Watt that the novelist Angus Wilson had a home in Suffolk, so Thistlethwaite suggested he be asked to lecture at UEA. As he needed time to write, Wilson accepted a part-time appointment, proudly commenting at the time that whilst it was not unknown for university lecturers to write novels: 'I am the first novelist to be appointed a university lecturer.'[26] He encouraged composition and went on to become a professor and co-founder (with Malcolm Bradbury) of a masters course in creative writing that would become one of the great jewels in UEA's crown. Thus was founded, almost in a fit of absence of mind, what would become one of the jewels in UEA's crown; a moment's inspiration was worth a year's planning.

In the planning and design of the School of Fine Arts and Music, which began in 1965, Thistlethwaite took advantage of his contacts and location to approach the composer Benjamin Britten at Aldeburgh about aspects of effectively combining music theory, musicology and performance. This connection with Aldeburgh was strengthened with the appointment of the Director of Music and Dean of the School of Fine Arts and Music, Philip Ledger, who was to become artistic director of the Aldeburgh Festival.[27] Imogen Holst, composer and co-director of the Festival, was drawn in as an external adviser and later as a lecturer; Peter Pears gave master classes in singing; and Britten himself became Honorary Music Adviser.

Men of their time, Thistlethwaite and his founding deans had great faith in planning and in their own ability to shape the future. But inevitably they were more successful in some areas than others, and it is perhaps ironic that, in many ways, they were most successful in the areas about which Thistlethwaite knew least – the sciences. Here again, success – when it came – was in part the product of the founding vice-chancellor's networks. The main scientific voice on the UEA planning committee belonged to Sir Solly Zuckerman, who came as close as anyone could to personifying the 'new men' who inhabited Snow's novels. For thirty years after the Second World War, Zuckerman served, 'often as Chairman, on every Whitehall committee in which issues of scientific concern were discussed.'[28] A scientist who had helped create the strategic bombing strategy during the war, by 1960 he was the government's chief scientific adviser and saw some strategic opportunities for Government planning with the creation of a new university in Britain's rural east. He also had a holiday retreat in Norfolk.

Zuckerman had been deeply involved in a 1961 Royal Society report on the biological sciences in view of the emergence of disciplines such as plant physiology, microbiology and biophysics. Cambridge was home of the government-funded Institute of Food Research, and Zuckerman saw the chance to hive off part of it to Norwich, where it could be co-located with the John Innes Institute for Horticultural Science. Whilst the two would not be part of a new School of Biological Sciences, whose dean was T A Bennet-Clark, a distinguished professor of Botany at King's College, London, who was lured to UEA in order to pursue the new Royal Society thinking,[29] their proximity would, Zuckerman was sure, lead to fruitful cooperation. Thus casually came about the creation of one of the biggest and most creative academic groupings at UEA. Zuckerman's sense of the shape of the future proved to be acute.

The same was true of the new grouping of physical and environmental sciences which became known as the School of Environmental Sciences. If Thistlethwaite was impatient at the old academic groupings in the humanities and social sciences, he found Zuckerman a kindred spirit when it came to the sciences. From his vantage point in Whitehall, Zuckerman could see that if Britain's 'age of affluence' was to continue, it needed good graduate scientists – and he knew what sort. He wanted meteorology, oceanography, geology to sit in a school alongside geography, economics, agronomy and ecology – he envisaged a school going from very hard science at one end, to a bridge to the Humanities at the other end as social and economic geography shaded into demographics and perhaps anthropology.[30] It proved difficult (yet again) to find a dean for such an experimental school, not least as most senior scientists kept wanting to replicate the structures they were comfortable with. But Zuckerman and Thistlethwaite refused to be deterred, even when six interviews failed to find the right man. It was in keeping with Thistlethwaite's unorthodox approach that he invited two of the likeliest candidates to dine with him. Unable to make up his mind which of them would do best, he asked them both to take on the job. Keith Clayton, who had been a reader in geography at the LSE, became the founding dean, and attracted scientists of great calibre, one of whom, Fred Vine, provided UEA with its first fellow of the Royal Society, and another, Hubert Lamb, who founded what would become the internationally famous (if not, in some circles, infamous) Climatic Research Unit, which, as Jon Agar shows in his chapter in this volume, would play a leading role in establishing the scientific reality of the theory of global warming.

The over-arching theme of interdisciplinarity ran into the proverbial stone wall in the setting up of the subject groupings of physics and chemistry. The intention of the APB, in particular of Sir Christopher Ingold FRS, a distinguished chemist at University College London and firm believer in interdisciplinary linkages, was the creation of a combined school of physical sciences. Counter to Ingold's thinking and to his choice of a physical chemist as dean was the whole approach of a rival candidate, Alan Katritzky, a thirty-one-year old organic chemist, initially from Oxford but then a fellow of Churchill College, Cambridge. One of his convictions that chemistry in itself was 'a universe of sciences' and that a school of chemistry must stand alone. He was reinforced in this view by what he observed had happened at Sussex in its creation of a school of physical science in which chemistry had been 'subsumed'. These opposing views led to a long-running conflict during 1962 between scientific supporters of the Ingold approach – the Cambridge tradition – and those of Katritzky's Oxford tradition. In the former camp was Zuckerman; in the latter Bennet-Clark. The outcome was the appointment of Katritzky as dean of a separate School of Chemistry – thus challenging Thistlethwaite's aversion to single-subject disciplines.[31] As far as physics was concerned, a separate School of Maths and Physics was created.

Defining interdisciplinarity, then, proved far from easy. Attractive as it was in the abstract to progressive educationalists like Thistlethwaite, it was less clear what students made of it. The fundamental aim of the UEA experiment was 'to prevent the isolation of different disciplines or departments and to procure the undoubted benefits that would accrue from dialogue between different disciplines and departments'.[32] Whilst it seemed clear enough in schools such as biological sciences and environmental sciences that the hopes of the Royal Society had been realized, and that breaking down disciplinary boundaries had produced new knowledge, it was far from clear that the same thing had happened in the humanities, where, contrary to predictions, the old disciplines proved intellectually tougher than imagined. This can be illustrated by debate within the School of English and American Studies about the attempt to restructure the school in the early 1970s. This School soon began to develop a distinctive offering in American Studies which caused some academics (including Malcolm Bradbury) to raise the question of whether it might be better to have a separate School of American Studies. The idea did not gain traction largely because the new professor of English Literature, John Broadbent (an import from King's College, Cambridge), was firmly of the view that such a direction would be inimical to the sorts of interdisciplinary work which the school was beginning to develop. He also doubted that a monoglot School of English Literature shorn of its links with history and with comparative literature could compete against other universities; and more importantly, at least for him and other senior colleagues, he felt it would be the betrayal of UEA's original mission – 'to separate any of these components would impoverish the whole complex'.[33] Even Broadbent, who in theory was all in favour of the new interdisciplinary way of doing things, could point to very little by way of actual scholarly achievement: his own claim to fame was an impeccably edited edition of Milton's *Paradise Lost*.

As one perceptive critic noted, the arts schools were 'inter-departmental rather than inter-disciplinary'. Relatively few students chose to do courses in other schools, and even in art history, which seemed the exception, only 12 per cent of students chose courses from outside their own school. On paper, the regulations showed a perfect structure to facilitate interdisciplinarity; in practice, students did the minimum required of them and often sought exemptions. 'UEA', the same critic concluded, 'was a bold experiment in education, but an experiment which has largely failed'.[34] The obvious answer would be to go where Warwick and York had gone from the start and revert to a university organized on departmental lines. Experience there had shown that not second-guessing the future had been a sensible option. Historians and literature scholars at both had, as scholars will, used whatever they found useful in other disciplines whilst remaining stalwartly in departments which applicants to universities could recognize. I can remember looking at the UEA prospectus in 1973 and quite missing the fact that it did history at all – a not uncommon experience, and one which led the historians to design a cross-school 'degree in history'.

The end of the utopian vision

The first challenge to the glad confident morning of UEA's foundation emerged during the late 1960s. Thistlethwaite was very much the driving force in the creation of the university, something which was quite obvious to some of the initial cohort of students, one of whom noted his propensity for 'talking about "my" plan, "my" colour-scheme, "my" architect", "my" advisory committee, and "what my architect calls."'[35] In time-honoured fashion, the innovative educationalist from one generation was seen as a 'typical Cambridge conventionalist' by the rising one.[36] Eventually the gap between the ideals of Thistlethwaite's generation and those of the students would lead to real trouble during the student revolt of the late 1960s, with the former never quite understanding how the latter, with all their privileges, came to see themselves as an oppressed group in need of liberation. Those who wanted to, in Timothy Leary's famous phrase, 'turn in, tune in, drop out', seemed foreign to the high seriousness of a mindset formed by the experience of the Second World War and the age of austerity. Malcolm Bradbury caught something of the baffled way with which the university responded to all of this when, during a sit-in in March 1971, he wrote to a colleague that they seemed to be presenting 'an atmosphere of considerable chaos and uncertainty' to the public. He lamented the lack of leadership and the fact that the secretarial staff seemed to have been left to themselves, as had the lecturers, with the result that the impression had been given that everything had ground to a halt: the university 'has the air of a place taken by surprise and left at a loss'. It suggested, he thought, that 'we do not seem to possess a coherence of value[s], which would allow us to say that our idea of the university as a community is something other than these protesting students want it to be – a bland enclave of the youth culture'. There was, he lamented, 'no clarity of effort or purpose, or a logical structure of command and responsibility'.[37] It has been argued, with some persuasiveness, that much of the mythology surrounding the 'troubles' is the product of self-congratulatory memorializing by some of the leaders and that, at least in the UK, its effects were more evolutionary than they were revolutionary.[38] As elsewhere, the university administration responded to pressures from the more vocal agitators by creating a 'Dean of Students' and eventually in 1972 allowing full student voting rights on both Senate and Council. But events had shaken the confidence of the mandarin mindset about its ability to deal with some of the new knowledge its universities were curating, if not creating.

It would be other factors, partly to do with the national higher education context, which contributed most to undermining the confidence of UEA's foundation years. Yet, in terms of the structures created, those established by Thistlethwaite proved to be strong ones, surviving the troubled late 1960s and early 1970s and outlasting even the economic stringencies of the 1980s. Nor are the reasons hard to find. One was that the school structure created a counter-balance to the centralizing tendencies inherent in Thistlethwaite's initial vision. The academics soon found that as an agglomeration of what elsewhere would be departments, schools had considerable autonomy, with a range of subjects allowing over-recruitment in one area to make up for a shortfall in others, with a consequent fire-proofing against turbulence in admissions figures. This helped make the deans powerful figures who could negotiate with the vice-chancellor rather in the manner of powerful medieval barons with the king – another reason why academics were reluctant to change the model – and why, eventually, in a later, more managerial age, vice-chancellors felt an urgent need to do so. That the strength of the structures of the 1960s proved, despite the criticisms made of them, resistant to change led one well-travelled literature professor to comment to me in the mid-1980s that he had thought Oxford and Cambridge the epitome of resistance to change – but they had nothing on UEA. This owed a great deal to the age structure of the staff in the schools. Thistlethwaite, who was in his early forties when he became vice-chancellor, was by some way the oldest senior figure. It might have been expected that men appointed to chairs in their mid-thirties would have moved on, but Norwich notoriously became known as a place from which people moved reluctantly: a combination of low house prices and a pleasant rural hinterland made it the graveyard of ambition. So it was that those who had created the structures and were invested emotionally in them were very reluctant to change. The discussions over the future of the School of English Studies were typical of all such attempts to change things; there was a great flurry of memoranda, but not much changed. In one sense it was a sign of the maturity of the 'new' university that pragmatic ways forward were found. It became a School of English and American Studies; subsidiarity was ensured through the creation of informal 'sectors' where groups of scholars in the same area would work together, but the dean remained more than *primus inter pares*. For some later recruits, it came to seem as though the early structures had become institutionalized; in fact what had happened was that academics had found a way of making things work and did not, even in the less benign climate of the 1980s, see the need to change things. Radical intellectual investigation sat, ironically for some, alongside an institutional conservatism.

Despite the headlines they garnered and the nostalgia they later created, the student disturbances of the late 1960s had far less effect on the development of the new universities than the political ethos at the centre of government. As Thistlethwaite noted in a memoir he wrote about his years at UEA, by the end of the 1970s the mindset of the post-war era, which had led to the UGC's 'beneficent policy of treating each university in the round and respecting the varying character and needs of each', had disappeared.[39] For a decade that system had been coming under strain, as it was apparently failing to deliver the economic prosperity necessary to sustain government spending at the levels required. In the public sector, university spending was particularly badly affected as costs were increasingly greater than the government had allowed for, partly due to the rate of inflation.[40] The experienced Whitehall warrior in Thistlethwaite identified the critical time as coming earlier than that – to the moment in the late 1960s when Whitehall responsibility for university funding passed from the Treasury (via the UGC) to the Department of Education and Science. He saw how this changed attitudes towards the universities. As a small part of a much larger budget, university funding via the UGC could proceed in the gentleman-like and professional manner he was used to: one set of professionals trusting another to do the right thing. As large part of a smaller departmental budget, at a time of economic constriction, spending on universities was bound to come under more scrutiny, and the UGC was hardly the creature to combat pressure from successive Governments to explain exactly why, at times of cuts elsewhere, universities should receive quinquennial grants guaranteeing them inflation-proof funding. The new universities, lacking the lobbying skills (and alumni in the right places) of the older foundations, found themselves particularly exposed, and Thistlethwaite noted with some distaste the proposal by one vice-chancellor of an established university to the effect that one of the new universities should be 'thrown to the wolves' in order to preserve funding for older ones such as his own.[41] The UGC's policy had been displaced by the neo-liberalism, associated with the Thatcher years. By the time of his retirement in 1980, to Thistlethwaite, the question was how to preserve the liberal idea of a university and to prevent it becoming nothing more than an engine of economic growth.[42] Perceiving with bleak clarity the world that was struggling to be born, it seemed to him that the trends of the previous decade had attenuated the idea of individual loyalty to an institution – something perhaps felt more keenly by a man who had devoted the last twenty years to one institution. He finished his last lecture at the university he had given so much to with a quotation from the bidding prayer of his own Cambridge college, St John's: 'Bless O Lord the work of this college … and grant that love of the brethren and all sound learning may ever grow and prosper here.' Even as he spoke those words, he was conscious of how 'remote' they sounded 'from our present day ideology and rhetoric.'[43]

Thistlethwaite's essentially Whiggish temperament, however, made him optimistic that clever people would find a way to hold on to all that was good about the past whilst adapting their talents for a society which apparently saw universities largely as a utilitarian tool. Nor were his hopes altogether in vain. He was right to see the longevity of UEA's founding fathers as an obstacle to the march of a more utilitarian ideology, and a guarantee of the preservation of a community spirit. Even during the long, lean years of the cuts of the 1980s, the school system survived and, as the Research Assessment Exercise of 1987 showed, enabled UEA to produce world-class research whilst maintaining a commitment to the seminar system, interdisciplinarity and teaching.

What can be measured and what brings in extra income has a natural tendency to preoccupy university management, and over time the effect of this would be to upset the balance between the twin functions of a university, teaching and research. In an era where money was in short supply, success in the RAE was valued for more than its reputational benefit. It not only added lustre, it added lucre. In such a climate, the professor who was never there to teach or administer anything, but brought in grants and produced plenty of high-class outputs, was a more valuable colleague than the school stalwart who was there for the students, carried a heavy teaching load, and was always available to carry out the increasing load of administration the new 'audit culture' was creating. Moreover, the way the RAE was managed was particularly inimical to interdisciplinary work, as it clearly favoured those scholars whose work fell into the neat brackets of 'English literature' and 'history' rather than the vaguer realm of 'studies'. Whilst there was much rhetorical commitment (especially in 1991 and 1996) to the virtues of interdisciplinarity, the fact that those new universities which had kept to the old disciplinary boundaries did better than the newer ones suggested where the money was to be had. It was precisely that line of thinking that led to the creation of a School of History in 1994, where the historians from the schools of English and American Studies, European Studies and Economic and Social Studies came together in a single-subject School. The advantages of Interdisciplinarity were no longer clear.

In some areas, most notably Climate Science and Development Studies, what had started as experimental interdisciplinary areas became enormously successful disciplines in themselves. It would be easy, but incorrect, to assume that the creation of single-discipline Schools at UEA in 2002 marked the end of an era. That would be to ignore the fact that the old subject boundaries had been so eroded that most scholars automatically used insights from cognate disciplinary areas without thinking anything of it. In that sense, UEA's model was successful and the abandonment of the old structures meant only that they were no longer necessary.

Thistlethwaite's university owed much to the classical model of a community of scholars dedicated to 'sound learning' and the 'good of the brethren'. Newman would have recognized it. In our own times, the reaction against the excesses of the neoliberal experiment has given rise to the 'slow university' movement,[44] which may well want to see what lessons can be learned from the example of the sustainable universities created in a quite different climate fifty years ago. In an age when so much of university income now comes from students, it may be time to rediscover the virtues of Thistlethwaite's model.

Notes

1 Michael Sanderson, *The History of the University of East Anglia, Norwich* (London: Hambledon, 2002). In 2013 UEA created a website documenting its 50 year history: http://www.uea.ac.uk/50years.
2 David Edgerton, 'C. P. Snow as Anti-Historian of British Science: Revisiting the Technocratic Moment, 1959–1964', *History of Science*, 53 (2005), 187–208.
3 'UEA: The Background Story' (17 May 1962), Frank Thistlethwaite Archive, University of East Anglia (UEA) Archives.
4 The APB comprised Professor (later Sir) Charles Wilson, vice-chancellor of the University of Leicester; Noel Annan (later Lord Annan), Provost of King's College, Cambridge; Professor (later Sir) Denys Page, Master of Jesus College, Cambridge; E. T. ('Bill', later Sir Edgar) Williams, Warden of Rhodes House, Oxford; Professor Solly Zuckerman (later Lord Zuckerman), chief scientific adviser to the Ministry of Defence; Professor (later Sir) Christopher Ingold, recently retired professor of Chemistry, University College London; Sir Walter Worboys, Chairman of ICI; Lincoln Ralphs (later Sir), chief education officer for Norfolk. For their discussions about the academic structures within UEA, see Sanderson, *History*, 60–8.
5 Thistlethwaite had been a Commonwealth Fund Fellow at the University of Minnesota (1938–40), and worked during the war at the British Information Service in New York before returning to work in the Cabinet Office. In 1946 he returned

to his undergraduate college, St John's, Cambridge as a fellow. See M. Sanderson, 'Thistlethwaite, Frank (1915–2003)', historian and university administrator, *ODNB*, https://doi.org/10.1093/ref:odnb/89466 (accessed 26 April 2019).
6 Frank Thistlethwaite, *Origins: A Personal Reminiscence of UEA's Foundation* (Colchester: Palladian Press, 2000), Frank Thistlethwaite Archive, UEA, 2.
7 Ibid., 3.
8 Quoted in Sanderson, *History*, 85–6. For an earlier account of UEA's origins, see Thistlethwaite, 'The University of East Anglia', in *New Universities in the Modern World*, ed. Murray G. Ross (London: Macmillan, 1966), 53–68.
9 'UEA: the Background Story', 5.
10 Frank Thistlethwaite, 'Doing Different in a Cold Climate', St John's College, Cambridge Lecture (1986), Frank Thistlethwaite Archive, UEA, 2.
11 Sanderson, *History*, 86.
12 Ibid., 83.
13 Robert Ashton, quoted in Thistletwaite, *Origins*, 14.
14 Ibid., 13.
15 Ibid., 14.
16 Peter Dormer and Stefan Muthesius, *Concrete and Open Skies: Architecture at the University of East Anglia, 1962–2000* (London: Unicorn, 2001).
17 Thistlethwaite, *Origins*, 56.
18 Ibid., 23.
19 Ibid. 24.
20 Ibid., 25.
21 For Robert Ashton (1924–2013), see my obituary of him, *Independent*, 13 March 2013, https://www.independent.co.uk/news/obituaries/professor-robert-ashton-historian-of-early-modern-england-8533191.html
22 Thistlethwaite, *Origins*, 18.
23 For James McFarlane (1920–1999), see Malcom Bradbury's obituary, *Guardian*, 30 August 1999, https://www.theguardian.com/news/1999/aug/30/guardianobituaries2 (accessed 6 August 2019).
24 Ross (1924–96) pops up in studies of Whitehall in the 1970s. For full details of his career, see *Inside the Bank of England. Memoirs of Christopher Dow, Chief Economist, 1973–84*, ed. Graham Hacche and Christopher Taylor (Basingstoke: Macmillan, 2013), 268.
25 Sanderson, *History*, 92.
26 *Eastern Daily Press*, 23 October 1963, 32–3.
27 For Philip Ledger (1937–2012), see Stephen Cleobury's obituary, *Guardian*, 20 November 2012, https://www.theguardian.com/music/2012/nov/20/sir-philip-ledger (accessed 6 August 2019).
28 Philip Ziegler, 'Zuckerman, Solly, Baron Zuckerman (1904–1993)', *Oxford Dictionary of National Biography* (Oxford University Press, 2004) http://www.oxforddnb.com/view/article/53466 (accessed 11 October 2017).
29 For the story of the move to Norwich of the John Innes Institute, see Sanderson, *History*, 101–2. For Bennet-Clark (1903–1975), see *Biographical Memoirs of Fellows of the Royal Society*, 23 (1977), 1–18.
30 Sanderson, *History*, 64–5; cf. Thistlethwaite, *Origins*, 15–16.
31 Sanderson, *History*, 105–9. Retrospectively, Thistlethwaite believed that Katritzky was the right choice: *Origins*, 34.
32 Memo by Martin Scott-Taggart (July 1971), UEA Archives, UEA Brad/4.

33 J. B Broadbent to the Vice-Chancellor, 11 May 1970, UEA Archives, UEA Brad/4. For John Broadbent (1926–2012), see the obituary by David Punter, *Guardian*, 27 November 2012, https://www.theguardian.com/books/2012/nov/27/john-broadbent (accessed 6 August 2019).
34 Memo by Martin Scott-Taggart (July 1971), UEA Archives, Brad/4.
35 Adrian Smith to his parents, 16 February 1964, UEA Archives, SMI/1/20.
36 Adrian Smith to his parents, 26 February 1964, UEA Archives, SMI/1/23.
37 Bradbury to George [Chadwick], 16 March 1971, UEA Archives, Brad/3/1.
38 Nick Thomas, 'Challenging the Myths of the 1960s: The Case of Student Protest in Britain', *Twentieth Century British History*, 13 (2002), 277–97.
39 Thistlethwaite, *Doing Different*, 12.
40 By the end of the 1970s maintenance grants were worth nearly 20% less than they had been in 1963: Gill Wyness, *Policy Changes in UK Higher Education Funding, 1963–2009,* Institute of Education, DoQSS Working Paper 10–15 (July 2010), 8–10 http://repec.ioe.ac.uk/REPEc/pdf/qsswp1015.pdf (accessed 4 June 2016).
41 Thistlethwaite, *Doing Different*, 11.
42 Ibid., 13.
43 Ibid., 14.
44 'The Slow University, a Seminar Series', http://www.celebyouth.org/slow-university-seminars/ (accessed 26 April 2019) and Maggie O'Neill, 'The Slow University', http://discoversociety.org/2014/06/03/the-slow-university-work-time-and-well-being/?utm_content=buffer6928b&utm_medium=social&utm_source=twitter.com&utm_campaign=buffer (accessed 4 June 2016).

6

Great expectations: Sloman's Essex and student protest in the long sixties

by Caroline Hoefferle

Great expectations for the new universities of the 1950s and 1960s gave birth to the University of Essex and shaped its destiny in many ways. Inspired by the discourse of modernization and liberalism, its founders, including Noel Annan and Albert Sloman, hoped that Essex would become a new global leader in higher education. Sloman's Reith lectures raised student and staff expectations that Essex would be a different kind of university, with a radical curriculum and architectural design more liberal, democratic and progressive than any before. When students demanded that it follow through with this ideal in the late 1960s, they encountered resistance from Sloman and the Essex administration, resulting in an escalating cycle of frustration and ever more extreme activism. Out of the great expectations inspired by the discourse of modernization and liberalism, 'red Essex' was born.

The creation of a new university

The foundations for the creation of the University of Essex were laid by Rab Butler and Winston Churchill in the 1940s. They had come to view higher and further education as an essential component of the nation's economic modernization programme. They argued that the whole education system would need to expand not only to produce the larger numbers of highly skilled and educated workers required by a modernizing economy, but also to provide access to higher education for children of the post-war baby boom. This economic and social vision for education was revealed in the Education Act of 1944, overseen by Butler who was Conservative Minister of Education and MP for Saffron Walden in Essex from 1929 to 1965. In a 1965 lecture at the University of Essex, Butler argued that the Education Act was critical to beginning the modernization and democratization of the entire education system in Britain and laid the groundwork for the reform and expansion of higher and further education in the 1950s and 1960s.[1]

Churchill was an MP for Epping in Essex from 1924 to 1945 and the adjacent Woodford constituency from 1945 until his death. He became convinced in the 1940s that the nation needed a British version of the Massachusetts Institute of Technology to compete with American and Soviet science and technology. Support for this plan and improved science and technology in higher education in general increased in the 1950s as a reflection of growing anxiety over Britain's reduced global influence at the height of the Cold War. By the late 1950s, they had convinced the Conservative government of Harold Macmillan and the University Grants Committee (UGC) to create this British equivalent to the MIT.[2]

Inspired by this discussion, the Essex County Council received permission from the UGC to locate this new university near Colchester in 1961. None other than Rab Butler became the university's first chancellor. Butler and the council then chose Noel Annan, Provost of King's College at Cambridge, to chair an Academic Planning Board (hereafter APB) to create the plan for the new university and select its first vice-chancellor. Annan was the most influential figure in creating the broad academic vision for Essex.[3] A historian, modernizer and leader of the liberal establishment, Annan was influenced by C. P. Snow's 1959 Rede lecture in Cambridge entitled *The Two Cultures*. Snow's main point was that the divide between science and technology versus literature and the arts was detrimental to both sides and to civilization in general. He argued that this great divide in Britain was caused by over-specialization in the British education system, and that the US and Russian education systems provided much broader education, which resulted in better engineers and scientists. His solution then was to reform British education to encourage more interdisciplinary study and more top-notch scientists, engineers, technicians and government managers.[4]

Snow's influence can be seen in the curriculum laid out in the first report of the Academic Planning Board, written by Annan and published in February 1962. The curriculum at Essex would specialize in the applied and social sciences, but require an interdisciplinary common first year of general education courses to bridge the two cultures. Like the other new universities envisioned in the Education Act of 1944, Essex would try to recruit a more diverse student body and try to meet the needs of industry and the economy in general. Annan proposed only two faculties: physical science (focusing on technology) and humanities (primarily social sciences and literature).[5] Annan and the APB also wanted Essex to join the other new universities in focusing more on innovative, quality teaching than on research. To break the traditional departmental focus on disciplines and research, they encouraged the new universities to replace academic departments with 'schools of study'.[6] Once Annan and the APB had laid down their blueprint for the University of Essex, they went about finding a vice-chancellor to fit their vision.

After a year-long search, in June 1962, Annan's APB selected 41-year-old Albert Sloman to be the first vice-chancellor of the University of Essex and the youngest-ever vice-chancellor in the UK. Sloman was born and raised in Cornwall and had received his DPhil in Spanish from the University of Oxford. He had joined the RAF during the Second World War and had briefly taught at Berkeley and Trinity College Dublin before beginning his distinguished career in 1953 at the University of Liverpool as professor of Spanish and later dean of the faculty of arts. With the first report of the

APB and Annan as his guide, Sloman would preside over the creation and evolution of the university in this position until his retirement in 1987. As Sloman recalled, the APB said to him, 'Look, here's what we have been thinking about, and broadly here is what we propose. But we accept that the first Vice-Chancellor will be the person who will actually shape the university; you have got our Report. It is now over to you.'[7]

Although Sloman was shaped by Annan's vision, he was determined to make his own imprint on the new university. Soon after he was hired, Sloman took a four-week trip to a dozen universities in the United States to gather ideas. He spent much time speaking with Clark Kerr, the chancellor of Berkeley, noting Kerr's organization of the university into departments and recruitment of the best staff, which Sloman would implement at Essex.[8] He agreed with the UGC and the APB's emphasis on interdisciplinarity and the overlapping nature of academic disciplines, but he did not support their insistence on a common first year and also felt that separation into discipline-based departments would give academics more ownership over their own disciplines and programmes, fostering research and teaching excellence. When he returned to Essex, Sloman blended the departmental system with the APB's ideas for a limited number of schools for study. He decided to start with eight departments or subjects (chemistry, economics, government, literature, mathematics, physics, sociology and the Language Centre), grouped into three schools: physical sciences, social sciences and comparative studies. To attract the very best scholars to be the chairs for these departments, he offered them the freedom to design their own programmes and all the resources they needed to make these programmes among the best in the world. Each school would also build its own common first year. Sloman also made Essex more explicitly international than other British universities, featuring not only an international-focused curriculum, but heavily recruiting students and staff from the Americas and Europe.[9] Roughly one-third of the Essex student population in the 1960s was international.[10] Sloman would profoundly shape the curriculum, structures and people of Essex for its first three decades.

As Sloman was hiring the best staff for Essex, he met with Kenneth Capon of Architects Co-Partnership (AC-P) to design their new university. AC-P was founded in 1939 and specialized in modernist educational buildings, such as at St John's College, Oxford and St Paul's Cathedral School, London. Influenced by the then popular new brutalist design, Capon favoured monumental concrete structures and compact mixed-use urban plans.[11] Capon had been Annan's pick since he had proposed four high-rise towers for Churchill College at Cambridge. Originally, Essex County Council had granted Wivenhoe Park, between Colchester and Wivenhoe, to the university for its main location, but Sloman and Capon decided to preserve the park and give the university a small foot-print in an adjacent valley. Capon tried to capture Sloman's vision for the university's central function, which was to connect, advance and transmit knowledge, within the design for the university.[12] To reflect the university's purpose, Capon placed the library in a central and prominent location in the plan. Sloman insisted that this library, unlike other universities at the time, would be open access, allowing students and staff to freely pursue knowledge outside their discipline.[13] Five pedestrian squares stretched out from the library, unified by a continuous zigzag of lecture halls, department spaces and food service options in a mixed use plan. There would be no student union or staff common rooms. All would share the same buildings and catering facilities, and they would be open twenty-four hours every day, with equal access. This design would force staff and students from different disciplines to mix in formal and informal settings, creating the modern, unified, democratic community that Sloman envisioned.[14]

To achieve the large departments and resources Essex needed to lead the world in the sciences, Sloman believed that the university would need to enrol at least 10,000 students, possibly expanding up to 20,000, making it the largest university in the country. To house these students in a compact way, Capon designed a series of twenty-eight residential towers, fourteen stories high, clustered around the zigzag core, resembling a small city, so that no lecture hall would be more than a five-minute walk from a residential tower. Although only six of these were built, and they were unfunded by the UGC, they were innovative in that they were designed to foster student community and independence. Sloman rejected the traditional *in loco parentis* university–student relationship, and believed that modern students should be treated as adults and have more independence than previous generations. Capon gave him a design that reflected this vision. Each tower floor was a same-sex residential apartment with its own kitchen, study room, washroom and thirteen bed-sits, bringing together students from different disciplines to live together and to bridge the gap between the 'two cultures'.[15] Controversially, male and female students would live in the same tower, but on different floors. They would set their own rules for themselves with little administrative interference. The architectural design would project strength, austerity and rigour, but also encourage student community and independence.[16]

To fund the initial phase of construction, the university began a public appeal fund in late 1963. In speeches and literature, the fundraising campaign emphasized how Essex would contribute to the nation's economic modernization initiative by providing the research and the trained specialists to meet the needs of the country's industrial and commercial interests. Churchill, who was a founding donor to this campaign, noted that 'it is of the greatest importance to increase our resources of higher education, particularly in the technological and scientific fields'.[17] With an exciting architectural plan and ambitious goal to educate upwards of 10,000 students, the fundraising campaign generated excitement for the endeavour and succeeded in raising over £1m.

Contributing enormously to the success of this public funds campaign was the Macmillan government's publication of the Robbins report in October 1963. The report formally set the national agenda for the expansion and reform of higher and further education for the next decade. The explicit purpose of this reform was to satisfy the labour needs of industry and government, but also to democratize access to higher and further education by providing more opportunities for the greater number of qualified baby boomers. It supported the creation of new universities and provided funds for adding facilities to existing colleges and universities to allow more students to enter higher education.[18]

To build upon the discussion sparked by the Robbins report, the BBC invited Sloman to give a series of six Reith lectures on the future of the universities.[19] With the help of Noel Annan and social policy professor Peter Townsend, Sloman would take this opportunity to share his concept for the University of Essex with the rest of the world and generate more public support for his endeavour. Asserting that the Robbins report had shown that the universities were threatened with the pressure of expanding numbers and knowledge, Sloman declared that this threat could only be solved by 'radical innovation'.[20] He expressed his vision for a high-tech modern university, which would offer the best education to students of all classes. Essex would be a model democratic university, not only admitting students of all classes, but making them full members of the academic community, where they would have full control over their personal lives and participate in university governance. The radical new curriculum would not only train more specialists in the sciences, but also be relevant to the problems that concerned modern students. These lectures repeatedly used the words: modern, radical, revolutionary, progress and planning. The whole point of Sloman's curriculum was to enable students to research and improve the modern world, to use their educations to solve the world's problems. These lectures helped to recruit a more forward-looking, egalitarian staff and student body, interested in being part of this non-traditional university and great progressive experiment. Interviews with sixty former staff and students show that many of them listened to the Reith lectures with excitement and intrigue. They were especially attracted by Sloman's vision of Essex as a place of progress and adventure, as a place that was 'in-tune with the feeling of the times'.[21]

Paul Thompson, the sociologist, who arrived at Essex in 1967, has collected an archive of oral histories of students, staff and administrators at Essex throughout the years that reveal much about Sloman and the Reith lectures. One of these is a 1998 interview with Peter Townsend, who was thirty-five when Sloman asked him to leave the LSE and be one of two founding professors of the university. Annan and Sloman selected him for his pioneering research in the field of social policy. Townsend became the first chair of an interdisciplinary Department of Social Sciences, including economics, political science and sociology. He believed that although Sloman was very precise, efficient and orderly, he was unapproachable and felt his status as vice-chancellor put him above the rest of the community. Townsend asserted that 'Noel Annan was, I think, the one who contributed more than anybody else to the seminal ideas for the distinct form that the University took. Sloman was not so imaginative. He was very efficient, he was concise, but he was a bit insecure about his own ideas'. Townsend helped Sloman in writing the Reith lectures because Sloman 'clearly felt ill-equipped to pronounce about the development of the whole clutch of new universities in the country'.[22] Other faculty and staff, however, believed that Sloman was the best vice-chancellor for the job, excelled at attracting the best people to work at the university, and ultimately was a great success at creating the university from scratch.[23] Whether the ideas behind the Reith lectures originated with Sloman, Annan or Townsend, they reflected the liberal modernization discourse spreading through the political, cultural and intellectual elite of the Western world in the 1960s.

High expectations

The public funds campaign, Robbins report and Reith lectures raised high expectations for staff and students who would come to Essex when it opened its doors in 1964.

That first year, the development of the architectural master plan had only just begun. Lacking the space for lecture halls and student accommodations, Essex enrolled only 122 students in the first year and 400 the next.[24] The first stage of building, including the library, a restaurant, the bases for the planned squares, and two residential towers, was not completed until 1966–7.[25] Students and staff, meanwhile, stayed in nearby Colchester and attended classes in temporary huts and draughty old buildings in fields of mud for the first few years. That first year, students still believed in that they would at least see some of the master plan completed before they graduated. They created their own newspaper, *Wyvern,* which was free from administrative oversight and whose editor was elected by the student body. They also had a student government council, which was also free from administrative oversight and elected by students representing each department. Their first few issues of *Wyvern* showed that students were generally satisfied with their lodgings, curriculum and staff. It was noted, however, that students 'were bombarded from the first by constant emphasis on the "challenging" aspects of their "unique position of responsibility."'[26] Even though Essex in its infancy was physically far different from the master plan envisioned by Sloman and Capon, students still had high expectations that the next year would be better.

By November of 1965, however, the student mood began to shift. Chris Piening wrote an editorial for the *Wyvern* complaining that 290 new students came to Essex that year believing the prospectus, which had promised that the students would have a new teaching building, restaurant, residential and study room tower, shops, and other amenities, but this did not happen. Troubles with contractors put the building far behind schedule. He asserted that the university officials knew of these troubles, but lied in the prospectus. 'Essex is a record of mismanagement and incompetence on the part of the contractors and the University alike,' complained Piening, suggesting that the new intake would have gone elsewhere, had they known the conditions they would face.[27]

These first generations of students not only came to the university with very high expectations about its physical plan, they also expected a radical curriculum and non-traditional organization for university governance and student accommodations. In a February 1966 questionnaire sent out to all first-year undergraduates, 70 per cent said that they had come to Essex because of its new, non-traditional nature and 57 per cent said they came to study their subjects very deeply. Relative to other universities in the study, Essex students were also more left wing, libertarian and non-religious.[28] Just as Sloman had envisioned, Essex was attracting forward-thinking, rigorous, non-traditional students with very high expectations. Just as importantly, however, Essex was not yet living up to their expectations and this would lay the groundwork for the student unrest to follow.

Things would only get worse in 1966 when the UGC visited the university to determine allocation of resources for the future. The UGC saw that the sciences were not attracting the numbers of students they had originally predicted and so scaled back their enrolment predictions from 3,000 to 1,700 by 1972. The House of Commons Public Accounts Committee then pressured the UGC to also scale back Essex's capital grant accordingly.[29] Sloman believed that the funding settlement actually resulted from some of his staff offending the committee by emphasizing Essex's commitment to research instead of only teaching. He saw this as a 'disaster', blaming the 1966–72 settlement as the reason for Essex's financial troubles for the next twenty years, and one reason why the university did not fulfil his vision for over 10,000 students until the twenty-first century.[30]

Student troubles

Student troubles at Essex, however, were not caused solely by the size and physical environment of the university. They emerged within the context of a global youth movement that had developed earlier in the decade. British students were especially aware of the massive student protests, which had shaken France and Berkeley since 1960. French students had begun protesting in 1960 against the military draft and in support of Algerian liberation, but in response to President Charles De Gaulle's crackdown on protest demonstrations in 1962–3, they broadened their demands to include educational reforms and an end to police violence.[31] While the French students won no concessions, students at Berkeley were more successful the next year. In the autumn of 1964, Berkeley students held massive protests over the right to political organization and free speech. The global news media spread word of the uprising and Berkeley students reached out to students in other nations, calling for solidarity and support for their protests. Student groups around the world responded with support, expressing a new feeling of generational identity that surpassed national boundaries and local university politics.[32]

Possibly because of its high numbers of French and American students, the Essex student newspaper reported on global student protests more than most other British universities. French and American students provided Essex students with role models for the potential power of student activism as well as a sense of a global student activist identity. In the second issue of *Wyvern*, editor Chris Piening saw the Berkeley crisis as a 'frightening example of what can happen in a democracy and at the same time an encouraging demonstration of the Berkeley students' determination to be allowed to exercise their basic rights … '.[33] He condemned the Berkeley administration's attempts to stifle their freedoms and supported the students in their struggle. The next year, the American Students for a Democratic Society wrote to student unions all over the world, including Essex, asking for support for a world student strike over the issue of the war in Vietnam. David Triesman (who would later go on to become a Labour member of the House of Lords and a minister in the Department for Innovation, Universities and Skills) was a twenty-three-year-old second-year member of the Socialist Society and assistant editor for the *Wyvern* that year. He responded with support for the SDS and outrage over the Vietnam War. While the British government was not fighting the war in Vietnam, it refused to condemn US actions there and this enraged many students like Triesman. He helped to organize several demonstrations against the war over the next year, but few students turned out for them. In an editorial, Roy Saatchi argued that an Essex strike over this issue would achieve nothing and that they had more direct issues to consider striking over, such as government grants and student affairs. Vietnam alone would not inspire 'red Essex'.

By November of 1965, Triesman was linking student financial and governance issues to a wider socialist critique of the British government and the Vietnam War, hoping to inspire more activism at Essex. That month he wrote about a government proposal to replace student maintenance grants with loans, arguing that this scheme would undermine the Robbins commitment to democratize education and that British students might strike if the government went ahead. He threatened, 'Crosland beware! The mood is militant! You cannot ride roughshod over Britain's student population.'[34] Using his base in the Socialist Society, Triesman won election to the student council the next month on a platform of making the Council more actively engaged in wider issues affecting Essex students. He and other radical socialist leaders in the student council gained supporters by linking global issues to local demands for increased freedom and democracy for students. In February of 1966, the student council clashed with the 'University Bureaucracy' over the creation of an autonomous student union. They called for more power for the student union, and a student union representative on Senate and on other powerful committees.[35] They also passed a resolution condemning the British government's support of the Vietnam War. By October of 1966, the Socialist Society had ninety members and was the largest society at the university.[36] They would provide the core of student radicalism at Essex through the end of the decade.

Global, national and local issues collided in February of 1967 for Essex's first formal protest on the issue of overseas student fees. In December of 1966, the British government announced that it would increase overseas student fees to help pay for the rising costs of higher education. This was opposed by many administrators, faculty and students. The Radical Student Alliance made the issue the focus of its campaign that year and was able to rally thousands of students across the country to its protests and strikes. Over 200 Essex staff and students participated in the nationwide strike against the overseas student fees on 22 February 1968, including a teach-in and protest march.[37] This was a peaceful demonstration with considerable support from staff. Despite the earnest pleas of staff and students from Essex and across the nation, however, the government went forward with the overseas fees increase.

Throughout the year, Essex students held a number of small protests on a range of issues including the price of food in the Hexagon restaurant, the price of rent in the towers, the lack of increases in student grants and the Vietnam War. The Socialist Society and student council stirred the pot of this discontent and suggested increased student power and participation within university governance as the solution. Sloman's plan for the university, integrating staff and students as well as the architectural environment, was included as inspirations for their dissatisfaction. Writing in January of 1968, the President of the student council, Colin Rogers, complained that 'we have no say in the academic affairs of this, *our* University'.[38] Rogers and two other members of the Socialist Society, Rick Coates and Mike Gonzalez, also published a pamphlet in January 1968 entitled 'There's something wrong with Essex'. Reflecting the influence of the American student movement, especially Mario Savio's 1964 speech at Berkeley, they insisted that 'we don't want to be trained to be new parts in an old machine'.[39] Essex students, who had been told by Sloman and other university officials that they were an important part of the university community and been promised 'integration' with the faculty, felt disillusioned by the university authorities' apparent betrayal of the spirit of this promise. Like their counterparts at Berkeley, Essex students wanted the university to live up to its educational ideals, to be truly a democratic learning community.

By February of 1968, Rogers and the Socialist Society were urging Essex students to protest and take more control over the direction of the university.[40] They saw their opportunity at the end of the month when the Colchester Conservative Association invited controversial MP Enoch Powell to speak on campus. About a dozen students set off fizzing bombs, turned on the fire alarms, poured beer on the chair of the meeting and shouted down Powell as he tried to defend his views on immigration. As he tried to drive off, they formed a chain of students to block his car and someone threw a pipe at the car.[41]

The newspapers gave heavy coverage to this incident and it sparked a debate and resolution in the House of Commons that month. Kenneth Lewis, MP for Rutland and Stamford, began the debate by making a motion 'that this House, while upholding the ancient right and ritual of students to protest, deplores militancy which indulges in activities of ill-tempered, disturbance, disorder or strike; and, supports any moderate student lead which recognises and seeks to improve the advantages provided by university and further education, paid for so largely out of the taxes of the general public'.[42] He argued that the current troubles at Essex and the LSE were caused by extremists, anarchists and Communists, and urged the House to take action. Antony Buck, MP for Colchester, who had been with Powell at the University of Essex meeting, believed that the university had become a great centre of learning and that Sloman was 'a highly admirable gentleman'. A hard-core minority of students, however, had impaired Essex's reputation with their actions. He was amused by the fizzing bombs, which were grapefruit blacked over and topped with a burning fuse, but he found the shouting of obscenities and letting off the fire alarms to be inexcusable. The attack on his new car in particular led him to plead with the House to support the resolution and draw the line at violent protest, which impeded the freedom of speech of others. In the end, however, the House rejected the resolution.[43]

Rogers, Triesman and other members defended their activism, arguing that it originated in 'a difference of opinion on the role of the university—the one we want and the one that is imposed on us from above'. Rogers argued that Sloman, the administration, the government and industry leaders created the new universities to satisfy the national need of the industries. They 'merely want a flow of ready-made technologists, managers and teachers with which to perpetuate our capitalist system'.[44] He asserted that students did not want to be slaves and civil servants. Rather, they wanted to study society and use their educations to oppose injustice and improve the world. Another Socialist Society student, Mike Gonzalez, expanded on Rogers's points in a *Wyvern* article in late April. He traced the origins of Essex's protests to the Robbins report and its emphasis on using the system of higher education to satisfy the needs of British capitalism. He asserted that Albert Sloman's Reith lectures confirmed Essex's role in supporting the needs of the government and capitalism's demands for trained workers, as opposed to focusing on the needs of the students. As a socialist student in a very strong and critical School of Comparative Studies, Gonzalez and other students like him had indeed been effectively taught to think critically about their society, government and educational system. As Sloman had envisioned, Essex students were independent thinkers and activists. Sloman had just not envisioned that they would direct their critique directly at him and his university.

In the aftermath of the Powell protests, the Dean of Students threatened to discipline seven of the students involved, but students and staff petitioned the administration to drop the charges and in April, students held several demonstrations in support of the accused students. In the end, the dean asked two of the students to pay the charges for the Fire Brigade and one of the students to pay for the pipe damage done to the car.[45] The administration then instituted a new set of disciplinary procedures to expel those who would prevent others from speaking at similar events to defend the right to freedom of speech.

While these disciplinary measures were being discussed in early May of 1968, the Socialist Society seized upon a new opportunity to challenge perceived injustices and flaws in their nation. The chemistry department had invited Dr Thomas David Inch from the government's chemical and biological weapons research facility at Porton Down to discuss his research and careers in toxic chemistry. David Triesman and the Socialist Society wanted to use this event to peacefully debate with Inch and to condemn the barbarity of chemical and biological warfare (CBW), and the sale of British weapons to the United States for its war in Vietnam. After a few minutes of listening to Inch speaking, Triesman stood up and spoke for five minutes, comparing scientists like Inch to Nazi scientists and demanding a ban on chemical weapons research. When Inch and members of the chemistry department shouted back, Peter Archard stepped in and threw mustard powder on Inch, shouting 'ban mustard gas'. After that, a number of other students blocked Inch from leaving. The police arrived thirty minutes later, hustled Inch out to a car and drove away.[46]

Sloman responded the next day by suspending Triesman, Peter Archard and Raphael Halberstadt, who was not even at the protest. These suspensions then set off a much larger occupation of the university, where students and sympathetic staff suspended all routine functions of the university until Sloman would agree to rescind the suspensions. They declared a 'free university' for a week, a continuous teach-in to discuss CBW, the suspension of the three students, and increased student participation in university governance. At the end of the week, Sloman finally agreed to meet with the assembled student and staff protesters to discuss the suspensions. Sloman defended his decision in the name of protecting the academic freedom of the community, but students and staff challenged him, arguing that they were the voice of the academic community and they wanted the suspensions lifted. Finally, on 17 May, Sloman agreed to allow a staff-student tribunal to investigate the matter more thoroughly and make a recommendation. Students returned to class the next Monday and prepared for final examinations.[47]

While it is easy to laugh at these student antics and view these protests as part of an insignificant youth rebellion, at the time many important people took these protests very seriously. Bertrand Russell, Jean Paul Sartre and more than forty scientists supported Triesman in his demands to stop chemical weapons research.[48] Student delegations from universities across the nation also gave Essex students their support. In mid-May, the founder of an anti-CBW group in Devon contacted the university to tell them that the Essex protest had prompted the *Guardian*, the *Sun* and the BBC to give more extensive coverage to the problem of CBW. They revealed that even though the British government had been researching CBW at Porton Down, it had not been manufacturing, selling or stockpiling them. At a resulting Parliamentary discussion, the government stated that it had no intention to pursue CBW further.[49]

The university administration took the protests seriously and conducted a full inquiry into these events. Thirty-eight student participants in the demonstration and twenty-two members of the chemistry department gave evidence. Chaired by the barrister Anthony Gifford, the tribunal of enquiry gave a fairly unbiased account of the events. The resulting report of the tribunal of enquiry, published on 24 June 1968, found that a breakdown in communications between students and administrators lay at the root of the protests. It recommended no punitive action be taken, the three suspended students be reinstated, and that the university reaffirm its statement of academic freedom and continue to work towards improved student representation on senate committees.[50]

In the autumn of 1968, Essex staff and students worked to restructure student representation on university committees. They established student representation on most university committees, including the senate general committee.[51] A core of Socialist Society students, however, continued to object to the lack of participatory democracy in Essex's governance structure and demanded a radical restructuring of the university's governance to include much more student participation and power within it, effectively creating a 'people's university'. They specifically wanted a 'university assembly' in which all members of the university could voice concerns and make decisions, and a 'community court' consisting of all members of the university to handle all disciplinary cases.[52]

'Red Essex'

Protests, occupations, mass meetings and disruptions continued at Essex through the mid-1970s, leading many newspapers to dub the university 'red Essex'. In the spring of 1969, protests resumed in response to the expulsion of two prominent activists for non-attendance. The House of Commons Select Committee on Education and Science visited the university in April 1969 in the midst of these protests as part of their nationwide tour to see how taxpayer's money was being spent and to discover the roots of student unrest. Student hecklers repeatedly interrupted its proceedings so much so that it had to move its final day of hearings to London. The committee questioned Sloman and other top Essex administrators as to the cause of student unrest, university discipline and governance. Sloman saw student activism as a demonstration of Essex students 'intense concern with the general quality of society' and part of the university's 'strength and vitality'.[53] He believed in protecting freedom of speech, but felt that only a small minority of students had recently begun challenging the limits of this by violently shouting down speakers. He asserted that they were in the process of revising their disciplinary procedures to prevent further such occurrences, but was not that worried about the unrest. The committee's final report condemned student radicalism, but supported improved student participation, teaching and facilities.[54]

A lower level of student activism continued at Essex through the 1970s against the Vietnam War, university investments in South Africa's apartheid economy and Margaret Thatcher's attacks on student union autonomy. Sloman believed that they had 'contained student unrest by firm methods coupled with judicious concessions'.[55] Essex, however, once again exploded in massive protests in 1973–4 in reaction to the Conservative government's reduction in student grants and rising rents and food prices. Although Sloman opposed the grant reductions, Essex students boycotted classes, formed picket lines and occupied administrative offices, shutting down the university. The NUS and students across the UK protested these issues, but at Essex the protests were more prolonged and intense, involving up to 1,000 students at times and even locking Sloman in his office to force him to meet with students. Eventually, the university brought in police to disperse and arrest student protesters.[56]

The UGC again visited 'red Essex' during the 1974 protests. The Student Union boycotted meeting with them, but protesters staged a number of actions such as begging for alms and showing up in sleeping bags all over the campus to make the point that they were impoverished by the insufficient grants. One student, Halford Hewitt, was caught painting 'UGC fuck off' on the vice-chancellor's house and later flung himself on top of a departing UGC car. Other students threw stink bombs at the UGC committee.[57] In the worsening economic crisis of 1973–4, when the UGC cut funding across higher education, this treatment almost guaranteed that Essex would receive more cuts to its budget. By 1974, Conservatives and even some liberals were asking for the resignation of Sloman and the closure of the university, arguing that its reputation was so bad that Essex students would face a stigma when seeking employment. Conservative MP for Brent North Rhodes Boyson argued that 'Essex should be closed down and law abiding students transferred to other institutions …. I think its problems have lasted so long that one might as well shut up shop now and start again …. Essex has become almost notorious in the country'.[58]

In the midst of this tempest of criticism, Noel Annan returned to Essex to chair a committee to investigate the protests and make a recommendation on the future of Sloman and the university. His report blamed the protests on a minority of professional revolutionary student activists in the International Marxist Group, International Socialists and the Communist Party.[59] Calling them 'wild men' and 'wreckers', the report argued 'for them the University authorities are not only the tools of the capitalist State but the weakest link in that State, and to strike at them is to promote the destruction of that State and achieve the political objectives nearest to their heart ….They want to provoke the university, to corner them, eventually to force them to capitulate'.[60] Privately, Annan predicted to Sloman, 'I believe that if this lot are really clobbered, you will never have trouble again. The experience at LSE proves just this'.[61] Students and staff disagreed with the Annan report and published their own analysis a few months later. They argued that the disciplinary procedures and power structure of the university, as well as the cutting of student grants, lay at the root of the problems of the university. They recommended that the university reform its decision-making and disciplinary structures to allow students a greater voice and some real power.[62] Many students were eager to work with Sloman, viewing him as a sympathetic supporter and criticizing the militants for their dishonesty and propaganda.[63] Sloman worked with students and staff the next year to implement some of these reforms, but smaller-scale protests continued around the issues of grants and rent prices.

Sloman continued on at Essex until 1987 when he retired as one of the longest-serving vice-chancellors in UK history, having held the position for twenty-five years. Weathering the storm of political and public condemnation due to the student troubles of the 1960s and 1970s, he saw enrolments and building grow very slowly over the years. He blamed the student protests and the UGC settlement of 1966–72 for the university's slow growth and failure to completely achieve his vision.[64] The university was threatened with budget cuts and closure several times in the 1970s and 1980s, but it survived with Sloman's design and curriculum intact and eventually witnessed new growth and achievements by the end of the century.

The University of Essex, with all of Sloman's plans and all of its student unrest, was a child of the nation's post-war dreams of economic modernization and democratization. Its clear mission, similar to that of all of the new universities of the 1960s, was to produce the trained workers to help the nation to compete in the global post-war economy and to provide places in higher education for the growing population of qualified school-leavers, giving them the opportunity to move into better careers. This dual purpose shaped its creation, but Sloman shaped its unique curriculum and structure. He created a remarkably innovative and successful curriculum, and the basis for a more democratic governance structure, allowing students much more freedom over their personal lives than at previous universities. Its innovative architectural plan, while stunted in its growth and aesthetically displeasing to some, helped to connect students and staff, and build an intentional community of learning and living. Sloman's vision for this innovative new university attracted critical-thinking students with very high expectations for their university and society. When the reality of Essex failed to live up to these great expectations, disillusioned students grew restive. Students in sociology and government directed their critical thinking skills towards Sloman and the university, as well as the wider world. Inspired by Sloman to be action-oriented participants in society, these students took direct action to express their critique and remake their world according to their ideals. In protest after protest, students reflected Sloman's vision for more participation, more democracy in decision-making, and more innovative problem solving over a wide variety of issues. Essex students were not alone in protesting these issues, but Essex protests were larger and more prolonged, and specifically referenced the unique ideals and vision created by Sloman and the universities founders. By the late 1960s, press and government criticism of this student activism had serious consequences for Essex in the form of budget cuts and threats of closure. The fact that Essex survived this period with its commitment to innovation, its unique curriculum and student participation intact is a testament to the strength of the original vision of its creators and its lasting relevance.

Notes

1. R. A. Butler, *The Education Act of 1944 and after: The First Noel Buxton Lecture of the University of Essex, 7 May 1965* (London: Longmans, 1966).
2. Jules Lubbock, 'University of Essex Vision and Reality' (University of Essex, 2014), https://www.essex.ac.uk/about/university/fifty/documents/exhibition-guide.pdf. Cf. *Something Fierce Exhibition*. https://www.essex.ac.uk/about/university/fifty/exhibition/ (accessed 26 January 2017); Allen Packwood, 'Churchill: MP for Epping and Woodford, 1924–65' (unpublished lecture, 2014), 2. The editors are grateful to Dr Packwood for letting us see a copy of this lecture.
3. Lubbock, 'University of Essex Vision and Reality'.
4. Robert Ruprecht, 'Forty Years Later: C. P. Snow's Two Cultures Revisited', *European Journal of Engineering Education*, 24 (1999), 231–3; Guy Ortolano, *The Two Cultures Controversy: Science, Literature and Cultural Politics in Postwar Britain* (Cambridge: Cambridge University Press, 2009).
5. Jules Lubbock, 'Building the New University: The Architecture of Essex and Its Context' (2014) https://soundcloud.com/uniofessex/buildinguniversity (accessed 26 January 2017).
6. Keith Tribe, *Economic Careers: Economics and Economists in Britain 1930–1970* (New York: Routledge, 2002), 22–8. For Sloman, see also A. Mackenzie, 'Sloman, Sir Albert Edward (1921–2012)', *Oxford Dictionary of National Biography*, http://www.oxforddnb.com.libproxy.york.ac.uk/view/10.1093/ (accessed 1 March 2019).
7. Sloman quoted in Tribe, *Economic Careers*, 227.
8. This paragraph draws on Tribe, *Economic Careers*, 229–32 and Lubbock, 'University of Essex Vision and Reality'.
9. *Wyvern*, 13 January 1967, 4. University of Essex, Albert Sloman Library, Special Collections, f LF 320.7.
10. Paul Thompson, 'University of Essex: Our 50th Anniversary', 'Creating a University' Podcast Series, http://www.essex.ac.uk/fifty/podcasts/ (accessed 6 July 2015).
11. Lubbock, 'University of Essex Vision and Reality'.
12. Albert E. Sloman, 'British Universities and Their Students', in *Universities: Boundaries of Change*, ed. Jack Straw et al.(London: Panther, 1970), 31.
13. Tribe, *Economic Careers*, 237.
14. Lubbock, 'University of Essex Vision and Reality'; cf. Jules Lubbock, 'Something Fierce: Brutalist Historicism in the University of Essex Library', in *British Design. Tradition and Modernity after 1948*, ed. Christopher Breward et al. (London: Bloomsbury, 2015), 127–40.
15. Marie Clossick, *Student Residence: A New Approach at the University of Essex: A Study of the Rayleigh Tower 1965–66* (London: Society for Research into Higher Education, 1967), 2–3.
16. Jules Lubbock, 'The Counter-Modernist Sublime: The Campus of the University of Essex', in *Twentieth Century Architecture 6: The Sixties*, ed. Elain Harwood and Alan Powers (London: Twentieth Century Society, 2002), 112.
17. Winston Churchill quoted in Lubbock, 'University of Essex Vision and Reality'. Churchill's £25 contribution was not publicized at the time: Anthony Montague Browne to Sir John Ruggles-Brise, 8 April 1964, CHUR 2/554, Churchill Archive, Churchill College, Cambridge. The editors are grateful to Dr Allen Packwood for this reference.

18 *Report of the Committee Appointed by the Prime Minister under the Chairmanship of Lord Robbins, 1961–1963, Parl. Papers* (1962–63) xi–xiv, Cmnd. 2154; Brian Simon, *Education and the Social Order, 1940–1990* (New York: St. Martin's, 1991), 229.
19 Albert E. Sloman, *A University in the Making* (London: BBC, 1963).
20 Ibid., 10.
21 Paul Thompson and Joanna Bornat, 'Myths and Memories of an English Rising: 1968 at Essex', *Oral History*, 22 (1994), 44; Paul Thompson, 'Creating a University' Podcast Series for the University of Essex 50th anniversary, http://www.essex.ac.uk/fifty/podcasts/ (accessed 6 July 2015).
22 Peter Townsend, interview with Paul Thompson, 14 September 1998, *Stories of Essex Sociology*, https://essexsociologyalumni.com/memories/tales-from-past-folk/extract-of-interview-with-peter-townsend-on-his-essex-days/ (accessed 26 January 2016).
23 Chris Garrington and Paul Thompson, 'Albert Sloman: The Man', *Creating a University* (Wivenhoe Oral History Project), http://www.essex.ac.uk/fifty/podcasts/ (accessed 6 July 2015).
24 *Wyvern*, 2 February 1966, 10.
25 Lubbock, 'University of Essex Vision and Reality'.
26 *Wyvern*, 5 February 1965, 4–5.
27 Ibid., 2 November 1965, 2.
28 Ibid., 2 February 1966, 10.
29 Michael Shattock, *Making Policy in British Higher Education 1945–2011* (Maidenhead: Open University Press, 2012), 193–4.
30 Cited in Tribe, *Economic Careers*, 235–6.
31 Mark Edelman Boren, *Student Resistance: A History of the Unruly Subject* (New York: Routledge, 2001), 130–1.
32 Ibid., 144.
33 *Wyvern*, 5 February 1965, 3.
34 Ibid., 19 November 1965, 1.
35 Ibid., 28 February 1966, 3.
36 Ibid., 28 October 1966, 10.
37 Ibid., 24 February 1967, 1.
38 Ibid., 19 January 1968, 7.
39 Colin Rogers et al., 'There's Something Wrong with Essex' (January 1968) reprinted in David Widgery, *The Left in Britain, 1958–68* (Harmondsworth: Penguin, 1976), 327; cf. David Triesman, 'Essex', *New Left Review*, 50 (July–August 1968), 70–1.
40 *Wyvern*, 16 February 1968, 2.
41 Ibid., 1 March 1968, 1–5.
42 HC Debs., 760 (15 March 1968), vol. 760, 1974–5.
43 Ibid., 1983–5.
44 *Wyvern*, 15 March 1968, 4.
45 Ibid., 26 April 1968, 1.
46 Ibid., 10 May 1968, 1, 8; 'University of Essex Report of the Tribunal of Enquiry on the Events of May 7th 1968 and the Events Leading Up to Them' (June 1968), 12, University of Essex, Albert Sloman Library, Special Collections, Foundation Papers Boxes 8–9.
47 Adrian Sinfield, quoted in Thompson and Bornat, 'Myths and Memories', 50; '1968: A Student Generation in Revolt, Ronald Fraser Interviews David Triesman' (transcript 17 April 1984), 61, 65, Columbia University Oral History Center, C896/05; cf. *Wyvern*, 21 June 1968, 8.

48 '1968: A Student Generation in Revolt, Ronald Fraser Interviews David Triesman', 65.
49 John R. Walker, *Britain and Disarmament: The UK and Nuclear, Biological and Chemical Weapons Arms Control and Programmes 1956–1975* (London: Routledge, 2016), 15.
50 'University of Essex Report of the Tribunal of Enquiry on the Events of May 7th 1968'.
51 'Progress Report on Developments in Student Participation in University Government', *In Brief* (Students' Council leaflet), 3 October 1968, University of Essex, Albert Sloman Library, Special Collections, f LF 320.7.
52 *Wyvern,* 31 October 1969, 1; 14 February 1969, 2.
53 'The Events of the Visit of the Select Committee of Education and Science to the University of Essex', *Parl. Papers*, 308 (1969), para. 1272.
54 Ibid., para. 1292.
55 Sloman quoted in *Report of the Disturbances in the University of Essex (The Annan enquiry)* (University of Essex, 1974), 8.
56 'Essex', *Oxford Strumpet* (21 February 1974), 4, Special Collections, Bodleian Library University of Oxford, PR 1/23/31/2; *Wyvern,* Fall 1973, 7.
57 Ibid., 84.
58 *Look East*, BBC television programme (20 May 1974), Transcript, University of Essex Special Collections.
59 *Report of the Disturbances in the University of Essex*, 3.
60 Ibid., 7.
61 Annan to Sloman, 19 March 1974, Annan Papers, The Archive Centre, King's College, Cambridge, NGA/5/920.
62 Executive Committee, University of Essex Students Union, 'The Great Double-Act: A Reply to Lord Annan' (September 1974), 3–7, University of Essex Special Collections, Collection of Material Relating to Disturbances, Foundation Papers, Box 13.
63 Andy Rollings, 'Opinions of a Silent Majority Student' (26 February 1974), University of Essex Special Collections, Collection of Material Relating to Disturbances, Foundation Papers, Box 13.
64 Tribe, *Economic Careers*, 236.

7

Science and the new universities

by Jon Agar

In this chapter I examine the teaching of, and research in, the sciences in the new universities of Sussex (1961), East Anglia (1963), York (1963), Lancaster (1964), Kent (1965), Essex (1965) and Warwick (1965). Since these were new foundations and had not passed through university college status, there was considerably greater freedom to design new disciplinary and interdisciplinary patterns. This opportunity was seized, to a greater and lesser degree, with implications for how subjects, including the sciences, were perceived, taught and to some extent researched. Scientific research at Sussex and East Anglia, as well as mathematics at Warwick, developed in comparatively important ways; this is less the case for the other four. But in terms of science teaching, the new utopianist universities were only part of a wider expansion of higher education in the sciences. Compared to other papers in this collection, I have to pay attention to the other great, contemporary wave of higher educational institution building of the period: the granting, following the advice of the Robbins report of 1963, of university status to the Colleges of Advanced Technology from the mid-1960s.

The chapter proceeds as follows. After a discussion of the variety of higher educational institutional possibilities for science and technology in the 1950s, I will examine the broad development of the sciences as taught and researched in the new universities. I will examine the rhetoric and record of interdisciplinarity, the distinctive developments in science and mathematics at Sussex, East Anglia and Warwick, and the relevance of 'long 1960s' social and cultural movements to the sciences on campuses. I will also note in passing the consequences of the contingent availability and use of high-speed electronic stored-program computers (perhaps the major, common, new object of investment for the sciences encountered by nearly all universities in this period).

Institutional experiments

Among the keywords recurrent in discussion of the new universities, 'community' and 'spontaneity' are noted by Muthesius. Another word – 'experiment' – is an interesting one in the context of the sciences.[1] It was used by some protagonists in early planning,[2] while the *Daily Mail* described the University of Sussex as 'the great experiment'.[3] 'Experiment' can convey novelty, a step into the unknown, but also an assay – a pitching and test of alternatives, to see what works and what does not. A variety of imaginative institutional arrangements for higher education were pitched in the 1950s and 1960s, especially concerning the provision of science and engineering teaching and research.

The expansion of higher education of the 1950s and 1960s 'took place against the background of a vigorous and continuing debate on the appropriateness of the courses on offer to a swiftly changing industrial society'.[4] In particular, the 'clear distinction between the universities and the technical colleges', confirmed in the language of the 1945 Percy Report on Higher Technological Education, in which 'industry must look mainly to Universities for the training of scientists, ... it must look mainly to Technical Colleges for technical assistants and craftsmen', continued to divide and shape the higher and further educational landscape.[5] The early post-war expansion of student numbers had marginally favoured the sciences. The Barlow report (1946) had recommended a doubling of the annual output of science graduates, which was achieved in five years.[6] In 1950, of 64,000 students studying in English universities, 20 per cent studied pure science (in 1939 the proportion had been 15 per cent), although only 11 per cent studied applied science.[7] The increase in numbers of science graduates took place in the existing universities and the university colleges (granted charters in the 1940s and 1950s). Following a White Paper on technical education in 1956, ten technical colleges were upgraded to become colleges of advanced technology (CATs).[8] In 1965 the CATs were given university status, while a year later a further group of technical colleges were designated polytechnics.

The vocational specialism of the technical colleges was sharply contrasted to the expectations of breadth in university studies, including the sciences. 'In general the university courses should be more widely based on higher standards of fundamental science and contain a smaller element of training related to immediate or special work in industry' as rightly found in the technical colleges, noted the University Grants Committee in 1950.[9] Here, alongside a desire to bolster community through shared intellectual life was one early contextual prompt for perhaps the strongest, distinctive feature of science in the new universities: interdisciplinarity. Interdisciplinarity was already being urged by the University Grants Committee in the 1950s as a response to the threat of over-specialization.

In 1956, the University Grants Committee was pressing for a decision on the extent of expansion in the universities. Harold Macmillan, then chancellor of the Exchequer, endorsed the Treasury view that one of the critical issues was how many more scientists and engineers would be trained. Already the planned increase from 84,000 students (1956) to 106,000 (1964) would be two-thirds scientists and engineers, but now, with the UGC saying that 168,000 students should be anticipated by 1968, it was hoped 'that the proportion of the increase which would be Science and Technology could be raised above two-thirds'.[10] Why was this? Partly it was expected that industrial growth would be in areas that relied on such expertise, but the Treasury's framing of this expectation also tells us about the visions of the future it held:

> In an age of atomic development and automation much larger numbers of University trained graduates in science and technology must be provided ... [The] broad conclusion is that output of scientists and technologists ... should be nearly doubled by the latter part of the 1960s.

The Treasury hoped that this expansion would take place in the (cheaper-per-student) Colleges of Advanced Technology. But as Roger Quirk, under-secretary at the Office of the Lord President of the Council, observed: This might be 'naturally attractive to the Treasury ... but will it not perpetuate the subtle bias by which engineering studies suffer under a heavy disadvantage from the snob point of view?'[11] He advised the Lord President of the Council, the Marquess of Salisbury, to 'strongly back the expansion in "higher" science and technology', drawing his attention to a recent article, 'New minds for old', by C. P. Snow.[12] This is a very early, and immediately influential, sighting of Snow's 'two cultures' thesis and associated widely discussed arguments supporting the education system must support more scientists and engineers and fewer arts graduates. (Snow's Rede lectures would be delivered three years later and published as *The two cultures and the scientific revolution*, in 1959.)[13] We can read the two cultures sentiment at work in A. J. P. Taylor's highly inaccurate comment:

> Of the new universities, Norwich will teach scientific subjects. The others will have no science departments at all. The students will not even receive a rigorous discipline in their own subjects. They will receive a general 'arts' education, a smattering of culture which enables you to do the crosswords in The Times newspaper.
> Scientists are to be provided for also, but they will go to Colleges of Technology in great cities. No gracious living here. No Latin prayers, no array of silver. Just good training and hard work. The future of our country depends on these scientists. But they are second-class citizens, with no gifts in the choice of port. Our rulers, our masters, will be men who can chatter about ideas and who cannot mend a fuse.[14]

A further argument that expansion in science and technology must be made in universities, not merely technical colleges, was offered by Solly Zuckerman, chief scientific adviser and, as we shall see below, a key figure in the sciences at one of the new universities:

> There can be no question of the universities expanding their engineering departments without a concomitant expansion in the basic sciences. To lose sight of this fact would be equivalent to increasing the power of an aircraft, without any commensurate adjustments in fuel supplies or air-frame.[15]

In other words, university engineering research and teaching was necessary and should be expanded, but because basic science fed engineering as fuel feeds an engine, so university science should be proportionately expanded too.

Therefore, there were several problems identified: the snobbish attitude to technical education, the apparently desperate need for experts for future industry, the intimate relation of basic science to engineering applications and so on. A variety of solutions were offered. One was the wish for a British MIT, a high-status, innovative, technological university. Ortolano dates the origin of the suggestion to a moment 'over a brandy in Sicily in 1955, a fortnight after his resignation as Prime Minister', when Winston Churchill 'lamented not having done more to promote science, technology and engineering', stated his admiration for MIT and 'regretted that Britain had nothing comparable'. In fact, the aspiration for British MIT began earlier than this, in the late 1940s. Churchill himself observed during an address at the Mid-Century Convocation held at MIT in 1949 that 'we have suffered in Great Britain by the lack of colleges of University rank in which engineering and the allied subjects are taught'.[16] A network of influential scientists and industrialists, coordinated by Churchill's wartime secretary, John Colville, rapidly raised funds, but 'it soon became clear that it would be impossible to establish a university on the scale of MIT'. The outcome instead was Churchill College, Cambridge, approved by the university's Senate in 1957 and, with New Hall, the first new Cambridge colleges since the 1880s,[17] Snow was a founding Fellow. This was expansion of science higher education that began as a proposal for a radical break but fell back firmly into the established pattern. Even then the idea of a British MIT did not die; it was quickly revived in public discussions following the surprise of Sputnik, for example.[18]

The language of experiment features prominently in the Robbins report of 1963. 'The organisation of higher education must allow for free development of institutions', for example:

> Existing institutions must be free to experiment without predetermined limitations, except those necessary to safeguard their essential functions; and there must be freedom to experiment with new types of institution if experience shows the desirability of such experiments.[19]

Snow's arguments were influential behind-the-scenes on Robbins. (Snow had been invited to sit on the Robbins committee, but declined the formal role; nevertheless, he provided considerable 'informal guidance'.)[20] One experiment proposed was the development of five Special Institutions of Scientific and Technological Education and Research (SISTERs).[21] 'In making this proposal,' the authors noted, 'we have been much influenced by the fact that there is as yet in this country too little that compares for both scope and scale with the great institutions abroad that we visited, such as the Massachusetts Institute of Technology'.[22] Robbins suggested three SISTER-style bodies were already being developed as existing institutions expanded (Imperial College of Science and Technology), Manchester College of Science and Technology, and the Royal College of Science and Technology at Glasgow. Two more were required, one from an existing CAT (in addition to all CATs being granted university status), and the other a new foundation, possibly, it was suggested, in the northeast.[23]

Robbins's recommendations were largely accepted by the government. The new universities of this volume, however, were already underway. The point of dwelling on the wider debate about science, technology and higher education is that the parallel expansion of science at older universities, CATs and would-be SISTERs was an important context for the location and style of teaching and research at the new universities, not least because it removed any expectations that the new universities should be specifically science-focused or tied to an agenda of modernizing industrial centres. The new universities were 'to be located not in the major centres and industrial conurbations, but in medium-sized, even smaller towns, preferably of the non-industrial, county-town type, and preferably those with national historical associations'.[24] Colchester, for example, was selected, not Chelmsford, where radio and electronics giant Marconi had its base. (Warwick, in the car industry centre of Coventry, was an exception). Arguments, such as that of B. V. Bowden (physicist and computer scientist, Principal of Manchester College of Science and Technology, soon to be Minister of Education under Harold Wilson), Nevil Mott (physicist, Master of Gonville and Caius College, Cambridge), and journalist Thomas Pakenham (and son of a Labour minister), that Harwell was a better location for a new university, being at the centre of Britain's atomic energy research programme, could be dismissed. But this alternative was significant. Pakenham summarized:

> If there is to be much science at a university, the savings on halls of residence may be more than outweighed by the cost of duplicating scientific facilities. The waste of human resources seems even more disturbing than the waste of money. Scientists are attracted to the best equipment like wasps to jam.[25]

The post-war period marks the decisive rise of 'Big Science', the organization of scientific research in managed teams working with large-scale and expensive, centralized instruments and facilities.[26] Paradigmatic examples include particle physics and radio-astronomy. Choosing Brighton, York, Canterbury, Coventry, Lancaster, Norwich and Colchester meant choosing to be at a distance from Big Science.

A second, contemporary sense of Big Science was given by Derek de Solla Price: the phase entered by science when the exponential growth in knowledge created qualitatively new problems, not least ones stemming from social and economic constraints on further expansion.[27] This analysis chimed with that of the first vice-chancellor of the University of York, Eric James, a chemist turned pedagogy maven. In speeches made from the 1950s through the 1970s, James argued that the most important factor – the 'dominant feature of the intellectual history of our time' – shaping education, especially higher education, was the growth in knowledge, especially scientific knowledge.[28] Profound questions were therefore raised, said James, for the teaching of scientists and non-scientists to live, work and continue to learn in such a world. James was alarmed by the cost of Big Science:

> If you go to McGill [in Montreal] today you can see the apparatus with which Rutherford made some of the fundamental discoveries in atomic physics. Compare that ... with the cyclotron, or whatever its successor is called.[29]

Moreover, if a university, argued James, was to pursue Big Science, then it had to receive funds from, and open its account books to, the state. 'What will be the effect of this on that vague but precious concept, academic authority?', James asked,

Does the prospect for man include the inevitable and complete control of the richest sources of knowledge and the highest peaks of education by an omnipresent State? And if such massive financial dependence even does not stifle any but conformist views, will it not lead to a distortion of curricula towards the manifestly "useful" as that great educator, Hutchins, feared in his University of Utopia?[30]

Therefore, for the leader of at least one of the new universities, Lord James, vice-chancellor of York, science, through its over-specialization, and Big Science, through its inevitable ties to the state, were threats to the utopian project.

In 1965, the date by when all the new English universities had opened their doors, and the post-war demographic bulge of students was at its peak, concern mounted about the flow of candidates into science and technology. The Dainton report, commissioned in 1965 and published in 1968, identified a distinct 'swing from science to arts and to social sciences' as sixth-formers voted with their feet. This 'relative decline' was not only 'potentially harmful' (ultimately because we live in a science-inflected world), but had occurred, surprisingly, at a time of 'remarkable progress in man's knowledge of the physical world'.[31]

So, in this context of a long-standing debate about the expansion of universities, of expansion of science and technology at existing universities (including the upgraded CATs), and evidence of a late swing away from science and technology, what was actually happening at the new universities, in the words of the Dainton report 'to attract the imaginative mind and fire the curiosity of the young'?[32]

Drawing a new map of scientific learning

If the 'most central of the key words' of the founders of the seven new universities was 'community',[33] then it is also the case that the sciences were the most difficult to integrate into a community, given its need for special disciplinary technical training. So how did it fit? One trend in partial favour of integration was the movement towards interdisciplinary formations within the sciences in the twentieth century. Several combinations were prominent, not least in Britain, in the post-war period, notably molecular biology and radio astronomy. So there was a set of intra-communal linkages here. Arts–sciences or sciences–social sciences combinations were rarer, although important examples of both will be discussed below.

Many, but not all, of the new universities expressed the desire to reshape and reconnect disciplines by organizing the subjects under broad 'schools'. The approach was noted in the Robbins report, again using the language of 'experiment':

> We are arguing that there should also be experiments in new combinations of subjects which have recognisably organic connections: technology, for instance, with some social studies showing the more general implications of the technologist's profession; philosophy and mathematics with the history of science.

However, Robbins also wanted to 'offer a word of caution', warning that 'undergraduates should not be made the guinea-pigs of experiments with totally new subjects without textbooks or a commonly accepted core of methods of thought'.

At Sussex, students took multi-subject honours courses, designed with the intention of 'breaking free from "excessive specialization"'; within the sciences this meant a 'study of the social context and application of science' in addition to the traditional topics.[34] Sussex set up a School of European Studies, a School of English and American Studies, a School of African and Asian Studies, and a School of Social Studies. These were planned early. 'The position in relation to the scientists was at first much less certain,' wrote Asa Briggs; indeed 'it was not clearly envisaged in the early discussions, before the arrival of the first academics, that science departments would be abolished and replaced by Schools'.[35] Nevertheless in 1962 Sussex opened a School of Physical Sciences, in which an undergraduate could major in physics, chemistry, mathematics and philosophy. 'The basic idea', explained Roger Blin-Stoyle, the founding dean, was that 'physical scientists and mathematicians educated within the University' would be 'complete', meaning they would study courses that gave them both a fundamental 'understanding' (an interdisciplinary grasp of the essentials of theory and method) and a 'spirit of inquiry' (a 'militantly questioning, doubting and inquiring frame of mind', felt to be largely erased by the grind of school study).[36] The pattern of compulsory majors and minors would continue for the undergraduate's three years. But the special demands of specialism made by the sciences could not be ignored completely. 'It is quite clear,' wrote Blin-Stoyle, 'that in any university institution there must be provision for extreme specialist study ... which has to be absorbed before any research can be embarked'; a fourth, optional year of such 'extreme' study was put in place.

Briggs, in his essay 'Drawing a new map of learning', argued that the 'map is seldom re-drawn' and cited the 'biological studies', which 'produce exciting new research which rests on cross-boundary thinking', yet have been 'controlled by independent potentates' of biology, botany and zoology, as a prime case in point.[37] (This line did not originate with Briggs, rather it was an accusation made by an influential report on biology by the Royal Society in 1961, concerned that some vital interdisciplinary areas, notably microbiology and molecular biology, were 'falling between the gaps of the existing Research Council structure'.)[38] In 1965, BIOS, a School of Biological Sciences, began with the appointment of a founding dean, the evolutionary biologist John Maynard Smith, an 'inspired choice', not least because of his interdisciplinary instincts: a 'naturalist trained as a mathematician and engineer, "prepared to play away games with philosophers."'[39] The pattern of undergraduate biology teaching at university, dominated by the nineteenth-century divisions of zoology and botany, underpinned by comparative anatomy, 'was broken by the establishment of the new universities,' states Maynard Smith, 'and in particular by the Schools of Biological Sciences at York and at Sussex'.[40]

Another innovation was the Arts/Science scheme, 'whereby arts students were exposed to some science, and vice versa'.[41] Neither arts nor science students could 'escape the fact that they are living in a scientific and technological age,' wrote Blin-

Stoyle. Courses were offered on 'Science and industry', 'Science and government', 'The moral responsibility of the scientist' and 'The impact of science on contemporary thought'.[42] The notable science writer John Gribbin was one product of this scheme and remembered it distinctly as creating 'a sense of alarm at the prospect of being forced to think outside the box, and the rewards, with hindsight, far exceeded my expectations'.[43]

The arts schools later opted out of the Arts/Science scheme in the late 1960s, a decision John Maynard Smith called a 'bloody disgrace'.[44] 'There were also difficulties in retaining interdisciplinarity in science', noted one of the founding lecturers in science at Sussex, Brian Smith, a 'multidisciplinary course, "Structure and properties of matter", the equivalent of an arts contextual course, was developed in time for the arrival of the first science students in October 1962 and served the early science schools well', but 'it was not popular with later arrivals and was eventually abandoned' by the early 1980s.[45] Likewise the science school split, forming separate Biological Sciences, MAPS (Mathematics and Physical Sciences) and MOLS (School of Molecular Sciences).[46] MOLS has been described as both 'genuinely interdisciplinary' and 'essentially a department of chemistry'.[47]

The Sussex tutorial system of teaching was also a better fit for arts than for sciences. In arts pairs of students met the tutors for two-hour tutorials, discussing essays; the scientists 'have rather more tutorials a week, but attend them in groups of five, and because of the nature of their subject their sessions tend to be used to clear up difficulties and increase understanding rather than for the more discursive approach proper to an arts tutorial'.[48]

'The freedom to work along new lines and the power to plan new combinations of subjects' initially proved to be, as Briggs noted in 1964, 'great attractions in recruiting academics from universities where curricula can be changed only with the greatest difficulty'.[49] But, over time, the radical interdisciplinary ambitions, led by a vision of teaching and the unity of the student experience, would be undermined by the specific disciplinary requirements of advanced research. 'Research, fascinatingly largely absent from the 1950s and early 1960s national debate about the expansion of higher education, became an increasingly important and defining characteristic of the University', notes Fred Gray, introducing his history of Sussex, and as 'disciplinary knowledge expanded, so contextual study and the schools themselves were often threatened'.[50]

The discussion and implementation of the Sussex 'schools' system influenced the other new universities. At Norwich, Noel Annan, Provost of King's College, Cambridge, and member of the Academic Planning Board (APB), praised 'the Brighton system' for the arts.[51] For the physical sciences, 'Sensibly conservative' and 'not inclined to flights of fancy', the eminent chemist Christopher Ingold proposed a single 'Faculty of Science', teaching a 'mix of interdisciplinary and single-subject' topics.[52] More broadly, Zuckerman, also a member of the APB, proposed what turned out to be a particularly distinctive feature of UEA. 'If one had in mind to do something absolutely new and fresh in science,' Zuckerman wrote to Ingold, 'I am wondering whether Norwich could not embark in its Faculty of Science on a Division of Environmental Sciences – meteorology, oceanography, geology, conservation, etc. If it were, I am quite certain nobody could ever be able to say that scientists were trained in too narrow a way.'[53] A School of Environmental Sciences ('ENV') was established, with Keith Clayton as dean, as was a School of Biological Sciences ('BIO') and a School of Mathematics and Physics (but proposals for schools in medicine, Management and Technology and Molecular Technology, and Engineering failed). Lancaster University also established a School of Environmental Studies in 1963, recruiting Gordon Manley, the renowned climatologist, from Bedford College University of London.[54] The new schools at UEA and Lancaster were the first of their name in the UK, paving the way for others such as Sussex.[55] Lancaster aimed to teach arts students some science, and vice versa. The UGC stifled attempts to teach agriculture (seen as a declining specialty, despite Norwich's location in the arable centre of England) or engineering (too expensive). Unlike at Sussex, there was little sympathy for teaching arts to science students or vice versa.[56]

At UEA the Senior Professor in Biological Sciences, Thomas Bennet-Clark, was 'driven' to leave King's College London by the 'narrowness of the study of biology' there.[57] The proposal for a broad school was not only devised by Bennet-Clark and Zuckerman, but also recommended by the Royal Society around the same time.[58] Bennet-Clark's explanation to the Vice-Chancellor, Frank Thistlethwaite, and to the architect Denys Lasdun is one of the most powerful invocations of interdisciplinarity expressed in the context of the new universities:

> Frontiers between the various scientific disciplines are arbitrary. Two centuries ago they did not really exist and the different named sciences of the 19th and 20th centuries were all grouped as 'Natural Philosophy.'
>
> Advances really depend on breaking down the traditional specialized frontiers between mathematics, physics, chemistry, botany and zoology. This is being achieved in the School of Biological Sciences.[59]

Bennet-Clark conceded that such a bonfire of the disciplines was less welcome in the physical sciences. Nevertheless, while 'it would be less reasonable to compel mathematics, physics and chemistry to form one school, as there is no notable decline in "classical" aspects of these sciences', it was

> however desirable and reasonable to try and return as closely as possible to the earlier situation of unified "Natural Philosophy". This implies location of all sectors as close to each other as possible. Mixing of Natural Philosophers (scientists) with others will be effected in Common Rooms and in the crescents of living accommodation.

In this way, interdisciplinarity – or even a return to unified natural philosophy – would be written into the architecture of the new university.

Researchers in interdisciplinary areas (biophysics, biochemistry, genetics, marine biology) were soon at work in well-provisioned laboratories (including a £156,000 Philips EM200 electron microscope), part of Lasdun's 'teaching wall', the central building spine of UEA. The Agricultural Research Council's Food Research Institute relocated nearby, as did the John Innes Institute at a slightly further distance. Finally, the Lowestoft Fisheries Laboratory was incorporated. Altogether this made for a distinct and 'powerful network of biological research all of whose components were particularly appropriate for the East Anglian region'.[60]

But interdisciplinarity at UEA could be, and was, challenged. Again we see the influence, this time negative, of the experience at Sussex. Alan Katritzky, a 'breathtakingly young [31], abrasively dynamic' chemist 'with the firmest ideas', was dead set against subsuming chemistry under a broader school of physical sciences.[61] As the first professor, he lobbied against Ingold's influence, mobilizing support from powerful outsiders, including John Cockcroft, Robert Robinson and Alexander Todd. At the critical meeting held at Senate House, London, on 25 August 1962, Katritzky won the argument, not least because he could cite the experience of the Professor of chemistry at Sussex, Colin Eaborn, that 'mistakes made at Sussex [must not be] repeated at East Anglia … an independent School of Chemistry was of primary importance'.[62] At an even later foundation, Warwick, the interdisciplinarity expressed in the schools structure of Sussex was rejected. Allen Griffiths, the founding professor of philosophy at Warwick, later declared: 'I wouldn't have wanted to go to Sussex because I wouldn't be starting my own department. I would have been fitting into an academic and intellectual strait-jacket.'[63]

Such tensions were largely disciplinary ones, expressed internally within the university. External competition could also shape the new map of scientific and technological learning. At UEA the presence of local technical colleges, such as Norwich Technical College and King's Lynn Technical College, was a factor in early discussions, while at Warwick there were concerns that there might be conflicts with the local technical college if the university also chose a 'similar technological bias'.[64] Warwick's novelty was to be found in such entities as a graduate school in business administration, something Essex also attempted.[65] Another external force was the UGC. When an oversupply of biologists was forecast by officials in 1966, the UGC's Biological Sciences subcommittee took a hard look at existing plans for expansion and foundation, including the new universities.[66] This was despite a recognition that biology 'provided … as good a general education as the arts subjects which traditionally served this purpose', not least because the 'recent revolutionary change in emphasis away from "Natural History" [towards lab-based yet interdisciplinary study] … had greatly enhanced biology students' adaptability and usefulness'. So now, in review, the UGC's biological subcommittee found UEA biology to be 'disappointing' and advised that it 'should be encouraged to add to existing staff but not start any new developments'. At Lancaster (where interdisciplinarity was also encouraged) buildings were 'already too advanced to be halted'.[67] At Sussex the problem was 'overreach' and going 'too fast'. At Warwick the study of life sciences 'should be discouraged'. Likewise, at York, while an interdisciplinary school in biochemistry and microbial genetics was noted, animal physiology 'should be discouraged'. Furthermore, at Kent, even though the natural sciences building was scheduled for occupation in 1970–1, there was a call for an enquiry into the need for the department.[68] Some CATs (Bath, Chelsea, Salford, Surrey, with Brunel as an exception) also received similar criticism, whereas the older institutions (such as Imperial, Kings, UCL) were marked as actual or potential 'centres of excellence'. The UGC's subcommittee's view was forthright and carried a clear sense that its top-down encouragement and discouragement would have effect.

Sometimes the interdisciplinarity of science at the new universities was present, if not vaunted. The case of biology at York, for example, illustrates how interdisciplinarity could hide behind apparent disciplinary singularity, while also showing how founding professors could have considerable leeway within the broad frame set by the academic planning board. In 1961, two years before the university opened, the UGC wrote to say that 'biology was a suitable subject'; the Academic Planning Board (chaired by Robbins) then discussed what 'biology' might mean, but its main decision was to confirm Eric James's view that 'you appoint a man whose interests are wide'.[69] In 1963, James wrote to Lady Ogilvie, member of the APB, seeking approval for the appointment of Mark Williamson as professor of biology. He stressed the 'absolutely first-class work in various fields', or, in other words, his wide interests.[70] Once in post, Williamson built up a Department of Biology that might have had a singular, disciplinary title but in fact was an expression of modern, interdisciplinary biological sciences including, for example, biochemistry, as well as a wide, non-traditional curriculum amongst which there was genetics, the diversity of living organisms and ancillary chemistry (Part One), and biochemistry, microbiology, biochemical genetics, plant and animal physiology, embryology, ecology, population genetics, evolution, behaviour, statistics, taxonomy, comparative anatomy, palaeontology, histology and electron microscopy (Part Two). Individual elements of this curriculum were traditional, but others were not, and nor was the interdisciplinary whole.[71] Williamson also secured top of the range, expensive equipment, such as an electron microscope and a Spinco-E ultra-centrifuge, 'critical' to the new department's 'research credibility', according to Rachel Leech, appointed by Williamson in 1966.[72] It is also clear that the sciences at York, which could not be housed in the colleges, strained the desired ideal of 'community' as envisaged by the vice-chancellor: 'Every teacher in *other than laboratory-based subjects* has a room in one or other of the colleges to which he is attached.'[73]

The organization of the sciences at the new universities therefore did indeed reveal 'freedom to work along new lines and the power to plan new combinations of subjects', and did so with imagination. But the early starters had more freedom to do so, while later developments were constrained, and existing traditional disciplinary forces could still overcome interdisciplinary best intentions. Academic Planning Boards, moreover, provided a broad framework for development, but individual professors still had agency to develop their subjects, albeit in often interdisciplinary ways. I will now turn to examine three specific case studies in more detail.

Sussex: Cognitive sciences and the Science Policy Research Unit

At Sussex the case of cognitive sciences shows how interdisciplinarity could overspill from the already interdisciplinary schools, while a second new development, the Science Policy Research Unit (SPRU), is another example of a distinctive and novel combination of approaches. Maynard Smith's plan for undergraduate biology teaching integrated study in molecular biology ('that hybrid between genetics and biochemistry that dominates so much biology today'), development, brain and behaviour.[74] Within this plan were soon seven majors. Six were – in order of increasing interdisciplinarity –

biology, biochemistry, neurobiology, geography (a combination of biology courses with mostly physical geography), human sciences and biology with European studies.

The seventh, though, shows how new interdisciplinary research could be generated from the school system. In 1965, Stuart Sutherland was appointed as professor of psychology and began a major in experimental psychology (neurobiology also overlapped with this development). But the mid-1960s were when the provision of electronic stored-program computers to UK universities reached full throttle. The Flowers report of 1966 confirmed that computers would be essential to research (and even teaching) and must be provisioned generously, while, also from the mid-1960s, the technology available diversified from the giant mainframes of the 1950s to smaller, cheaper, more networked machines.[75] There was therefore space for imaginative exploration of new possibilities. Sutherland brought from a couple of spells as a visiting professor at MIT a strong and influential 'belief that artificial intelligence and computational modelling provided new and powerful ways to tackle the problems of cognitive psychology'.[76] In the same year Margaret Boden was appointed as a lecturer in philosophy, and swiftly became a leading interpreter of AI. Sutherland first planned a 'Brain Institute' and then a new 'School of Cognitive Studies', encompassing 'artificial intelligence and computer science, psychology, linguistics, logic and possibly, on an experimental basis, English language and literature'.[77] This plan became, not a school but a programme in 'cognitive studies' in 1972. Based fundamentally on the computer-as-mind and mind-as-computer models, this project survived the mid-1970s dip in funding support for artificial intelligence and flourished in the 1980s.

Discussions about another interdisciplinary endeavour, this time at the intersection of the sciences, social sciences and humanities, took place between 1961 and 1964 between, amongst others, Stephen Toulmin (a professor of philosophy at Leeds) and Asa Briggs.[78] When Toulmin decided that levels of funding were inadequate, Briggs appointed Christopher Freeman to lead a new outfit. The Science Policy Research Unit (SPRU) was entrepreneurial as well as interdisciplinary: nearly all of the staff and projects were supported by contracts, with funds coming from governments (UK and foreign), research councils, foundations and business; the 'balance between longer-term, more fundamental research and shorter-term contract research', noted Freeman, was 'one of most difficult issues'.[79] Indeed, SPRU here was wrestling with relationships with sponsors that would become more widespread, as the encouragement of user-relevant, applicable interdisciplinary research became a common concern for UK universities.

East Anglia: Environmental interdisciplinarity

UEA's Climatic Research Unit is a good example of how the institution created space for influential interdisciplinary research even as organizing and funding teaching was the main concern in the early years. Its origins were quite contingent. In 1966, Hubert Lamb, who had been employed for thirty years at the Meteorological

Office at Bracknell, was browsing the brochure that came with his daughter's UCCA application when he noticed a proposal to start a school of environmental sciences.[80] Intrigued, Lamb wrote wondering if there might be space for his work on climatic forecasting (as distinguished from short-term weather forecasting). 'My research on what affects climatic development and trend has started to lay the foundations for climatic forecasting ... the endeavour must involve coordinating research in different disciplines ranging from history, archaeology and botany to satellite meteorology and solar physics.' With opportunities at the Meteorological Office restricted (it had a service focus on short-term weather forecasting), Lamb saw expansion of such highly interdisciplinary research as only likely in an academic context. The university responded by saying that his proposal would be put to the dean of the new school, once one was appointed.[81] Lamb in turn suggested his name as a possible candidate.[82]

After Keith Clayton was appointed, Lamb eventually returned with his first proposal renewed. With the Met Office 'basically unwilling to divert staff to these ultra-long-range problems (as they appear to most meteorologists)', and hinting that other universities were interested, Lamb told Clayton that there was 'now a new element in the situation in that I now have a firm offer of financial backing from Shell to contribute towards the setting up the proposed Centre'.[83] Clayton this time responded with interest: 'I would certainly agree with you about the timeliness of the development, and it could certainly fit in very well with some of the inter-disciplinary studies that we are trying to foster here.'[84]

Much of UEA's financial resources were being devoted to developing teaching, but since the 'inter-disciplinary scope of Mr Lamb's research would fit particularly well into our range of interests', there was a determination to move forward.[85] There was nevertheless a protracted period during which Lamb and his UEA supporters sought promises of funding, largely from industry. Shell offered £10,000, less than expected (£80,000 was needed) but partly so because it thought that such was the 'usefulness' of the Centre other stakeholders would be willing to contribute.[86] Burmah Oil, Esso, BOAC, Unilever and Birds Eye, and organizations such as NATO, all refused. Behind the scenes, Zuckerman worked his contacts in academia and government.[87] Eventually, with funds primarily from Shell, but supplemented by BP, the Central Electricity Generating Board, the Electricity Council and potentially from the Nuffield Foundation, Lamb's Climatic Research Unit opened in 1971. Lamb retired in 1977, but before then his unit's work had established precisely the broad-ranging, statistical understanding of past climate variability that enabled it to be a counter-weight to the Met Office's scepticism towards climate change.[88] In particular, Lamb's arguments were part of the mix of climatological theories and evidence that informed central government in the mid-1970s that climate change was a genuine phenomenon of political significance. (Interestingly, John Ashworth, the chief scientist, Central Policy Review Staff, who reviewed this evidence, had been recruited to the government's think tank from another of the new universities, Essex.) In the 1990s and 2000s, UEA continued to be a centre of the worldwide expert investigation of climate change.

Warwick: Catastrophe and sudden change

Like almost all his fellow founding professors, Erik Christopher Zeeman, known as Chris Zeeman, who arrived in 1964, wanted to establish a research-oriented, largely single-disciplinary subject department.[89] Zeeman contributed to the course 'Enquiry and change', which all Warwick undergraduates took. But this was not a display of interdisciplinarity; rather, it was a beauty contest of disciplinary competition ('it was about each professor trying to convince a general audience that his subject was an exciting thing to do', recalled one seminar leader).[90] Zeeman built up critical masses (this deliberate, focused growth of nuclei of scholars approach was influential) of staff in the pure mathematics specialties of topology, algebra and analysis. He also established in 1965 a Mathematical Research Centre to house symposia held by world-leading specialists, not least René Thom in catastrophe theory, a subject Zeeman would significantly further develop at Warwick. Thom had applied catastrophe theory (which in its essence is a specialty in pure mathematics) to embryology. But Zeeman's extension of the subject was distinctively an interdisciplinary conversation. A flavour of this can be seen in his reminiscence:

> Amongst my applications of catastrophe theory I particularly liked buckling, capsizing, embryology, evolution, psychology, anorexia, animal behaviour, ideologies, committee behaviour, economics and drama.[91]

In language strikingly similar to Bennet-Clark's language of a rebirth of older, more connected scholarship, Zeeman also spoke of 'natural philosophy':

> Ever since the disappearance of natural philosophy from our universities and the fragmentation of mathematicians into pure and applied, our canvases have steadily been growing smaller and smaller. At least catastrophe theory marks a revival of natural philosophy, to be enjoyed once again for a while at any rate.[92]

Indeed this extraordinary range can be found in Zeeman's *Catastrophe Theory: selected papers, 1972–77*. What is striking is how this research at one of the new universities drew on and rationally analysed the sudden, seemingly irrational problems of the 1960s and 1970s, from prison riots and stock exchange instabilities to anorexia. All this analysis took place in a university which went through its own abrupt spasm of protest, as E. P. Thompson described in *Warwick University Ltd* (1971). The resonances between seemingly arcane mathematics and the era of radical change partly explain why catastrophe theory reached a surprisingly popular audience (in many ways it was a precursor of late 1980s popular interest in 'chaos' theory). Zeeman himself not only became a public scientist (he presented, for example, the Royal Institution's Christmas Lectures in 1978), but also encouraged people to 'make and play with' their own 'catastrophe machine', a device of '2 elastic bands, 2 drawing pins, half a matchstick, a piece of cardboard and a piece of wood', thereby showing 'how continuous forces can cause catastrophic jumps'.[93] This toy is the materialization of 1970s hopes and fears: the fears of abrupt change, the hope for rational explanation and that they might be achievable through self-experimentation.

The long 1960s

Elsewhere I have asked whether the category of the 'long 1960s', the period of social and cultural change from roughly the mid-1950s to the mid-1970s, has any use for the historian of science.[94] My answer was that amongst the continuities and noise, a common pattern, formed of the interference of three waves, could sometimes be usefully discerned: the proliferation of experts produced by the expanded post-war educational systems, powerful social movements that could stage conflicts, not least between experts, and, thirdly, an orientation towards the self, in diverse ways. Since the new universities of the UK were created during the long 1960s, it is worth asking whether these waves marked the sciences found within them.

In the United States, student-led social movements led to protests on campuses in which Cold War military research and development were a focus for dissent. At Princeton military R&D research and development was fiercely debated in 1967. At Stanford the much more extensive secret contracts and classified research at the Applied Electronics Laboratory and the nearby Stanford Research Institute were protested by students and faculty in 1966. Additionally, at MIT, which received more defence research and development grants than any other university, the Lincoln and Instrumentation laboratories, specializing in electronics and missile guidance technologies respectively, were the target of a strike intended to 'provoke "a public discussion of problems and dangers related to the present role of science and technology in the life of our nation."'[95] In the UK, while defence research contracts were placed with university departments, they were at a much lower level, and, despite fraught public debate about the implications of Sputnik for education (mostly school rather than higher education), post-war British universities are not best described as 'cold war campuses', except as part of the broader context.

Nevertheless, at the new UK universities – as Harold Perkin put it, a 'laboratory in staff student relations' – unrest was also evident.[96] The disruption at Warwick has already been noted. At Essex, the most affected by the events of 1968, it was indeed *science* that sparked the major incident. Dr Thomas Inch, a scientist from Porton Down, the UK government's chemical and biological warfare laboratory, arrived to give a talk at the Chemistry department on 4 May 1968. A witness recalled how

> a young woman with long black hair, stood up and started reading the indictment in a strong, loud voice. She described in some detail the activities of Porton Down; in particular, the development of CS gas which was being used in Vietnam and on the streets of Paris.[97]

It is not known who this student was, or whether she was a scientist or arts student. At Essex, all arts students were taught some science in the first year, as well as a 'paper in social arithmetic or elementary statistics'.[98] Inch received for his pains a tin of Colman mustard powder emptied on his head, accompanied with the cry of 'Ban mustard gas. Ban mustard gas'.

Conclusion

This chapter has reviewed some of the features of the sciences in the new universities of the 1950s and 1960s. The upgrading of the colleges of advanced technology, as well as earlier expansion at older universities and new establishments such as Churchill College, meant that there were lower expectations, or demands, that the new universities must focus on science and technology than might have been expected. The 1950s and 1960s, after all, were the highpoint of technocratic modernism. Nevertheless, the largely free hand given to the designers of curricula, research programmes and laboratories – especially the earlier of the new foundations (Sussex and East Anglia), where all was fresh – meant that interesting experiments in the provision of university science could be and were conducted. In particular, I have argued that interdisciplinarity, while not without problems for the consciously disciplined sciences, was a hallmark. Partly, interdisciplinarity was a reflection of longer, deeper trends (evident throughout the twentieth century), but also it could be specifically trialled in new forms at the new universities. Whilst interdisciplinarity in the sciences was not uniform, and partially retracted over time, it stands as the most significant characteristic of science at the new universities.

Notes

1   Stefan Muthesius, *The Post-War University: Utopianist Campus and College* (London and New Haven: Yale University Press, 2000), 5, 101.
2   Ibid., 3.
3   *Daily Mail*, 24 May 1962, quoted in *Making the Future: A History of the University of Sussex*, ed. Fred Gray (Falmer: University of Sussex, 2011), 11.
4   Roy Lowe, *Education in the Post-War Years: A Social History* (London: Routledge, 1988), 159.
5   Quoted in ibid., 58.
6   Michael Sanderson, *The Universities and British Industry, 1850–1970* (London: Routledge and Kegan Paul, 1972), 351.
7   Lowe, *Education in the Post-War Years*, 60–1; W. A. C. Stewart, *Higher Education in Post-War Britain* (London: Macmillan, 1989), 61.
8   Stewart, *Higher Education in Post-War Britain*, 83–4. Tyrell Burgess and John Pratt, *Policy and Practice: The Colleges of Advanced Technology* (London: Allen Lane, 1970). Peter Venables, *Higher Education Developments: The Technical Universities, 1956–76* (London: Faber and Faber, 1978).
9   Quoted in Lowe, *Education in the Post-War Years*, 159.
10 'The Universities Building Programme', enclosed in Macmillan to Lord Salisbury, 23 August 1956, TNA CAB 124/2040.

11 Quirk, 'Universities Building Programme', 28 August 1956, TNA CAB 124/2040.
12 Quirk to Lord Salisbury, 'The University Building Programme', 3 October 1956, TNA CAB 124/2040; C. P. Snow, 'New Minds for the New World', *New Statesman*, 8 September 1956, 279–82.
13 Guy Ortolano, *The Two Cultures Controversy: Science, Literature and Cultural Politics in Post-War Britain* (Cambridge: Cambridge University Press, 2009).
14 *Sunday Express*, 7 October 1962, 16.
15 Zuckerman to E. H. Boothroyd, 25 October 1956, TNA CAB 124/2040.
16 Ortolano, *The Two Cultures Controversy*, 110; http://libraries.mit.edu/archives/exhibits/midcentury/mid-cent-churchill.html (accessed 15 February 2019). For the proposals for a Technological University, see Quirk to F. R. P. Vintner, 1 October 1957, TNA CAB 124/2040. Michael Shattock, 'Introduction', in *Making a University: A Celebration of Warwick's First 25 Years*, ed. Shattock (Coventry: University of Warwick, 1991), 9.
17 James Jackson Walsh, 'Postgraduate Technological Education in Britain: Events Leading to the Establishment of Churchill College, Cambridge, 1950–1958', *Minerva*, 36 (1998), 147–77.
18 *Financial Times*, 10 October 1957, 6.
19 Committee on Higher Education, *Higher Education. Report of the Committee Appointed by the Prime Minister under the Chairmanship of Lord Robbins 1961–63* (London: HMSO, 1963) (hereafter *Robbins report*), 9.
20 Ortolano, *Two Cultures Controversy*, 107.
21 *Robbins report*, 128.
22 Robbins and his team also visited ETH Zurich and Delft Institute of Technology (now Delft Technical University).
23 *Robbins report*, 129. 'Tees-side's Claim To Have a SISTER', *New Scientist*, 19 December 1963, 718.
24 Muthesius, *Postwar University*, 96–7.
25 *Observer*, 21 January 1962, 20.
26 James H. Capshew and Karen A. Rader, 'Big Science: Price to the Present', *Osiris*, 7 (1992), 2–25; Peter Galison and Bruce Hevly (ed.) *Big Science: The Growth of Large-Scale Research* (Stanford: Stanford University Press, 1992).
27 Derek de Solla Price, *Little Science, Big Science* (New York: Columbia University Press, 1963).
28 'Democracy and Authority in Education', undated speech (*c.* 1969). Borthwick Institute for Archives (hereafter BIAUoY), University of York, JAM/2/1/1.
29 James, 'The Universities and the Prospect for Man', undated speech (*c.* 1975). BIAUoY JAM/2/1/2.
30 James, ibid.
31 Council for Scientific Policy, *Enquiry into the Flow of Candidates in Science and Technology into Higher Education, Presented to Parliament by the Secretary of State for Education and Science by Command of Her Majesty* (hereafter *Dainton report*), Cmnd. 3541 (London: HMSO, 1968). See also Richard Layard, John King and Claus Moser, *The Impact of Robbins* (Harmondsworth: Penguin, 1969).
32 *Dainton report*, 35.
33 Muthesius, *Post-War University*, 101.
34 Asa Briggs, 'Drawing a New Map of Learning', in *The Idea of a New University: An Experiment in Sussex*, ed. David Daiches (London: André Deutsch, 1964), 61.
35 Ibid., 61–2.

36 Roger Blin-Stoyle, 'The School of Physical Sciences', in ibid., 121–3.
37 Briggs, 'Drawing a New Map of Learning', 72.
38 Peter Collins, *The Royal Society and the Promotion of Science since 1960* (Cambridge: Cambridge University Press, 2016), 46.
39 David Harper, 'The JMS Legacy', in *Making the Future*, ed. Gray, 183.
40 John Maynard Smith, 'Integrating the Biological Sciences', in *The Sussex Opportunity: New University and the Future*, ed. Roger Blin-Stoyle (Brighton: Harvester Press, 1986), 126.
41 Brian Smith, 'Life through a Plate-Glass Window …', *Times Higher Education Supplement* (10 August 2001), 22.
42 Blin-Stoyle, in *Idea of a New University*, ed. Daiches, 27.
43 Gribbin, quoted in Jonathan Bacon and Benedict du Boulay, 'Overview: Doing Science', in *Making the Future*, ed. Gray, 164.
44 John Maynard Smith quoted in ibid., 165.
45 Smith, 'Life through a Plate-Glass Window', 22.
46 Colin Eaborn and Ken Smith, 'The Heart of the Matter: Physical Sciences at Sussex', in *The Sussex Opportunity*, ed. Blin-Stoyle, 111.
47 Bacon and du Boulay, 'Overview: Doing Science', 165. cf. John Murrell, 'Life in MOLS', in *Making the Future*, ed. Gray, 208.
48 Walter James, 'Purposeful Design: Britain's New Universities', *Texas Graduate Journal*, 7 (1964), 208.
49 Briggs, 'Drawing a New Map of Learning', 60.
50 Fred Gray, 'Introduction', in *Making the Future*, ed. Gray, 12. Cf. Smith, 'Life through a Plate-Glass Window', 22.
51 Michael Sanderson, *The History of the University of East Anglia* (London: Hambledon and London Books, 2002), 60.
52 Ibid., 63.
53 Quoted in ibid., 64.
54 Marion McClintock, *University of Lancaster: Quest for Innovation. A History of the First Ten Years, 1964–1974* (Lancaster: University of Lancaster, 1974), 136.
55 Eaborn and Smith, 'The Heart of the Matter: Physical Sciences at Sussex', 113.
56 McClintock, *University of Lancaster*, 131; Sanderson, *History of the University of East Anglia*, 66, 103.
57 Sanderson, *History of the University of East Anglia*, 99.
58 Biological Sub-Committee (22 March 1966), TNA UGC 8/85.
59 Bennet-Clark to Frank Thistlethwaite and Denys Lasdun, 6 January 1963, UEA Archives. FT/C3D1. Bennet-Clark's text was reproduced in Lasdun's *UEA Development Plan*.
60 Sanderson, *History of the University of East Anglia*, 103.
61 Ibid., 105.
62 Katritzky to Thistlethwaite (September 1962), quoted in ibid., 107.
63 Quoted in Robert G. Burgess, 'Working and Researching at the Limits of Knowledge', in *Making a University*, ed. Shattock, 95.
64 Minutes, E (61) (8 May 1961), TNA ED 188/74. Cf. Shattock, 'Introduction', in *Making a University*, ed. Shattock, 21.
65 James, 'Purposeful Design: Britain's New Universities', 207.
66 Minutes, UGC Biological Sub-Committee (16 November 1966), TNA UGC 8/85.
67 McClintock, *University of Lancaster*, 134.
68 Minutes, UGC Biological Sub-Committee (22 March 1966), TNA UGC 8/85.

69 'Record of Meeting of the Academic Planning Board of the University of York Held at the Offices of the UGC at 10.30 on 23 March 1961', BIAUoY UOY/F/APB/1/2.
70 James to Ogilvie, 8 August 1963, BIAUoY UOY/F/APB/2/1.
71 *A History of the First Fifty Years of Biology at York*, ed. Mark Williamson and David White (York: Department of Biology, 2013), 9-13.
72 Ibid., 17.
73 James, 'The University of York', undated manuscript. BIAUoY JAM/2/1/2 (my emphasis).
74 Maynard Smith, 'Integrating the Biological Sciences', 128-9.
75 F. Verdon and M. Wells, 'Computing in British Universities: The Computer Board 1966-1991', *The Computer Journal*, 38 (1996), 822-30; Jon Agar, 'The Provision of Digital Computers to British Universities Up to the Flowers Report (1966)', ibid., 39 (1996), 630-42.
76 Nicholas Mackintosh, 'Obituary: Professor Stuart Sutherland', *Independent*, 18 November 1998, 6.
77 Alistair Chalmers, 'Thinking about Thought: The Cognitive Sciences at Sussex', in *The Sussex Opportunity*, ed. Blin-Stoyle, 178.
78 Christopher Freeman, 'Policy Research for Science and Technology', in ibid., 191.
79 C. H. Geoffrey Oldham, 'Twenty-Five Years of Science and Technology Policy Research at the Science Policy Research Unit', in Hiroshi Inose, Masahiro Kawasaki and Funio Kodama (ed.), *Science and Technology Policy Research. 'What Should Be Done?'. 'What Can Be Done?'* (Tokyo: Mita Press, 1991), 31-41. Freeman, 'Policy Research for Science and Technology', 200.
80 Lamb to Secretary to the University of East Anglia, 5 October 1966, UEA Archives, UEA/Jones/40.
81 Osborne to Lamb, 24 October 1966, ibid.
82 Lamb to Osborne, 29 December 1966, ibid.
83 Lamb to Clayton, 26 March 1969, ibid.
84 Clayton to Lamb, 2 April 1969, ibid.
85 R. G. Chalkey (Shell), 22 May 1969, ibid.
86 Chalkey to Lamb, 31 October 1969, ibid.
87 Zuckerman to Clayton, 31 May 1970, ibid.; 'Industrial Sources of Research Funds', undated (1970), ibid.
88 Jon Agar, '"Future Forecast – Changeable and Probably Getting Worse": The UK Government's Early Response to Anthropogenic Climate Change', *Twentieth Century British History*, 26 (2015), 602-28.
89 Burgess, 'Working and Researching at the Limits of Knowledge', 102-3.
90 Rolph Schwarzenberger, quoted in ibid., 98.
91 J. J. O'Connor and E. F. Robertson, 'Erik Christopher Zeeman', http://www-history.mcs.st-and.ac.uk/Biographies/Zeeman.html (accessed 15 February 2019).
92 E. C. Zeeman, 'Catastrophe Theory: A Reply to Thom' (1974), in Zeeman, *Catastrophe Theory: Selected Papers 1972-77* (Reading, MA: Addison-Wesley, 1977), 622.
93 E. C. Zeeman, 'Catastrophe Theory: Draft for a Scientific American Article', in ibid., 8.
94 Jon Agar, 'What Happened in the Sixties?', *British Journal of the History of Science*, 41 (2008), 567-600; cf. *Groovy Science: Science, Technology, and American Counterculture*, ed. W. Patrick McCray and David Kaiser (Chicago: University of Chicago Press, 2016).
95 Stuart W. Leslie, *The Military–Industrial–Academic Complex at MIT and Stanford* (New York: Columbia University Press, 1993), 233.

96 Quoted in Muthesius, *Postwar University*, 179.
97 Chris Ratcliffe, 'May Days at Essex', http://www.essex68.org.uk/may68-e.html (accessed 15 February 2019); cf. 'A Very Essex Protest', *Scenes from Student Life*, BBC Radio 4 (28 April 2016), http://www.bbc.co.uk/programmes/b07881l1 (accessed 15 February 2019).
98 James, 'Purposeful Design: Britain's New Universities', 207.

8

The new and the old: The University of Kent at Canterbury

by Krishan Kumar

I

It all began with a name. A new university was to be founded in the county of Kent. Where to put it? In the end, the overwhelming preference was for the ancient Cathedral City of Canterbury. But that choice of site was by no means uncontested. There was no lack of alternatives. Ashford, Dover and Folkestone all put in bids. The Isle of Thanet in East Kent was for some time a strong contender, and its large seaside town of Margate, with its many holiday homes, undoubtedly seemed a better prospect than Canterbury for the provision of student accommodation, in the early years at least. Ramsgate too, with a large disused airport, was another strongly urged site in Thanet.

But as far back as 1947, Canterbury had been proposed as the site of a new university in the county. That proposal got nowhere, but the idea of Canterbury being the site of a new university had caught the imagination of many in the county. Thanet lacked the cultural significance and the wealth of historical associations of the city of Canterbury. Moreover the 'University of Thanet', or the 'University of Margate' or Ramsgate, did not have quite the same ring to it as the University of Kent or the University of Canterbury. That at least was the strong opinion of the two bodies that brought the University into being, the steering committee of Kent County Council and the group of the great and good in the county that came together as 'Sponsors of the University of Kent' – most prominent among them being its Chairman, Lord Cornwallis, Lord Lieutenant of Kent and former chairman of Kent County Council.[1]

I joined the University of Kent, as a lecturer in Sociology, in 1967, and left it in 1996. Much of what I have to say in this chapter is based on my own observation and experience of developments during this period, eked out with regular visits thereafter, especially when I was a visiting professor there (2006–11). I have found helpful Graham Martin, *From Vision to Reality: The Making of the University of Kent* (Canterbury: The University of Kent at Canterbury, 1990), for the early history of the University. Thanks also, for the provision of much useful material, to the University Archivist Ann MacDonald, and to Triona Fitton, whose *Hidden History: Philanthropy at the University of Kent* (Canterbury: The University of Kent, 2015) contains many unexpected gems. Conversations with Jan Pahl and Colin Seymour-Ure, old friends and veterans of the university, have not only been highly enjoyable but immensely useful in putting together this picture of Kent.

Canterbury having been chosen as the site, what to call the new foundation? That proved trickier. The newly founded universities of Sussex and Essex suggested one model, the 'county' designation, and that found favour with some of the Kent county representatives. But the predominant view among the Sponsors – the most influential body – was that it should be called the University of Canterbury in the time-honoured European tradition of naming a university after the town in which it was situated, as with the universities of Oxford and Cambridge, and, somewhat more recently, the new University of York, also in a venerable Cathedral City. Sussex and Essex might be acceptable in some places (for what were the alternatives? Falmer? Wivenhoe?), but they were unnecessary novelties in Kent when one had so obvious a choice as Canterbury.

So the University of Canterbury it was to be. It was therefore something of a shock when news came from the University of Canterbury in Christchurch, New Zealand that it took a dim view of the new English university taking the same name. The vice-chancellor pointed out to the University Grants Committee that having two universities of Canterbury would cause confusion at meetings of Commonwealth universities. There was, in fact, a precedent for one of the new English universities taking the same civic name as that of a Commonwealth university – York, England and York, Canada.[2] On the whole the 'Sponsors' felt they should not insist, although some were displeased enough to resign.[3] A compromise name was hastily devised. The new university would now be the University of Kent *at* Canterbury, thus appeasing the New Zealanders while at the same time retaining the favoured city of Canterbury in the title. It was a somewhat cumbersome title, without English precedent – though plenty in the United States – but the abbreviation UKC made it easier to say and soon became the familiar form of reference.

In the event, it turned out to be a happy choice for a number of reasons. It made it clear that the university did indeed represent the county as a whole, not simply the people of Canterbury. But more importantly it symbolized the marriage of tradition and modernity that came to be the hallmark of the new university. Kent, the county, close to the metropolis of London and with strong ties to it, stood for the modern, the forward-looking aspect of the new university. Canterbury, the city, looked back: to Roman times, when St Augustine founded the first Christian church in the country there and made Canterbury the head of the English church; to the glories of the middle ages, when Canterbury flourished as the site of the shrine of Thomas Becket, one of the greatest pilgrimage destinations of medieval times (Canterbury was fourth in terms of popularity after Jerusalem, Rome and Santiago de Compostela). Then there was of course Chaucer's *Canterbury Tales*, which furnished the university with its banner and its motto: 'and gladly wolde he lerne, and gladly teche'. It would have been unthinkable for the university to ignore the ancient heritage deriving from the city; the problem rather might be how to throw it off, how not to find itself sucked back into a past that had such agreeable associations. That it did not do so had a lot to do with the make-up of the chief founding members of the university.

For there was – perhaps by chance rather than design – a mix of the old and new not just in the initial conception of the university but in the earliest of its principal Officers. The first, and the university's longest-serving, vice-chancellor was Geoffrey Templeman (1962–80), formerly Registrar at the University of Birmingham.[4] He brought with him his Assistant Registrar, Eric Fox, and made him Registrar of the new university. Also from the Birmingham Registry came the first Finance Officer, Dennis Linfoot (later succeeding Fox as Registrar). All served long terms, of nearly twenty years each, giving the new university an unusual degree of continuity. Here were men well versed in the running of a modern university, one of the new civic universities that originated in the late nineteenth century.

But Templeman had started life as a medieval historian, and, it turned out, had a profound belief in the virtues of the ancient universities' collegiate system. There was a similar feeling for old forms in David Edwards, the Buildings Surveyor who had responsibility for the initial layout of the buildings and the physical development of the site. Edwards had worked both at the University of Oxford and the City of London, and had much experience with ancient college buildings and old City structures. Most important of all perhaps was the fact that the external chairman of the Academic Planning Board (APB), which was responsible for curriculum structure and academic organization, was Sir Derman Christopherson, vice-chancellor of the University of Durham and on the APB of the University of Sussex.[5] Christopherson played a key role in the design of the university, staying on after the Planning Board was wound up to be chairman of its successor, the Academic Advisory Committee. In all he served in these pivotal positions from 1962 to 1971. Such an unusually long association meant that he had ample opportunity to shape the university. In this the example of Durham, as a collegiate university modelled to some extent on Oxford and Cambridge, clearly influenced much of his advocacy. Oxford, where he had been an undergraduate, and Cambridge, where he had been a lecturer and where he was later to become Master of Magdalene College, also added their considerable weight to his thinking. Yet he was by no means a dyed-in-the-wool traditionalist, partly due to his having been an innovative mechanical engineer in his days of active research. During the war years he had worked with Solly Zuckerman, the government's principal scientific adviser on military planning. At Durham he had overseen a substantial programme of reform of the colleges, and he was enthusiastic about many of the new developments that had taken place at Sussex. He was, therefore, a fitting proponent of the fusion of old and new.

The old certainly got its due, perhaps more than it deserved. Oxbridge by way of Durham: for many of the new professors and lecturers that took up their positions in the late 1960s and early 1970s, this seemed to express quite accurately the ethos of their new place of work. Most of them found it highly congenial. That was hardly surprising, since so many of them had been educated or taught in those institutions. From Durham came Alec Whitehouse, professor of theology and first Master of the first college, Eliot College; Graham Martin, first dean of the natural sciences and later deputy vice-chancellor; Reginald Foakes, professor of English and the second dean of the Faculty of humanities; Stephen Darlow, the first University librarian.[6] Taken with Christopherson's chairmanship of the APC, it is not surprising that talk of a 'Durham mafia' was common in those early days. From Cambridge came Walter Hagenbuch, professor of economics and first dean of the social sciences. From Oxford came Guy Chilver, professor of classics and first dean of the Faculty of humanities; Bryan Keith-Lucas, professor of politics and first Master of Darwin College; and Maurice

Vile, Reader, later professor of politics and deputy-vice chancellor; and (via Leicester) Patrick Nowell-Smith, the first professor of philosophy.[7]

All of these had been teaching members of their respective institutions. But they brought in their wake a whole galaxy of young lecturers who had been educated there. Here Kent followed closely a pattern very visible more generally. There was a marked sentiment among many recent graduates from the older universities that the exciting new developments, and the interesting challenges, were taking place in the new universities of the 1960s, whose planning and establishment they followed closely. The older universities remained highly resistant to change – for example, the very late acceptance of sociology at Oxford and Cambridge. Particularly if one was interested in new disciplines, such as the social sciences, or new experiments in teaching, the new universities were the places to be. When, after graduating in history from Cambridge and sociology from the London School of Economics, I looked around for a job, I applied only to the new universities. I had no desire to be anywhere else.[8]

As with all of the new universities, Oxbridge was particularly well represented at Kent. The Department that I joined in 1967, the Department of Sociology and Social Anthropology, was headed by an Oxford-classicist turned anthropologist, Paul Stirling. His lieutenant was the Cambridge-educated (English literature) Derek Allcorn. They recruited heavily from Oxbridge, such that more than half of the initial department of about a dozen lecturers had been educated at Oxford of Cambridge. Keith-Lucas's Politics department contained no less than four members from the very same Oxford college, Nuffield, where Keith-Lucas had been Fellow and Bursar. The third entity here was the London School of Economics, where Stirling had taught Anthropology and where many of the Oxbridge-educated lecturers had received their graduate education in the social sciences. The only university outside the magic triangle to make any kind of showing was Hull, where Allcorn had been teaching and from which he recruited two outstanding sociologists David Morgan and Frank Parkin. The latter's sparkling contributions in particular did much to enhance the department's reputation.[9]

The fact that it was in the social sciences as much as in the humanities – where one might have expected it anyway – that Oxbridge figured so heavily in the early years is a striking testimony to its importance in the early history of the university. Tradition here was in its element. It showed itself in many other ways: the adoption of highly traditional subjects such as theology and classics (rejected by many of the other new universities – Essex even spurned history); the elaborate graduation ceremony in the grand setting of Canterbury cathedral, with orations (though not in Latin) in classic style; the traditional structure of governance, with a Council overseeing a Senate, which in turn oversaw the various faculty boards (humanities, natural sciences, social sciences); above all in the adoption of the collegiate system.

II

The Vice-Chancellor Geoffrey Templeman was the most ardent and tireless campaigner for a full-scale college system. The new campus would be 'very different from the civic universities and also from the new ones ... eg: Sussex, York and even Keele', Templeman explained to the Sponsors in 1962.[10] Equally typical it was not knowledge or experience of Oxbridge but – apparently – a tour of some Australian universities where the college system operated that most convinced Templeman of the virtue of colleges.[11] The colleges of course lacked the munificent endowments of Oxbridge colleges and were in no sense financially independent. Nor did they select students themselves, like the Oxbridge colleges, though a proposal of this kind was mooted at various times, one as late as 1975. Each was to be a microcosm of the university as a whole, each a community of 'masters and scholars'. Not only were students resident in the colleges, but there was provision for staff residences too, and staff and students dined together nightly in the grand dining halls designed for each college (complete, in the case of the first college Eliot at least, with a raised dais for high table dining for staff). In the early years, staff actually processed to their separate tables through standing ranks of students – an echo of old Oxbridge that not surprisingly later disappeared.[12] The most striking departure from the practice of those other new universities that also adopted colleges – York and Lancaster – was that the colleges did not simply house separate departments, as in those universities, but were representative of each department (the laboratory-based sciences to some extent excepted). Staff members from each department, for example, were distributed across all the colleges, not concentrated in a college that housed their department. Sociologists, economists, historians, mathematicians, physicists, and teachers of literature and languages were to be found more or less equally distributed in all colleges. Unlike Oxbridge colleges, they were not teaching institutions – for their classes and lectures students moved between colleges depending on courses – but they were where teaching staff taught. Students came to your college room for their classes, or as far as possible to seminar rooms in your own college. All staff members were also tutors (Oxford's 'moral tutors') to students – not necessarily only within their own subject – within their college. An effort was made to identify staff as well as students with colleges, and it worked to a remarkable degree – as was made clear when this principle came under attack in later years. The colleges quickly developed their own identities. Eliot and Rutherford, the two oldest colleges, came to stand for ritual and tradition; Keynes and Darwin, the younger members, for daring breaks (e.g. no high table, senior common rooms and junior common rooms close to each other to allow for greater interaction of staff and students). Staff as well as students were aware of these differences and chose accordingly. When, after a year of arriving, I moved from Eliot to Keynes, I was acutely conscious of a change of culture.

A fair attempt to match this emphatic gesture to tradition was made in the naming and design of the colleges. If the college system was old, at least their names should be new. No Anselms, Beckets, Chaucers or Marlowes – historic names which were already abundantly distributed around the city of Canterbury. The decision was made to name all the colleges after outstanding men and women of twentieth-century Britain.[13] The very year – 1965 – that the first college was opened, the poet T. S. Eliot died, thus making it easy to name the first college after him. Proper due having been paid to the humanities, the next college had to be named after a scientist, hence Rutherford (1966). That left social science, so the third college was Keynes (1968). The fourth was Darwin (1970). This of course broke the twentieth-century rule. The first choice had been Russell, after the great twentieth-century philosopher. But Bryan Keith-Lucas, the

Master-designate of the fourth college, disapproved of Russell's politics and his well-known advocacy of 'trial marriages', which Russell himself had energetically put into practice. This distaste was shared by several members of the University's Council and Senate. After much hand-wringing and canvassing of alternative names it was decided – following Keith-Lucas's suggestion – to call the new college Darwin, of course after a great nineteenth-century Kentish luminary. But it spoilt the symmetry and consistency of the twentieth-century rule, and raised suspicions as to Kent's equal commitment to the modern as well as to the ancient. Women were conspicuously absent from this act of recognition and commemoration, so the fifth college – opened much later, in 2008 – was named Woolf, after Virginia. Later still, reflecting an even greater daring, the sixth college, opened in 2015, was named Turing, after Alan Turing, the great mathematician and computer scientist who was also a homosexual and who became an icon for the promoters of gay rights. So at least there has been a clear re-assertion of the twentieth-century principle, within the context of greater inclusivity and an acceptance of newer norms. It should also be pointed out that all colleges, from Eliot on, and all subsequent ones, were mixed: Kent apparently has the distinction of being the first university in the country to have introduced the principle of mixed residences. Common as this became later, it was certainly a radical move at the time and was so noted (student newspapers had a good deal of fun with the privacy rule that futilely tried to keep men and women apart in their own areas during the night hours between 11 pm – '12 midnight on Saturdays' – and 9 am).

The promotion of the new within the framework of the old is further shown in the physical design of the colleges. The University of Kent sits on a hill just outside Canterbury, amidst fields that had originally been part of a sheep farm. It commands splendid views over the city and the Stour valley – the big windows in the dining halls of Eliot and Rutherford were indeed carefully designed to frame Bell Harry, Canterbury Cathedral's magnificent central tower, making the dining halls the best places from which to view the Cathedral. The view from between the two colleges (now rather spoilt by in-building) is equally stunning.

The first architect, Lord Holford, working closely with his associate Anthony Wade and the Surveyor David Edwards, had the idea of placing the first two colleges – Eliot and Rutherford – on the hill directly facing the city, thus mimicking the traditional Kentish forts similarly placed in various parts of the county.[14] The colleges do indeed have a fort-like appearance, as though guarding (or possibly intimidating) the city. Holford wished to show the local in the new built environment, as much as was shown in the conversion of the original farmhouse, Beverley Farm, into first an administration building, later a student residence. But the architecture of the colleges was to be resolutely modernist. All the colleges have clean, functional lines, with no decoration, following a pattern typical of the other plateglass universities in being airy and full of light, with any local and traditional features incorporated being subjected to an overall uncompromisingly modern design.[15] The later colleges, under different architects, gave up any attempt to follow local traditions, being modern in the style of modern buildings everywhere. The same is true of most of the other buildings on the campus, the older Library and Gulbenkian Theatre as much as the shining new music building, the Colyer-Fergusson Building, with its state-of-the-art concert hall, and the Jarman Arts Building (named after the gay artist and filmmaker Derek Jarman), liberally sprinkled with contemporary works. Abstract modernist sculpture is to be found at various points throughout the campus, as in most of the new universities. Kent may be in an old city, it seems to be saying, and respects the old traditions, but don't mistake it for anything but a modern university of the most forward-looking, progressive, kind.

III

If the colleges were a mix of the modern and the traditional, the original curriculum was innovative in the highest degree. The enemy, as Christopherson and the APC saw it, was departmentalism, the division of disciplines into self-contained compartments which fought each other for resources and jealously policed their borders. Christopherson himself favoured the introduction of schools, on the Sussex model, but reluctantly yielded to the preference for faculties (though school re-emerged with the School of Mathematical Studies).[16] But the faculties – of humanities, social sciences and natural sciences – were not composed of departments, as in traditional universities. Instead, disciplines were organized in boards of studies, so that sociology, for instance, came under the Board of Studies in Sociology and Social Anthropology. The very fact that the two disciplines could be linked together like this showed that the 'board' designation was not simply a trivial concession to the reformers but had genuine implications for a broader definition of disciplines and the possibility of interdisciplinary work.

Probably the most radical departure was in the organization of the degree structure. There was a Part One, to be taken by all undergraduates, which lasted not just the usual first three academic terms of the first year but, going beyond the long vacation, included the fourth term of the second academic year. This was to make clear that Part I was not simply a flourish, a concession to outside expectations and conventions, but an integral part of the student's three years at the university. Nor was the Part I simply an introduction to the various disciplines – economics, politics, etc. – as was conventional but was instead composed of 'topics' all of which were interdisciplinary, taught by a team which included staff from several disciplines, and which met regularly to discuss method and substance. I myself was involved in a topic, 'Exploring reality', which included teachers in sociology, philosophy, English literature, film studies, art history, theology, biochemistry and physics.[17] Other topics included 'The idea of the university' and 'Cities and the built environment'.

Equally bold, though destined not to last long, was the provision of a 'long vacation term' between the second and third years (something of course that already existed in some Oxbridge science degrees). Like the four-term Part I, this was intended to break up the usual divisions of the academic calendar. The economics of student life – many needed to work in the long vacation – finally put paid to this, but it is a good indication of the determination to re-think and re-imagine the structure of degrees. The four-term Part I in particular meant that the disciplinary concentration of Part II was never as large a part of the student learning experience at Kent as was normal elsewhere, where Part I is something to be hastily got through – and often much resented by subject teachers – before one proceeds to the real business of subject specialism. There was even, more radically, a proposal for a continuation of the Part I syllabus as Part IA, which would allow a student in the second and third years to pursue a range of topics all of which were interdisciplinary. This, which would have brought Kent closer to an American generalist pattern of undergraduate education, never made it beyond the planning stage, but it is once more a tribute to the adventurousness of the APC, and its determination to break out of departmentalism.

Generally, the APC put itself firmly behind the new, as shown in several ventures. If the humanities reflected tradition, with strong support for classics, theology, history, philosophy and English literature (which though also included American, African and Caribbean literature), there was no lack of commitment to the newer social sciences, sociology, social policy, anthropology, politics, psychology and law.[18] It was indeed in these disciplines that the university made some of its strongest showings, as expressed by the rankings in the various research assessment exercises. There was also the establishment of joint degree programmes, such as that in sociology and English. This was genuinely joint in that staff from both English and sociology contributed in equal measure to the several bridge courses that were the core of the degree, and students took an equal number of their other courses in both English and sociology. No other university, it appeared, which had tried similar ventures – Essex was one – was ever able to sustain such a strong and lasting commitment from both disciplines.

The degree in sociology and English was a good example of how a fully fledged college system, with staff distributed evenly across the colleges, could generate new initiatives in the curriculum. It was entirely owing to the fact that the two instigators of the degree, Steven Lutman from English and myself from sociology, found ourselves frequently together in the Senior Common Room of Keynes College, that we came up with the idea of the joint degree to reflect our common interests. We rapidly recruited two other members of the college – Louis James from English and David Morgan from sociology – and the degree was born. A similar initiative, this time emanating from Eliot College, was the joint degree in anthropology and computing, again arising out of Common Room conversation.

One further bold initiative, this one the brainchild of the second Dean of the Social Sciences, Maurice Vile, was the institution of more than half a dozen interdisciplinary lectureships, in such areas as development, the city, race and ethnicity. Remarkably – and unthinkable in these more parsimonious days – the lecturers appointed to these positions were given three years in which to develop their courses, during which time they had a very much reduced teaching load in the Boards of Studies to which they were loosely attached. For them, and for other enterprises of a similar kind in the social sciences, there was established the Centre for Research in the Social Sciences, one of a number of multidisciplinary centres, such as the Unit for the History, Philosophy and Social Relations of Science (later the Centre for History of Cultural Studies of Science). The University of Kent in those days did not merely pay lip service to multidisciplinarity – the word was not in any case much in vogue then. It was simply part of the working assumption of the people who set up the university. The new universities had to be different. Here was an unprecedented and probably unrepeatable opportunity to throw the pieces up in the air and let them settle in new ways. With remarkable unanimity, the university put itself behind these initiatives – even as it willingly paid its dues to the call of the past that beckoned from the ancient city at the foot of the hill.

IV

The German sociologist Max Weber wrote memorably of the 'routinization of charisma', when the innovations of great prophets and leaders were often transformed in the bureaucratization that must inevitably follow if the enterprise is to last. Whether that was absolutely true in the case of the new universities is a moot point. Undoubtedly, however, they were all responding to powerful new forces in a changed environment. To account for this change of climate in the 1970s and 1980s is a complex business. It can partly be seen as a widespread reaction against the radical calls of the 1960s for new practices and experimentation in virtually all areas of society – economic, political, cultural. The May events of 1968 in Paris might be regarded as the high point of the radical current, after which there was a drawing back everywhere. Kent had its own local version of radicalism. In June 1968 students called for better representation in university governance, and in 1970 they occupied university buildings in protest over secret files allegedly held on them by the Registry: 'a real exercise in university students running their own affairs' observed the *Times Educational Supplement*. In 1968 Templeman, wary of student militancy, responded with a perhaps misguided paternalism, saying 'the best point of contact was between university teachers and the individual student'.[19] Thereafter, the reaction that set in especially hit the higher educational sector, hardened by the growth of neoliberalism, dominant in Margaret Thatcher's Britain and Ronald Reagan's United States. What that meant was an attack on progressive education, and a tightening of state control over universities, together with more stringent demands that they justify the public money that was spent on them. The overall effect was highly inhospitable towards experimentation and led many universities to re-assert traditional forms and methods of teaching and academic organization. The retreat was across the board, shown no less, as other chapters in the volume highlight, in Keele and UEA than in Kent.

The first to go were the new Part I structure and the long vacation term, which had ended by the mid-1970s, to be replaced by the more traditional three-term, one-year Part I, followed by a two-year specialized Part II. But it was only in the late 1980s and 1990s that the interdisciplinary topics of Part I were given up for more conventional introductions to Part II subjects. Slowly too the radical interdisciplinary lectureships were abandoned, and staff in those positions were forced to give up their special designations and seek a home in one of the regular boards of studies. Some, unhappy with the situation, such as Hannan Rose, interdisciplinary lecturer in race relations, simply left, joining the politics department at Leicester.

But boards of studies also disappeared, to be replaced by the 1980s by conventional departments. A sign of what that might mean – the end of some imaginative couplings – was the break-up of the board of studies in sociology and anthropology into two separate departments. Now students of sociology would not be exposed to studies of non-Western societies, with all that might contribute to their understanding of society in general. The subject would be even more Eurocentric, or at least 'Westocentric', than it had already begun to show a tendency towards in the earlier part of the century, replacing the broader vision of its nineteenth century founders. Correspondingly, anthropologists were cut off from sociological approaches that would have much to teach them as their traditional subjects – 'primitive', non-industrial societies – disappeared in the general modernization and development of Third World, non-Western, societies. The passing of the interdisciplinary lectureship in development studies, held by Henry Bernstein (who departed soon after to Wye College and later SOAS), was a further blow in this direction.[20]

What was happening, of course, was the return of departmentalization, the original enemy. The conviction gained ground at the highest levels of the university that the best way to be 'productive' – the new mantra emanating from powerful quasi-governmental outside bodies – was to be organized into departments on the traditional model. That would allow for a concentration of energies that – it was alleged – were lost when staff were attached to too many multidisciplinary projects and spread out over too many sites. This, it has to be said, was a matter of faith, rather than based on any hard evidence. Impressionistically at least Kent authors were at least as productive before the changes as they were after, as shown by the many high rankings of subjects in the early years of the research assessment exercises – not matched again until quite recently.

Perhaps the most significant casualty of the new turn was the college system as it had been originally conceived. Not that students did not continue to choose colleges and to reside in them; and the addition of two new colleges, Woolf (2008) and Turing (2014), showed that colleges remained the principal social units. What was undone however was the diffusion of staff throughout the colleges, as representatives of their subjects: the key part of the original, Oxbridge-derived, model, unique at Kent among the new universities. Now, as at York and Lancaster, each department was to be housed in a particular college and staff were to be concentrated there. All sociologists were to be in Eliot, all political scientists in Rutherford, all economists in Keynes. This was supposed to provide increased synergy: staff would bump into each other in the corridor or the Senior Common Room and share ideas or talk over departmental business. The kind of synergy that, in the old model, had produced the joint degree in sociology and English was apparently of the unproductive kind.

With the loss of this central element of the college system went a general weakening of the system as a whole. A centralized body, 'UKC Hospitality', took over the arrangements for dining and accommodation. There would now be dining only in Rutherford; dining halls in all the other colleges were closed and turned over to other functions. Provision was made for other dining outlets – suitably differentiated as to style and cuisine – in other parts of the campus and students were encouraged to pick and choose; they might by chance bump into a staff member doing the same. Colleges lost their function as dining units bringing staff and students together – one of the principal elements of the original vision. Even the residential principle was weakened, with the construction of student residencies – Park Wood – at some distance from the colleges and separate from them. Kent remained a collegiate university in name, but much of what had made its system distinctive – from, say, York or Lancaster – no longer applied.

Once the representative principle – each college a microcosm of the university as a whole – had been broken, the parts could be treated as more or less modular, readily moveable, units. Departments found themselves being bodily shunted from one place to the next, like parcels of goods. Sociology, starting in Eliot, woke up one day in Darwin, and then, in a further move, was bundled out of the colleges altogether and placed in a corner of the Cornwallis Building, home to a number of social science research units. Even more unceremoniously, in one of those acts of 'rationalization' that was all the rage in the 1980s and 1990s, sociology lost its independence altogether and was merged in the joint super-department named the School of Social Policy, Sociology and Social Research.

In fact, in a further twist, departments themselves were abolished, and subjects came to find themselves in schools: the School of English, the School of History, the School of Physical Sciences. Some of these are basically single-discipline schools, like the old departments; some, such as the School of European Culture and Languages (which includes classics as well as modern European languages and literatures), or the School of Anthropology and Conservation, corral subjects that are not thought to be viable as free-standing disciplines. Needless to say, such schools have little in common with the Sussex prototypes that Christopherson hoped to introduce at Kent. They are basically cost-cutting exercises devised to disguise the fact that many disciplines are no longer supported as they once were. As such they can be the effective prelude to extinction, if and when the need arises.

Ironically, all this reorganization did not have the hoped-for result: increased research productivity, or, more precisely, increased national recognition and higher rankings. Quite the opposite. From the high positions previously enjoyed, many Kent subjects, especially in the humanities and social sciences, slumped dramatically in the rankings produced by the Research Assessment Exercises in the 1980s and 1990s. The situation was not helped by the appointment in 1980 of a Vice-Chancellor, David Ingram, a physicist who had previously been principal of Chelsea College (formerly a College of Advanced Technology) in London, who felt that Kent's main problem was that science had been sacrificed to the humanities and social sciences. His efforts – as for instance in an attempted merger with Mid-Kent College, defeated by a rattled Senate – to turn Kent into an Imperial College-in-the-shires predictably ended in disaster. Kent had neither the structure nor the resources to establish world-class scientific and technological schools (though it had always had some scientific departments with a high reputation, such as the Department of Electrical Engineering, and it developed a highly enterprising Faculty of Information Technology, under Brian Spratt). The main result of this effort was to starve the humanities and social sciences, on which Kent had built its reputation, of much-needed resources at a time when cold winds were blowing from the new Conservative government of Margaret Thatcher. It also played its part in the departure of a number of Kent's best-known academics, many to overseas universities, dismayed by this turn of policy as well as by the new political climate. The 1980s and 1990s, it is fair to say, were a low point for Kent, when much of what had been achieved in the previous thirty years seemed to be unravelling.

V

UKC was always more than just a place for teaching and research; there was also a vibrant cultural and intellectual life. It began with the lucky fact that the first college was named after T. S. Eliot, for that led to the endowment by Eliot's publisher Faber and Faber, and the patronage of Eliot's widow Valerie Eliot, of the T. S. Eliot Memorial Lectures, which have always been the most prestigious lectures at the university. The poet W. H. Auden, in 1967, established the high ambition and standing of the lectures, and ensured the continuance of distinguished scholars and artists for these lectures. Among the memorable ones in my experience were those given by Seamus Heaney, George Steiner, Jonathan Miller, Anthony Burgess, Julia Kristeva, Helen Vendler, John Carey and Edward Said.[21] Audiences for these lectures have always been large, including many people from the surrounding region. The Eliot lectures quickly established a certain cultural identity for the new university, as a place that could attract some of the best intellectuals in the world, who in addition to giving the public lectures resided in the colleges and interacted with staff and students over a period of nearly a week. Several – Julia Kristeva, John Carey and Edward Said stand out in my mind – were notable for the amount of time they spent with various members of the university community. In turn the Eliot lecture series inspired the Keynes seminar, launched by the first Master of Keynes, Robert Spence, and continued by his long-time successor, Derek Crabtree who were responsible for bringing a number of famous economists to the university, including Roy Harrod, Joan Robinson and Nicholas Kaldor.[22]

Other cultural ventures included the Gulbenkian Theatre, home not just to student productions but also to those by visiting companies from London and elsewhere. The Gulbenkian Theatre since 1969 has also housed the Gulbenkian Cinema, Canterbury's regional branch of the British Film Institute, and a venue for a wide variety of English and foreign language films. It was one of the most successful initiatives of the early years, again as a centre for both town and gown, and has continued to flourish. It has recently been complemented by the Colyer-Fergusson Building, opened in 2012, with its splendid concert hall and practice rooms. Generally music and the arts have been one of the most successful areas of Kent's activities, one where in recent years it has been able to attract a considerable amount of private outside funding – over 6 million pounds in the case of the Colyer-Fergusson Building – which contrasts with a relatively thin record of private philanthropy in the early years.[23]

VI

We end where we began, with naming. On 1 April 2003, the University of Kent at Canterbury formally changed its name to the University of Kent. This abbreviation signified an important change in the way the university saw itself, and how it wished to present itself to the world. The dropping of 'UKC' does not mean that the university is turning its back on the city of Canterbury. That would be impossible for both symbolic and practical reasons. If anything, ties have increased since the early years when there was some tension between the old established city institutions of King's School and the Cathedral, on the one hand, and the new upstart on the hill. Relations were smoothed as each got to know the other better, as for instance in the very successful joint venture between the university and the cathedral, the Centre for the Study of Religion and Society. The Archbishop of Canterbury is in any case *ex officio* the university's Visitor (which means he is the final court of appeal), and the cathedral has become the irreplaceable focus of the university's graduation ceremony. The university without the city would be hard to imagine.

But the university has to some extent outgrown the city, overflowed it. It is now not just the University of Kent at Canterbury but also the University of Kent at Medway and the University of Kent at Tonbridge, with various new combinations on the horizon.[24] It has broken out of its east Kent fastness to move more to the centre and the north, to be nearer the large centres of population of Maidstone and the Medway towns, nearer to London, the great metropolis, linking up indeed with the University of Greenwich in the Medway venture. There are now partnerships now not just with Canterbury Christchurch University but also with West Kent College, South Kent College, and MidKent College. In 2008 Wye College, specializing in agricultural science and a constituent part of the University of London, came under Kent's remit jointly with Imperial College London. The University of Kent at Canterbury clearly would now truly be a misnomer, however much many regret the loss of the familiar UKC.

There is more than this, signified by the university's informal logo, the 'UK's European university'. Kent's proximity to the Continent has recently been heavily accentuated with large numbers of continental students, many under the Erasmus programme, arriving at the university, as is readily clear from the languages spoken in its bars and cafés. Along with this has been a move across the Channel, Kent putting out roots on the continent. The first of these – set up in 1999 – was the Brussels School of International Studies, a postgraduate institution offering master and doctoral degrees in several social science subjects, as well as History, drawing on the parent Kent departments. Later came a postgraduate centre in Athens, partnered with the Athens University of Economics and Business, and specializing in heritage Management. A postgraduate centre in Paris now offers year-long MAs in several of the Humanities subjects, with Kent staff regularly making the journey to the French capital to teach. The newest postgraduate centre, specializing in the classics and Roman history, is attached to the American University of Rome. Europeanization rides high at Kent, and it is not surprising that its spokesmen – including successive vice-chancellors – have spoken out volubly and vigorously against a British withdrawal from the European Union. For Kent this is not just a matter, as for many other British universities, of valuable exchange programmes and joint research projects, but of the very image it now promotes.

To walk across the campus now is to be struck by the enormous changes that have taken place since those heady days of the 1960s. Six colleges instead of four, together with a host of new buildings that have filled in many of the agreeable green spaces that allowed for walking, thinking and viewing. A student body that has swollen to nearly 20,000, from the 5,000 that began the experiment, and of that larger number 22 per cent from overseas, from the continent and further afield. As many as 600 academic and research staff compared to the barely 200 in the first decade. New subjects – a new School of Architecture, for instance – and new combinations of subjects proliferate. We are in a new world. But then I stand between Eliot and Rutherford – admittedly a more cramped space now than before – and see Canterbury Cathedral rising gloriously out of the old city: the iconic view on every piece of publicity produced by the university. Some things don't change. Kent may no longer be the University of Kent *at* Canterbury, but its centre, and perhaps its heart, is most certainly *in* Canterbury. The old continues to nourish the new.

Notes

1 Lord Cornwallis (1892–1982), the second Baron Cornwallis, came of an old Kentish family whose ancestors included the Cornwallis who surrendered at Yorktown in 1780. He was active in Kent's public life, having served as chairman of Kent County Council in 1935–6, as well as later being Lord Lieutenant of the county. Apart from chairing the 'Sponsors' committee, he was later Pro-chancellor of the University (1960–71) and remained to the end of his long life one of the University's strongest supporters. Julie West (ed.), *Debrett's people of Kent* (London: Debrett's Peerage Limited, 1990), *sv.* 'Cornwallis', 40.
2 *Times*, 18 July 1962, 8.
3 Ibid., 17 October 1962, 15.
4 For an outline of his career, see his obituary: *Guardian*, 25 February 1988, 37.
5 For Christopherson, see Duncan Dowson and Gordon Higginson, 'Sir Derman Guy Christopherson: 6 September 1915–7 November 2000', *Biographical Memoirs of the Fellows of the Royal Society of London*, 50 (2004), 47–59.
6 For Whitehouse, see his obituary: *Times*, 29 April 2003, 32. For Foakes, see his autobiography, *Imagined Places: A Life in the Twentieth Century* (Bloomington: Xlibris, 2005). For Darlow, see his obituary: *Kent Bulletin*, 34 (Spring 2000), 6.
7 For Hagenbuch, see his obituary: *Times*, 22 January 1981, 16. For Chilver: ibid., 11 September 1982, 8. For Keith-Lucas, see George W. Liebmann, *Six Lost Leaders: Prophets of Civil Society* (Oxford: Lexington Books, 2001), ch. 6. For Vile (b. 1927), see Joaquín Varela Suanzes-Carpegna, 'Politics and the Constitution in British and American History: Interview with Professor Maurice Vile', *Libros y Revistas Historia Constitucional*, 10 (2009), 559–81. For Nowell-Smith, see his obituary: *Guardian*, 22 February 2006, https://www.theguardian.com/news/2006/feb/22/guardianobituaries.obituaries (accessed 1 September 2019).
8 For some reflections on this situation, particularly as regards the development of sociology in the 1960s and 1970s, see my 'Sociology and the Englishness of English Social Theory', in *The Idea of Englishness: English Culture, National Identity and Social Thought* (Farnham: Ashgate, 2015), 165–98; see also Mike Savage, *Identities and Social Change in Britain since 1940* (Oxford: Oxford University Press, 2010), 124–34, who quotes the sociologist Ray Pahl (Cambridge and LSE) on his 'excitement' and 'enthusiasm' at being at the new University of Kent (ibid., 129).
9 For Stirling, see David Shankland, 'An Interview with Professor Paul Stirling', *Turkish Studies Association* Bulletin, 23 (1999), 1–23. For Allcorn, see his obituary:

Anthropology Today, 3 (1987), 19. For Parkin, see my obituary of him: *Guardian*, 9 November 2011, 11.

10 'Notes of a meeting of members of the architectural subcommittee of the Sponsors of the University of Kent', 3 November 1962, Holford papers, University of Liverpool, D147/C75/4.
11 Martin, *From Vision to Reality*, 109.
12 For the merging of 'living and learning' at Kent, see *Times*, 27 April 1964, 7.
13 For the tussle over this, see Martin, *From Vision to Reality*, 122–6.
14 Holford later fell out with the university, resigning his consultancy in October, 1965. Although reticent in public about the episode, he later criticized Templeman for setting his 'face against grand designs and long-term planning': Holford to Templeman, 5 August 1970, Holford papers, D/147/C5.
15 Holford, 'Memorandum on the planning and development of the University of Kent at Canterbury' (April 1965), UKC Archives, Registry E/F 00001–10.
16 Christopherson, 'Preliminary Proposals for an Academic Structure' (April 1962), Academic Planning Board, UKC Archives, Registry E/F 00047–63; Templeman, 'The Structure and Development of First Degree Studies' (1 January 1963), ibid.
17 So successful was this that it was turned into a book of essays by the course teachers: *Exploring Reality*, ed. Michael Irwin (London: Allen and Unwin, 1987).
18 For the original discussions around the curriculum, see 'Memoranda submitted by members' (1961–2), Academic Planning Board (1961–3), UKC Archives.
19 Senate, 19 June 1968, UKC Archives. For the occupation, see *Times Educational Supplement*, 13 March 1970, 6; Registry E/F, 1969–70, UKC Archives, 00220–00325.
20 For Bernstein (b. 1945), see G. Capps and Liam Campling, 'An Interview with Henry Bernstein', in 'Introduction' to *The Political Economy of Agrarian Change: Essays in Appreciation of Henry Bernstein*, ed. Liam Campling and Jens Lerche, *Journal of Agrarian Change*, 16 (2016), 370–89.
21 Nearly all the lectures are published as books by Faber, including such famous ones as Seamus Heaney's *The Government of the Tongue* (1988), George Steiner's *In Bluebeard's Castle* (1971), and Julia Kristeva's *Proust and the Sense of Time* (1993).
22 Most of the seminar proceedings have been published, starting with *Keynes: Aspects of the Man and His Work: The First Keynes Seminar Held at the University of Kent at Canterbury, 1972*, ed. D. E. Moggridge (London: Macmillan, 1974). Highlights include Geoffrey Keynes on his elder brother, republished in *Keynes and the Bloomsbury Group: The Fourth Keynes Seminar 1978*, ed. Derek Crabtree and A. P. Thirwall (London: Macmillan, 1980).
23 Fitton, *Hidden History*, 16.
24 The University of Kent at Medway, the largest extension since the University's foundation, was launched in 2001 at the site of MidKent College; in 2005 it moved to a new campus at the old Chatham dockyards. For its purpose and development, see Fitton, *Hidden History*, 81–3.

9

Social history comes to Warwick

by Carolyn Steedman

'Isn't your father Vice-Chancellor of Wolverton University?' asked the women from the Department of the Environment, and initiated a long conversation about new British Universities ... Frances struggled, and then gave in. It was going to be boring, after all.

Margaret Drabble, *The Realms of Gold* (1975).

Old provincial society had its share of ... subtle movement .... Some slipped a little downward, some got higher footing .... Settlers, too, came from distant counties, some with an alarming novelty of skill, others with an offensive advantage in cunning.

George Eliot, *Middlemarch*, 1871–2, Book 1, Chapter 11.[1]

1970: An introduction

Why might the story of social history at Warwick be worth telling, yet again? Is it historically significant in a way that social history at other new universities of the period are not? 'The Warwick Centre for Social History', 'E. P. Thompson', and 'The Files Affair' are caught together in a perpetual *ménage à trois* of the historical imagination; that's why. The story goes something like this. At the brand new University of Warwick, a Centre for Social History was established. This had something – or a lot, depending on your perspective – to do with Edward Thompson, who had joined the Warwick History School in 1965, two years after the publication of his enduringly influential *Making of the English Working Class*. Then, early in 1970 there was a series of student occupations of the Warwick Registry. They were protesting at the continued reluctance of the university to grant them autonomous control of the Students Union. During the second occupation in February, students opened administrative files in one of the Registry offices and found letters and other documents suggesting that surveillance was being carried out against certain university staff, and that there had been political profiling of prospective Warwick students. Thompson was telephoned at his house in Leamington Spa; he drove in; the papers were copied and distributed among university staff. Thompson published a burning piece of polemic on the background to 'the files affair' in *New Society*.[2] There were rapid negotiations with Penguin Books: *Warwick University Ltd,* famously written in a week, was an edited collection of participants' testimony, published in March 1970. Now the university's broken promises to students about management of the Union were linked to surveillance at Warwick. Questions were asked about the university's relationship with the Midlands automotive industry and its own system of governance. All the factors were now in place to allow one to imagine Warwick as Thompson's dystopian eighteenth-century state, with Oliver the Spy in its employ; as William Cobbett's 'It', or 'The Thing'; as William Blake's 'The Beast'.[3] If you had a mind to do so, 'the new mid-Atlantic world of the Midlands Motor and Aircraft Industry' might be mapped onto the great landed aristocracy of the eighteenth century, 'exerting their power by manifold exercise of interest, influence, and purchase'. The new lords of 1970 'infiltrate the command-posts of our society, including our educational institutions, not through any transparent democratic process, but quietly, in unnoticed ways', wrote Thompson. 'They apparently share with their precursors the same assumptions that it is their world, to dispose of by ownership and by right of purchase.'[4] That particular 'social-history-at-Warwick' story ended with Thompson resigning from Warwick. He went in September 1970, but left behind his legacy in a Centre for the Study of Social History. Told that way, with that end-stop, the story is about the confrontation between a bureaucratic institution and a radical historian.

A primary source for anyone pursuing this history is the reissued *Warwick University Ltd. Industry, Management and the Universities* (2014). Its Introduction is particularly valuable for identifying university personnel connected to the Midlands automotive industry. Valuable too is the perspective that has emerged from Warwick's Fiftieth Anniversary Oral History Project. Forty years on, some former student participants in the Registry Occupation now remember how their own politics of everyday life – attempts to change the food available in the Students' Union, domestic arrangements in halls of residence – were appropriated to a much grander narrative of the secret state, the inner workings of local capitalism, and political and industrial espionage.[5] The story can then be used to understand Thompson's obsession with finding the roots of the surveillance state. None of these versions have much to say about the new social history with which Thompson's reputation is so closely linked, or about disciplinary formation in one utopian university, between about 1965 and 1975. This chapter provides a different story, to add to all the ones we possess.

Settlers, too, …

First, a confessional prolegomenon, of three items. The first is that from 1997 to 1998 I was director of the Centre for the Study of Social History (CSSH) at the University of Warwick, the graduate teaching centre which Edward Thompson and many others laboured so assiduously to establish between 1964 and 1968. In 1998 it was closed (or disestablished, or dissolved, or whatever the term is for disbanding part of a university with formal departmental status). In my wilder and guiltier moments I believe that one day I shall be held to account for this – not saving it, being responsible for its demise. I am involved in the story told here. The second point is about *Middlemarch* (1871–2). I had read it at least ten times by the time I came to live in Warwickshire in 1976. 'Loamshire' is Warwickshire in George Eliot's historical novel, and 'Middlemarch' is Coventry. It had been the set text for the literature part of my history degree at the University of Sussex, 1965–8, though then I hadn't known these things. I *had* been on a school trip to Coventry from London in the fifth form at my grammar school. In 1963 we girls were shown the industrial dereliction of the West Midlands through the train window and then walked to the new Cathedral – the new city – rising from the bombsites of the old. When I arrived in Loamshire in 1976, I found *Middlemarch* an invaluable guide to the habits and manners of this little bit of provincial England. *Middlemarch* continues to shape my vision – comic, tragic, historical – of Warwick and Warwickshire.

The final introductory point is to do with Utopia and utopianism. The fifteen-year-old on the train to 'Middlemarch' in 1963 had an idea of utopianism – had heard the term – though like the rest of the fifth form, probably, had not yet *read* a word on the topic. I think we all believed that we were living a kind of utopianism: that as the clean, well-fed, well-educated children of the 1944 Education Act, of the National Health Service – of the welfare state – we were lucky, lucky girls. Did anyone on that trip mention that land had been acquired for a new university three miles from Coventry Station? We certainly heard a lot about the new universities throughout the early sixties, not just from schoolteachers but from the press as well. At Sussex in 1965 I didn't exactly think 'I'm here – at university – because of Lord Robbins'. But much later I came to connect the Robbins Report on the provision of post-school education, to the idea of reconfigured human and class relations in a situation of relative material comfort; to the modest, social democratic utopianism in which I had been born and educated.[6] Paragraph 810 of the Report can still make me cry, at the way in which it interpellates all the children of the post-war years, the mildness with which the last plank of the welfare state is laid in place; for all that is now lost: 'The young people who will be seeking to enter higher education in the years 1965/6 to 1967/8 were born in the period when the population of this country was beginning to return to normal life after the upheavals and separations inevitable in war. The trials that their parents had to undergo are in themselves sufficient reason for the country to exert itself to meet the needs of their children….'[7] To add to that lullaby of social justice is the observation that in 1969 the 20 per cent of eighteen-year-olds who had made their way through the narrow gate to the golden city of higher education became new kinds of social subject under the Representation of the People Act (1969), which extended the franchise to their age group. In 1970, after appeal to the High Court, students were permitted to vote in parliamentary elections in their university constituency instead of that of their parental home.

Coming to Middlemarch

Did social history come to Warwick, sometime between 1963 and 1975? The origins and progress of social history were frequently discussed in the post-war period. 'The "Cinderella of English historical studies" only a few years earlier, it was by the mid-1960s the "in" thing, the queen of the ball ... It is still [2000] often regarded as the main innovation of the period.'[8] The *Annales* school in France has been described as one point of origin, as has the labour history undertaken in economic history departments from the 1930s onwards. An Oxbridge axis of social history has been outlined, in which Sir Lewis Namier's concerns with the psychological as well as economic factors underpinning social structure and social distinction ('what lay beneath', in Miles Taylor's nice depiction) influenced the work of historians far beyond the golden triangle, from the 1930s through to the 1950s.[9] David Feldman and Jon Lawrence have described a baggy, protean concept – 'social history' – which for fifty years has stood in, more or less, for 'everyday life'.[10]

Most modern commentators on the genesis of social history have indicated the importance of G. M. Trevelyan's *English Social History: A Survey of Six Centuries: Chaucer to Queen Victoria* (1942) for popular understandings of what kind of history it was.[11] Others remind readers that J. R. Green's *Short History of the English People* (1874) had underpinned many early-twentieth-century histories for children, teaching not 'of English Kings or English Conquests, but of the English People'.[12] Most young children educated in the state system during the first half of the twentieth century learned more about the furnishings of a Viking long-house, the way food was cooked by medieval people and clothes made 'through the ages' than they did about war and high politics.[13] When Victor Neuberg discussed Edward Thompson's phrase 'history from below' as indicative of a new way of doing social history, he actually reviewed eight works of social history for young children, including the emblematic *Life in England* (1968–70) series, 'a vivid, panoramic view of how men lived'.[14] Social history for children was targeted at parents as well as schools: in 1962 the new *Look and Learn* children's magazine promised 'Art, Literature, Science, People, World, and Social History – These and More are Presented in a Vivid, Authoritative Way'. Readers of the broadsheets and the literary magazines had been prepared for understanding E. P. Thompson's 1966 proclamation that 'social history lives in the language of the commons', reported in *The Times* under the title 'history's widening frontiers'.[15] The vocabulary of 'social history' was made familiar to parents and children of the post-war era. Everyday understandings of what social history was, and what it did, may well have accompanied the first intakes of students at the new universities, whether or not they studied for a history degree.

Thompson distanced himself from popular social history in one of the many committee papers he wrote in the long run-up to the formal establishment of the Warwick Centre for Social History: 'The very considerable advances in recent years in this country in social studies and in sociology has been accompanied only by a piecemeal and disconnected advance in the study of social history,' he wrote in 1965. 'In some places "social history" has scarcely advanced beyond the status of belles lettres.'[16] University social history was different from the middlebrow version, and also from the kind 'advanced within the aegis of departments of economic history'. Progress there had been impressive, but 'distorted by the tendency to limit study to quantitative or institutional problems'.[17] Ten years later (Thompson long departed from Warwick) CSSH Director Royden Harrison acknowledged the turn from labour history to social history and the special relationship between the two. But academic social history had still not lost the accent of its somewhat philistine origins out in the wider society: social historians must guard against 'a return to the "everyday," the residual and the tedious'.[18] Labour history remained the more respectable endeavour well into the 1970s.[19]

Thompson had been appointed as a reader in the History of Labour in the Warwick School (Department) of History, at a university that placed great emphasis on the study of industrial society and industrial relations.[20] Thompson used the university's ties with the Midland's automotive industry (the target of his later polemic in *Warwick University Ltd*) to persuade his many committee audiences of the value of 'the social history of industrialisation, and of mature industrial societies, with special emphasis upon the "History of Labour," or the social history of the working classes'. However, he added, 'The "History of Labour" should not be taken to indicate only the history of the institution of the "labour movement" and of the trade unions, as developed by the Webbs and G. D. H. Cole.' In the proposed Centre 'the unorganised and unpolitical majority' would be paid as much attention as 'traditional labour history'. What *was* social history? The historical questions to be pursued in the Centre were the best answer to that question: 'social tensions during the transitions to industrialism'; 'urban growth and popular demoralisation'; 'the social history of leisure'. In the future there would be 'the history of poverty and of social welfare work: of family relations, sexual codes' (there existed no 'serious social history of illegitimacy' for example) and popular attitudes to children and education. There would be 'industrial studies'; there would be work on 'racial and ethnic assimilation or antagonisms'; on 'sub-political' attitudes and crowd behaviour and on 'popular attitudes (and superstitions) in medical matters'. The 'serious history' of crime had 'scarcely emerged from infancy'; a Centre for Social History would help it grow up. Comparative study was imperative, but the Centre must first establish 'a reputation for sound work in British studies'. He pointed to a series of journals in the United States, France and Italy that showed the way.[21] Appointments should be made at junior level of young specialists in the history of industrialization in America and Europe: 'Ultimately it will be from comparative work (a comparison of labour conditions in Hamburg, Liverpool and Boston, for instance) that the unique function of this Centre will be recognised.' And, he observed, 'A new university is exceptionally well-placed to commence such a Centre'; 'the location of the University of Warwick and the proposed work of the Centre of Industrial Studies, suggests that it will be ideally situated to initiate this work.' By such endeavours and in this environment the 'prolonged debate between historians and sociologists' might be resolved, in a sustained effort to bring 'historical techniques to bear upon problems which sociologists have indicated'. He ended his paper with a discussion of 'The Centre and the Undergraduate who goes into Industry'.[22]

Building Middlemarch

Construction work had started in 1964 on 400 acres of wood- and farmland granted by Warwickshire County Council and the City of Coventry. The ancient conflict between county and city (between Loamshire and Middlemarch) in the disposal of its capital (cultural and monetary) has been described elsewhere, as have the principles underpinning the architecture of the new university.[23] In his own accounts of the university's history the first Vice-Chancellor Jack Butterworth made much of its location, and its connections with William, 1st Baron Rootes (1894–1964). Rootes was a Coventry motor manufacturer and in 1961 chair of the promotion committee set up to found a university near the city. His vision of Warwick's close academic links to industry was not unique, but many accounts make it so. In the early years Butterworth told many audiences about Billy Rootes being driven up the M1 assiduously studying architectural and curriculum plans as his chauffeur sped into Warwickshire.[24] It had been intended that he should be the first Chancellor of the new university, but he died in late 1964. In his first Report to University Court in November 1965 Butterworth eulogized Rootes telling how 'versed as he was in the ways and history of Warwickshire and Coventry, and conversant with nearly all the notable personalities in the area, it was almost inevitable for [him] ... to emerge as a leader of the new venture ... He made no pretensions to an academic background ... [but] quickly appreciated the potential of a University taking a new look at Engineering and the Sciences in the heart of the industrial West Midlands'.[25] In publicity directed at prospective students and their parents there was less about the white heat of manufacturing burning in the lands bordering the M1 south of Birmingham; for them Loamshire was described. There were lyrical descriptions of a 'typical Warwickshire field and hedgerow pattern ... intermingled with woodland planting', of the 'most central county in England, with its wealth of historical and cultural traditions ... all that is characteristic of the English countryside – meadows, fine woods, quiet rivers, charming villages and many notable churches and lovely old manor houses'.[26]

In the Report eulogizing Lord Rootes, the vice-chancellor discussed the curriculum of the new university. He was sure that there would be interest in the history degree, which included 'an unusual arrangement of the periods studied, and the even rarer feature of a term of academic work at an American University, and another in Florence'. Between the appointment of an Academic Planning Board in 1961 and the granting of the University Charter in 1965, 'an academic "blue-print" of the University been devised', he said.[27] The first professors were appointed in the summer of 1963; they joined the Planning Board to work on the curriculum to be submitted for approval to the University Grants Committee. John Hale was the founding professor of History and it was his view of the undergraduate curriculum that entranced Jack Butterworth: he frequently repeated Hales's proclamation that unlike universities 'elsewhere', history at Warwick would be characterized by 'breadth rather than chronological completeness'. There would be 'no medieval history, except in the form of an introduction to [second year studies] and no detailed study of the period 1650–1776'. Thus would be provided 'a sound and uninsular historical training which appeals to the imagination'. Modern history should be taught before early modern (the 1960s descriptor was 'renaissance').

The reasoning behind this reverse chronology was that the modern period was more familiar to most students and would particularly suit those taking a joint degree or options in other departments. The aim was not to produce historians, but to 'promote reflection on the past' through discussion of the evidence on which knowledge of it was based. There would be discussion of economic and sociological approaches; reading the work of 'great historians' would be interspersed with the 'latest articles' from the history journals. The terms spent abroad (in Italy, in the United States and, for a brief period in the late 1960s, at some African universities[28]) were planned because 'history is most meaningful if it is studied where it is made. Italy was the most significant country in the Renaissance, America in the Modern world'. The function of a history department was to explore 'man's changing attitudes to sex, family and class; gaining a livelihood; the state, religion and morals; environment, natural and intellectual'. There were indeed 'horrid chronological gaps' in the programme, but the advantages were breadth, and a 'world-wide' history 'in some depth'. It was a modern history curriculum for a modern world.[29]

'Social history' did not figure in these very early plans; or rather core courses and options were not entitled 'social history', but a social history approach was implied in the descriptions of many of them, and of special subjects (final year) offered from 1966 onwards. The special subject might be 'studied under the supervision of the Reader in the History of Labour', said the Prospectus for 1966–7. Edward Thompson offered 'The Transition to Industrialisation' (using 'English sources as a basis, but introducing comparisons with Lyons, Silesia and Ireland') and 'Sans-culottes and socialists, 1791–1848' ('the artisan Jacobin tradition in France, England and America; millenarian movements; St. Simon, Baboeuf, Fourier, Owen, "Utopian" Communitarianism in America and Britain') until 1968, when the Centre for Social History became an independent entity.

Labour history? Social history?

Appointed in 1964, Thompson arrived at Warwick at the beginning of the academic year 1965–6. His memos and letters, promoting, haranguing, persuading about social history are preserved in administrative records for 1964–5, but he did not move to Leamington Spa and Warwickshire until the autumn of 1965. He did not need to be present to promote the cause of social history and it did not have a single point of origin in him, or in his work teaching adult learners for the University of Leeds Extra-Mural department, or his experience of working on *The Making of the English Working Class*: all of that was important, but really, social history was already at Warwick. It was made at the meeting point of the experiences and expectations of teaching staff and students and administrators, in a highly specific locale. It actually arrived as 'labour history', with its connotations of academic credibility described above.

In February 1968, in a long memo to his colleagues, Thompson told his version of the tale: 'I was appointed in June 1964,' he wrote; 'the Vice-Chancellor ... [and] Professor Hale and Professor Sargent [founding chair of the Department of Economics] were present at the interview, and large plans were sketched for developments in comparative labour history.'[30] Then, he recalled, from June until December 1964 'a protracted dream-sequence ensued, in which the plans were put on paper, endorsed, and the Librarian and I visited Milan to make a bid for the £1,000,000 Feltrinelli Institute Library'. Then 'the dream-sequence ended abruptly in January/February 1965, when communications dried up with Warwick, when I ceased to be kept informed of the Feltrinelli negotiations, and when I came tardily to realise that a) the university had no funds for a Centre, b) fund-raising was not being pursued with vigour or system'.[31] All of his story – and mine, here – pivots on that economic point: *independent* funding for a centre of labour or social history, funding that did *not* come through the School of History; a constitutional setup that allowed an independent Centre to earn its own living from the fees for charged for graduate study.

In February 1965 Thompson wrote to the vice-chancellor (he was keen to assure him that he only did this because his Head of Department John Hale was out of the country) reminding Jack Butterworth that at his appointment 'some rather grandiose plans were put forward as to a Centre for Comparative Labour History. Those were splendid (although I thought at the time that they might be over-ambitious). It was not made clear to me at the time that these were dependent upon successful fund-raising activities … Whether there is any real acceptance by the university of the idea of the proposed Centre – whether there is any expectation of an early decision … – I haven't the least idea … I don't want to come to Warwick if the whole thing has been based on fantasies; if in fact this is to be a determined Business University which really has very little interest in my kind of Non-Mathematical Humanology; if in fact the Centre never looks like materialising … I think you will agree that it would be much better to arrange a dissolution of engagement than a subsequent divorce … I may very well be the wrong kind of person to get into a new University at this stage'.[32] The vice-chancellor's reply had been reassuring ('but not quite specific enough'), telling Thompson that the Centre was 'one of the central projects in our work in Social Studies'. Thompson recalled that he had replied enclosing plans for a Centre that he and John Hale had worked out back in October 1964. He also asked to withdraw from the university. He had not yet moved from Halifax; the vice-chancellor invited him down to Warwick: 'I underwent an interview (the first of several similar experiences repeated in the past four years) which I commenced with anxiety and concluded in euphoria, flattered by assurances as to the central importance of my work to the university and be-dazed with the glint of the vast sums of money liable to descend in the next week or fortnight into the lap of Social History. As a result of this interview I withdrew my proffered resignation and redrafted the plans for the Centre, from Labour to Social History.'[33] In June 1966 the Senate determined that a 'Centre … be established forthwith' and that 'Mr E. P. Thompson be appointed to act as Director … in the initial period'; in the vice-chancellor's report it was noted that 'Mr Thompson was appointed as Director of the Centre for the Study of Social History in June 1966'.[34]

The Centre had an advisory committee (responsible to the Board of Arts) drawn from the history, economics, English, French, and politics departments and the library; but it did not really exist in administrative and financial terms. Its director was a member of the School of History; the graduate students who came to work with

Thompson were registered in the School; Thompson's publications were returned as history publications. A series of pronouncements by the vice-chancellor, made in a different rhetorical space from the one that produced Thompson's trance-like euphoria, made it clear that the Centre would have to earn its own income. Through John Bruce Lockhart, Warwick's planning and development officer, John Hale approached the Trades Union Congress (and many other organizations) without success.[35] When new posts were proposed for Social History late in 1966, Thompson reminded the Registrar that the Centre depended entirely on the number of graduate students recruited, and recommended that one of the new appointments should 'be of a Director for the Centre ... He could supervise the programme of the existing graduates. He will be heavily concerned also in planning the M.A. course'.[36] This was an attempt, not at resignation from Warwick University, but from directing an entity that didn't really exist. 'Did we ... acquire Edward Thompson because of a clear prior commitment to Labour History, or did our enthusiasm for Labour History arise when it was found out we might get him? Which came first ... ?,' mused the vice-chancellor in 1964.[37] The idea came first: a centre for social history was on someone's mind when Thompson attended his job interview, perhaps on John Hale's mind, as he devised the curriculum for undergraduate history students.

There were material factors to consider in the development of a modern labour history: an economics department, a planned Centre for Industrial Studies and interdisciplinary co-operation with both – but also Hale's belief that in historical terms, America *was* modernity. There was also an anonymous donation of £215,000 for building a hall of residence in which Anglo-American understanding might be fostered, announced in 1965.[38] It was to be used for exchange students. The plan for a term of academic work at an American university for History undergraduates had already emerged. John Hale had been in the United States earlier in the year, and now reminded the director of the American studies programme at the American Council of Learned Societies (ACLS) that 'in New York in the Spring, you suggested that I should send you details of our projected Centre ... We are particularly anxious to establish ... the study of the History of American Labour ... the Council might find it possible to give us a senior post'. The person appointed would be 'work from the outset within the School of History itself ... increasingly as the Centre developed, within the Centre ... we are ... stressing American History at Warwick'. From 1969 onwards history would be offering both undergraduate and higher degrees in American history, he said.[39] In December 1965 Hale asked the university to make a bid to ACLS 'to give us a Lectureship for 5 years at a cost of £1,625 p.a., coupled with a sum that would enable us to invite an American specialist in the History of Labour to visit us for one term in each year'. This would be 'a sound foundation from which to build up the American Labour History side of the work of the Centre for the Study of Social History': the most 'responsible and useful way for Warwick to contribute to the study of American history in Britain: by teaching it with a social emphasis'.[40] 'It was the appeal of such a Centre which enticed the ... [ACLS] to establish the lectureship in the History of American Labour and which brought me to Warwick for a two-year visit', wrote ACLS Senior Lecturer in the History of American Labour, David Montgomery – 'at considerable sacrifice to my own income', he noted.[41]

Both Montgomery and Thompson taught in the School of History whilst organizing the seminar programme of the notional Centre and supervising PhD students (there were no MA students until 1969). Students (undergraduate and graduate) were value; the monies allocated to departments for teaching them determined workloads of staff. What Thompson wanted was for the Centre to be allocated the value of the graduate students its staff supervised and taught and to have the right of its disposal. He did some numerically febrile accounting (familiar to anyone labouring under a workload points system) to demonstrate what the effect of current arrangements was: 'My teaching ratio is made up of: 8 full time and 2 part-time graduates equals 9 counts as 18. 15 special subject students count as, say, 7. Directing, examining, lecturing and conducting some seminars on industrialisation: about 40 students first term, about 30 second & third: say 36 divided by one-quarter equals 9 divided again by one half (to account for graduate help in seminars) count as 4 ½. Basic One [core course] lectures &c say 2. Total: 31 ½ ... I have worked harder for [this university] than I have ever worked for any institution in my life ....'[42]

Thompson offered his resignation again halfway through the year 1965–6. This was withdrawn at the last minute after a subvention from the head of another department helped him obtain some secretarial support for graduate work. The vice-chancellor released funds donated by the Cadbury Foundation ('with a letter signifying the benefactor's interest in social history'). All of this was recognized in a Senate resolution formally instituting the Centre in June 1966; Thompson agreed to serve as acting director for the year 1966/7. But it was not enough. 'Independence' inscribed an economic rather than a political or ideological principle. 'Everything turns upon the disputed point of independent budgeting,' wrote Thompson to the Registrar in February 1968; 'constitutional questions rest on that point.' The Head of History John Hale and all history staff endorsed 'the independent solution', he said.[43] Members of the Planning and Executive Committee (PEC) thought that if Thompson and Hale 'could regulate conflicts of interest' between themselves, no other constitutional remedy was necessary. In Thompson's words, 'the Committee has refused to accept independent budgeting for the Centre'. It had recommended the semi-independent solution: 'that Social History should be treated as a semi-independent division of History, entitled to submit financial and other estimates, through History, to the appropriate University committees'. 'Through History' meant no independence at all.[44] The way through Warwick's complicated constitutional and financial structures was to give social history the status of a subject, so that it might become, under the Warwick statutes, a unit with departmental status, able to receive funding in its own right. This is what was proposed at a meeting of the Board of Studies for Arts in March 1968, when the Board recommended to Senate that 'The Centre for the Study of Social History have the status of an independent subject for the purposes of estimates and government, within the Board of Studies for Arts'; 'graduates supervised by staff appointed to the Centre, or following courses within the Centre, should count for purposes of estimates towards staffing the Centre'. 'Social History Independent' announced the University of Warwick *Bulletin* in April.[45]

John Hale was worried: staff in the Centre whose time and teaching he no longer had the disposal of *must* contribute to undergraduate teaching in History; when *his* staff taught or supervised social history students, their efforts must be recognized through the financial 'matrix' that governed teaching time and costs across the university (both Centre and School of History should get the money its staff 'earned' by teaching students). Thompson, now effectively a head of department with a budget, was also concerned about these questions.[46]

At the same meetings that declared social history's independence, the MA course in 'Labour history, 1867–1926' was approved. It comprised taught courses in comparative labour history, labour in England and the United States, and 'States and industrial relations' ('the system of industrial relations in Britain and its central problems. An introduction to relevant economic and sociological tools of analysis, and to the legal basis of the system'). Introduced in 1969, it was the only MA offered by the Centre until the late 1970s.[47] By December, with David Montgomery in post, it had become clear that comparative labour history must be based on survey courses in both British and American labour history. 'This is being undertaken this term by all candidates. Next term all candidates will take the comparative course.' But uncertainty in the staffing position made it difficult to 'guarantee the continuation of the M.A. in 1969–1970 … in certain circumstances it might be necessary to introduce new options (for example, in the comparative study of early popular movements, c. 1760–1850, in France and Britain, to be taught by Dr G. Lewis and Mr E. P. Thompson').[48] There are nineteen of the MA dissertations produced by CSSH students between 1969/70 and 1975 in the Warwick history department's own archive. All of them focus on the period 1850–1920; nearly all concern the organized working class, employed or unemployed, the political organization of labour (the SDF, the Communist Party, syndicalism); there were dissertations on workers' associational life, and strike action in the east and west Midlands. The first women's history appeared in 1972 with 'Sylvia Pankhurst and the Workers' Socialist Federation – the red twilight 1918-1924'. The students were after all working for an MA in Comparative Labour History with the specific date-boundaries of 1867–1926.[49]

What Thompson himself did by way of research during these years as he worked towards 'The crime of anonymity' and *Whigs and Hunters*, together and apart from his PhD students, was shift the gears of social history in consideration of crime and the criminal law in the long eighteenth century.[50] For both works, substantial amounts of time were spent (or wasted) by him and his Research Assistant Ernest Edward Dodd in pursuing what legal historians think of as social history's criminal fix. Mesmerized by the criminal law, by the judges making it, the magistrates administering it, the ordinary people become felons by breaking it, we too have made the same kind of pursuit as Thompson, who longed to find some writers of anonymous threatening letters *actually prosecuted* quite as much as he longed for the actual trials of men suspected of deer-poaching and bark stripping under the Black Act (1723). He sent Mr Dodd haring after trial records in the Public Record Office and Surrey County Record Office; he went tearing through all the country record offices of southern England and the East Midlands. We were (are still) mesmerized by the criminal law say some, because the 'pre-eminent … inspirational sources' for social-history research remain 'the classic studies of the "Warwick school"' dating from the 1970s, 'which emphasized oppressive features of the administration of justice'.[51] In *Warwick University Ltd* Thompson made a notorious connection between Warwick University and the eighteenth-century state and legal system, as we have seen.[52]

Though his undergraduate teaching in the History School continued, the last prospectus to offer special subjects taught by Thompson was for 1967–8. I can find no trace of the enticing-sounding 'Politics and Poetry in England in the Age of the French Revolution, 1791–1801', which was a new special subject almost certainly offered by Thompson for the academic year 1970–1. He had left Warwick by the autumn of 1970: he wanted to write; to get back to his 'trade as writer'; not to be overwhelmed by undergraduate teaching; he wanted out.[53] Thompson was discreet about his reasons for leaving Warwick; sometimes 'he just … didn't answer the question when it was put to him … his reasons were personal', says Bryan Palmer.[54] Perhaps the business of the 'files affair' and 'Warwick University Ltd' had little to do with his final, long-threatened departure. In October 1969 with the Centre just a year old (officially speaking), the Board of Arts resolved to recommend that it 'should continue as an independent graduate school, [and] that Mr Thompson's successor should be appointed as soon as possible'.[55] His resignation was in, and arrangements for the future being made four months before the occupations of February 1970.

'Only an independent centre can attract students', reiterated the new head of History in 1969; 'it has been remarkably successful … the Centre is one of the prominent features of the University. Its demise could scarcely fail to demoralize within the University and to impair the latter's reputation without'.[56] Successful? Yes, said Thompson a year after independence, in May 1969. He provided the figures to demonstrate it, showing an increase of four to eleven students in the period 1965–8. And then again, the vast panorama of a future was laid out: exchange of North American, Australian, Panamanian and Filipino researchers in social history, not to mention the European, Latin American and African. 'This is the plan: but of course it won't happen. What is happening at present is a boot-strapped operation, with frayed boot-laces; or, more exactly, a gamble upon upward revision of UGC arts graduate ceiling.' He prophesied that 'monies will be fought for between the Schools, and then within them … the boot-straps will snap'. In which case, why not take the Centre back into History 'as a substantial component of its regular graduate school'?[57] The only loss would be interdisciplinarity and its status as an international centre. He was working on eighteenth-century threatening letters at the time: this and his many letters of resignation suggest how much he learned from them.

Leaving Middlemarch

The story of social history at Warwick reflects the social history of Middlemarch and Loamshire; a materiality and a culture intertwined, with the important structural factors of Whitehall, government policy on higher education and the UGC to be taken into account in the as-yet-unwritten social history of this time and place. In the novel, no one leaves; no one gets out. But you could (and can) get out of Warwick: the large majority of all universities' populations are those birds of passage, its students. Did Warwick's history undergraduates know of utopia? From 1969 second year undergraduates might choose 'The Social and Intellectual History of England, *c.* 1500–1660' and read 'the utopias of More and Winstanley and selected texts of Francis Bacon'. A student could have encountered utopianism on Thompson's special subject 'Sans-culottes and socialists, 1791–1848' which ran from 1966. Anna Davin, undergraduate in the history school from 1966–9, recalls a different kind of utopia. She remembers the walk to the university from her home in Canon Park, though the sweetly lost lands of content, up to 'where the administration was, maths and administration … at Gibbet Hill cross road, more or less … to go to the teaching block where History was, the Library, the students union … between them, you went down across a stream, and there was a lovely, lovely woodland, in spring all blue with flowers. I was quite sad when I went back many many years later for the first time, and saw the concrete and lots more buildings, and none of that there'.[58] She was a young mother, living with her three small children and partner in a family home in a pleasant Coventry suburb. She didn't *live* in the place of desolation. 'Desolate, sometimes this place was desolate,' recalled Will Fitzgerald. 'I mean when the wind blew and the rain rained and the mud mudded and the tiles fell, the weekends particularly a few of us had to for domestic reasons had to stay here but god it was bleak sometimes and just nothing was open. And just think about it, there was Rootes Hall and most of the main catering, even with the caff closed down so it was minimal catering at weekends.'[59] Mike Shattock came for his interview for the post of Registrar in July 1969: 'The whole place struck me as strange. I mean you know there were very, very few buildings and a lot of space,' but 'it was a lovely summer's day' when the wind didn't blow and the empty seas of mud didn't have to be negotiated.[60]

Anna Davin met Dorothy Thompson, Edward Thompson's wife, at a party. 'We were talking and I'd said how I was worried about not knowing how to write an essay. I hadn't been in formal education … for eight years or something, and she said we could pick a subject for me to write an essay about, and then she would read it for me and make comments … I went on the bus to Leamington … and I remember the incredible sense of freedom, going on a journey on my own, without the three children in tow … seeing the hinterland of Coventry and the University, going through Kenilworth, and on to Leamington ….'[61] Half a century later, the heart still lifts as you leave; the land and sky open up, whether you take the No. 12 bus that still maps the terrain between Gibbet Hill and North Leamington, or drive more briskly along what is now the A46/A452. On both routes you may glimpse, through all the houses in between, on a low hill, on a wooded incline – in Loamshire, the place where you are – a figure watching all of it; all that has been and will be: 'Destiny stand[ing] by sarcastic with our dramatis personae folded in her hand.'[62] In his interview for the Voices of the University project, Will Fitzgerald remembered Mike Shattock saying to him sometime in 1970 ('if I'm wrong Mike I do apologise') that 'actually, if we got closed down for whatever reason or we suddenly just didn't happen, the leaves would blow very quickly through here, the buildings would fall down and it … [would be] as though it never existed'.[63] Destiny inclines her head in agreement. This is her kind of historical analysis.

Notes

1. Margaret Drabble, *The Realms of Gold* (1975) (Harmondsworth: Penguin, 1977), 234; George Eliot, *Middlemarch. A Study of Provincial Life* (1871–72) (Harmondsworth: Penguin, 1965), 122.
2. 'The Business University', *New Society* (19 February 1970), 301–4.
3. E. P. Thompson, *The Making of the English Working Class* (London: Gollancz, London, 1963); idem., *Witness against the Beast. William Blake and the Moral Law* (Cambridge: Cambridge University Press, 1993).
4. E. P. Thompson (ed.), *Warwick University Ltd. Industry, Management and the Universities* (1970) (Nottingham: Spokesman, 2014), 17.
5. Grace Huxford and Richard Wallace, 'Voices of the University. Anniversary Culture and Oral Histories of Higher Education', *Oral History*, 54 (2017), 79–90.
6. *Report of the Committee Appointed by the Prime Minister under the Chairmanship of Lord Robbins, 1961–1963, Parl. Papers* (1962–63) xi–xiv, Cmnd. 2154.
7. *Report of … Lord Robbins*, 257; Carolyn Steedman, 'Young Women at University in the 1960s and 1970s' (unpublished paper, University of Warwick, 2015); Nicholas Barr (ed.), *Shaping Higher Education. 50 Years after Robbins* (London: LSE, 2014).
8. Jim Obelkevich, 'New Developments in History in the 1950s and 1960s', *Contemporary British History*, 4 (2000), 125–42.
9. Miles Taylor, 'The Beginnings of Modern British Social History', *History Workshop Journal*, 43 (1997), 155–76.
10. David Feldman and Jon Lawrence, 'Introduction: Structures and Transformations in British Historiography', in *Structures and Transformations in Modern British History*, ed. Feldman and Lawrence (Cambridge: Cambridge University Press, 2011), 1–23; Ben Highmore (ed.), *The Everyday Life Reader* (London: Routledge, 2002), 1.
11. Laura Carter, 'The Quennells and the "History of Everyday Life" in England, c. 1918–69', *History Workshop Journal*, 81 (2016), 106–34.
12. J. R. Green, *A Short History of the English People* (London Macmillan,1874), v.
13. David Cannadine et al., *The Right Kind of History. Teaching the Past in Twentieth-Century England* (Basingstoke: Palgrave Macmillan, 2011), chs 1–2. Carolyn Steedman, 'Battlegrounds: History in Primary Schools', *History Workshop Journal*, 17 (1984), 102–12.
14. *Times* (22 June 1968), 20.
15. *Times* (9 April 1966), 9; *E. P. Thompson and the Making of the New Left. Essays and Polemics*, ed. Cal Winslow (New York: Monthly Review Press, 2014), 290–306.
16. University of Warwick Modern Records Centre (MRC), 'Meeting of the Planning and Executive Committee' (9 March 1966), Agendum 3C, MRC UWA/FICHE/REF/R1724.
17. 'A Centre for the Study of Social History at the University of Warwick. The Scope of the Subject' (9 July 1965), MRC UWA/FICHE/REF/R1724.
18. Royden Harrison, 'From Labour History to Social History', *History*, 60 (1975), 236–9.
19. Alexander Hutton, '"Culture and Society" in Conceptions of the Industrial Revolution in Britain, 1930–1965' (unpublished PhD thesis, University of Cambridge, 2014), 131–73, 207–310.
20. MRC UWA/PUB/8/1/1, University of Warwick, *Bulletin*, October 1964, No. 1.
21. *Economic Development & Cultural Change* (University of Chicago, 1952–), *Comparative Studies in History and Society* (The Hague, 1958–), *Annales* (Paris,

1929–), *Le Mouvement Social* (Paris, Institut Français d'Histoire Sociale, 1961–) and *Movimento Operaio* (Istituto Feltrinelli, Milan, 1947–1955) all showed 'how far this work has already advanced'.

22 'A Centre for the Study of Social History at the University of Warwick. The Scope of the Subject' (9 July 1965), MRC UWA/FICHE/REF/R1724; also 'Meeting' Appendix A (20 April 1966), MRC UWA/M/PES/1.

23 Henry Rees, *A University Is Born. The Story of the Foundation of the University of Warwick* (Coventry: Avalon Books, 1989); Michael Shattock, *Making a University. A Celebration of Warwick's First Twenty Five Years* (Coventry: University of Warwick, 1991). For architecture and building in the British new universities, see Elain Harwood, *Space, Hope and Brutalism. English Architecture, 1945–1975* (London: Yale University Press, 2015).

24 J. B. Butterworth, 'Rootes, William Edward, first Baron Rootes (1894–1964)', rev. G. T. Bloomfield, *Oxford Dictionary of National Biography*, Oxford University Press, 2004 (accessed 10 February 2016).

25 'First Report of the Vice-Chancellor to the First Meeting of the University Court' (17 November 1965), MRC UWA/PUB/1/1.

26 University of Warwick, *Prospectus* (1965/66), MRC UWA/PUB/4/2.

27 'First Report of the Vice-Chancellor', MRC UWA/PUB/1/1.

28 'Memo' from VC to Registrar and Vice-Chairman, Board of Arts (undated, c. 1964–5) MRC, UWA/F/FD/3/2, R.1038.

29 '1st Degree Courses in History' (undated), MRC UWA/F/FD/3/2, 'History at Warwick, (memo) (October 1964).

30 'Graduate Work in History' (28 February 1968), UWA/M/BFA/1/2; Tony Mason, 'The Final Curtain, Centre 1968–98', unpub. Ms., 1998, MRC 1040/7/2/1: 'Edward Thompson once said that the idea for the Centre was roughly sketched during his job interview.'

31 Thompson to History Staff, 'Another Thrilling Instalment', 19 February 1968, MRC UWA/FICHE/REF/R1724.

32 Letter quoted in Thompson, 'Another Thrilling Instalment'.

33 Thompson, 'Another Thrilling Instalment'. Later Thompson remembered being 'told that one file which attracted a good deal of sardonic interest was a bulky file full of my own fatuous and long-winded attempts at resignation'. *Warwick University Ltd*, 157.

34 Extract from minutes of the meeting of Senate (15 June 1966), MRC FICHE/REF/R1724; 'University of Warwick. Report of the Vice-Chancellor 1965–1966', UA/PUB/1/2.

35 J. M. Bruce Lockhart to John Hale, 3 May 1966, MRC UWA/FICHE/REF/R1724.

36 Thompson, 'Memo' (21 November 1966), extract from minutes of the meeting of Senate (30 November 1966), MRC UWA/FICHE/R1724.

37 'Memo from VC to Registrar and Vice-Chairman, Board of Arts', N.D. (October 1964), MRC UA/F/FD/3/2, R.1038.

38 'First Report of the Vice-Chancellor'.

39 John Hale to Richard W. Donar, 21 October 1965, UA/FICHE/REG/R1724.

40 John Hale to Registrar (9 December 1965), MRC UA/FICHE/REF/R1724; also 'Extract from minutes of the meeting of Senate 30 Nov 1966 for the appointment of the selection panel for a Lectureship in American Labour History', UA/PUB/1/2 UA/FICHE/REF/R1724.

41 Thompson, 'Another Thrilling Instalment'.
42 'Extract from Minutes of the Meeting of Senate 15 June 1966', MRA/UWA/FICHE/REF/R1724.
43 Thompson to the Registrar, 13 February 1968, MRC UWA/M/PES/3.
44 Thompson, 'Another Thrilling Instalment'; 'Minutes of a Meeting held on 28 February 1968', Board of Studies for Arts, MRC UWA/M/BFA/1/2. 'The Centre Bites the Dust (for the third time)' was a subheading to Thompson's Thrilling Instalment.
45 University of Warwick, *Bulletin* 2 (April 1968), MRC UWA/PUB/8/2/5.
46 Hale to the vice-chancellor, 29 May 1968; vice-chancellor to Hale, 29 May 1968; Thompson to the vice-chancellor, 31 May 1968, MRC UWA/FICHE/REF/R1724.
47 'Minutes of a Meeting held on 28 February 1968', Board of Arts, UWA/M/BFA/1/2; Minutes of the Meeting held on 13th March 1968', Senate, MRC UWA/M/S/7; University of Warwick, *Bulletin*, 2 (April 1968), UA/PUB/8/2/5.
48 David Montgomery and E. P. Thompson, Agendum 11, Report of the Board of Studies for Arts for the Meeting of the Senate (4 December 1968), MRC UWA/M/PES/4. For Gwynne Lewis (1933–2014), see Colin Jones and Robin Okey, 'Gwynne Lewis Obituary', *Guardian*, 25 January 2015, http://www.theguardian.com/education/2015/jan/25/gwynne-lewis (accessed 30 September 2019).
49 My warmest thanks to Warwick History for allowing me to examine the MA dissertations it holds (and which I packed away at the Centre's demise in 1998).
50 After much moving about of material (on anonymous letters, riots, poaching) between the two, both works were published in 1975: E. P. Thompson, *Whigs and Hunters. The Origins of the Black Act* (London: Allen Lane, 1975); Thompson, 'The Crime of Anonymity', in *Albion's Fatal Tree. Crime and Society in Eighteenth-Century England*, ed. Douglas Hay et al. (London: Allen Lane, 1975).
51 David Lemmings, 'Introduction', in *The British and Their Laws in the Eighteenth Century*, ed. Lemmings (Woodbridge: Boydell Press, 2005), 3–5; John H. Langbein, 'Albion's Fatal Flaws', *Past and Present*, 98 (1983), 96–120; Carolyn Steedman, 'At Every Bloody Level. A Magistrate, a Framework Knitter and the Law', *Law and History Review*, 30:2 (2012), 387–422; idem., 'Threatening Letters. E. E. Dodd, E. P. Thompson and the Making of "The Crime of Anonymity"', *History Workshop Journal*, 80 (2016), 50–82.
52 Thompson, *Warwick University Ltd*, 17.
53 Thompson to Dodd, 13 December 1967, MRC, MSS 369/1/13; Thompson to Dodd, 29 September 1969, MSS 369/1/13.
54 Michael Merrill, 'An Interview with E. P. Thompson', *Radical History Review* (12), 1976, 4–25; Bryan D. Palmer, *E. P. Thompson. Objections and Oppositions* (London: Verso, 1994), 113.
55 J. J. Scarisbrick to the Registrar, 24 October 1969, MRA/UWA/FICHE/REF/R1724.
56 Ibid.
57 'Social History Ten Year Planning', MRA/UWA/FICHE/REF/R1724.
58 'Voices of the University', University of Warwick Oral History Project, MRC, 1069/47/1. Interview with Anna Davin (4 February 2015). She went on to study for her PhD with Eric Hobsbawm: 'We had a very prickly and difficult relationship … Warwick seemed like golden days by comparison.'
59 'Voices of the University', MRC 976/5/1/1-4, Interview with Will Fitzgerald (15 March 2013). Will Fitzgerald was secretary of the Warwick Students' Union in 1969 and President 1970–1.

60 'Voices of the University', MRC 974/1/1, Interview with Michael Shattock (29 November 2012).
61 Interview with Anna Davin.
62 Eliot, *Middlemarch*, 122.
63 Interview with Will Fitzgerald.

10

Innovation and evolution: Lancaster learning, 1964–74

by Marion McClintock

Lancaster was the seventh and last of the so-called plateglass universities to be announced, but the fastest in formation. Only three-and-a-half years separated the initial bid by Lancashire County Council to the UGC (12 May 1961) to the moment when the first students settled into temporary teaching accommodation at St Leonard's House in the city centre on 6 October 1964. By the time of Lancaster's announcement in November 1961 the Robbins Committee was already sitting and so the origins of placing a university in northwest Lancashire must be seen, like its peers, as the result of deliberations leading back to the nation's recovery from the Second World War and a determination in the late 1950s to create a stronger higher education system.[1]

The stated criterion for the allocation of seven new institutions was an expectation that each one should be part of 'the community in which it lives', and so evidence of local support was always required.[2] Decisions about their designation and location were however taken at national level, thus responding to local bids, but serving national needs.[3] Lancaster's chances were enhanced by the advocacy of the much larger, pre-1974, County Palatine of Lancaster in its promotion of a bid to the UGC and, with the University of Manchester, the provision of resources for the preparation of the new institution. Some key local families, notably Arthur Peel (the 2nd Earl Peel) and Edward Stanley (the 18th Earl Derby), were also highly influential in garnering support and raising funds, while the contributions by Lancaster City Council in making an approach to the County Council, and in offering a greenfield site at a peppercorn leasehold two miles to the south of Lancaster, were essential.

Following the precedents set at East Anglia, Essex, Sussex and York, Lancaster obtained full university powers at the moment of the implementation of its royal charter.[4] The initial decisions over curriculum, organization and governance had been determined by an Academic Planning Board (APB), reporting direct to the University Grants Committee, and chaired by Sir Noel Hall, principal of Brasenose College, Oxford, and founder of the Administrative Staff College at Henley. Five of the other six board members had first degrees from either Oxford or Cambridge, and all had held posts at leading UK universities. The Board's primary function was to 'consider the arrangements by which the Universities [i.e. the sector as a whole] may be assured of the maintenance of satisfactory standards at the university' and only secondly to 'consider the range of subjects to be studied at the University during the first years of its existence and the length and general character of the undergraduate courses'.[5] Significantly, postgraduate programmes were not included in its scrutiny.

Planning

The APB worked closely with the 'Executive Council for the Establishment of a University at Lancaster', a body run and chaired from the County Council, most notably in the appointment of the vice-chancellor. Their first choice, Jack Butterworth, went to Warwick, but they were more successful in their selection of Sir Charles Carter.[6] A Quaker, educated at Rugby School and St John's College, Cambridge, Carter's studies were interrupted by the Second World War, part of which he spent in Strangeways Prison while his credentials as a conscientious objector were examined. He graduated with first class honours in Economics in 1944 and, after a spell at the Queen's University, Belfast, became the Stanley Jevons professor of Political Economy and Cobden lecturer at the University of Manchester. He was already known to Hall, having served with him on the APB for the University of Kent, and their shared interest in business studies and economics was helpful in setting Lancaster's priorities. A shy man, by his own account lacking social graces, aristocratic birth or inherited wealth, nonetheless, as Christopher Driver correctly noted, 'Everything at Lancaster, from its constitution to its colours (Lancaster red and Quaker grey), has a touch of Carter.'[7] The effectiveness and energy that he injected into the process brought into being an institution that to this day bear the print of his influence and of his values, as he summarized on the occasion of Lancaster's tenth anniversary:

> First, a living concern for the quality of civilisation – and this includes an interest in the creative arts and the physical and social environment, as well as in the traditional learned subjects. Second, a willingness to do better in extending educational privilege, and in giving special help to the deprived. Third, a continuing readiness to experiment and to seek an informed understanding of educational problems. Fourth, a continued and developed service to the schools and to other institutions of higher education; for we care, not only about our own little bit of the educational system, but about the whole.[8]

Carter was concerned with productivity and value for money. An early action was the appointment of a research fellow in teaching methods. Other start-up initiatives included an enquiry for the Committee of Vice-Chancellors and Principals into the use of academic staff time; an examination of a four-term year; a scrutiny of whether two-year degrees were feasible, and a proposal that the new University's Senate and Council might be merged into a single body.[9] Carter was quick to welcome national moves to include students on the university's governing bodies. He also initiated commercially sponsored residences at Lancaster, when the UGC was unable to continue capital funding for the colleges.[10]

An early task for the APB was to appoint site development architects. Interviews held at the RIBA headquarters in London led to the choice of Bridgwater, Shepheard and Epstein. The beautiful pasture land at Bailrigg with which they were to work lies some 300 feet above the River Lune estuary, consisting of windswept and clay land, with rain soaking in at the top of a hill that oozes out in springs and that slopes down steeply to the west.[11] The decision was made to position the buildings on the plateau at the top of the hill, and to construct a north–south spine as a determining principle, with pedestrians above and services below. Single-function buildings such as the library, the administrative headquarters (University House) and the Physics building were interspersed with colleges which also housed academic departments, all designed for human occupation, with the sole exception of the residential Bowland Tower that encased the boiler house chimney. Construction started at the centre and worked outwards north and south, with an underpass below the main square to take traffic from west to east. Staff and students were able to live and work in the completed spaces without interference from the builders at the periphery. Lancaster was, said Gabriel Epstein, 'a very special job, a key job, a favourite child', although compromised by budgets set too low for the higher-quality finishes that would have better stood the test of time.[12]

One exciting and perhaps unexpected building was the multi-faith Chaplaincy Centre, built in three segments like a clover leaf, and sweeping upwards to a tripartite spire. The Roman Catholic diocesan authorities had taken the initiative in approaching the University with offers of funding and also with an introduction to Cassidy and Ashton of Preston, architects who specialized in religious buildings. Financial support also came from the Jewish community. The UGC was persuaded to help with fees and fittings for the social areas of the centre. Boldly distinctive on a visitor's first approach to the university, with its resident chaplaincies and its public rooms available to all members of the university, the Centre opened in April 1969.[13]

Lancaster was designed by the ABP to be organized on a collegiate basis. The form and location of the founding colleges were pivotal in early design discussions. They were built around open courtyards, with social, dining and administrative accommodation on the ground floor. Academic departments occupied much of the first floor, alongside student rooms that continued onto the second floor, topped off by roof-level flats for members of staff. Four colleges were completed by 1970: Bowland (named after the nearby forest) and Lonsdale (after Lonsdale South of Sands, an old county division) formed mirror images of each other, linked by shared refectories, then Cartmel (named after the local village) and finally the County College, funded by the County Council.[14] All staff were allocated to a college and each student – whether resident on campus or not – was a member of a Junior Common Room that underpinned the college's social life and its governance.

Students were represented on the 'syndicates', that is, the colleges' governing bodies which in turn reported to the University Senate, and also on the Federation of Junior Common Rooms which fed into the Student Representative Council. Students therefore had a very direct influence on student affairs through their colleges, although, as with other new UK campuses, there was no central Students' Union. There was a zest and liveliness about early student cohorts, who thrived on being part of an embryonic institution. Numerous societies were set up: rowing, speleology, ballroom dancing, mental health, 'Third World First' and crown green bowling. There were two rival student publications, *John O'Gauntlet* and *Carolynne*. There was a debating society, an exclusive dining club and lively rag processions through the city of Lancaster that caused disquiet for local residents, unaccustomed to being spectators at wrestling matches between students involving flour and water. In a university where many junior staff were not much older than the students, there was a common identity in being pioneers.

Lancaster was however the least visible of the plateglass universities. Partly this was because the announcement about its foundation came last, but also because of its relative geographical isolation. The submission to the UGC was for the 'promotion of a university in north-west Lancashire' rather than for Lancaster.[15] Emphasis was placed in the bid on the absence of any university west of the Pennines for over 200 miles, from Manchester to Glasgow, whilst the principal illustration in the bid document was of open landscape with stretches of water, uninterrupted by roads or habitation. Commentators placed their own interpretations on that blank canvas. For example, the *Observer* contrasted Lancaster with its White Rose neighbour, epitomizing York as being 'the brave old world [while] Lancaster looks like being the bleak new one … [Charles Carter] seems almost a personification of the new technocratic Britain. Economist – the applied kind, with statistical background; social commitment – good adult education man, involved in the North-West – ex-Manchester Professor of Political Economy … Verdict on Lancaster: the dark horse'.[16] Wrongly, as we shall see, Lancaster was perceived as traditional. Teaching and learning seemed to be based on existing precedents of the early 1960s: a prescribed curriculum for each subject with little choice; student use of primary texts at a time when widespread development of textbooks was in its infancy and electronic media unknown; a reliance on students themselves to make the adaptation from secondary to tertiary teaching and learning; and at first assessment that relied principally on three-hour written examinations. Students sat quietly at desks arrayed in academic gowns, their teachers also wearing academic dress.

The emerging curriculum

With Carter in post, the first academic appointments quickly followed. Most of these were older men, from London, Oxford, Cambridge and in some cases the civic universities. Often with wartime experience, they were forceful and pushy. They were appointed as founding heads of departments and came with strong views about the direction of teaching and research that their disciplines should take. For example, Gordon Manley, the climatologist, arrived from Bedford College, London to head up the new Department of Environmental Studies. Chemistry was led by John Bevington, who moved from Birmingham.[17] As Alfred Sloman, the first vice-chancellor of the University of Essex, noted, 'Appoint a professor and establish a department for whatever reason, and you have probably accepted the subject for all time. The professor will die but his department may well live on for ever.'[18] In their turn, these founding professors appointed younger and less experienced staff to take forward their initial ideas, many of whom had chosen to come to a new university in the expectation that their role would be to shape the institution rather than simply work in it. Martin Blinkhorn, one of the early appointments in history, later recalled:

> They [the heads of departments] had to deal with young, obviously ambitious and often quite spirited academics, who by virtue of the fact that they were taking jobs in, you know, in the wilderness, as it were, it told you something about them, or about us, and that is to say we had a sense, something of a sense of adventure, and we weren't necessarily going to be easy people to boss around or control ... So it was a very distinctive environment and I think most of us enjoyed it, even if we didn't think we were going to stay a long time.[19]

Turning to the curriculum, the UGC had accepted in its entirety the recommendations of the APB's interim report about the need for three broad disciplinary areas. These were the natural sciences, the humanities (including languages), and management and business studies. Some general principles were laid out: all undergraduate programmes would be at honours level; there would be no fourth year, as at Keele, however desirable, as finances did not permit; and all students would be required to take a subsidiary subject, or 'distant minor', of which more below.

In January 1964, the first meeting of the shadow Senate took place, symbolically marking the moment when the process of curriculum development and delivery passed from the external APB to the university, in particular to a 'Development Committee', the powers of which would later be challenged.[20] By the time the charter and statutes were approved, and the first students started arriving the following Autumn, the new curriculum was ready. How innovative was Lancaster? Harold Perkin, who moved to Lancaster's new history department from Manchester, later claimed that the new British universities 'enjoyed from the beginning both the capacity and the liberty to experiment in every aspect of university education'.[21] Although that freedom was quickly circumscribed in practice, Lancaster's development nevertheless marked the beginning of a major shift in the understanding of subject areas. While at a national level the UGC of the early 1960s delegated curriculum innovation to fixed subject-based subcommittees and panels, a much more fluid and dynamic pattern emerged in individual universities, as the example of Lancaster demonstrates.[22]

Lancaster's sciences included biology, chemistry, environmental studies, mathematics and physics.[23] Despite Carter's considerable efforts to have medicine included, he was unsuccessful.[24] Physics also featured theoretical and applied topics, and the department later became renowned for its work on ultralow temperatures and particle physics. Mathematics developed considerable strength in statistics, while chemistry honed expertise in polymer science. Biology was the greatest chameleon, with additions of ecology and biochemistry. In the early 1970s applied science subjects were added: engineering including design and mechatronics (the combination of robotics and engineering), computing and psychology.

The arts and social sciences were more protean. English led the way in innovation. The curriculum devised by the founding professor, Bill Murray, formerly of the

University of Khartoum, and colleagues such as the medievalist Stanley Hussey, who came from Queen Mary College London, required study of three branches: English literature, Anglo-Saxon language and literature, and modern language.[25] From English emerged theatre studies, linguistics and creative writing. History included history of science and medieval studies and later on American studies. Lancaster also pioneered social history, Harold Perkin becoming the first chair in the subject in the UK in 1967.[26] French Studies was another founding subject, with Russian and German quickly following, along with Central and South-Eastern European studies, and, for a brief period, Arabic and Islamic Studies, an externally funded programme.[27] Another new subject was politics and international relations including peace and conflict studies, then not part of the school 'A' level curriculum. Philosophy burgeoned to include a choice of joint degrees with social studies, the natural sciences or languages, with an ambition for the continuation of all three strands at Part Two. After much debate, classics was introduced, initially requiring students to tackle both Latin and Greek texts in the original.

Lengthy discussions took place, starting at the shadow Senate, about whether theology should be offered. In October 1964, the Canon Theologian of Blackburn Cathedral called for a department that examined basic questions of humanity within a divinity faculty. By December, the founding Professor of Philosophy, Frank Sibley, who had come from Cornell University, had broadened the discussion, influencing Senate's decision to run a religious studies programme 'not to convince students of a revealed truth ... but, rather, to examine in a rigorous and scholarly manner the variety of human religious belief and experience, and the impact of religions upon culture and civilization'.[28] The founding chair was duly advertised as open to candidates of all faiths or of none. Come the day of interviews, the professors of Mathematics and Philosophy, on their way to lunch in 'The Stork' pub at Condor Green, agreed to lobby the vice-chancellor to change tack and approach the external assessor for the chair – Ninian Smart from the University of Birmingham, a champion of 'religious studies' over theology – and invite him to be considered for the chair instead. Once another external assessor had been found, Smart's appointment was confirmed. Smart put in place a challenging curriculum, including religious and atheistic thought, scientific approaches to religion, New Testament studies and the history of Indian religion, all at Part One.[29] As well as founding Religious Studies, Ninian Smart turned out to be an innovator in academic planning. In 1967 he drew on his North American connections to bring undergraduates from the United States to Lancaster for their junior year, taking transferable credits back to their home institutions, an initiative that was later extended to students from Australia.[30]

Whilst the UGC did not encourage Sussex, York and UEA to include vocational subjects, Carter was persistent about adding Law to the curriculum at Lancaster.[31] He was vindicated when the UGC approved a second wave of departments at Lancaster: Law was introduced alongside other applied subjects, namely Visual Arts, Educational Research, Social Administration and Management Teacher Development. The creative arts proved even more malleable. Bill Murray in English and Tom Lawrenson, founding Professor of French Studies (who arrived at Lancaster from Aberdeen, via the University College of the Gold Coast), obtained funding from the Nuffield Foundation for an experimental studio enabling the practical investigation of world theatre, for example, by staging seventeenth-century French dramas in their original form.[32] This studio, equipped appropriately and supported by workshops for fine arts, music and dance, formed a bridge with local public arts provision, demonstrating how the humanities could benefit from practical demonstration facilities, conventionally seen as relating only to the sciences.

However, it was in Business and Management Studies, initially at postgraduate level, that the shared enthusiasms of Carter and Noel Hall were most fully realized. Operational Research was the star department here. Carter gained funding from the Institute of Marketing. The new department was established in 1965, signalled as 'a new subject as an academic discipline ... so little organised research into marketing problems has been carried out, except on behalf of companies and their product and brand policies, that it would be premature to take a narrow view of what constitutes a suitable research topic'.[33] Carter persuaded Patrick Rivett to move from the National Coal Board, where he had led on operational research and come to Lancaster as founding professor. Rivett set up a department that relied on consultation work with industry and government, 'a model of a business situation in the same way that his military counterpart does in a battle situation ... [as] a set of mathematical equations which predict how the business will perform under alternative conditions'. Rivett only stayed three years, enticed away (to his later regret) by Sussex.[34] Other innovators came to Lancaster as well. Gwilym Jenkins was brought in from Imperial College, University of London, to establish Systems Engineering, advertised to the aspiring 'A' level entrant as 'the study in their totality of complex systems ... it is the optimization of the overall behaviour of the system that is the main object of Systems Engineering [and] requires collaboration between persons with different backgrounds such as engineers of different types, mathematicians, computer experts, chemists, economists and accountants'.[35] From this 'hard' systems programme emerged 'softer' systems led by Peter Checkland who came from ICI. This was a field that sought to engage with 'situations involving everyday human decision making' as Checkland later summarized the discipline.[36] Closely allied to soft systems was the Department of Behaviour in Organisations, headed by Sylvia Shimmin, appointed from University College, London as one of Lancaster first female chairs, its mission identified (with no intended irony) as the study of the psychology of 'man-made systems, in human aims, aspirations and expectations'.[37]

The Lancaster degree schemes had the strengths and weaknesses of wide choice. Students could register for combinations of cognate or contrastive disciplines, such as Physics and Music, or Economics and French Studies. However, joint programmes were more demanding, particularly where the students had to make the intellectual connections between two or three major subjects. The APB stated to the UGC in March 1963 that its members wished to avoid 'excessive specialization in a single subject or, indeed, in facets of a single subject. We equally seek to avoid undisciplined spread over several subjects more or less at the option of the original undergraduate'.[38] By 1966 there were serious debates about how to ensure that the almost infinitely variable choices were combined into coherent schemes of study, and about how much student choice to permit. Some academic staff favoured a single route through the curriculum, for example, physics argued that progression builds on previous knowledge, so a prescriptive pattern of units was required.[39] At the other end of the spectrum, others favoured even greater freedom for student choice, including the opportunity to dispense with the apparatus of an agreed and uniform course structure, and to make the individual student a producer of his own curriculum rather than a consumer of one designed for students in general.[40] This initiative became the School of Independent Studies, approved in 1973, whereby students could devise their own schemes of study, with guidance, each scheme being individually approved and assessed. Such an approach could be a high risk, but it appealed to some students.

Despite the involvement of students at Lancaster in the university's governance, encouraged by Carter himself, it was the academic staff who determined the curriculum.[41] In some cases, learned societies such as the Royal Institute of Chemistry and the Law Society provided advice. Mostly, however, new courses came from within. For example in 1969 in response to an invitation by the university's Development Committee seeking new ideas, thirty-six proposals for new subject areas were promptly delivered for discussion at a conference of the Senate.[42] The APB had set the new university a further challenge, which was to require all students to cross the so-called 'Snow line' (coined to refer to C. P. Snow's influential critique of 1959 of the unhealthy separation of the 'two cultures' of science and the humanities). This would be achieved by taking a 'distant minor' subject. Part of the inspiration here came from APB member Jack Ratcliffe, the radio physicist and director of the Radio Research Station at Slough, who had earlier tried to introduce cross-fertilization at Cambridge and was keen to use Lancaster as a test bed.[43] Thus, Lancaster's first prospectus proclaimed that 'an educated man [sic] should have some idea of the habits of thoughts both of science and the humanities, and to this end particular attention is to be given to studies that cross traditional faculty boundaries'.[44]

What is striking is the seriousness with which the University took the commitment to a 'distant minor', and also how much investment senior academics and university managers put into its planning, implementation and successive reviews. Despite the many weighty issues to be decided before the university opened its doors, the shadow Senate discussed the distant minor on three spate occasions in 1964, agreeing that it should occupy one of the nine Part Two units and normally be taken in the second year. A special subcommittee on the 'breadth subject', as the 'distant minor' became known, met on a regular basis, its membership including senior staff.[45] Some courses proved very successful. The 'History of Architecture' by Basil Ward was so popular that a ballot was required. Similarly, 'Biology and Man', 'Principles of Physics' and the 'Oil Industry' were destined for longevity.

Even more significant was the level of controversy that the 'distant minor' initiative generated. In a fractious and divisive debate that ensued, Carter proved a keen advocate of contrastive study, seeing a course such as the 'Economics of Engineering Projects' as being capable of depth and rigour.[46] However, students sought courses in cognate departments that more closely complemented their major subjects. Some staff entered fully into the spirit of making their subject attractive and accessible across disciplinary chasms, but others subverted the process rather than design attractive offerings.[47] Some staff put effort into demonstrating how subjects were so broad that students were already experiencing opposing habits of thought, while others thought the requirement could be satisfied by contrastive subjects in the first year. It was also suggested the distant minor should be given less weighting in the assessment. 'Distant minor' students could be at a disadvantage. There were no prerequisites for 'distant minor' units such as 'Introduction to Business and Control' or 'French Biology in the 18th and 19th Centuries', although the assessment from such courses counted equally with the other units.[48]

Unusually for the time, postgraduate-taught programmes were developed concurrently with the first degrees, although with less emphasis on subject coverage. Indeed, the first degrees to be awarded, at Lancaster's Ashton Hall in December 1965, were masters degrees. Doctoral students were admitted from 1964, initially in the sciences, and made do with temporary laboratory accommodation at St Leonardsgate, before moving to Bailrigg in 1968. Staff began seeking research grant funding, particularly in the sciences. Physics was notable for its success in grant income generation.[49]

The university and the region

One early responsibility that Lancaster assumed was for the accreditation, assessment and award of certificates and degrees at teacher-training institutions in the north-west. With the approval of the Department of Education and Science, which delegated 'Area Training Organisation' status from Manchester to Lancaster, a School of Education was set up. It was headed by Alec Ross, an educationalist from Exeter, who came in 1966 as head of a new Department of Educational Research and a year later became founding director of the school.[50] Early college partners included Edge Hill College of Education at Ormskirk, St Martin's College at Lancaster, and three smaller colleges at Chorley, Blackpool and Ambleside. The shared headship of school and department benefitted both. The School quickly gained credibility, while the Department, in which research spanned the range from nursery schooling to university education, enabled Lancaster to make significant contributions to national policy debates. Ross was also a close confidant of Carter and his successor as Vice-Chancellor, Philip Reynolds.

Given the important regional imperative in the decision to found a university at Lancaster, an extra-mural studies programme for 'mature' students or lifelong learners might have been anticipated. The reality was more complex, for while the merits of extra-mural studies were debated at the shadow Senate, there was already competing provision from the universities of Liverpool and Newcastle that extended into Lancashire and Cumbria.[51] Additionally, the 1973 enquiry led by Sir Lionel Russell failed to provide direction, with no clarity given to perplexing questions of what might be grant-aided, and how the universities, especially the new institutions, would fit in with the Workers' Educational Association and local education authorities.[52] Nevertheless, the internal pressures for extra-mural studies continued. In December 1973, Carter announced to the Court the appointment of an organizer for extra-mural studies, and that discussions would reopen with the University of Liverpool about Lancaster taking over responsibility for the northern part of its adult education area.[53] So lifelong learning did eventually come to Lancaster, growing from a modest beginning of afternoon and evening sessions into an extensive network of individually accredited tutors and a customised curriculum across Cumbria and north Lancashire.

Pedagogy

Lancaster developed a distinctive style of teaching. The founding staff, predominantly recruited from long-established universities, relied on familiar precedents: small group discussion in the humanities and closely supervised practical classes in the sciences, with generous support from technicians. The first lecture theatres were proportionate to the size of the student body. Academic staff offices could accommodate seminar groups of up to twelve people. Sometimes, for example, in French Studies, joint teaching took place. Originally, open 'mixing bays' were left for informal student meetings, and the larger seminar rooms were interspersed with the staff offices, allowing for the ethos of a discipline to travel down corridors.

Lancaster went further, however, not only proving innovative in its curriculum and teaching style, but also investing in monitoring and improving its programmes. The influential Hale Report of 1964 approvingly noted that, along with Keele, Essex and Manchester, Lancaster was undertaking research into university teaching.[54] The first Senior Research Fellow, Wacek Koc, funded by the Leverhulme Trust, was appointed in 1963, joining two other new colleagues in a mini Department of Higher Education established the following year, in order 'to conduct research into processes of education at universities and similar institutions, and whose special task is to gather information on the functioning of the University of Lancaster'.[55]

One of the first tasks that Koc set himself, when plans for the new Open University were forming, was to examine how television might be used in university teaching, and how offices could be designed to maximize teaching opportunities. However, his next project, analysing students' teaching and learning experiences, proved controversial. The findings Koc published were unflattering to academic staff, including senior professors. As Christopher Driver noted, 'The scholar hired to examine learning situations was soon given to understand that he would be wiser to conduct his researches in universities other than his own.'[56] Looking back now at Koc's questions, in the light of subsequent peer observation and student satisfaction surveys, it is difficult to understand why the responses generated such fury that the reports were suppressed. Some heads of department were however not yet ready for first-year students to make such comments as 'English is the least satisfying because I dislike the study in linguistics, and most lectures and tutorials on literature have been dull and off the point.'[57] An abrupt end was made to this line of work. John Heywood, the research fellow on the project, moved to Dublin, and Koc was sidelined into trivial tasks. Similarly, a survey in 1971 by Operational Research of their past MA students elicited sharp criticism of the casework and practical project approach of which the department was so proud, and this too was kept under wraps.[58]

Despite these episodes, Lancaster increasingly embraced active participation by students in the curriculum and teaching methods. Staff were open to change, including the use of the final year for more research-based units. This approach is well captured in an annual introduction to the university, published annually from 1964 to 1971 by a member of the English staff, David Craig. He invited students to regard themselves as 'joining a guild of specialists, who wish to teach you their craft ... The staff are also constantly learning themselves, partly by trying to add to knowledge and partly by trying to become better at communicating what they know. In this quite literal sense the senior members of a university are themselves students ... and part of their learning takes place in teaching you'.[59]

In retrospect

By 1974 most of Lancaster's main disciplines were in place. Carter retired in 1979, having seen the new university through the teething troubles of an ambitious programme of expansion and some difficult student discipline, against a backdrop of an erratic national policy. His response to an early student occupation of the Senate chamber had been to wish the occupiers a good night's sleep.[60] However, the extent and ferocity of subsequent student troubles, particularly within the Department of English in 1972 and a far-reaching rent strike in 1975, affected him deeply and were widely understood to be amongst the main reasons he left Lancaster after sixteen years of service to the university.

During the 1970s building provision for teaching on the Bailrigg site grew as student numbers expanded. The open spaces within the perimeter road became densely filled. A large area of student accommodation, divorced from teaching, was built at Alexandra Park. Existing student residences, remodelled into larger and taller blocks, replaced the commercially sponsored residences behind the Cartmel, Furness and Fylde colleges. Rather than lose the college ethos, however, Lonsdale and Cartmel were transferred to Alexandra Park. Nowadays, undergraduates continue to take their degrees by college and the centrality of college life means that there is still no separate students' union building.

In 2019 the wide student choice of Lancaster's first decade of life has dwindled. There is now less flexibility. Fewer departments are prepared for potential Part II major students to follow three separate Part I disciplines, thereby limiting the original freedom enjoyed by students, able to change their major after their first year. After a few years the 'distinct minor' became a free ninth unit, before the reduction of Part II to eight units extinguished the option altogether. Other pioneering developments at Lancaster fell away too. Extra-Mural Studies grew into a short-lived Department of Continuing Education, opened in 1994. But the high costs of bought-in tutors and the problems around transferring credits into major degree programmes led to closure in 2010. Similarly, the School of Independent Studies folded in the late 1990s, having proved too resource-intensive.

A backbone of founding subjects continues, including economics, English, history and physics. Some areas have merged into larger clusters, such as the Lancaster Environment Centre (est. 2003) that now includes biology, environmental science and geography. The Lancaster Institute for the Contemporary Arts (est. 2005) hosts art, theatre and design. Management became an integrated school, the first non-science subject at Lancaster to have its own building, although individual departments were retained, with cross-disciplinary offerings at both BA and MA level. By way of contrast, operational research and systems engineering did not survive as free-standing units, but were subsumed into management studies. The spirit of Rivett and Checkland lives on. The most recent innovation in the School of Management has been the Institute for Entrepreneurship and Enterprise Development.

In the arts and social sciences, classics and music have been discontinued. Russian and Central and South-East European Studies were closed down during the era of cuts in the 1980s. However, Italian and Spanish Studies were started up, and, for a while, Japanese and Chinese Studies too. As elsewhere across the UK higher education sector, whilst foreign language study has declined, linguistics remains strong, in Lancaster's case in the field of corpus linguistics, brought by Geoffrey Leech who was appointed in 1969, becoming head of the new Department of Linguistics in 1974.[61]

Early in the new millennium Carter's aspiration for a medical school was realized in the form of a department and later a School of Medicine. Its five-year degree programmes were initially accredited by the University of Liverpool, but are now fully under Lancaster's jurisdiction. Other schemes that took root in the Carter era have flourished too. Lancaster's collaborative partnerships in the north-west expanded to the point where equal numbers of first degrees were being awarded to associated institutions as to Lancaster graduates. From these partnerships emerged Edge Hill University in 2006 and the University of Cumbria in 2007.

Such adaptation and innovation is perhaps inevitable for survival in any institution. The Lancaster story is a little different however, for a reflexivity and flexibility around its curriculum, organization and governance was built into its operations from the start. All the new universities of the 1960s were given exceptional opportunities to make their own choices and set their own paths in learning, teaching and assessment, with light-touch, permissive APBs. Lancaster flourished in this environment. Out of the media spotlight in its early years the University freely experimented, its initiatives informed by staff and by students, for better and worse. In this sense it was perhaps the most utopian campus of all.

Notes

1 University Grants Committee (hereafter UGC), *University Development 1957–1962* (London: HMSO, 1964), paras. 193–222, Cmnd 2267.
2 Ibid., para. 280.
3 John Carswell, *Government and the Universities in Britain: Programme and Performance 1960–1980* (Cambridge: Cambridge University Press, 1985), 14.
4 'Letters Patent of the University of Lancaster, Approved by H.M. the Queen in her Council', 14 September 1964, University of Lancaster Archive (hereafter UA) 1/16/4/2.
5 APB Minutes, March 1962 to March 1964, UA 1/15/1/3; Executive Council Minutes 1962–64, UA 1/9/7/12. The other members were Professor Fred Dainton

(Nottingham); Sir Malcolm Knox (St Andrews); Kathleen Major (Oxford); Professor R. J. Pumphrey (Liverpool); Jack Ratcliffe (Radio Research Station, Slough); Professor B. R. Williams (Keele).

6   'Staff Appointments: C. F. Carter', UA 1/29/7/12.
7   Christopher Driver, *The Exploding University* (London: Hodder and Stoughton, 1971), 179. For Carter, see *Economic Careers: Economics and Economists in Britain 1930–1970*, ed. Keith Tribe (London: Routledge, 1997), 145–55; Roger Middleton, 'Carter, Sir Charles Frederick (1919–2002), Economist and University Administrator', *Oxford Dictionary of National Biography*, https://doi.org/10.1093/ref:odnb/77032 (accessed 12 April 2019).
8   Carter, 'Speech for Tenth Anniversary University Dinner, 30 September 1974', UA l/29/7/12)
9   *Fulcrum*, 32 (November 1977), UA 1/9/1/2.
10 'Commercially Sponsored Residences 1964–69', UA 1/27/3/18; Tony Birks, *Building the New Universities* (Newton Abbot: David and Charles, 1972), 33, 120–2.
11 'John O'Gauntlet Interview with Peter Shepheard', *John O'Gauntlet*, 4 (March 1965), UA 1/9/3/1.
12 Gabriel Epstein, 'An Architect Gives His Verdict', *Lancaster Graduate*, 7 (Autumn 1989), 9–21 (UA.1/12/2).
13 Marion E. McClintock, *The University of Lancaster: Quest for Innovation* (Lancaster: University of Lancaster, 1974), 83–6.
14 Shadow Senate Minutes, SS 64/5/8 (8 July 1964), UA 1/16/2/1.
15 'Submission to the University Grants Committee', Council for the Promotion of a University in North-West Lancashire (8 May 1961), UA 1/15/1/7.
16 *Observer*, 14 June 1964, 12.
17 *Lancaster Guardian* (Supplement), n. d. [October 1964] (UA 1/9/2/3. For Manley, see Christopher P. Green, 'Manley, Gordon (1902–1980), Geographer and Meteorologist', *ODNB*, https://www-oxforddnb-com.libproxy.york.ac.uk/view/10.1093/ref:odnb/9780198614128.001.0001/odnb-9780198614128-e-62248 (accessed 30 September 2019). And for Bevington, see his obituary, *Guardian*, 24 November 2007, https://www.theguardian.com/news/2007/nov/24/obituaries.guardianobituaries (accessed 30 September 2019).
18 Albert E. Sloman, *A University in the Making* (London: BBC, 1964), 23.
19 Interview with Martin Blinkhorn (18 February 2015), UA 1/29/9/2. In fact, like so many, staff, Blinkhorn worked out his entire career at Lancaster.
20 For later complaints, see 'Proceedings of a Conference of the Board of the Senate, Windermere, 1970', GM/70/435, UA 1/15/2/15.
21 Harold Perkin, *New Universities in the United Kingdom* (Paris, OECD, 1969), 17.
22 UGC, University Development, 1957–1962, paras. 21–8; P. Trowler et al., 'Academic Practices and the Disciplines in the 21st Century', in *Tribes and Territories in the 21st Century: Rethinking the Significance of Disciplines in Higher Education*, ed. Trowler et al. (London: Routledge, 2012), 257.
23 'University of Lancaster, Prospectus 1963–64', UA 1/1/1/1/1.
24 Shadow Senate SS 64/3/9 (vii) (9 May 1964); SS 64/5/18 (8 July 1964), UA 1/16/2/1; Senate S 65/1/24 (7 January 1965), UA.1/16/2/7.
25 For Murray, see *Steps. Alumni Magazine* (Summer 2006), 7 and for Hussey, see his obituary: *Independent*, 14 December 2004, https://www.independent.co.uk/news/obituaries/professor-stanley-hussey-686592.html (accessed 30 September 2019).

26 For Perkin, see Jeffrey Richards, 'Perkin, Harold James (1926–2004), Social Historian', *ODNB*, https://www-oxforddnb-com.libproxy.york.ac.uk/view/10.1093/ref:odnb/9780198614128.001.0001/odnb-9780198614128-e-94388 (accessed 30 September 2019), and his autobiography, *The Making of a Social Historian* (London: Athena Press 2002).

27 Arabic and Islamic Studies ran from 1973 to 1982, led by Professor Walid Arafat, a Jordanian, appointed from a seven-year funding agreement between Lancaster and Kuwait (another 1960s campus). Kuwaiti funding ended in 1979: GM/81/22, GM/81/29, Senate Minutes S 81/35 (25 January 1981), UA 1/16/2/7.

28 Senate Minutes S 64/4/5 (3 December 1964) UA 1/16/2/7.

29 Interview with Dr C. R. Claxton (16 February 1972) UA 1/29/6/10). See also the obituary of Smart, *Guardian*, 2 February 2001, https://www.theguardian.com/news/2001/feb/02/guardianobituaries (accessed 26 February 2017).

30 Senate Meeting (20 December 1967), UA 1/16/2; Senate Minutes S 81/156 (21 November 1981), UA 1/126/2/7.

31 Shadow Senate, SS 64/2/12 (7 March 1964); SS 64/3/9 (viii) (9 May 1964), UA 1/16/2/1; Senate S 64//1/38 (18 September 1964), UA 1/16/2/7.

32 T. E. Lawrenson, 'The Place of the Theatre in the Modern Humanities', in *Inaugural Lectures, 1965–7* (Lancaster: University of Lancaster, 1967), 63–79.

33 University of Lancaster, Graduate Studies Prospectus, 1967–8, 34, UA 1/1/1/1.

34 *Lancaster Guardian* (Supplement) (October 1964), 13, UA 1/9/2/3. For Rivett, see his obituary, *Independent*, 18 August 2005, https://www.independent.co.uk/news/obituaries/professor-patrick-rivett-306560.html (accessed 30 September 2019).

35 University of Lancaster, Graduate Studies Prospectus, 1967–68, 57 UA 1/1/1/1.

36 Peter Checkland, *Systems Thinking, Systems Practice* (Chichester: John Wiley, 1981), 284.

37 Sylvia Shimmin, *Behaviour in Organisations: Problems and Perspectives* Inaugural Lecture (Lancaster: University of Lancaster, 1970), 14; 'Sylvia Shimmin 1925–2004', *The Psychologist* (July 2004), 375.

38 McClintock, *Quest,* 107. This was also the position as understood by the UGC: *University Development 1957–1962*, para. 309.

39 'Notes of the Senate Conference held at Ullswater' (19–20 March 1966), 14, UA 1/15/2/9; 'Proposals for Courses, Appendix vii', Shadow Senate SS 64/2/19 (7 March 1964), UA 1/16/2/1.

40 McClintock, *Quest*, 112.

41 'Revision of Statutes', G.68/334 (September 1968), UA 1/16/4/5 (Carter).

42 McClintock, *Quest*, 159–63.

43 J. A. Ratcliffe, 'Can Science Courses Educate?', *Advancement of Science*, 53 (1957), 421–6.

44 University of Lancaster, Prospectus, 1963–64, 11, UA 1/1/1/1.

45 'Committee on Breadth Courses', 2 vols. (1968–70), UA 1/18/1/14.

46 McClintock, *Quest*, 124.

47 An example of how seriously some staff took these experiments can be found in John Heywood and Hubert Montagu-Pollock, *Science for Arts Students: A Case Study in Curriculum Development* (Guildford: SRHE, 1977).

48 GM/71/58, UA 1/18/1/14.

49 Of Lancaster research grants awarded at 30 September 1968, £134,239 of £379,330 went to Physics: 'Fourth Annual Report of the Vice-Chancellor to the Court' (17 December 1968), UA 1/16/3/2, 8.

50 'Associated Institutions Origins', UA 1/1/4/7.
51 McClintock, *Quest*, 297–302.
52 Department of Education and Science, *Adult Education: A Plan for Development* (London: HMSO, 1973), paras. 211–23.
53 'Minutes of the Tenth Annual Meeting of the Court' (15 December 1973), 13, UA 1/16/3/2.
54 University Grants Committee, *Report of the Committee on University Teaching Methods* (London: HMSO, 1964), 107.
55 University of Lancaster, Graduate Studies Prospectus 1967–68, 31, UA 1/1/1/1.
56 Driver, *Exploding University*, 226–7; 'Student Questionnaires and Analysis of Responses', UA 1/1/3/7.
57 W. T. Koc, 'Some Results of the Investigation into the Work-Load of First Undergraduate Intake 1964–65', Department of Higher Education (March 1965), UA 1/1/3/7.
58 Department of Operational Research, University of Lancaster, 'Results of Survey of Past Graduates and of their Views on MA course, 1971', UA 1/1/3/7.
59 David Craig, 'An Introduction to the University' (University of Lancaster, 1969), 15–16, UA 1/16/2.
60 'Interview with Joseph Thornberry' (10 February 2015), UA 1/29/9/3.
61 For Leech, see 'In Memory of Geoffrey Leech' (2004), http://wp.lancs.ac.uk/geoffreyleech/ (accessed 30 September 2019).

11

Failed utopia? The University of Stirling from the 1960s to the early 1980s

by Holger Nehring

Introduction

In early April 1968, Lord Robbins was installed as the first chancellor of the University of Stirling. In his inaugural address, he confessed to have had 'quite a special admiration for the Scottish University tradition, both for the achievements and for the principles on which it rests'. That tradition, he argued, combined specialism with 'common values', and it emphasized that positive sciences had to be taught 'in conjunction with the social values they are supposed to serve'. Robbins hoped that experimentation, links to the local community and a 'universal outlook' would characterize the work of the new university.[1] In his last speech as chancellor at the graduation ceremony in June 1978, Robbins took stock. He highlighted to 'the most enchanting setting for a campus anywhere on this island'. But he also pointed out that the University's importance went beyond purely aesthetic achievements. Because of what the institution had achieved in the field of higher education reform, 'to have visited Stirling [was] a must for all serious students of recent developments in Higher Education in this country'.[2]

A year later, Stirling's staff newspaper struck a less optimistic chord: under the heading 'A chilling prospect', the lead article discussed the increase of cost in fuel supplies and their uncertainty and asked colleagues to cut back on their heating. Another op-ed focused on the perennial issue of lack of car parking on campus.[3] Meeting the new Principal Sir Kenneth Alexander at Stirling, a *Times Higher Education* journalist in January 1982 noted that the offices at Stirling were 'barely heated'.[4] Successive cuts to university grants over the 1970s and early 1980s had affected Stirling especially badly: its economics, sociology and biology departments in particular were threatened, and to some the survival of the institution itself appeared at risk.[5]

Many thanks to Karl Magee, the archivist of the University of Stirling, for providing me with access to much of the material on which this chapter is based and for his help and support in tracking down sources. Not least, my great thanks go to David Bebbington and the late John McCracken, colleagues of (almost) the first hour at Stirling, and Ewen Cameron (Edinburgh) for reading a draft chapter so carefully and making a number of excellent suggestions for improvements.

This chapter tells the early history of the University of Stirling from its foundation in autumn 1967 to the multiple crises of the 1970s and early 1980s by embedding it in the history of government and higher education in Britain more generally. Specifically, this chapter is interested in investigating whether, and if so, how Stirling, the latest of the new foundations in the UK, was still part of the utopian moment that characterized the founding of the other new universities. Superficially, this is a story from 'dreams to disillusionment', during which the dominant perception of universities in British society 'shifted sharply from golden places of uplift, aspiration and modernity, to very tarnished places of conflict, rebellion and, above all, difference – a space apart from society rather than one integrated into it'.[6]

Instead of accepting this narrative of decline and demise at face value, this chapter seeks to highlight the ways in which this history from seeming progress to decline was part of a broader reconfiguration of government in the UK: interpretations of progress and decline were not just mirror images of each other but structurally connected. This had several dimensions: the ambitions that lay at the foundation of the University, which then led to a series of administrative problems; the timing of the opening that came to fall at the beginning of a financial and economic crisis; as well as the hope for liberal values that were not accompanied by an expectation that these values themselves might be open to challenge.

This chapter offers a case study of a history of 'unintended consequences of purposive government action',[7] in particular its assumption of a linearity of progress, that was, in reality, never reliable and settled.[8] The crisis of the University during the 1970s and especially during Margaret Thatcher's government thus occurred as part of a longer history of conflicts in higher education governance and funding in the UK connected to the constitutional question of government in the UK.[9] On the one hand, universities 'represented the ultimate realization of the concept of autonomy, the integration of the disciplines and the high degree of academic participation in governance which had been the ideal prerequisites of university status […]'.[10] On the other hand, universities became part of a 'centralized educational system', in which the 'supply is basically controlled by the government'.[11] Seventy per cent of annual expenditure and 90 per cent of capital investment now came from government, and the university grant (and university spending) consequently became part of a regular parliamentary audit in 1965.[12] Paradoxically, the Scottish universities' defence of academic freedom meant that they accepted rule from the London-based UGC in preference to devolution of higher education to the Scottish Office: academic self-government from London, even though ultimately managed at arms length by the Treasury, seemed to be better than control by civil servants at the Scottish Office in St Andrews House in Edinburgh.[13]

Scotland's 'fifth University'

The arguments surrounding the foundation of the University of Stirling encapsulate in a nutshell the complexity of government in Britain and emphasize certain features of the University's later development and the conflicts around it: they highlight an element of local competition and emulation and the interplay between central and local government; they show the variety of motives and intentions attached to the founding of a new University; and they underline the complexity of the work of the UGC and Whitehall to plan higher education while at the same time serving – and being seen to serve – Scottish as well as academic interests.[14]

The first modern attempt to found a university in Stirling had not been a direct response to the work of the Robbins Committee; it had come as a response to the Barlow Committee on Scientific Manpower which published its report in 1946.[15] Discussions about a fifth university in Scotland, which would complement the four 'ancient' universities of Aberdeen, Edinburgh, St Andrews and Glasgow, began immediately after the Second World War. Five Scottish burghs, including Stirling, applied. The University Grants Committee (UGC) concluded at the time that 'present demands for admissions cannot be regarded as evidence of a need for further expansion'.[16]

Robbins's recommendations in the early 1960s were only successful politically, because his demographic argument for expansion and equality was tied up with quite different arguments – 'the need, in a modern Britain, for much greater numbers of scientists and technologists' and the move away from assumptions about manpower planning to an approach that highlighted the democratic implications of the equality of opportunity in higher education.[17] By the early 1960s, a 'bulge' of students was expected, and Scotland was deemed to be especially affected. Falkirk and East Stirlingshire, Stirling and Inverness Councils as well as Perth, Dumfries, Ayr and the 'new town' Cumbernauld launched campaigns to locate a fifth Scottish university in their communities.[18]

The specific arguments for the creation of a new university in Scotland still contained some assumptions related to manpower planning, but combined these with references to regional development, industrial regeneration, unique setting and local heritage, nationalism and Unionism. Inspired by similar campaigns around the country, especially the ones in Norwich and York, the local branch of the National and Local Government Officers' Association began to lobby the Secretary of State for Scotland for a university in the Falkirk area on the basis that 'Scotland badly lacks a University of the residential type'.[19] This remit was especially noteworthy as the traditional Scottish universities had not been residential and accommodated primarily local students.

In autumn 1962, a 'National Committee for a fifth University in Scotland' was formed in Glasgow, with sponsorship from the Association of University Teachers and the Educational Institute of Scotland, the Scottish teachers' union, the Church of Scotland and the Scottish branch of the National Union of Students (NUS).[20] While awaiting the outcome of the report of the Robbins Committee, the Scottish Education Department used the opportunity of the existence of the grassroots campaigns to make its own case for expansion in Whitehall. It highlighted two factors which would make the creation of another university in Scotland 'inevitable': the importance of demand for science and technology and the current relative disadvantages for Scottish students in the university system as a whole. It was more difficult for Scottish students to go to English universities than the other way round because of their perceived 'qualification gap' and specifically the system of five-year secondary education with much less specialization at the higher levels, with final qualifications that were difficult to compare with the system in England. Only the foundation of a new university in Scotland could address this imbalance.[21]

The government published the findings of the Robbins Committee in September 1963, and its recommendation to create another Scottish university immediately gained traction north of the border.[22] The UGC subsequently began formally to accept expressions of interests from Scottish local communities by the beginning of 1964.[23] This process brought to light issues that were to return repeatedly, though perhaps more subtly, in the discussions about the new University of Stirling: the first of these were concerns in the Treasury about how to pay for the expansion of higher education, in particular in Scotland – in an internal memorandum, a civil servant complained that the UGC did not 'give details why their estimate [of demand for student places] vis-à-vis 1960 [had] changed' and concluded, in line with the official policy of the 'promotion of growth poles' at the time, that only economic considerations and regional planning 'would be essential for determining the location of the new university'.[24]

The second key question discussed was about who had control and authority over Scottish educational matters. While the responsibility for schools had been moved, as part of a process of administrative devolution, to the Scottish Office quite some time ago, university education remained a London matter. Some were therefore concerned that it would be the London-based UGC that would make the decision for a new university site. The Scottish National Party (SNP), buoyed by coming second after Labour in the West Lothian by-election of June 1962, felt 'very disturbed by the country's educational, cultural and constitutional fibre'. As the Scottish educational system was fundamentally different from that in England, the Secretary of the SNP argued, the UGC required Scottish representation lest it become 'virtually a committee of foreigners' and its work 'destroy yet another aspect of Scotland's national life in the interests of the uniformity of the United Kingdom'.[25] This assessment is remarkable because the UGC tended to have a good level of Scottish representation, not in a statutory sense, but in regularly appointing Scottish academics to its panels.[26]

In late spring 1964, on two separate visits, a delegation from the UGC visited the various sites, from Ayr and Dumfries in the south, over Cumbernauld, Stirling and Falkirk in the Scottish central belt via Perth to Inverness in the north.[27] In the end, it appears that the choice for Stirling was a compromise decision, though it was unanimous.[28] When Stirling was announced as the site of the new University on 17 June 1964 in a response to a parliamentary question, there was a significant degree of surprise and disappointment within the Scottish press and some of Stirling's competitors. Critics regarded the choice for Stirling as a violation of Robbins's principles. The East Stirlingshire Campaign Committee, in particular, kept pressing for a reconsideration. Alexander Duncan, the chairman of the East Stirlingshire Campaign Committee, described Stirling as 'just a recognised snob town', 'a county town like Colchester or Canterbury'.[29]

Setting up the University

The Stirling Sponsoring Committee went straight to work, however, and began planning for the new University. The UGC created an Academic Planning Board that was to appoint the first principal and vice-chancellor, map out the academic programme and guide the process for the submission of the application for a Royal Charter.[30] It also started a local and national fundraising campaign, following the example of the other new universities.[31] The UGC appointed its former Chair Lord Murray as head of the Stirling's Academic Planning Board. Other members included Lady Ogilvie, the principal of St Anne's, Oxford who was appointed in particular to share her view on welfare and residential questions; the economist Professor A. T. Peacock from York and the Principal of the University of Glasgow, Dr Charles H. Wilson, who had been involved with the foundations of Sussex, East Anglia and Strathclyde and who now served on the Committee of Vice-Chancellors and Principals and whom Murray knew well.[32] The appointment to the board followed discussions with the Scottish Office. Lord Wolfenden, UGC's chair, was keen to avoid 'impression of official control' from London.[33]

Stirling's Academic Planning Board (APB) mainly rehearsed the objectives that Robbins and his team had identified in their report, that is, the education of 'cultivated men and women' by the 'example of teachers who are themselves learning', the 'transmission of common culture', the importance of skills and vocations and the 'advancement of learning'. However, there is little evidence of a transfer of ideas and concepts from the other new universities.[34] Specifically, the APB wanted to create a 'forward-looking institution' that prepared students 'for a vigorous life in a free society'. It wanted to do this through 'instruction in specific vocational skills' that were of 'of practical use', but taught 'consonant with the development of a disciplined mind'. In doing so, the APB wanted to 'commit common standards of citizenship' and play a 'more active role in the development of the community, particularly of the local community, in which it works and lives'.[35]

The APB's first appointments reveal the educational agenda they had in mind in a bit more detail. It recruited Dr Tom Cottrell as the first principal. He took up his position part-time for a year in August 1965. His biography appeared to fit the bill for a new Scottish University perfectly: he was Scottish; he was, at forty-two, still relatively young; he was a chemist at the University of Edinburgh, but with an interest in working in and with industry and interdisciplinarily, as a previous stint as a research chemist with ICI demonstrated. And he was known to be concerned about student welfare and interested in innovative teaching and learning methods.[36] He had also been a member of the rival East Stirlingshire Campaign Committee. His appointment may therefore have been, at least in part, a diplomatic act.

Cottrell was joined, in October 1965, by Harry Donnelly as the first University secretary. Donnelly had been the civil servant in the Scottish Office working on the foundation of the new University, so that there was probably no-one with a better insider knowledge of Scottish higher education policy with respect to the new University at the time.[37] In early 1966, they were joined by the thirty-year-old R. G. Bomont, then assistant finance officer at Lancaster, as University accountant and the thirty-four-year-old J. F. Stirling, then deputy librarian at York, as University librarian.[38]

In line with Scottish practice, Stirling opted for a four-year degree. But it introduced some innovations: a semester system and periodic assessment in the form of class tests, similar to, but not necessarily following the example of, UEA and Lancaster.[39] This was to guarantee high academic standards by forcing students to avoid cramming and learn continuously.[40] As at the other new universities and perhaps closest to the examples of Essex and Lancaster, teaching (and potentially research) were to happen in an interdisciplinary environment: departments existed but were supposed to play a relatively small role.[41]

The planners decided that Stirling should leave out the capital-intensive subjects of engineering and medicine, and instead offer programmes in four key areas: arts and humanities, social sciences, pure sciences and education.[42] Principal Cottrell was especially keen on two innovations in terms of the subjects represented at the university[43]: he made a strong case for education in order to allow teacher training at the University (which at that time existed at no other Scottish university) and in direct conversation with the relevant subject areas; and also to enable research on education to take place in schools and universities. Cottrell also regarded universities as an important site for further education and envisioned the Education department in particular to offer refresher courses for teachers: the 'increasing complexity of the cultural, social and economic life of this country and the rapid rate of growth in knowledge' meant that education did not end at age twenty-one, especially in the light of the recommendations on the emphasis on vocational training and research in higher education as part of the recommendations of the Hale Committee on university teaching methods.[44] Cottrell's second innovation was the creation of a School of Technological Economics, similar to existing programmes at Lancaster and Manchester: 'Leadership in government and in industry is beginning to pass to those whose knowledge and skills are in the fields of technology and the social sciences, especially economics.'[45]

The first appointees to the academic positions were mostly promising younger scholars, many of whom had Commonwealth connections. Most of the appointments were male; among the batch of three or four dozen initial appointments, there were only nine women, with only one female professor. Just over 50 per cent of new staff came from Scotland, and a remarkable 40 per cent had biographical or academic backgrounds from outside the UK.[46] It is noteworthy that Stirling decided to move away from classic subject-specific norms: English Studies shied away from compulsory courses on Anglo-Saxon and Middle English and instead focused on modern and contemporary literature, including Commonwealth literature. Tommy Dunn (founding professor) and encouraged the study of African writers from the start. One of Dunn's first appointments was of Alastair Niven, a pioneer in the field and later Literary Director of both the British Council and the Arts Council.[47] History was, initially, taught primarily as *modern* history since the French Revolution, and its emphasis on geographical areas outside Europe, including courses on African and Japanese history, along certain themes, was noteworthy in a context where global history was still taught in the context of 'Britain and the wider world'. Biology focused on human ecology and whole organisms rather than more traditional approaches in isolated subfields of the discipline.[48]

Conflicts

The conflicts that characterized Stirling's early history reveal a broader pattern of government in Britain: despite this local involvement, the new universities were not local universities, as they were subject to the rules and regulations – and ultimately benevolence – of central state funding.[49] Stirling opened its doors not only during a period of financial and economic contraction, but also at a time when the huge public and political support and approval of the early 1960s were gradually giving way to a more challenging environment: the student protests at LSE, Essex and Sussex and across the world, appeared to bring 'students', with their seemingly distinct and 'permissive' lifestyle, as a distinct social group into focus: they became symbolic markers in debates about what had gone wrong in British society since the end of the Second World War. Student funding was expensive, but students' understanding of citizenship clashed with more conservative notions of civic duty. Moreover, despite a significant public investment in higher education, universities did not appear to have delivered the affluence that many had connected with the project, and the number of science students had not increased as planned. There therefore seemed little merit in funding in what contemporaries pejoratively called a 'meritocracy'.[50]

Similar to the other new universities, Stirling was a 'victim of its own success'. As William Whyte has argued: the 'size and scale of project' 'helped create a counter-reaction which was impossible to contain.'[51] But because the funding cuts started at a time when expansion was still vital – and innovation crucial – this pattern of development was perhaps more significant for Stirling than for the other new universities, such as York, East Anglia or Essex. This highlights the problems at the heart of the systemic approach to planning the higher education reforms implied.

Stirling operated in an increasingly challenging environment: one in which it was supposed to grow in student numbers in order to finance itself, when student numbers had already reached their peak nationally; one in which, against the background of a crisis in UK public finances, funding for universities began to dry up, especially as far as equipment and capital investment were concerned; and one in which projections of future student numbers in the capital-intensive natural sciences proved to be vast overestimations of actual demand, leading to severe imbalances of staffing and student numbers.[52]

The number of 'justified serious complaints by students about lodging' at Stirling in the late 1960s was still small. But they already pointed to some of the conflicts over accommodation and rents that would characterize the university's history in the 1970s and into the 1980s. Rather than live in the somewhat bare study bedrooms, the 'point had been taken that many of the students would prefer to share flat accommodation if such accommodation was available.'[53] As at Warwick, the first sustained student protests at Stirling emerged over discussions about a lack of a common social building for staff and students.[54]

When the University's Development Committee, against its original plans to avoid separate social spaces for staff and students, decided in 1970 to award 4,000 sq ft of social space 'for use by staff only', the student newspaper screamed 'social apartheid' and registered the 'widespread anger, sorrow and frustration among the whole student body'.[55] As the University went ahead with the decision to create a separate staff space, around a hundred of the 640 students went on strike and occupied Airthrey Castle. Attendance at tutorials and lectures was 'minimal'; some teaching was cancelled.[56]

These issues around social space were exacerbated by the quality and cost of residences during a period of rising inflation and economic and financial crisis. Unfortunately for Stirling, the building of residences began just when the UGC norms had been reduced to save costs, although its capital allocation per student (£894) was 'at least on par with' those of the other new universities.[57] But the UGC's position took little account of specific local conditions of recent foundations. It pointed out that the 'allocation of resources between [...] competing claims [was] a proper and indeed inevitable function of Government.'[58]

All these issues came to a head when Stirling became the national focus for everything that seemed wrong with British universities. When the Queen visited the campus on 12 October 1972, Stirling students and the University became the 'national symbol for undergraduate anarchy'. Protests against the situation in the residences and the social spaces as well as the cost of rent and food had been going on for several days, and the students had occupied some of the administrative buildings. The Council of the Students' Association had voted in favour of a sit-in that would alert the Queen to the continued cuts to student grants and budgets, while significant costs had been associated with her visit.[59]

As the Queen emerged from Murray Hall, she was shielded by a heavy police presence as many protesters were drunk.[60] Students 'chanted, jeered, sang lewd songs' when royal party passed them. A mature student was photographed drinking from a bottle in front of the Queen. There were reports in the press that he swore at her and that the Queen's Lady in Waiting had to manhandle several protesters to push them out the way of Her Majesty.[61] With the monarchy as key for symbolic integration of the UK and for public decency and propriety, the debate about the Queen's visit brought to light fundamental conflicts that had accompanied the higher education reforms.[62] At one level, the debates saw a clash between a rhetoric of hard-working normal people who complained about the lack of decency and civility of a highly privileged class of students or 'drunken, ill-bred louts who merely used tax payers' money to indulge themselves.[63] Shopkeepers in nearby Bridge of Allan refused to serve students and buses refused to stop at the University as a sign of outrage, porters and cleaners refused to continue to serve students, and Stirling Town Council distanced itself from the behaviour of the students in a public statement to the Queen.[64] The example of Stirling seemed to suggest that education would not serve society at large, but foster the belief in some students that they had 'a duty to manipulate education as best they can for their own ideological purposes', all at taxpayers' expense.[65]

The incident proved highly divisive within the University, as severe disciplinary action was taken against some of the students involved. Some members of staff came out in their support, resigning from the University Court and other committees in protest. Students occupied the principal's offices in protest, and a series of legal challenges and counter-challenges ensued. It was Stirling's misfortune to have embarked on fulfilling Robbins's utopia of higher education at a time when the utopian moment that had enabled it was over.

Shock and transformation

Principal Cottrell's death from a heart attack in his residence on campus in the wake of these debates in 1973 and the continued divisions between and among students and staff on campus were the symbolic markers of the end of optimism at

Stirling. In 1970, the UGC and the Department of Education and Science developed a system that would help universities plan for inflation, an issue that had repeatedly thrown a spanner in Stirling's building projects: the value of the quinquennial grant was now guaranteed, including some compensation for inflation, and there would no longer be an 'endless haggle' over numbers.[66] But early in December 1973, the chancellor of the Exchequer announced significant cuts: he cancelled half of the agreed inflationary compensation for university costs and made another reduction within weeks of that, amounting to 10 per cent of overall contributions across the board.[67] By the mid-1970s, the UGC annual report registered 'a deep and damaging sense of uncertainty which can only be removed by the restoration of a longer-term planning horizon'.[68] But that 'planning horizon' did not appear. When the last quinquennium ended on 31 July 1977, there was no longer any room 'in the now uniform public expenditure survey system for exceptions of the old-fashioned sort'; and there was no longer any general norm against which expectations could be planned.[69]

Stirling suffered especially hard from these changes to the funding system: the building stock was 'out of phase', observed the internal planning document, and 'growth to 3,000 vital to resolve present difficulties'. This, in turn, posed significant problems for the availability of existing residences. Student complaints about rising rents and rent strikes were therefore a perennial feature of Stirling campus politics.[70] There were several acute accommodation problems over the course of the 1970s, either because building work had not yet finished when the students arrived or because of larger-than-expected student intake.[71]

Throughout this period, University planners still assumed the continuation of a steady growth of student numbers by 10 per cent per year from 1977 to 1982 with a final target of 3,420 students, and a target ratio of 75:25 non-science:science students.[72] But this strategy of growth with the aim of maintaining the UGC's recurrent grant by keeping up with the student projections came with significant capital costs, both in terms of residences and in terms of staff–student ratios across the institution and therefore potential staff costs.[73] In February 1976, Deputy Principal Professor James Trainer wrote a desperate letter to Sir Frederick Dainton, UGC's chair, mentioning that his university was now running with a 'small accumulated deficit, no general reserves, [and a] loan-financed residence problem'. What made this difficult to solve was a 'serious imbalance in the sphere of academic staffing': there were comparatively too many staff in the sciences (with high levels of capital and equipment costs), and too few in the arts and social sciences.[74]

The early 1980s did not see a revival of quinquennial planning and a move towards the control of cash flow in the system. Against this backdrop, the UGC decided to make cuts by differentiating between subjects at particular institutions on the basis of academic merit rather than by differentiating between institutions.[75] The idea was to encourage science degrees and discourage social science by using institutions' grant income and the quality of their students as yardsticks.[76] As a consequence, Stirling suffered the sixth largest reduction of its budget in the UK: it was asked to make cuts of 27 per cent from 1981 to 1983/4. This equalled a 22.7 per cent reduction of its projected income in 1986/7 vis-a-vis 1981.[77]

It might well have been the case that the cuts in 1970s occurred primarily due to international financial and economic pressures, whereas they were, in the 1980s, primarily ideological.[78] But there is a broader continuity here in terms of what they tell us about the governance and government of universities in the context of the Union. The Thatcher government's understanding of unionism was based on the idea of untrammelled individual freedom – this meant that it was unitarian and assimilationist, wishing to implement the same standards across the Union and cutting back on redistribution.[79] The irony is that higher education campaigners in the 1970s had argued against the inclusion of higher education in the debates about devolution in the 1970s, and principals came out strongly against higher education becoming devolved to the proposed Scottish Assembly, regardless of whether or not they were in favour of political devolution more generally.[80] To many in Scotland, Scottish national institutions in the context of the UK still appeared 'levers […] by which they might enjoy full equality' in the Union.[81] The seeming self-government of universities through the UGC, with its emphasis on institutional autonomy, thus worked as an 'antidote to nationalism' in the debates of the 1970s.[82]

Conclusion

This chapter, then, has shown that the 'failed utopia' that many contemporaries began to see in Stirling's history of the 1970s and 1980s was not simply an episode in a story of decline. The debates surrounding Stirling highlight especially well a set of issues that pertains not just to the Robbins Report, but also shines a powerful light on the fundamental problems in post-1945 UK government. Unlike the other new universities, Stirling was never part of the utopian moment in UK higher education history that saw the stars of ample funding, appropriate political environment and ideologies of progress perfectly aligned. Stirling was the expression of utopian thinking, but that thinking was never fully realized in the first two decades of its history.[83]

The first issue is perhaps an obvious one that historians often take it for granted: time – and timing – matters. The early history of the University of Stirling highlights just how much theories of path dependency are insufficient to account for problems of policy implementation – and conflicts around it.[84] Stirling opened its doors just after a time was coming to an end which Sir Maurice Dean, the permanent secretary at the DES in 1963, the joint permanent secretary at the DES from April to October 1964, and the permanent secretary at the Ministry of Technology from 1964 to 1966, called 'the great plastic period' of higher education.[85] The whole system had been designed to allow for expanding (or at least stable) income, but not for its contraction.[86] The fact that Stirling was founded just when the system began to contract had significant implications for its subsequent development: residences were completed late or at a lower quality than expected; positions remained vacant; and yet student recruitment had to be maintained to secure a grant that allowed the University to continue its operations.

Perhaps more than at the other new universities in mainland Britain, Stirling's malaise revealed especially clearly the difficulties involved with combining the ideal of academic freedom with tightly controlled public funding and direction.[87] During the

1970s, these conflicts were not (yet) primarily framed in terms of Scottish nationalism. The conflicts over funding, equipment and expansion at Stirling highlight the broader contradictions in British higher education policy. From north of the border, these issues began to appear differently as 'Scotland' offered an ideological category as a way of seeing the political world.[88] This meant that the welfare state – and the place of higher education within it – was identified 'in the popular imagination as the fundamental British institution that cemented Scottish loyalty to the Union'.[89] This consensus, which reached well into the Labour and Conservative Parties, appeared as a 'pillar of the modern Union [and] an expression of Britain-wide solidarity and redistribution'.[90] The 'alteration in the size of the state, or of the employment and benefits which it purveys' therefore had direct implications for the arguments for union.[91] This constellation had particular repercussions for Stirling. On one level, the foundation of the University in 1967 was part of a trend whereby public expenditure in Scotland was much higher per capita than in England and Wales.[92] On another level, the 'quasi-democratic'[93] aspects of Robbins's proposals appeared to fit neatly into what Ewen Cameron has called the 'myths of egalitarianism and social mobility'[94] – myths that formed an important part of Scottish self-perceptions in education.[95]

Accordingly, the President of the Student Association Jack McConnell, who was to become the First Minister of Scotland after devolution, during his tenure in the early 1980s expressed a solidaristic conception of the Union. He was certainly not a nationalist at the time, but he vehemently criticized the Thatcher government policies in terms of Scottish interests.[96] Margaret Thatcher and her government, by contrast, tried to present themselves, however misguided, as true guardians of Scottish interests by highlighting their idea of fostering individualism away from government control, with some significant support from Scottish Conservatives.[97] This meant that for those opposed to the reforms, 'Scotland' could now emerge more broadly in public discourse as a dream space against which general aims of welfare and redistribution could be measured – precisely as the solidarity of the Union appeared to unravel against the backdrop of economic crisis and fiscal austerity.[98] The failure of one utopia gave rise to another.

Notes

1 Inauguration Address, Stirling, 5 April 1968, British Library for Political and Economic Science [BLPES], Robbins Papers, ROBBINS/8/1/4.
2 Stirling Graduation Ceremony, June 1978, 3, 4, BLPES, ROBBINS/8/1/4.
3 *Campus. The Newspaper of Stirling University*, 48 (November 1979), BLPES, ROBBINS/8/1/4.
4 William Whyte, *Redbrick. A Social and Architectural History of Britain's Civic Universities* (Oxford: Oxford University Press, 2015), 309.
5 (Principal) Sir Kenneth Alexander to Robbins, 11 September 1981, with an enclosed letter by Alexander to the UGC and a letter to the *Guardian* newspaper by the economists C. V. Brown and B. J. Loasby, BLPES, ROBBINS/8/1/4; HC Debs., 13 (18 November 1981), 319–20.

6   Peter Mandler, 'Educating the Nation II: Universities', *Transactions of the Royal Historical Society*, 25 (2015), 1–26; Glen O'Hara, *From Dreams to Disillusionment. Economic and Social Planning in 1960s Britain* (Basingstoke: Palgrave, 2007). For students, see Caroline Hoefferle, *British Student Activism in the Long Sixties* (London: Routledge, 2013), 87.
7   Glen O'Hara, *Governing Post-War Britain. The Paradoxes of Progress* (Basingstoke: Palgrave, 2012), 2.
8   O'Hara, *Governing*, 200.
9   The longer-term continuities are highlighted for higher education by Harold Perkin, 'University Planning in Britain in the 1960s', *Higher Education*, 1 (1972), 111–20; also I. G. C. Hutchison, 'The Scottish Office and the Scottish Universities, c. 1930 – c. 1960', in *Scottish Universities: Distinctiveness and Diversity*, ed. Jennifer J. Carter and Donald J. Withrington (Edinburgh: John Donald, 1992), 56–66; and his *The University and the State: The Case of Aberdeen* (Aberdeen: University of Aberdeen, 1993). On the general issue of government in the context of Unionism and devolution, see Paul O'Leary, 'States of Union: Modern Scotland and British History', *Twentieth Century British History*, 27 (2016), 124–43.
10 Michael Shattock, *Making Policy in Higher Education 1945–2011* (Maidenhead: McGraw Hill/Open University Press, 2012), 43.
11 Richard Layard, John King and Claus Moser, *The Impact of Robbins* (Harmondsworth: Penguin, 1969), 17.
12 Harold Perkin, *New Universities in the United Kingdom, Case Studies on Innovation in Higher Education* (Paris: OECD, 1969), 221; John Carswell, *Government and the Universities in Britain. Programme and Performance 1960–1980* (Cambridge: Cambridge University Press, 1985), 86.
13 Hutchison, 'Scottish Office', 64–5.
14 My account comes to different conclusions from a recent detailed survey of the arguments around setting up Stirling from the perspective of social geography: Michael Heffernan and Heike Jöns, '"A Small Town of Character": Locating a New Scottish University, 1963–1965', in *Geographies of the University*, ed. Peter Meusburger et al. (Dordrecht: Springer, 2018), 219–50.
15 On the background, see Richard Saville, 'The Industrial Background to the Post-War Scottish Economy', in *The Economic Development of Modern Scotland, 1950–1980*, ed. Richard Saville (Edinburgh: Edinburgh University Press, 1985), 1–46.
16 Scottish Education Department to Montmorency, UGC, 28 April 1947, TNA UGC7/237.
17 Carswell, *Government*, 43.
18 See the expressions of support from autumn 1960 in TNA UGC7/237.
19 *Scotsman*, 5 August 1960.
20 *Glasgow Herald*, 11 September 1962; *Times*, 22 November 1962.
21 *Scotsman*, 19 February 1962; Arbuckle, SED, to Syers, UGC, 20 February 1963, TNA UGC7/239; M W Graham, SED, to Pottinger, SDD, 21 May 1965, National Archives of Scotland, Edinburgh (NAS) ED26/1600.
22 David Waddell, 'The University of Stirling 1967–1992. The First Twenty Years of Innovation', MS Stirling 1992, 19–20. Also, R. G. Bomont, *The University of Stirling. Beginnings & Today* (Stirling: University of Stirling 1995).
23 For Stirling, see Town Clerk Norman to UGC, 15 August 1963 and 24 October 1963, TNA UGC7/240.

24 C. S. Hennett, 'New University in Scotland', n.d. (Spring 1964), TNA T227/1629. Michael Keating, *The Independence of Scotland. Self-Government and the Shifting Politics of Union* (Oxford: Oxford University Press, 2009), 50; Jim Tomlinson and Ewan Gibbs, 'Planning the New Industrial Nation: Scotland, 1931–1979', *Contemporary British History*, 30 (2016), 584–606;
25 Gordon Wilson, Secretary of the Scottish National Party, to Wolfenden, 1 May 1964. TNA UGC 7/240.
26 Many thanks to Ewen Cameron for pointing this out to me. Lord Murray, its chair until the early 1960s, is perhaps the most prominent example of this practice.
27 For the different arguments, economic, technocratic and meritocratic, see Minutes of the UGC Sub-Committee on New Universities, 28 February 1964, NAS ED26/1600. Cf. Heffernan and Jöns, 'A small town of character'. On the regional planning aspects, see the fascinating case study by Ewen A. Cameron, 'University Realities: The Inverness Campaign to Establish Scotland's Fifth University', *Transactions of the Gaelic Society of Inverness*, 67 (2013–15), 9–50.
28 On the discussions in UGC and the final decisions: 'Confidential Note of Discussion with SED and SDD', Edinburgh, 28 April 1964; 'Confidential Report, New University in Scotland', 12/13 May 1964 TNA UGC7/244. On the decision in favour of Stirling, see Baird to Sinclair, SED, Confidential Telegram, 16 June 1964, NAS ED26/1600.
29 *Scotsman*, 20 July 1964.
30 Extract from Cmnd 2264 on the remit of Academic Planning Boards, TNA UGC7/245.
31 Meeting of Sponsoring Committee, 25 January 1965, University of Stirling Archives [UA] A11/5/1.
32 Attempts to recruit the historian Norman Gash from St Andrews did not go further as he was seen as too conservative to have ideas of use for the foundation of a new university. Wolfenden to Donnelly, 4 September 1964; Confidential Letter Donnelly to Coppleston, 9 September 1964, NAS ED26/1485.
33 Donnelly to Secretary of State, Memorandum, 'New University at Stirling', 19 August 1964, NAS ED26/1485.
34 Claus Moser, 'The Robbins Report 25 Years after – and the Future of the Universities', *Oxford Review of Education*, 14 (1988), 5–20.
35 First Report of the Academic Planning Board to the Sponsoring Committee (14 December 1965), UA/A11/5/1, 3.
36 Minutes, Meeting of Sponsoring Committee held in the Municipal Buildings, Stirling (18 June 1965); The University of Stirling. First Report of the Academic Planning Board to the Sponsoring Committee (14 December 1965) UA A11/5/1.
37 Waddell, 'University of Stirling', chs 2, 4.
38 Minutes of Meeting of Directors, 4 February 1966, UA A11/5/1, 2.
39 Perkin, *New Universities*, 184.
40 First Report of the Academic Planning Board to the Sponsoring Committee (14 December 1965), UA/A11/5/1, 9.
41 Ibid., 3. On Essex, see Stefan Muthesius, *The Postwar University. Utopianist Campus and College* (London and New Haven: Yale University Press, 2000), 151. On Lancaster see Perkin, *New Universities*, 86.
42 First Report of the Academic Planning Board to the Sponsoring Committee (14 December 1965), UA/A11/5/1, 4.
43 Meeting of Sponsoring Committee, 18 June 1965, UA/A11/5/1.

44 First Report of the Academic Planning Board to the Sponsoring Committee (14 December 1965), UA/A11/5/1, Appendix I, 2 and Appendix II, 1.
45 First Report of the Academic Planning Board to the Sponsoring Committee (14 December 1965), UA/A11/5/1, 3.
46 Waddell, 'University of Stirling', 12.
47 Many thanks to the late John McCracken for sharing this information.
48 Waddell, 'University of Stirling', 9.
49 Muthesius, *Postwar University*, 98. See, for example, H. H. Donnelly, 'Note of Record of a Meeting with Mr Steele and Mr Roberts Morrison', 18 November 1965, NAS HH101/3879; cf. Harding, 'Note for Record, Proposed University at Stirling' 16 July 1964; and Harding to Sir Richard Clarke, 'New Scottish University', 17 July 1964, TNA T227/1629.
50 Cf. Nick Thomas, 'Challenging Myths of the 1960s: The Case of Student Protest in Britain', *Twentieth Century British History*, 13 (2002), 277–97.
51 Whyte, *Redbrick*, 267.
52 Perkin, *New Universities*, 112.
53 Meeting of Directors, 13 November 1967, UA/A11/5/1.
54 For Warwick: Hoefferle, *Student Activism*, 136.
55 *Brig*, 27 February 1970, 1.
56 *Brig*, 13 March 1970, 1.
57 Young Enquiry, UA, 85, 109.
58 UGC, *University Development 1967–1972, presented to Parliament by the Secretary of State for Education and Science by Command of Her Majesty*, Cmnd 5728 (London: HMSO, 1974), 51 at UA/A/1/11/1.
59 Letter by Principal Cottrell to Linda Quinn, CSA President, 3 October 1972, printed in *Brig*, 12 October 1972.
60 Kenneth Roy, *The Invisible Spirit. A Life of Post-War Scotland 1945–75* (Glasgow: ICS Books, 2013), 444.
61 *Times*, 13 October 1972, 1; *Scottish Daily Express*, 13 October 1972, 1, 10; *Glasgow Herald*, 13 October 1972; *Daily Record*, 13 October 1972; *Stirling Observer*, 18 October 1972.
62 Alvin Jackson, *The Two Unions. Ireland, Scotland, and the Survival of the United Kingdom* (Oxford: Oxford University Press, 2012), 153–63.
63 *Stirling Observer*, 18 October 1972; *Times*, 16 October 1972, 13; *Scottish Daily Express*, 27 October 1972. On the broader background for this trope, see Jon Lawrence and Florence Sutcliffe-Braithwaite, 'Margaret Thatcher and the Decline of Class Politics', in *Making Thatcher's Britain*, ed. Ben Jackson and Robert Saunders (Cambridge: Cambridge University Press, 2012), 132–47.
64 *Scottish Daily Express*, 16 October 1972.
65 *Times Educational Supplement*, 20 October 1972; *Stirling Observer*, 17 October 1972. On the general background, see Michael Freeden, 'Civil Society and the Good Citizen: Competing Versions of Citizenship in Twentieth-Century Britain', in *Civil Society in British History. Ideas, Identities, Institutions*, ed. José Harris (Oxford: Oxford University Press, 2003), 275–91.
66 Carswell, *Government*, 139; A. E. L. Parnis, UGC, to Norman Hardyman, DES, 1 March 1971; I. R. M. Thom, DES, to R. H. Seebohm, Treasury, 7 September 1971, SNA ED188/312.
67 Carswell, *Government*, 145–6. Forsyth, Treasury, to Hardyman, DES, 25 January 1972. SNA ED188/312.

68 UGC Annual Survey for 1975–76, Cmnd 6750 (London: HMSO), para 24.
69 Carswell, *Government*, 157.
70 *Brig* 7, 15 September 1975; *Brig*, 12 (1981).
71 *Brig*, 25 September 1970, 1; *Campus*, 31 (Spring Semester 1977).
72 Agenda, Committee of Academic Development, Report from Academic Council, CAD (75)19, UA/A/1/11/1.
73 Paper, 'Staffing Implications for Student Growth', 15 May 1975, CAD (75)12 UA/A/1/11/1.
74 Trainer to Dainton, UGC, 2 February 1976, UA/A/1/11/1/2. Cf. *Campus*, 20 (November 1975). On the over-projection of the need for science places in the context of the 'declinism' of the debates, see UGC, University Development 1967–1972, presented to Parliament by the Secretary of State for Education and Science by Command of her Majesty, September 1974, Cmnd 5728 (London: HMSO, 1974), 24 at UA/A/1/11/1. On the background: Jim Tomlinson, *The Politics of Decline. Understanding Post-War Britain* (Harlow: Pearson, 2000), chs 5–6.
75 Shattock, *Making Policy*, 126.
76 Whyte, *Redbrick*, 291.
77 Shattock, *Making Policy*, 128–9.
78 Anderson, *British Universities*, 169.
79 Richard Finlay, 'Thatcherism, Unionism and Nationalism: A Comparative Study of Scotland and Wales', in *Making Thatcher's Britain*, ed. Alvin Jackson and Robert Saunders (Cambridge: Cambridge University Press, 2012), 165–79; cf. Jackson, *Two Unions*, 265–8.
80 Ewen Cameron, *Impaled upon a Thistle. Scotland since 1880* (Edinburgh: Edinburgh University Press, 2011), 301.
81 Jackson, *Two Unions*, 136–7.
82 Cameron, *Impaled upon a Thistle*, 299.
83 Heffernan and Jöns, 'A Small Town of Character' highlight the importance of more traditional thinking for the early history of the University. The design of the campus, the assessment and semester systems and the choice for subject systems do not bear this out.
84 For a conceptual discussion of this problem, see Paul Pierson, 'Not Just What, but When: Timing and Sequence in Political Processes', *Studies in American Political Development*, 14 (2000), 72–92.
85 Carswell, *Government*, 52, 62. Carswell takes the term 'plastic period' from J. G. Lockhart's *Memoirs of Sir Walter Scott* (1900).
86 Carswell, *Government*, 88.
87 Keating, *Independence*, 6–8.
88 Catriona Macdonald, *Whaur Extremes Meet. Scotland's Twentieth Century* (Edinburgh: John Donald, 2009), citing David McCrone, 193.
89 Richard Finlay, 'Thatcherism, Unionism and Nationalism', 167.
90 Keating, *Independence*, 53.
91 Jackson, *Two Unions*, 172.
92 Iain G. C. Hutchison, 'Government', in *Scotland in the 20th Century*, ed. T. M. Devine and R. J. Finlay (Edinburgh: Edinburgh University Press, 1996), 46–63; G. C. Peden, 'The Managed Economy: Scotland, 1919–2000', in *Transformation of Scotland. The Economy since 1700*, ed. T. M. Devine et al. (Edinburgh: Edinburgh University Press, 2005), 233–65.
93 Mandler, 'Educating the Nation', 2.

94 Cameron, *Impaled upon a Thistle*, 223.
95 Cf. Macdonald, *Whaur Extremes Meet*, 151–9; Lindsay Paterson, 'Liberation or Control: What are the Scottish Education Traditions of the Twentieth Century?' in *Scotland in the 20th Century*, ed. Devine and Finlay (Edinburgh: Edinburgh University Press, 1996), 230–49.
96 *Brig*, 11 (May 1980); 'Protest and Survive', *Brig* (October 1981).
97 Cf. Jackson, *Two Unions*, 268. For a case study in another field, see Jim Phillips, *The Industrial Politics of Devolution. Scotland in the 1960s and 1970s* (Manchester: Manchester University Press, 2008).
98 See Ben Jackson, 'The Political Thought of Scottish Nationalism', *The Political Quarterly*, 85 (2014), 50–6.

12

The New University of Ulster and the Northern Ireland crisis

by Thomas G. Fraser and Leonie Murray

Introduction

In the autumn of 1968, the first intake of students arrived at the Coleraine and Magee campuses of the New University of Ulster, established as the result of the recommendations of the Lockwood Committee three years earlier. The Coleraine campus was newly minted, like the seven universities recently founded in Great Britain, while the Magee campus thirty miles away in Londonderry, or Derry, inherited a history stretching back over a century. Only four days after the university's opening on 1 October, the political history of Northern Ireland changed forever when a civil rights march that had assembled in Derry, in defiance of a government ban, was forcibly dispersed by the police, followed by rioting in the city. These events were an unpromising backdrop to the university's official opening in Coleraine on 25 October by the Northern Ireland Prime Minister Captain Terence O'Neill. They had been foreshadowed by the unprecedented mass protest in Derry provoked by the decision not to locate Northern Ireland's second university in the city. From its controversial origins in 1963–5 to its merger with the Ulster Polytechnic twenty years later to form the University of Ulster, subsequently Ulster University, the fate of the university was part of the history of Northern Ireland as political protest gave way to violence.[1] The 1960s was a revolutionary decade in many parts of the world, both for higher education and politically, and nowhere more so than in Northern Ireland where by the end of the decade the existing political structures had been brought to the point of collapse. The years that followed were ones of sustained violence, euphemistically known as the 'Troubles', which ultimately claimed the lives of

---

The authors are grateful to their former colleagues in history, Alan Sharp and Ken Ward, for their helpful recollections and the provision of material. Professor Lester C. Lamon, formerly of Indiana University South Bend, sent his observations based on his period as Leverhulme Fellow in 1978–9. Our thanks must go to the staff of the Public Record Office of Northern Ireland (PRONI) and the Library and Archive staff of Ulster University, particularly Sarah O'Deorain and Fiona Clyde, for their unfailing help. Dr Brendan Lynn assisted with the material on the University's CAIN website. Any opinions expressed are entirely those of the authors. Crown copyright material is reproduced by permission of the Controller of Her Majesty's Stationery Office.

over 3,500 people.² This chapter will have as its focus how the attempt to create a new university in line with the rest of the UK both affected and was in turn affected by the dramatic political developments of the time.

Northern Ireland: The political and educational context of the New University of Ulster

Established by the Government of Ireland Act of 1920, 1960s Northern Ireland was an anomaly within the UK in having a devolved government, colloquially known as 'Stormont', based upon the Westminster model.³ Northern Ireland had been created to give a political home within the UK to Ireland's Protestant and unionist minority concentrated as it was in parts of the historic province of Ulster. Derry was a different case, holding a distinctive place in the traditions of the Protestant community since 1689 having held out for 105 days against the army of the Catholic King James II, but by 1851 Catholics had become a clear demographic majority.⁴ Protestant political control of the city's affairs was only sustained by a carefully contrived construction of ward boundaries whereby in 1967 8,781 Protestant voters elected twelve unionist councillors to Londonderry Corporation while 14,429 Catholic voters elected eight non-unionists.⁵ It was a city set to combust. The nature of its political complexion was to impact upon the university issue.

In the early 1960s, higher education in Northern Ireland was dominated by Queen's University Belfast, which in its wide range of faculties and subjects was broadly akin to the English civic universities.⁶ There was, however, another provider of university-level education in Northern Ireland: Magee University College in Derry. Founded in 1865, it bore the name of Mrs Martha Magee of Dublin. Excluded from the National University of Ireland in 1908, Magee students completed their studies in Trinity College, graduating from the University of Dublin. With 27 academic staff and 245 undergraduates in 1963–4, the college was by any reckoning modest compared with Queen's or other third-level institutions in Great Britain or Ireland.⁷

Such, in brief, was the position in the city of Derry, of higher education in Northern Ireland, and of Magee University College when on 25 March 1963 Terence O'Neill became prime minister. From the first, O'Neill was determined to move Northern Ireland forward. On 5 April, he informed his party's ruling body, the Ulster Unionist Council, that his government's mission was to transform Ulster.⁸

The Lockwood Committee: Origins and purpose

With this dynamic in mind, the Stormont government set in hand a series of modernizing measures similar to those experienced in post-war Britain.⁹ Higher education followed a pattern mirroring Great Britain. The Committee on Higher Education in Great Britain, chaired by Professor Lord Robbins, reported in October 1963, recommending the expansion of many existing universities and the immediate creation of six new universities.¹⁰

Northern Ireland had not been part of Robbins's brief but officials in Belfast had been anticipating his findings and the following month appointed a committee under the chairmanship of the classicist Sir John Lockwood, Master of Birkbeck College London, charged with reviewing the facilities for university and higher technical education in the light of Robbins, and making recommendations.[11] He was joined by three other experienced British academics: Sir Willis (later Lord) Jackson FRS, Sir Peter Venables and Miss (later Dame) Rosemary Murray, tutor-in-charge and founder of New Hall Cambridge and later the first female vice-chancellor of the University of Cambridge. Murray was familiar with Derry having spent part of the war there as an officer with the Women's Royal Naval Service.[12] The four Northern Ireland members were Major J. A. Glen, former assistant secretary at the Ministry of Education, A. B. Henderson of Ulster Television, W. H. Mol, headmaster of Ballymena Academy, and the Managing Director of Harland and Wolff shipyard, Dr Denis Rebbeck. The secretaries were W. T. Ewing of the Ministry of Education and Zelda Davies of the Ministry of Finance. Also attached to the Committee were (later Sir) Robert Kidd of the Ministry of Finance and Dr A. T. Park.[13]

The question of location

While the government's instructions to the committee referred simply to recommendations, it was clearly to work within Robbins's parameters, one of the principal proposals being the creation of new universities. Certain authorities were anticipating this, preparing their case for a second university to be located in their area. Two such possibilities were Armagh, the ecclesiastical capital of Ireland, and the embryonic new city of Craigavon, but neither of these developed any traction, despite the fact that there was some government support for the latter.[14] Londonderry County Borough's bid analysed both the city's claim to house the university and the place that Magee might occupy within it, arguing that it would be inconceivable to ignore the possibilities of developing Magee in order to establish a university de novo somewhere else. The document addressed the criticism that Magee did not award its own degrees by arguing that this could be remedied either by granting it full university status or by becoming a constituent college of a new University of Londonderry or University of North-West Ulster, the preferred option. The city's rich history was pointed to as conducive to the atmosphere for learning, as well as the reputation of its four grammar schools and the Municipal Technical College, its industrial base, its sporting facilities and two hospitals. The council was prepared to offer a contribution of a 3d rate for the institution's first ten years, a sum amounting to around £5,000 per annum. On two other matters there was less precision. Acknowledging the problem of finding student accommodation, the council claimed that they saw little difficulty in meeting the demand. Taking as its starting point the view of the University Grants Committee that a single site should be some 200 acres, the council argued that a single site was not the only option. In either case, the council offered the assurance that several sites existed.[15]

Derry's chief rival was Coleraine, a small borough located in a predominantly unionist area at the other end of County Londonderry. In 1963, Coleraine came together with the seaside resorts of Portrush and Portstewart, with a total population of 20,000, to prepare a case. Their submission pointed to the area's sporting and cultural amenities, instancing its good schools, sea bathing, golf courses and amateur dramatic societies. Its principal thrust was the availability of housing accommodation, the local Hoteliers Association claiming that 2,600 places of a standard approved by the Tourist Board could be used in term time. The submission claimed that for a resident student population of 2,500 there would have to be a capital requirement of £3,500,000. The conclusion was that capital could initially be directed to academic purposes, with residences coming at a later stage. The case was reinforced by the new University of Sussex, which found available lodgings in Brighton. No specific sites were identified, although it was claimed that two were available.[16]

The Lockwood Committee at work

Lockwood and his colleagues held their first meeting on 6–7 December 1963 beginning a series of seventeen which concluded on 26–27 September the following year.[17] They initially agreed that they would not recommend an actual site for a new university unless the development of Magee was to be their recommendation which meant by implication that it would be Derry, a course of action seemingly recommended to save them from the pressures to which they would be subjected.[18] Magee's future surfaced at their meeting of 31 January–1 February 1964 when it was decided that they should visit the College on 21 February. At this stage they felt that after so long an existence it would be difficult to close the College, but that its present form of government inhibited development.[19]

The visit to Derry on the 21st fell into two parts. There was a formal meeting with civic leaders in the city's Guildhall. The substantive part of the visit was at Magee where they toured the facilities and met the trustees, president, heads of department and other academic staff. In discussion it became clear that there was real dissatisfaction with the way that the college was being governed through the Board of Trustees. Relations with Trinity were felt to be satisfactory. There was acceptance that a university in Derry could not be based on Magee as it currently existed, but that it should be established on an autonomous basis with a new constitution and degree-awarding powers, including postgraduate provision. The main teaching focus should be on the Arts, Social Sciences and teacher training. With the possible exception of chemical technology, technology should not be included. Some teaching of chemistry had been started, but as a temporary measure they were using the laboratories of the Technical College and a girls' school. There was little support for strengthening the link with Queen's. When the committee met the following day to review their impressions of the visit, it was clear that Magee had failed to impress. The leadership, they concluded, lacked dynamism, while the trustees, faculty and academic staff had failed to offer a clear vision for future development of the College or 'what shape any future university development at Londonderry should take'.[20] The visit had dealt Magee a mortal blow.

At its next meetings on 12, 13 and 14 March, the committee concluded that 'because of its circumscribed mental outlook and its cramped physical situation the College could not be adjudged the best nucleus for a major expansion'.[21] The ingenuous admission that use was being made of schoolgirl laboratories had clearly not impressed a body containing members from a scientific and engineering background, since this was judged a willingness to accept low standards. This dismissal of Magee, however, did not seem to foreclose on Derry as a possible location for a new university, since it was felt that the submission of Londonderry County Borough showed greater vision. While it was considered unwise to raise Magee's expectations, the door still seemed open for the city to give oral evidence about a second university. A possibility might be an autonomous body with Magee as part of an arts faculty.[22] This thinking did not last long, since opinion in the committee was evidently turning against Derry as the location of the new university.

In Henderson's later account, the committee felt that the creation of the new university should point to the future, offering generations of young people the opportunity to break free from the past.[23] Such thinking, of course, chimed with the O'Neill government's claims to modernization. While acknowledging the anomalous position of Magee and the problems that a divided site would create, it seems that the visit to Derry had convinced the committee, including its English members, of the deeply divided nature of its politics and their continuing tensions. It was this perception, Henderson argued, that led the Coleraine option, which did not carry the legacy of the past and could offer a green-field site, to become increasingly attractive.[24] On 4 March, the secretary to the committee had raised the critical issue of whether it should, contrary to its initial decision, be given greater freedom in recommending a site.[25]

The matter was addressed on 1–2 May, setting a target date for the admission of students at the new university in 1968. Armagh and Craigavon were deemed unsuitable, while beyond a discussion of how Magee might relate to the creation of an institution elsewhere, Derry did not feature. The closure of the College was seen as a possible option. The attractions of the Coleraine bid were the provision of residential accommodation and the feeling that its amenities could be attractive to staff. Thinking appeared to 'point inescapably' to a particular location. Aware of the political implications, Lockwood promised to consult the ministers of education and finance.[26] The location was explored with the rival claimants by the committee in meetings on 14–16 May, their contributions adding little of substance to their written submissions.[27]

Since a decision on location could not be long delayed, Davies, Kidd and Park met two senior officials of the University Grants Committee on 4 June to explore its criteria. The advice that for a new university of 5,000 students there should be a population of 75,000 was difficult for anywhere in Northern Ireland outside Belfast, but far beyond that of the Coleraine area. More positive for the Coleraine bid was the information that exchequer money was now being directed entirely at academic buildings, not for halls of residence. They were advised that the critical criterion was the availability of lodgings, together with the opinion that if the new university were to be based on different criteria from those used in Great Britain, then these would have to be clearly justified.[28] The green light for the Coleraine bid had been shown.

Over the following two days the committee reviewed the question of location in the light of UGC criteria applied in Great Britain, Lockwood advising his colleagues that the chosen site should enable the smooth progress of the university, that the university should reflect the region's needs, that it should attract quality staff, that it should address the question of lodgings and residences, that there should be a suitable site, whether there was the possibility of financial or other support, and whether the area had industrial and research facilities. It was felt that none of the sites satisfied all these criteria, and it was confirmed that Magee's role as a university college should end. Neither the Derry nor the Coleraine promotion committees were judged to be impressive. In Derry's favour was that it could provide a site, its location favoured agriculture and biology, being described somewhat fancifully as lying 'on the coast', it had a sizeable population, some cultural activity and its choice would ameliorate the decision on Magee. On the other hand, its housing provision was bad, necessitating expensive halls of residence, and Magee's existence was a complication. Worse still was the perception that Derry's industrial development was doubtful and that 'it gave the impression of a frontier town and had never lost the siege mentality'. Those in charge of the new institution might have to expend their energies on the city's political divisions.[29] While Coleraine was devoid of cultural amenities, it was felt to have an adequate population, some industry, could provide housing for the staff, was supported by Londonderry County Council and could probably find a site. The key arguments appeared to be that its suitability for agriculture, marine biology and biology reinforced the academic case and that lodgings were immediately available. By then it was clear that Coleraine's case was effectively won, the committee believing that 'its proximity to Londonderry would smooth hurt feelings, while easing the demise of Magee'.[30]

The Lockwood recommendations and the origins of the New University

On 27 November 1964, Lockwood submitted his committee's report. The document examined the future size of Queen's University, the role of the teacher training colleges and of further education. The future of higher technical education was to be addressed through the proposed creation of an Ulster College combining various existing colleges. This development, in time, impacted significantly on the new university.[31] Overriding almost everything else in Lockwood's recommendations were its conclusions regarding a second university and, critically, its proposed location. That the creation of a second university should be pursued as a matter of urgency was its core conclusion. By 1973–4, 8,000 or 9,000 university places would be required, 6,000 of which could be met by Queen's, leaving between 2,000 and 3,000 for the new university. By 1980, it was envisaged that the new institution would have between 5,000 and 6,000 full-time undergraduates and postgraduates. The committee also suggested certain broad parameters for the university's provision, placing particular emphasis on the biological sciences. The physical, social and behavioural sciences would be needed in a supporting role. Two other areas were recommended. The first was a sound basis in the humanities, and the other was the creation of an education centre with facilities for around 1,000 students as an integral element of the new institution. These subject areas were to form the basis of the new university's initial undergraduate provision. Just why the Committee was so dirigiste in its recommendations remains unclear.[32]

The other major, and infinitely more controversial, recommendation was the location of the new university. In presenting its judgement the committee was careful to refer to the criteria that had already been used in Great Britain by the UGC. The first concerned the site which it felt should be of 300 acres, more being needed if agriculture were to be included. The report emphasized the need for good housing and amenities for the staff. The availability of lodgings for students would enable finance to be concentrated on academic buildings. The university would have to be located where there were good communications but be sufficiently far from the attractions of Belfast. The target date of October 1968 meant that a site would have to be acquired quickly if buildings were to be in place by then. As a result, the committee recorded its recommendation that these criteria were best satisfied by the Coleraine area. It pointed to the willingness of the local sponsoring committee to support the university's establishment. The reasons for rejecting Craigavon were spelled out but why the cities of Armagh and Derry had been rejected was left unstated, beyond the oblique, if significant, observation that the development of the university should be 'unaffected by political considerations either at local or at central government level'.[33] The position of Magee could not be ignored, its claims being dismissed on the grounds that its existing buildings and site could not be the basis for a university of 5,000–6,000 students, apart from which 'we do not consider that the College provides a suitable base for expansion'; hence, they saw 'no alternative to its discontinuance'.[34]

The government's response to the Lockwood report

The committee's recommendations that Coleraine should be the location of the second university and that Magee be closed were clear, but the government, quick to spot the likely political ramifications, had the Christmas parliamentary recess to compose its response. When the cabinet met on 21 December, it had to consider a memorandum by the Minister of Education, Herbert Kirk. Recommending acceptance of the Coleraine decision, Kirk had reluctantly accepted that Magee should cease to be a university-level institution. Nevertheless, cautioning of the political dangers if the government were seen to have deliberately killed it off, he advocated some other function, perhaps as a teacher training college.[35] While there was broad agreement on the choice of Coleraine, the future of Magee could not be ignored, with the Minister of Home Affairs counselling caution while wondering if the premises could be used by the new university. Kirk feared that the Treasury would not sanction a second university while continuing to finance what he dismissed as a 'relatively derelict institution'.[36] When discussion of the issue was resumed on 6 January 1965, it was clear that the cabinet was marking time, searching for a formula that would preserve Magee in some form. O'Neill was still inclined to the teacher training option.[37]

With some kind of rescue mission now in mind, a white paper was drafted to accompany the report, indicating the government's reluctance to accept the discontinuation of Magee, and proposed 'to ask the Academic Planning Board of the new university to consider whether it would be possible to incorporate the College as part of the new university'.[38] Aware of rumours that were circulating in Derry, on 13 January, with minor wording amendments, this formula was adopted by the cabinet.[39] When the report was at last published on 10 February, the accompanying white paper, while accepting the Coleraine decision, made it clear that it could not endorse the discontinuance of Magee as a university institution and proposed 'to investigate whether the College can be incorporated in the new university'.[40]

The 'University for Derry' campaign

Attention in Derry crystallized when unsettling rumours hinted that the city was to be by-passed in favour of the Coleraine tender; these took more coherent shape when, in December 1964, the influential former unionist mayor of the city and a current Trustee of Magee, Sir Basil McFarland, commented publicly that he did not believe that the Lockwood report would 'do us any good'.[41] This fear was confirmed by the President of the City of Londonderry and Foyle Unionist Association, Major Gerald Glover, who stated days before the report was due: 'Strong rumours have it, that we may have failed in our efforts.'[42] John Hume, a young school teacher who had made a name for himself through his Credit Union activities, had watched earnestly as these events unfolded. Hume and a group of fellow concerned citizens voiced their alarm in a statement on 29 January calling for the formation of an action committee in order to press Derry's claim.[43]

In early February, church leaders from across the city answered the call, issuing a joint declaration endorsing the Londonderry bid and establishing a University for Derry Committee (UDC). Leadership was assumed by Hume by popular resolve with members drawn from across the two traditions, and with three Magee staff in an advisory role. Support came from MPs, trade unions and representatives of almost every industrial group in the city.[44] A public meeting, at which Hume gave the closing speech, was held at the behest of both the UDC and Londonderry Corporation on 8 February at the city's Guildhall to 'give the people of the city an opportunity' to voice their support and 'to show that they would fight for the university as one people'.[45] The citizens of Derry thronged the aisles 'in one of the largest [public meetings] seen in the city in recent years'.[46] The publication of the Lockwood report and the accompanying white paper two days later was met with astonishment in the city. Hume's rallying speech pointed to the unity of the city's two communities.[47] The next day Hume led a deputation to Stormont to discuss the issue with O'Neill who was 'sympathetic' to the Derry claim, but 'non-committal'.[48]

The UDC and the Corporation designated 18 February 'University for Derry Day', calling upon employers in the city and citizens alike to treat it as a public holiday and were answered by the mass support of the people of Londonderry. Businesses issued supportive advertisements in the local newspapers and most had at least one representative in the motorcade of some 25,000 people that wound its way to Stormont, led by Mayor Albert Anderson and the nationalist MP for the city and leader of the official opposition, Eddie McAteer, who shared the mayoral car in an unprecedented show of unity.[49] The motorcade was joined by many more on its journey and by local supporters, including from Queen's which had a strong supporters' committee.[50] It was, by all accounts, an impressive sight and could not have failed to make a positive impression on the politicians. The same could be said of the cogent arguments made by the UDC in questioning by MPs and the response to Lockwood which was circulated to MPs and published province-wide in the days following the motorcade.[51]

The UDC was impressive in its organization, intellectual underpinning and its cross-community, civic, economic and political backing, but the government's mind was made up and it was not to be. Making the issue one of confidence, despite reservations held by some unionist MPs, the government's majority was assured, but only by eight votes. The failure of the campaign marked a critical point in Northern Ireland's political development. The 'loss' of the university was the final blow that left the people of Derry teetering on the brink. It took precious little to tip them over the precipice. The campaign also marked a decisive stage in Hume's political evolution. He was to become one of the leading figures in the subsequent civil rights movement and, after a long political career, a principal architect of the peace process.[52]

The creation of the New University of Ulster

With the target date set for 1968, planning for Coleraine had to begin if staff and buildings were to be in place. A 300-acre greenfield site outside Coleraine was secured and a Development Plan drawn up by Sir Robert Matthew, Johnson-Marshall and Partners. The name New University of Ulster was chosen. Under the direction of the Ministry of Education an Academic Planning Board (APB) was established under the chairmanship of Sir James Cook FRS, vice-chancellor of the University of Exeter and subsequently of the University of East Africa. His background in organic chemistry made him a neat counterpoint to the classicist Lockwood, although it immediately became clear that their views on the way forward were broadly similar. In its first meetings on 25–27 October 1965, the Board concluded that although they were not bound by the academic structures that Lockwood had proposed, they nevertheless took the view that the path set out in the report, albeit as modified by the government, should be followed. Taking this decision as its starting point, the Board set about establishing four schools: Biological and Environmental Studies, Physical Sciences, Social and Economic Studies, Humanities (historical and cultural studies) and the Education Centre. The initial student intake for 1968 was to be 445, leading to a student population of 3,575 by 1973, at least of quarter of whom would come from outside Northern Ireland, chiefly Great Britain. The search was undertaken for a vice-chancellor, the choice falling on Dr N. A. Burges. An Australian by birth who had served in wartime Bomber Command in which he was mentioned in despatches, his profile as an eminent botanist and administrative record as senior pro-vice-chancellor of the University of Liverpool being ideally suited to the academic direction that

Lockwood had set out for the new institution. Ewing was invited to become Registrar. The Board was all too conscious that the government's decision over Magee made their work different to that of any equivalent body in Great Britain. Its initial decisions were that the college should not be the location of a single school, that the office of president should be retained and that the six existing professors be recognized as professors of the New University and that at least the current level of provision should be maintained.[53] Aware of student restlessness in Great Britain, the Board sought to prevent this by providing joint facilities for staff and students, ensuring frequent staff–student contact and flexible course provision.[54] The latter was encouraged by the implementation of an imaginative unit of study structure that foreshadowed the later modularization policies of the wider university sector. Burges and Ewing were followed by founding professors in English, history, Russian, geography, economics, chemistry, physics, mathematics, social organisation, education and two in biology. It was an academic profile that broadly followed the template that Lockwood had set out.[55]

The University at work: Opportunity and challenge

The founding professors brought broad experience of the university sector and were open to new ideas. Innovative courses in, for example, biological and environmental studies, education, and history were set in place. Non-professorial staff members were mainly recruited from Great Britain.[56] They were largely unaware of the controversies that had accompanied the university's birth, and were excited at the prospect of developing a new institution and with the accompanying sense of collegiality.[57] Even as the university was starting, there were danger signals. The initial intake of 340 undergraduates at Coleraine and 60 at Magee had been cut back. The issue of numbers was to eat at the university's problems throughout its existence, but few predicted the political storm that was to wreck the surface calm of life in Northern Ireland, impacting on student demand. In August 1969 Northern Ireland finally erupted with rioting in Derry, the 'Battle of the Bogside'. A cycle of violence and failed political initiatives only came to an end with the cease fires of 1994 and the Good Friday and St Andrews Agreements of 1998 and 2006. In short, Northern Ireland endured a level of conflict unique to contemporary western Europe, a backdrop shared with Queen's but with nowhere else in the UK university sector.[58]

With an experienced professoriate in mid-career and junior staff recruited fresh from postgraduate work, the New University developed an impressive research profile. In the ten years 1972–82, over 1,800 books, articles and research papers were recorded, while staff attracted £2,674,876 in research grants.[59] The quality of its research is evidenced by the fact that in the area of pure mathematics Professor Ralph Henstock enjoyed an international reputation for his pioneering work on integration. Nor were staff unaffected by the violence; witness James Simmons's moving song 'Claudy', a memorial to the nine people, including a young girl, who were killed in an explosion there.[60]

Problems were foreshadowed almost from the start. Burges's first report to the University Court identified problem areas. The original target intake for 1968 had been 445. Because of the disturbances in Derry, it had been difficult to find staff for Magee; its proximity to the Bogside where there were recurrent clashes between republican youths and the security forces made it worrisome.[61] With the introduction of internment in 1971 and the events of 'Bloody Sunday' in Derry on 30 January 1972 in which soldiers killed fourteen unarmed men and youths, the situation in Northern Ireland, and in the city in particular, impacted the university significantly. By 1971, it was evident that the target of 300 Magee students would be missed. In early 1972, when the low level of applications for that year's entry was confirmed, Senate voted to withdraw undergraduate studies from the campus, replaced by an Institute of Continuing Education. Over the next decade the Institute undertook pioneering work in access courses and professional development in Nursing and Education.[62] On the political front, 'Bloody Sunday' was quickly followed by the suspension of the Northern Ireland parliament, which meant that responsibility for higher education passed to a secretary of state and ministers from London unbeholden to the Northern Ireland electorate.

The university's abiding problem was numbers, the Lockwood committee's projected campus of 5,000–6,000 hanging like the proverbial albatross around its neck. In posing the question 'where have all the students gone?', Burges inevitably pointed to 'the Troubles', which had resulted in an exodus of students to universities in Great Britain and fewer coming to Northern Ireland. In October 1972 a target of 2,200 was agreed with the UGC but the actual enrolment for that year was 1,632, 452 of whom came from Great Britain.[63] With violence showing no sign of abating, the latter figure was especially vulnerable, falling over succeeding years, touching a low point of eighty-four in 1979–80 before recovering somewhat. The university's overall figures remained stubbornly low, with its total enrolment of 1,723 in 1979–80 having shown little improvement over the decade.[64] In fact, in addition to the rival attractions of cross-channel universities and Queen's, a new factor had entered the equation as the Ulster College, originally suggested by Lockwood, morphed into the Ulster Polytechnic. Under the vigorous leadership of (later Sir) Derek Birley, the institution based at Jordanstown outside Belfast developed a broad provision of degree courses under the auspices of the Council for National Academic Awards (CNAA). By 1980–1, 54 per cent of its students were pursuing degree-level work, the total enrolment for 1978–9 being 7,446.[65] These figures inevitably gave rise to anxieties about the future. A visiting professor from the United States, who spent the academic year 1978–9 as a Leverhulme Fellow, recollected his overarching impression as being that of insecurity, with the university conscious of the comparisons with Queen's and the Polytechnic and 'a sense that NUU was not appreciated for the quality of what it offered'.[66]

The Chilver report and the creation of the University of Ulster

Burges retired in 1976, succeeded by Dr (later Sir) Wilfred H. Cockcroft, a noted mathematician and experienced administrator from the University of Hull who had served on the UGC.[67] Even so, the institution's prospects showed little sign of improvement. In December 1978 the government appointed the Higher Education Review Group for Northern Ireland, chaired by Sir Henry Chilver, vice-chancellor of

Cranfield Institute of Technology, charged with reviewing the current provision and making recommendations in the light of the community's needs in the 1980s and 1990s. When Chilver reported to the minister on 21 January 1982, his group's conclusions on the New University made depressing reading, especially on the critical issues of student numbers and entrance qualifications. Student numbers, Chilver observed, in 1980–1 had not risen above the 1970–1 level, noting the Lockwood target of 6,000. The number of students entering with top A level grades was 9.5 per cent, compared with 15.0 per cent five years before, and 56.8 per cent at Queen's. The inescapable conclusion was that the university 'has consistently attracted only a small proportion of Northern Ireland full-time students'. While acknowledging the impact of the disturbances, the report criticized the institution for failing to provide courses that were attractive to school leavers, noting that Queen's and the Polytechnic had been able to recruit heavily in the areas provided by the New University. In numerical terms, Chilver concluded, the higher education needs of Northern Ireland could be met without the New University. On balance, it was felt that it should not be closed, but retained as a small institution of some 1,000–1,500 students with a focus on courses other than traditional degree-awarding. Emphasizing that they saw 'no prospect of N.U.U. becoming a strong and cost-effective institution within the limitations of the traditional university model', they concluded that if it could not succeed in its new role then closure would have to be considered.[68]

The government endorsed Chilver's bleak analysis of the New University but not his recommendations over its future. In an accompanying white paper, it judged that the role Chilver had envisaged would make it even less attractive to students and lead to it further running down. The status quo was not an option. Instead, what was proposed was the radical idea of a merger between the New University and the Polytechnic to form a new university institution in which Magee would be a significant element. While the university authorities sought to rebut many of Chilver's conclusions, and there was inevitable heart searching in both institutions, the government was adamant that if negotiations between them were to fail then the only alternative was the New University's closure.[69] The process by which the Steering Group chaired by Sir Peter Swinnerton-Dyer created the University of Ulster, which under Birley's vice-chancellorship came into existence on 1 October 1984, lies beyond the scope of the present discussion.[70] After much discussion, the Senate and Council of the New University agreed to the proposal, and on 16 September 1983 Court endorsed the merger.

Conclusion

The New University of Ulster's story had come to an end, but with the creation of the University of Ulster new opportunities for Coleraine and Magee, now joined with campuses in Jordanstown and Belfast, were opening up. The fledgling New University of Ulster had proved to be an uneasy partner of its sister utopian universities in Great Britain, but given Northern Ireland's distinctive place in the United Kingdom, and the nature of its contested society, perhaps that was no surprise. The controversy over its location helped bring into sharp focus the tensions lying not so far under the surface of Northern Ireland, not least in the city of Derry where the university decision unlocked a mass popular protest that anticipated what lay ahead. While the other utopian universities created in the UK could tackle the inevitable challenges and upheavals facing higher education in the context of civil peace, the New University of Ulster had no such opportunity, however hard it tried.

Notes

1. See Gerard O'Brien and Peter Roebuck (ed.), *The University of Ulster: Genesis and Growth* (Dublin: Four Courts Press, 2009).
2. The authors are grateful to Dr Martin Melaugh, Director of CAIN (Conflict Archive on the Internet), Ulster University, for the above figure.
3. See R. J. Lawrence, *The Government of Northern Ireland: Public Finance and Public Services 1921–1964* (Oxford: Clarendon Press, 1965).
4. See T. G. Fraser, 'The Siege: Its History and Legacy', in Gerard O'Brien (ed.), *Derry and Londonderry: History and Society: Interdisciplinary Essays on the History of an Irish County* (Dublin: Geography Publications, 1999), 379–403.
5. *Disturbances in Northern Ireland. Report of the Commission Appointed by the Governor of Northern Ireland*, Cmnd. 532 (Belfast: Her Majesty's Stationery Office, 1969), ch. 12.
6. *Higher Education in Northern Ireland. Report of the Committee Appointed by the Minister of Finance*, Cmnd. 475 (Belfast: HMSO, 1965) (hereafter *Lockwood Report*), 19–21.
7. Ibid., 35.
8. *The Autobiography of Terence O'Neill: Prime Minister of Northern Ireland 1963–1969* (London: Rupert Hart-Davis, 1972), 44–6. Marc Mulholland, *Terence O'Neill* (Dublin: University College Dublin Press on behalf of the Historical Association of Ireland, 2013).
9. Frank Curran, *Derry: Countdown to Disaster* (Dublin: Gill and Macmillan, 1986), 25, 40. Cf. Robert Gavin et al., *Atlantic Gateway: The Port and City of Londonderry since 1700* (Dublin: Four Courts Press, 2009), 265–97.
10. *Report of the Committee Appointed by the Prime Minister under the Chairmanship of Lord Robbins 1961–3*, Cmnd 2154 (London: HMSO, 1963), 284.
11. *Lockwood Report*, vii.
12. Alison Wilson, *Changing Women's Lives: A Biography of Dame Rosemary Murray* (London: Unicorn Press, 2014), 102–5, 162–3.
13. *Lockwood Report*, vii–ix. See also Gerard O'Brien, '"Our Magee problem": Stormont and the Second University' in *Derry and Londonderry*, ed. O'Brien, 647–96; cf. John Draper, 'A University Gerrymander? Coleraine and the Lockwood Report', *History Ireland*, 22 (2014), 44–7.
14. 'The Lockwood Report', Cabinet meeting (21 December 1964), Public Record Office of Northern Ireland (henceforth PRONI), PRONI CAB/4/1286; W. T. Ewing to Kenneth Bloomfield, 5 February 1965, enclosing 'The Lockwood Report. Some Possible Questions and Answers', PRONI CAB 9 D/31/2; O'Brien, 'Our Magee Problem', 672–3.

15 *Submission to the Government of Northern Ireland Presenting the Case for the Promotion of a University at Londonderry* (City and County Borough of Londonderry, 31 December 1962).
16 'A Second University for Northern Ireland: Proposals from the Coleraine/Portrush/Portstewart Area' (February 1964), PRONI CAB/9/D/31/2.
17 Lockwood report, Minutes of Committee Meetings (6–7 December 1963) and (26–27 September 1964), PRONI ED 39/9.
18 Note by Secretary, 'Siting of a New University' (4 March 1964), PRONI FIN 58/8/14937.
19 Lockwood Report, Minutes of Committee Meeting (31 January–1 February 1964), PRONI ED 39/9.
20 Ibid. (21–22 February 1964), PRONI ED 39/9.
21 Ibid. (12–14 March 1964), PRONI ED 39/9.
22 Ibid.
23 Brum Henderson, *Brum: A Life in Television* (Belfast: Appletree Press, 2003), 105.
24 Ibid., 104–7.
25 Note by Secretary, 'Siting of a New University' (4 March 1964), PRONI FIN 58/8/14937.
26 Lockwood Report, Minutes of Committee Meeting (1–2 May 1964), PRONI ED 39/9.
27 Ibid. (14–16 May 1964), PRONI ED 39/9.
28 Note by R. H. Kidd (4 June 1964), PRONI CAB 9/D/3112.
29 Lockwood Report, Minutes of Committee Meeting (5–6 June 1964), PRONI ED 39/9.
30 Ibid.
31 *Lockwood Report*, 93–100.
32 Ibid., 66–75.
33 Ibid., 76–81.
34 Ibid., 76–81.
35 'Memorandum by the Minister of Education. The Lockwood Report' (11 December 1964), PRONI CAB/4/1286/6.
36 'The Lockwood Report', Cabinet discussion (21 December 1964) PRONI CAB/4/1286.
37 'The Lockwood Report', Cabinet discussion (6 January 1965), PRONI CAB/4/1287.
38 'Memorandum by the Minister of Education' (8 January 1965), PRONI CAB/4/1288.
39 'The Lockwood Report', Cabinet discussion (13 January 1965), PRONI CAB/4/1288.
40 *Higher Education in Northern Ireland. Government Statement on the Report of the Committee Appointed by the Minister of Finance*, Cmnd 480 (Belfast: HMSO, 1965).
41 *Londonderry Sentinel*, 6 January 1965, 17.
42 Quoted in ibid. (20 January 1965).
43 *Derry Journal*, 29 January 1965, 1.
44 *Londonderry Sentinel*, 3 February 1965, 1.
45 John Hume, quoted in *Derry Journal*, 2 February 1965, 1.
46 *Derry Journal*, 2 February 1965, 1; ibid., 9 February 1965, 5. Paul Routledge, *John Hume: A Biography* (London: HarperCollins, 1998), 50.
47 Barry White, *John Hume: Statesman of the Troubles* (Belfast: Blackstaff Press, 1984), 38–9.
48 Curran, *Derry: Countdown to Disaster*, 32; *Derry Journal*, 12 February 1965, 1.
49 Curran, *Derry: Countdown to Disaster*, 33; *Derry Journal*, 19 February 1965, 1; *Londonderry Sentinel*, 24 February 1965, 14.

50 *Derry Journal*, 19 February 1965, 1, 9. Queen's also had a smaller University for Coleraine Committees.
51 Ibid., 19 February 1965, 1, 9; ibid., 23 February 1965, 6–7.
52 Routledge, *John Hume*, 50–4.
53 Arthur P. Williamson, 'The New University of Ulster', in *University of Ulster*, ed. O'Brien and Roebuck, 41–51; 'The New University of Ulster. Interim Report of the Academic Planning Board', May 1967, Ulster University Archive.
54 'The New University of Ulster. Report to the University Court 1968–1970', Ulster University Archive.
55 *The New University of Ulster, University Handbook, 1970–71*, Ulster University Archive.
56 Personal information.
57 Personal information.
58 See T. G. Fraser, *Ireland in Conflict 1922–1998* (London: Routledge, 2000).
59 'The Chilver Report. The New University Comments', May 1982, Ulster University Archive.
60 Peter Bullen, 'Ralph Henstock an Obituary', *Scientiae Mathematicae Japonicae Online* (2007), 773–5, www.jams.org.jp/scm/contents/e-2007-8/2007-71-pdf (accessed 19 January 2017); James Simmons, 'Claudy', in *The Selected James Simmons* (Belfast: Blackstaff Press, 1978), 78–9.
61 'The New University of Ulster. Report to the University Court 1968–1970', Ulster University Archive.
62 Ibid.; T. G. Fraser, *The History of Magee* (Coleraine: University of Ulster, 2016).
63 'The New University of Ulster. Report to the University Court 1972–1973', Ulster University Archive.
64 'The New University of Ulster. Report to the University Court, 1979–1980', Ulster University Archive.
65 See Don McCloy, 'Ulster College to Ulster Polytechnic', in *The University of Ulster*, ed. O'Brien and Roebuck, 69–94.
66 Professor Lester C. Lamon to the authors, 9 August 2016.
67 A. J. Crilly, 'Cockcroft, Sir Wilfred Halliday (Bill) (1923–2009)', *Oxford Dictionary of National Biography*, Oxford University Press, September 2015) http://www.oxforddnb.com/view/article/72961 (accessed 7 March 2017).
68 *The Future of Higher Education in Northern Ireland* (Belfast: HMSO, 1982).
69 *Higher Education in Northern Ireland: The Future Structure* (Belfast: HMSO, 1982).
70 Rosalind Pritchard and Peter Roebuck, 'The Merger and after, 1982–1991', in *The University of Ulster*, ed. O'Brien & Roebuck, 95–118.

13

The new British campus universities of the 1960s and their localities: The culture of support and the role of philanthropy

by Jill Pellew

There was much that was novel, even utopian, in the thinking about the physical and intellectual shape of the new universities in Britain. They were to be centrally funded and planned on an unprecedented scale, as part of the post-war state welfare system. The new institutions would be national – even international – residential universities, on greenfield sites in cities with a cultural tradition, offering new cross-disciplinary academic disciplines. Above all, they were to be autonomous, in the sense of awarding their own degrees from the start. Yet the very idea of locating them on campuses on the peripheries of cities without experience of a university and needing the support of the surrounding county society meant that this was not yet the end of a hundred-year era in which the founding of English universities depended on local enthusiasm, energy and money. That history, partly bound up with the strength of English local government and the need for private endowment, was well understood by the University Grants Committee (UGC), established in 1919 with responsibility for making recommendations to the Treasury about the funding of universities. This understanding was strongly to influence the choice of locality for the new institutions. Close local cooperation was needed in securing a welcome for complex organizations destined to make a substantial impact on the local environment. While Treasury finance was intended to provide for capital costs and basic recurrent needs, there were gaps in funding that would have to be filled from other sources. Despite the new age of socialism and central planning, the historic importance of local involvement remained a key factor in achieving the overall vision for these new universities. The UGC had a history of believing that universities depended on a diversity of sources of interest and support – particularly from their localities – in order to maintain their independence. This was a constant theme in the late 1950s and early 1960s of its then Chairman, Sir Keith Murray.

While the expansion of university places was being discussed, word was spreading in towns and cities eager to promote their case as civic hosts for a new university. In 1946 and 1947 delegations from Brighton, York and Norwich presented their cases to the first full-time Chairman of the UGC, Sir Walter Moberly, whose committee's post-war role had been enhanced in order to ensure that universities were 'fully adequate to meet national needs'.[1] The petitioners included civic dignitaries, educators, senior churchmen, members of Parliament, professionals and businessmen – many of the same social elements that had been instrumental in founding civic university colleges in the half-century before the First World War. Again, in the 1940s, civic pride played an important part. Historic, cultural and geographical rationales were displayed. York based its claim for a 'full and founded University' on its historical significance as an important Roman encampment, archiepiscopal seat, and centre of commercial, artistic and archaeological activity.[2] Norwich stressed its cultural heritage as well as its distance from any other university centre.[3] Moberly understood these arguments. An erudite and deeply religious Anglican theologian, who had been vice-chancellor of the University of Manchester, he had thought deeply about higher education in England, which he believed had lost its moral compass. Much as he admired the role that 'redbrick' universities played in their urban communities, he dreamed of a location such as 'Canterbury or York ... cities rich in historic buildings and traditions', that was [sic] not 'drabtown'.[4] While not closed to new ideas – such as those that underlay the experiment at Keele – Moberly was part of the old guard and did not breathe the 'spirit of post-war optimism'. Above all, he was nervous about creating new autonomous institutions.[5]

A crucial source of momentum came from a new, professional quarter: the cadre of local education officers, backed by their city and county councils. These officials, pushing for the expansion of higher education opportunities, differed from the late-nineteenth-century industrial and commercial giants (such as the Ashton, Roscoe and Rathbone families of Manchester and Liverpool) who, in an age of liberal 'laissez-faire', had inspired the creation of university colleges in their cities. Endowment provided by these founders had remained a notable financial boon for those universities throughout the inter-war period. For example, in 1938–9, the income from endowment of the universities of Bristol, Manchester, Birmingham and Liverpool respectively was 20.7 per cent, 19.9 per cent, 16.3 per cent and 14.3 per cent. By 1950 this had been seriously eroded and public support from central government had become far the largest element of British university income.

Responsibility for secondary education lay with counties and county boroughs. Administration was led by professional local education officers, a group of enlightened and powerful men, concerned about a new trend: the increasing number of able children

Table 13.1 Sources of university income in Britain, before and after the Second World War, 1938/9 and 1950/51[6]

Date	Parliamentary grants	Tuition, exam, etc. fees	Endowments + donations	Local authority grants	Other
1938/9	35.8%	29.8%	18.0%	9.0%	7.4%
1950/1	64.9%	16.6%	7.3%	4.3%	7.0%

entering grammar school sixth forms with ambition and the capability of pursuing higher education.[7] Where were the places for these children going to come from; and how could their families afford to pay for university education? Particularly striking for his dogged persistence in working to resolve this issue was the Director of Education for the Borough of Brighton, W. G. 'Bill' Stone, who succeeded Herbert Toyne, leader of the unsuccessful delegation to the UGC in 1946. A visionary, with a keen sense of history, Stone revived the project and led the charge for his locality becoming the key liaison between it and the UGC.[8] In 1958 he arranged a series of lectures by distinguished academics to an invited local audience on 'The idea of a university'.[9] Stone later became secretary both to the University of Sussex Council of Governors and to its early appeal; and in 1964 the University awarded him an honorary degree. In Norfolk the real visionary was Dr (later Sir) Lincoln Ralphs, chief education officer for Norfolk County Council who had long foreseen the importance of a local university.[10] His equivalent at Kent was Dr John Haynes, County chief education officer from 1959 to 1974.[11] Local government teamwork was evident in the support for these officers from their town clerks. At Norwich this was Gordon Tilsley, Town Clerk from 1959 and 'a leading figure in the movement' to found a university.[12] The city of Coventry had a particularly dynamic local authority led by the Town Clerk, Charles (later Sir Charles) Barratt, regarded by some as the 'linchpin' of the effort to create what became the University of Warwick.[13] The initiative in Essex came from Alderman Charles Leatherland (later Baron Leatherland of Dunton) who, in July 1959, convincingly argued the case to his fellow councillors and to local rate-payers. Leatherland saw to the setting up of an Education Committee that played a critical role in persuading the UGC to promote a new university in Essex.[14] Education officers were not the only leading inspiration. At Warwick the perseverance of a headmaster, Dr Henry Rees, was also significant in the university project.[15] York's inspiration came from an independent group of historically and culturally minded individuals associated with the York Academic Trust which became a 'committed nucleus of university promoters'.[16] The first Registrar, John West-Taylor, boasted that 'while the claims for universities elsewhere all mainly depended … on the initiative and co-ordination of local authorities … the initiative in York was held … in the hands of an independent group'.[17] At Lancaster the drive for a new university came from senior figures in the county, imbued with local pride, rather than from local education officers concerned about opportunities for able sixth formers.[18]

The appointment of Keith (Sir Keith from 1955) Murray as chairman of the UGC heralded a new dynamic in Whitehall.[19] From the mid-fifties his Committee became increasingly convinced that sufficient opportunities for qualified sixth-formers could not be met without the creation of new institutions. This focused the UGC's attention on possible locations.[20] In 1958 it was announced that a new university would be founded in the Brighton area. Two years later, UGC calculations of further expansion requirements enabled Murray to convince his Treasury masters of the need for two further new universities – announced in April 1960 as being York and Norwich, to be opened in 1963. No total number of new institutions was yet specified, but the following year another four locations were announced: Essex, Kent, Warwick and Lancaster.[21] By this time the possibility of new universities for Scotland and Northern Ireland was being discussed, although their actual realization as the University of Stirling and the New University of Ulster emerged from the Robbins Report of 1963.[22]

While undoubtedly a believer in fresh thinking, Murray held to traditional values that underpinned British universities – not least the protection of academic freedom through autonomous status. These values were to be reinforced in 1963 by Robbins. The great expansion of university places was predicated on a huge and continuing Whitehall subvention. At that time few seriously anticipated potential problems – such as threats to academic freedom or the lack of long-term sustainability – that might lie in the creation of new universities far more reliant than ever before on a single paymaster: central government. After all, it was an optimistic post-war socialist society that believed that taxes were the only realistic way of paying for a major educational experiment. However, Murray may well have sensed the very same danger that the first secretary to the UGC had done. Back in its formative days in 1918, its first permanent official, Alan Kidd, had lamented the fact that several universities were deriving up to 50 per cent of their revenue from the state, urging that 'adequate endowment lies at the root of free development for a university'.[23] Murray's last official quinquennial report made clear the UGC's approval of universities' efforts to raise funds from non-government sources: for 'they are anxious not to become solely dependent on state funds'.[24] It was part of constant UGC iteration of the government's dependency on strong local initiatives in the creation of the new institutions.[25] To underline this Murray encouraged local 'promotion committees' for each of the new universities. The Promotion Committee at East Anglia gives an idea of the sought-after range of support. Its full complement consisted of seventy-six individuals representing local government (twenty-four), industry and business (sixteen), academics (thirteen), landed gentry (nine), churchmen (four), women's organizations (three), legal and medical organizations (three), MPs (three) and trades unions (one).[26] The successors of these promotion committees, once the universities were up and running, were their councils whose external membership of the local great and good gave them public authority and legitimacy. The perceived need for this can be seen to hark back centuries, as Clark Kerr observed:

> Even in the Middle Ages, emperors and popes, dukes, cardinals, and city councils came to authorize or establish the universities to make them legitimate – the guild alone was not enough.[27]

While influenced in the choice of new institutions by cultural and demographic considerations, Murray was, after all, a Treasury civil servant and hard-headed about financial input. What the UGC required from contending bidders was evidence of material support in the form of land for sites (150–200 acres, preferably with room to expand); financial commitment towards recurrent expenditure; and the potential to raise funds, particularly for student residences, in an early appeal.

The universities were founded on the peripheries of cities that, apart from Warwick, lacked any nearby major industry, located on 'green field' sites because – as Noel Annan put it – 'the cost of acquiring sites in the centre of cities would have been prodigious, the delays and frustrations intolerable'.[28] By and large, only city and/or county councils could find the means and had the legal authority to purchase and sanction the development of appropriate land; and it was the business of the promotion committees, with local government representation, to concern themselves with site location and purchase. That is not to say that there was no private involvement in this process. At York a revered city father, J. B. Morrell, chairman of the York Civic Trust (the original sponsor of the university project), made possible the diversion to the university project not only the nucleus of its 184-acre site but also Heslington Hall which he had formerly owned.

All possibilities were vetted by the UGC – indeed, usually by Murray himself. In Essex there arose a dispute between forces at Chelmsford – first off the starting block with a possible site – and those at Colchester who alleged that they had been kept in the dark and, moreover, had an option on a better 204-acre site, Wivenhoe Park. Difficulty arose when the owner of the latter hesitated to sell it to the county for fear of offending the owner of the Chelmsford site. At this point, Murray intervened to resolve the issue in a quiet exchange with the respected and emollient Lord Lieutenant (and Chairman of the Chelmsford Promotion Committee), Sir John Ruggles-Brise. Murray made it clear that the UGC preferred Wivenhoe Park – partly because he was concerned that the proximity of Chelmsford to London might detract from the community life of the university – and gently implied that if agreement could not be reached in Essex, the whole university project there could be over-ridden by other contenders.[29] This veiled threat did the trick. The County Council not only promised to consider purchase of the Wivenhoe site for about £100,000, but also pledged a capital grant for the same amount as well as a generous recurrent grant.[30]

Negotiations preceding the eventual choice of site, involving local sensitivities, were often complex. City and county authorities that were committing rate-payers' money not surprisingly sought some ownership in the naming of 'their' institution. At Kent the 270-acre site chosen for the new university was on the northern edge of its cathedral city and both Canterbury and the County Council agreed to contribute £100,000 each towards its purchase – hence the somewhat awkward early name of the University of Canterbury at Kent to which Krishan Kumar also refers in this volume. The designation of the 'University of Warwick' emerged as a compromise, apparently proposed by the Bishop of Coventry, when tensions arose between the city of Coventry and the county of Warwickshire about the nature and funding of the site. In the event, city and county authorities were content to match each other's £92,000 contributions.[31] At Norwich, complexities were augmented by the regional concept of East Anglia and the need for recurrent financial support from the wider region. The term 'University of East Anglia', a nomenclature formally and successfully proposed by the Ipswich member of the Executive Committee of the Promotion Committee, became current and, incidentally, foiled the UGC's tidy Shakespearean approach to the new universities: York, Essex, Lancaster, etc.[32]

Contributions towards the new universities from local authorities were a crucial component of their early income since the UGC would not sanction any Treasury funding of recurrent costs (including staffing) until a university had received its charter and was into its first intake of students. The number of borough and county contributing authorities varied considerably from one university to another – from eleven in Yorkshire down to two in Kent and Warwick. There were major differences in the amount granted depending on what an authority felt it could justify from the rates. Annual grants to the University of York in 1962/3, totalling £36,000, varied from £16,000 from the County Borough of York down to under £250 from the poorer county boroughs of Rotherham and Hull. That year grants totalling £59,000 to UEA, from eight authorities, varied from Norwich's £15,000 down to £2,000 from Huntingdonshire and the Isle of Ely. Essex, in its early years, had support for recurrent costs from only four authorities, one of which – Essex County Council – was notably generous in granting £107,000, the major portion of the total of £122,000.[33]

As each of the new universities became established and the numbers of their students increased, the essential early support of their local authorities substantially declined. Meanwhile, annual income from the Exchequer – always the mainstay of funding in this period, through quinquennial grants – and from additional sources, such as research income, tended to increase. By 1973/4 Essex, Kent, Lancaster, Stirling, Sussex and Warwick had ceased to receive any regular grant from their local authorities, and in the case of the other three this amounted to only 1 per cent.

This was because the focus of local authorities on higher education was changing. Following the implementation of the Anderson Committee report in 1962, they had become the source of scholarships for increasing numbers of university students from within their catchment areas, wherever in the country those students were studying.[34] Then, from the 1964 Labour government's introduction of the 'binary system' of higher education, local authorities became responsible for the administration of 'public sector' (non-university sector) forms of higher education including polytechnics and colleges of further education. Thus, inevitably, they had less direct contact with local universities.

Part of the UGC's attempt to broaden the base of financial support was its encouragement of appeals to local industry, commerce and individuals. But its thinking about this was muddled, led to difficulties for the fundraisers and may have limited success in securing funds. One problem lay in its early view that the cost of

Table 13.2 Recurrent grants from public funds (Exchequer and local authority) for three universities over early ten-year period[35]

University	Year	Total income (£)	From Exchequer	% from local authorities	Total grant from public funds
Sussex	1963/4	686,441	77%	56%	83%
	1973/4	6,748,647	72%	Nil	72%
York	1963/4	205,014	71%	18%	89%
	1973/4	3,896,177	78%	1%	79%
East Anglia	1963/4	240,870	63%	26%	89%
	1973/4	4,717,824	83%	1%	84%

building residences should lie with the new institutions themselves. This was partly for ethical reasons.[36] Earlier civic universities had expected students to live at home and therefore provided relatively little residential accommodation. The vice-chancellor of Birmingham had pointed out the difficulty of older universities becoming national (let alone international) institutions without adequate residential accommodation.[37] It was a perceived lack of 'fairness' in directing major government funding towards accommodation for the new universities that led to the UGC's early pressure to direct appeals towards provision of accommodation. As far as industry and commerce were concerned, this attitude seemed a diversion from their own interests and an unsubtle plea for donors to make up for a gap in government funding. Albert Sloman also pointed out the UGC's lack of vision in not encouraging universities to raise valuable unrestricted funds that would enhance their academic freedom to determine their own academic programmes.

> It would certainly have helped Essex … if the use of government funds had been unrestricted … could have been for all aspects of its development … It is far easier for an electronics firm, for example, to justify to its shareholders' donations towards the study of electrical engineering than for the building of a study bedroom.[38]

At Lancaster it was felt that the absence of an engineering faculty accounted for the lack of interest from Leyland Motors.[39] The leadership at UEA was disappointed in the early days that the UGC did not support agriculture as a core subject since it was felt that this would bring in substantial support from East Anglian farmers.[40] Thus, although in theory Murray believed in the value of philanthropic income as an enhancement to a university's autonomy, he showed no understanding that donors will respond to their own interests rather than to a need dictated by government.

In 1963 the Robbins report, in applauding the response of industry and private donors to the appeals, supported the 'enhanced sense of freedom and flexibility that comes from the possession of funds that are not dependent on government'.[41] It also touched on the wider issue of incentives or disincentives to donors at a period when wealthy individuals were subject to relatively high levels of income and inheritance tax, urging that in no way should public subsidy be reduced on account of generous donations. But it glossed over the fundamental dichotomy about tax and benefactions that arose from the government's exhortation to individuals and the corporate sector to make generous contributions to the new universities without offering serious financial concessions. The only tax incentive then offered was a seven-year covenant scheme whereby, if a major donation was paid in annual instalments over a period of seven years, the recipient of the gift (the university) could recover income tax each year. Several of those involved in these appeals ruefully noted the very different approach to individual philanthropy in the United States where, for example, the graduated tax deduction system allowed donors to pay relatively little tax on major gifts.[42] Such initiatives were not forthcoming in Britain. The spirit of socialism that had spawned a generation of new, state-funded universities was not about to encourage private financial support for them through tax concessions that would erode Treasury control.

The fundraising appeals tended to last from two to five years. Although the new universities were planned as national – indeed, international – in their intake and their research activity, it was assumed that the appeals would be community led and the fundraising prospects, broadly speaking, locally targeted. The hope was that local industries would want to invest in a new educational industry that would enhance the local economy, while private money would come from individual stakeholders in the area, proud of their university and the potential benefit to local society. This had, after all, been the pattern of fundraising for new universities for the last hundred years. The leadership pattern also resembled appeals for earlier civic university institutions. Chairmen and appeal committee members – preferably including an aristocrat, bishop or JP or two (predominantly white Anglo-Saxon men) – tended to be closely linked to the local promotion committee. At Lancaster the active chairman was the 18th Earl of Derby, whose great-grandfather had been a major donor to University College, Liverpool and whose grandfather and father had been chancellors of the University of Liverpool. At East Anglia the appeal chairman was Timothy Colman whose family had been associated with the region around Norwich since the seventeenth century and whose knowledge of local farming, industry and business was extensive.[43]

What was not in the traditional university fundraising pattern was the early use of professional fundraisers to manage the running of the appeal. In the heyday of civic university foundation appeals – for example, at Manchester, Liverpool, Bristol, Reading – this work had been undertaken by prominent local businessmen – for example, the Ashtons at Manchester, the Rathbones at Liverpool, the Wills at Bristol – often actively supported by leading academics. British universities had not until this point followed the American practice of using professional fundraisers who were developing techniques for bringing in substantial support especially for private universities, although British public schools and Anglican cathedrals were beginning to do so. At the new universities it was often retired service officers who filled this role. At East Anglia, Group Captain G. R. Montgomery (former Air Attaché in Tokyo) was 'Appeal Secretary'; his equivalent at Essex was Captain G. C. Phillips (a former submariner). The universities of East Anglia, Essex, Lancaster and Stirling made use of a young, professional firm, Hooker Craigmyle, run by Michael Hooker (1923–2004), a pioneer of institutional fundraising. Educated at Marlborough and Oxford, a high-principled Anglican, he had gained fundraising experience working with an American firm, the Wells organization, in Australasia and Britain. Both Montgomery and Phillips went on training courses run by Hooker.[44] The resident appeal directors at Lancaster and at Stirling, Colonel W. B. Shine and Major R. Ogilvie respectively, were in fact working as Hooker Craigmyle consultants. Some of the fundraising techniques that they absorbed had been used in earlier university appeals in the nineteenth century: dividing up the 'prospect' area into regions, headed by local voluntary chairmen, supported by 'stewards', who took responsibility for approaching likely donors and hosting fundraising functions. But these more professional campaign managers began to pick up American refinements such as the effectiveness of securing lead financial commitments before the launch of a campaign in order to stimulate lesser but more numerous donations.

The appeals used the nature and identity of their localities to their advantage. Lancaster's Promotion Committee pointed out the economic disadvantages suffered by the north west of England, urging potential donors to help redress the unfair geographical balance.[45] East Anglia played to local pride in its own geographically distant region by adopting 'do different' as its motto, particularly targeting farmers from a region where they had recently prospered.[46] Kent used its proximity to the continent of Europe as a marketing point. Ulster made a strong pitch to the northern Irish diaspora in North America and Australia.[47] Personalities among the leadership were critical. Lord James, the inspiring educationalist vice-chancellor of York, was a powerful figure who engaged closely with the city's business-cultural environment.[48] At East Anglia Lord Mackintosh, a tough and shrewd businessman, 'was quite blatant in suggesting to firms that supplied [him] with raw materials that they should support UEA'.[49] At Warwick the 'intimate relationship' of the chairman, Lord Rootes, with the motor industry and his 'abiding loyalty to the Coventry community' enabled him very nearly to realize his flamboyant promise to raise £1 million for the University's first residences before his death deprived it of its 'most potent fundraiser'. But the spectacular campaign continued under Warwick's founding Vice-Chancellor, Jack (later Lord) Butterworth, also a forceful personality.[50] Conversely, at Kent neither the campaign chair, the former Chairman of Kent County Council, Lord Cornwallis, nor the founding Vice-Chancellor, Geoffrey Templeman, exuded the passion required for successful fundraising. Kent's appeal – in a challenging socio-economic environment – got off to a slow start and never gained momentum. East Kent had few wealthy individuals and no corporate ethos. By contrast, Stirling's slightly later appeal, chaired by the industrialist, banker and politician, Lord Polwarth, successfully galvanized local pride in the first new Scottish university on what was generally acknowledged as a breathtakingly beautiful 300-acre site at the former Airthrey Castle estate.

The *Financial Times* in January 1967 printed a comparative list of the total amounts raised by various new universities at a moment when most were well under way.[51] A year later, the *Glasgow Herald* announced the three-year result of the campaign for the University of Stirling which had started later.[52] This makes a rough comparison of success rates possible as shown below in the descending order of philanthropic funds raised by that time (see Table 13.3).

Reasons for the different outcomes have already been suggested. A glance at the lead and second-tier donors is also instructive. Stirling's lead gifts of £250,000 and £100,000 respectively came from two Scottish trusts (MacRobert and Gannochy). York's major donors, who pledged £100,000 over ten years, were from the dominant local business trusts (Rowntree and J. B. Morrell) that had been involved in the university plan from the start. At UEA similar pledges were made by John Mackintosh and Sons, and by the Norwich Union – which were also engaged in the project at an early stage. At both York and East Anglia these lead gifts set the bar for good second tier donations. Lancaster's relatively high out-turn was partly due to generous initial local authority donations; and also to the inclusion in the Appeal total of a grant of over £500,000 from the County Council earmarked for a local 'County College'.[53] Kent and Essex were encouraged by good lead gifts – £100,000 to Kent from Queen Elizabeth the Queen

Table 13.3 Appeal campaigns in the new universities

University	Appeal launched	Raised by 1967	Final target
Warwick	1964	£2,750,000	£4,000,000
Stirling	1966	£2,203,000 by '68	£2,000,000
Lancaster	1964	£2,200,000	£2,500,000
York	1962	£1,850,000	£2,000,000
East Anglia	1962	£1,400,000	£1,500,000
Essex	1963	£1,300,000	£1,000,000
Sussex	1960	£1,000,000	£1,000,000
Kent	1965	£ 600,000	£2,000,000

Mother, and from Princess Marina of Kent (the first Chancellor); £200,000 to Essex from the Ford Motor Company (whose base was at Brentwood). But these were not followed up with equivalent second-tier donations.

Overall, the University of Warwick came first by a substantial margin. Its position in the heartland of contemporary British industry – combined with its leadership personalities – gave it the edge over its peer group in terms of the appeal outcome. Its three lead gifts were £360,000 from an individual, £150,000 from the Nuffield Foundation and £104,600 from the Ford Foundation with significant second-tier contributions, including eleven of £75,000. The result illustrated 'the remarkable cross section of the industrial, commercial and private wealth contained in the West Midlands in the early 1960s'. Warwick, in fact, never closed its appeal, and by 1990 it had facilitated building projects worth at least £11 million.[54]

Another non-governmental source of income was growing, particularly where some of the new universities forged links with local industry. Asa Briggs in 1967 lauded the commercial benefit that accrued from Sussex's Science Policy Research Unit (SPRU) 'which mainly on a contract basis is examining matters of direct relevance to our national and international well-being'.[55] Sloman encouraged such links at Essex where close contact was maintained with Plessey (whose former chief scientist became professor of Electrical Engineering), Marconi, GEC and Bell Telephone Laboratories. At Lancaster the Vice-Chancellor, Charles Carter, cooperated with 'Enterprise Lancaster' to offer university facilities to new science-based industries brought to the town. Above all, Warwick soon realized the intended objective of West Midlands industrial proximity through collaboration on many fronts – business studies, management education and a wide variety of engineering subjects (including the automotive industry).[56] By the early 1970s Sussex, York, Essex and Warwick were doing well compared with the national average in terms of their income from research grants and contracts as a percentage of their income.

The UGC began to relax its dictum about appeal funding being directed to student residences and universities started to realize other objectives: scholarship endowment; enhancement to library, sporting and cultural facilities. An important aspect of the appeals coincided with Murray's objective that the universities should act 'as a stimulus ... to the

Table 13.4 Income from research grants and contracts to Britain's new universities in their early years showing its percentage of total income[57]

	Sussex	York	UEA	Lancaster	Essex	Kent	Warwick	Stirling	Average all UK universities
1969-70	£786, 875 20%	£420, 365 18%	£122, 748 6%	£166, 858 7%	£264, 264 15%	£112, 239 6%	£277, 892 13%	£60, 218 6%	12%
1973-74	£1,259, 437 19%	£421, 269 11%	£334, 758 6%	£288, 857 6%	£443, 144 12%	£245, 063 7%	£501, 644 11%	£148, 353 5%	11%

life of the area'.[58] A direct impact on local communities was made in some cases through centres for the creative arts, intended primarily to link the practical study of fine arts, drama and music to academic courses but also to provide a 'cultural bridge' with the wider community. Sussex pioneered this with its Gardner Centre – named after a major donor, Dr T. Lyddon Gardner, a local businessman who was chairman of Yardley Cosmetics, and supported by the Gulbenkian Foundation, the Arts Council and the UGC. Asa Briggs was also proud of the fact that the university library was open to the wider public and an important feature of two-way support between the university and its local community.[59] Kent put the case for a university theatre to the Gulbenkian Foundation which responded with £35,000 towards the cost, matched by the UGC and the university's Foundation Fund. The university opened its doors to an audience that proved 'mainly town rather than gown' for an 'Open Lecture' series with distinguished outside speakers.[60] A founding mission of the University of Stirling was 'to play an active role in the development of the community'.[61] An important part of its early appeal funding went towards a new MacRobert Arts Centre (again supplemented by the Gulbenkian Foundation and the Arts Council), opened in 1971, which presented a broad cultural range of performances whose audiences included the general public.[62]

Two outstanding cultural ventures were enabled through individual philanthropy. Warwick's largest individual donor, Miss Helen Martin, whose Ohio family made a fortune out of Smirnoff vodka, was struck by the way that theatres and concert halls at American universities invigorated campus life and integrated wider regional communities. Butterworth even visited Dartmouth College in New Hampshire which provided a model for what became the Warwick Arts Centre, part of Miss Martin's beneficence.[63] Opened in 1974, this facility developed into a renowned public centre for the performing and visual arts. Meanwhile, at the University of East Anglia, the Vice-Chancellor, Frank Thistlethwaite – through links with the director of the Yale University Art Gallery – was developing a relationship with the businessman and art collector Sir Robert Sainsbury, who was seeking a permanent home for his remarkable art collection. The outcome was not only the munificent donation to UEA of the Robert and Lisa Sainsbury collection in 1973 but also (with their son, David) finance to house it on campus in Norman Foster's striking building, the Sainsbury Centre for Visual Arts.[64]

Such cultural outreach no doubt helped to break down barriers between the new universities and their host cities. Yet, once these universities began to acquire inner momentum and expand beyond their initial pioneering communities, the consequences of their growth were startling to many who had originally welcomed their foundation. The work of building up research capability, along with its capacity for commercialization, may have appeared as self-absorption to outsiders. As yet, there was little fruitful outcome of early enthusiasm – from idealists such as Asa Briggs and Keith Murray – for continuing education courses which could have reached out to local residents. This was rejected at Lancaster, and only came (in the shape of a Department of Continuing Education) at UEA in 1991.[65] The growth of a relatively sizeable university community may, in due course, have helped the local civic economy, as studies conducted at the University of Lancaster suggested.[66] But in the later 1960s there appeared to be an increasing lack of understanding between town and gown, not helped by geographical distance. A local survey at Norwich showed that two-thirds of townspeople had never been to UEA and 'barely a fifth had any idea of the size of the University'.[67] Moreover, their basic function of nurturing fundamental enquiry and challenging assumptions – something that students were cottoning onto – were 'activities which were not always welcomed and did not make a university a comfortable neighbour', observed the operational research scientist Patrick Rivett in a lecture delivered to the Lancaster Town and Gown Club.[68]

It was student unrest, starting in 1967–8 and lasting for up to five years, that led to serious disenchantment with the universities on the part of their communities. This echoed what was happening in the United States, in German and French universities and at some other English universities, not least the London School of Economics. Forms of activity and the causes upheld by protesting students – from lofty political hostility to South African apartheid to more mundane local objections to the non-representation of undergraduates on university committees – have been addressed elsewhere in this volume. Vice-chancellors often had to face dissent among the staff and students within their institutions, while trying to reassure the outside world in the days before they had professional public relations teams to help them. All these 1960s universities were affected to a greater or lesser extent by student protest and its wider ramifications. Michael Sanderson, historian of UEA, suggested that the seriousness of the events affected the seven new English universities in the following descending order: Essex, Lancaster, Sussex, Warwick, York, Kent and UEA.[69] Even at UEA the vice-chancellor was deeply embarrassed by a student attempt to disrupt a visit by the Queen, Visitor of the University, in May 1968 (a tactic repeated at Stirling in 1972). There was little sympathy with the students from the press, or from the local citizenry, and there were financial repercussions at a time when there was, anyway, a growing downturn in the national economy that was affecting Treasury budgets for all universities. Particularly close to home were members of university councils whose lay membership included local dignitaries and respected citizens. At UEA a decision of the vice-chancellor to terminate the university's bank account with Barclays, a bank with significant local connections – in response to protest about its South African investments – led to resignations from the Council of the Treasurer, Lord Cranbrook, and eight local MPs, and also threats of withdrawal of local authority funding. In the wider world of Norfolk residents, there was general suspicion that the university was 'a hotbed of chaos'.[70] This reaction was widespread among equivalent communities around the country, exacerbated by their historic unfamiliarity with university culture. Concerns about ungrateful, irresponsible students wasting hardworking people's taxes and rates were widespread.[71] Not surprisingly this was to affect philanthropic support. At Lancaster, in 1972, the Vice-Chancellor, Charles Carter, received a letter from a former generous donor to the university's earlier appeal who was 'disturbed' by reports of 'the unsavoury happenings at the University', declaring that he would not be giving again. 'Tighten your belts,' cautioned Carter to his heads of department.[72]

These storms – which paled in comparison with those at some continental and American universities – were ridden out. Compromises were made by university authorities in the face of the demands of a new generation with political power, a social change signalled after the Family Reform Act of 1969 which lowered the age of majority from twenty-one to eighteen years of age meaning that universities were no longer *in loco parentis*. These years were a watershed in the narrative of English universities' relations with their local communities. The deference of the generation born during the war, those who had worn jackets, ties and gowns to lectures, who naturally respected their teachers, had gone. It had been a deference appreciated by those local societies – headed by the hierarchy of lords lieutenant and local dignitaries – that had sought in the late 1950s and early 1960s to enhance their localities with new universities. The optimism in which they had been founded disappeared. A new climate of opinion emerged in which government financial cuts of increasing severity to higher education budgets became publicly acceptable. Not only were university students looked on with suspicion but, in consequence, the academic profession began to lose status.

Let us try to analyse the nature and significance of early local support for these institutions. Initially, a significant impetus was brought to bear in the choice of location by the local civic establishment. This was reinforced by a cadre of highly committed local education officers who successfully articulated the need for new local universities in order to meet the urgent needs of able sixth formers benefitting from their success at grammar schools. These officers were backed by colleagues on borough and county councils that worked hard to secure large-scale sites which might even include a historic building that could serve as an early administrative facility. Thereafter, for a few years, these local authorities provided an essential source of income until the UGC's contributions to recurrent spending began to kick in when students arrived. Meanwhile, those great and good who had originally promoted these universities spearheaded fundraising appeals that were helpful in generating support for specific projects that could not be justified to the public exchequer. In terms of securing long-term institutional endowment, the appeals were virtually irrelevant. By the mid-1970s none of the new universities had an endowment that yielded more than 1 per cent of total income except Warwick's which was 2 per cent.[73] (This was to lead, in due course, to the growth of professional university fundraisers.)

By this time local communities were becoming more detached from their universities. It was partly the affront created by student unrest. Also, as has been noted, the changing higher education responsibilities of local authorities diverted their attention from the university sector so that by 1973–4 none of the nine new universities were receiving more than 1 per cent recurrent income from local authorities while funding from exchequer grants ranged from 87 per cent down to 72 per cent of their income.[74] But above all, it was the increasing complexity and growth of these universities – located away from urban centres – that made them seem to many locals something like impersonal businesses. The University of East Anglia, for example, in line with UGC thinking, anticipated a major expansion of students to 5,000 by 1981–2. This forecast alarmed Sir Edmund Bacon, chairman of its Council (and Lord Lieutenant of Norfolk), who was concerned that the university 'was going to unbalance Norwich by having too many people employed by one industry … and this risked a certain amount of anti-UEA feeling'.[75] In less than a decade, there was little if any sense of the early relationship between Whitehall, local authority and local benefactors (whether corporate or individual) that had been effective in enabling the new universities of the 1960s to take off. Keith Murray's dream of even a modicum of independence through diversity of funding sources had vanished.

Notes

1. Christine Shinn, *Paying the Piper: The Development of the University Grants Committee 1919–46* (London: Falmer Press, 1986), 279.
2. Memo to the UGC, 27 March 1947, TNA UGC7/218.
3. City of Norwich Town Clerk to Chairman of UGC, 17 June 1946, TNA UGC7/212.
4. Sir Walter Hamilton Moberly, *The Crisis in the University* (London: SCM Press, 1949), 243. Also Bruce Truscot, *Red Brick University* (London: Faber and Faber, 1943).
5. For example, draft letter, 'Proposed University of York', Moberly to Lord Mayor of York, May 1947, TNA UGC 7/218. See also Robert Anderson, *British Universities Past and Present* (London: Continuum, 2006), 126.
6. 'Returns from Universities and University Colleges in Receipt of Treasury Grant', 1938–9, TNA UGC/3/20 and 1950–1, UGC/3/32. These aggregate figures mask large variations in endowment, notably between the ancient universities and more recent foundations.
7. Maurice Kogan, *County Hall: The Role of the Chief Education Officer* (Harmondsworth: Penguin, 1973), 13.
8. Stone to Murray, 6 March 1956, TNA UGC7/215.
9. 'Proposal for a University of Sussex, 1956–58', Stone Papers, Box 4, University of Sussex Special Collections, The Keep Archives Centre, Sussex. See also W. G. Stone, 'University Commentary from Brighton', *Universities Quarterly*, 12 (1957–8), 223–7.
10. Interview with Michael Sanderson (30 September 2008).
11. Graham Martin, *From Vision to Reality: The Making of the University of Kent at Canterbury* (Canterbury: University of Canterbury, 1990), 15.
12. Michael Sanderson, *The History of the University of East Anglia, Norwich* (London: Hambledon, 2002), 22.
13. Michael Shattock, *Making a University: A Celebration of Warwick's First 25 Years* (Warwick: University of Warwick, 1991), 14.
14. D. N. Bungay (Deputy Chief Education Officer, Essex County Council), 'The First Three Years: The Development of the University of Essex' (August 1963), Foundation

Papers, Box 3, University of Essex Collection/University of Essex Archives (held in Albert Sloman Library, Colchester).
15 For Rees (1916–2006), see his obituary: https://www.theguardian.com/news/2006/apr/24/obituaries.mainsection (accessed 31 December 2018).
16 Christopher Storm-Clark, 'Newman, Palladio and Mrs Beeton', in *York 1831–1981: 150 Years of Scientific Endeavour and Social Change*, ed. C. H. Feinstein (York: William Sessions, 1981), 294.
17 Quoted in Katherine A. Webb, *Oliver Sheldon and the Foundations of the University of York (*York: Borthwick Institute, 2009), 63.
18 Interview with Marion McClintock (March 2014).
19 Geoffrey Caston, 'Murray, Keith Anderson Hope, Baron Murray of Newhaven (1903–1993)', *Oxford Dictionary of National Biography*, Oxford University Press, 2005, https://www.oxford.dnb.com/view/article/52309 (accessed 27 June 2019).
20 University Grants Committee (UGC), 'University Development 1957–1962', *Parl. Papers*, xx (1) (1963–4), Cmnd 2267, paras 261–4.
21 Ibid., para. 290.
22 Report of the Committee Appointed by the Prime Minister under the Chairmanship of Lord Robbins (hereafter *Robbins Report*), 1961–63, *Parl. Papers*, Cmnd 2154, paras 701–10.
23 Shinn, *Paying the Piper,* 37–8.
24 UGC, 'University Development 1957–1962', para 157.
25 For example, Memo from Cecil Syers to R. W. B. Clarke, 9 March 1960, 'New University Institutions', TNA, UGC1/7.
26 Sanderson, *History of the University of East Anglia,* 23. There was a much smaller executive committee.
27 Clark Kerr, *The Uses of the University* (Cambridge, MA: Harvard University Press, 1963), 24.
28 Noel Annan, *Our Age: Portrait of a Generation* (London: Weidenfeld and Nicolson, 1990), 373.
29 Ibid., 372–3.
30 Correspondence (1960–1), 'Proposed University of Essex', TNA, UGC7/232. See also Annan, *Our Age*, 372–3.
31 Shattock, *Making a University*, 16–17.
32 Sanderson, *History of the University of East Anglia,* 28.
33 *Returns from Universities and University Colleges in Receipt of Treasury Grant* (1963–4), TNA, UGC3/45.
34 *Report of Committee on Grants to Students,* Parl. Papers, xxi (1959–60), Cmnd 1051, 481.
35 *Returns from Universities and University Colleges in Receipt of Treasury Grant* (1963–4), TNA, UGC3/45. Figures also taken from *Statistics of Education, 1974, Vol 6, Universities*, TNA, UGC3/56.
36 See, for example, Cecil Syers to R. W. B. Clarke, 9 March 1960, TNA, UGC7/170. Also, Note: 'Visit of Keith Murray to Lancaster, 14 June 1961', University of Lancaster University Archive, UA.1/15/1/12.
37 'Note on Need for Halls of Residence' (Vice-Chancellor, University of Birmingham), May 1960, Sir Frank Bowers's Survey, University of Lancaster UA.1/123/1/5/2.
38 Albert E. Sloman, *A University in the Making* (London: BBC, 1963), 16–17.

39 Colonel Shine, 'Factors Affecting the Campaign' in his 'Final Report on the University of Lancaster's Appeal' (August 1963), University of Lancaster, UA.1/13/1/4/1.
40 Interview with Michael Sanderson (30 September 2008).
41 *Robbins Report*, paras 658–9.
42 For example, W. E. Wade (Finance Officer) to vice-chancellor at Essex, 29 April 1964, University of Essex Collection, Foundation papers, Box 2, Special Collections, University of Essex Archives.
43 Sanderson, *History of the University of East Anglia,* 46.
44 For Hooker, see https://www.telegraph.co.uk/news/obituaries/1455508/Michael-Hooker.html (accessed 31 December 2018).
45 'The University of Lancaster: The First Seven Year Plan: 1964–1971, Appeal Brochures, 1963–71', University of Lancaster, UA. 1/13/1/9.
46 Sanderson, *History of the University of East Anglia,* 48.
47 'Interim Report of the Academic Planning Board for the New University of Ulster' (May 1967), para. 73, TNA UGC7/258.
48 Roger Young, 'James, Eric John Francis, Baron James of Rusholme (1909–1992)', *Oxford Dictionary of National Biography*, Oxford University Press, 2004, http://www.oxforddnb.com/view/article/51145 (accessed 2 February 2014).
49 Sanderson, *History of the University of East Anglia,* 52–4.
50 J. B. Butterworth, 'Rootes, William Edward, first Baron Rootes (1894–1964)', rev. G. T. Bloomfield, *Oxford Dictionary of National Biography*, http://www.oxforddnb.com/view/article/35825 (accessed 16 June 2016).
51 Cited in Shattock, *Making a University*, Appdx.
52 Cited in R. G. Bomont, *Beginnings and Today* (Stirling: University of Stirling, 1995), Appdx.
53 Colonel W. B. Shine, 'Final Report on the University of Lancaster Appeal' (August 1965), University of Lancaster, UA.1/13/1/4/1. Also: Maureen McClintock, *The University of Lancaster: Quest for Innovation* (*a History of the First Ten Years, 1964–1974*) (Lancaster: University of Lancaster, 1974), Appdx ix.
54 Shattock, *Making a University*, Appdx: *The University Foundation Appeal*.
55 Vice-Chancellor's Address to Court, 8 December 1967. University of Sussex Special Collections, The Keep Archives Centre, Sussex.
56 Michael Sanderson, *The Universities and British Industry 1850–1970* (London: Routledge and Kegan Paul, 1972), 370–1.
57 Figures taken from *Statistics of Education, 1970, Vol 6, Universities, TNA UGC 3/52* and from *Statistics of Education 1974, Vol 6, Universities, 1974, UGC 3/56*.
58 University Grants Committee, 'University Development 1957–1962', *Parl. Papers*, xx (1) (1963–4), Cmnd 2267, para. 280.
59 Vice-Chancellor's Address to Court, 8 December 1967. University of Sussex Special Collections, The Keep Archives Centre, Sussex.
60 Private information.
61 Bomont, *Beginnings and Today*, 17.
62 *Principal's Annual Report, 1971–2* (Stirling: University of Stirling), 3.
63 Michael Shattock and Roberta Warman, *The Martin Family and the University of Warwick: The Contribution of Private Giving to a University's Success* (Warwick: University of Warwick, 2000).

64 Sanderson, 263, 266; Witold Rybczynski, *The Biography of a Building: How Robert Sainsbury and Norman Foster Built a Great Museum* (London: Thames and Hudson, 2011).
65 For Briggs, see Miles Taylor, 'Introduction: Asa Briggs and Public Life in Britain since 1945', in *The Age of Asa: Lord Briggs, Public Life and History since 1945*, ed. Taylor (London: Palgrave Macmillan, 2015), 10–11. Murray encouraged continuing education at Stirling: Murray to John Wolfenden, 8 December 1964, TNA UGC 7/245. For Lancaster, see M. McClintock, *University of Lancaster: The First Thirty Years* (Lancaster, 1994), 297. For UEA, see Sanderson, *History of the University of East Anglia*, 373.
66 McClintock, *University of Lancaster*, 294–6.
67 Sanderson, 189.
68 McClintock, *University of Lancaster,* 319–20. Rivett, then at Sussex, cited the experience of two (unidentified) new universities.
69 Sanderson, 219.
70 Ibid., 198–9, 219.
71 In 1969–70 total government expenditure on UK universities was £196,657,000. £13,129,000 went to the new universities. *Statistics of Education 1970, vol. 6, Universities* (HMSO), TNA UGC 3/52, Table 40.
72 Vice-Chancellor to Heads of Department, 18 May 1972, University of Lancaster Archive, Vice-Chancellors' Circulars (CFC), UA 1/9/3/6.
73 *Statistics of Education, 1974, vol. 6, Universities* (HMSO), TNA UGC3/56.
74 Ibid.
75 Sanderson, 224, 272.

240   *Utopian Universities*

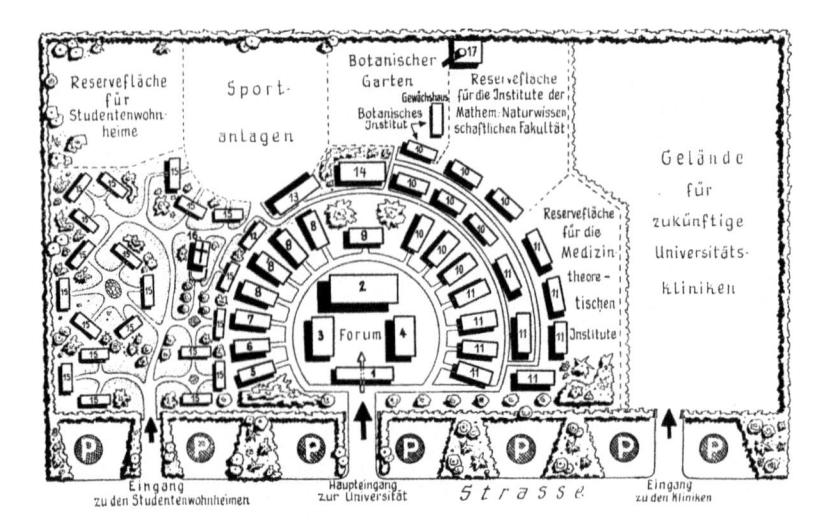

Figure 1 Hans Werner Rothe's ideal campus (1961). Rothe, *Über die Gründung einer Universität in Bremen* from Rolf Neuhaus (ed.), *Dokumente zur Gründung neuer Hochschulen. Empfehlungen und Denkschriften auf Veranlassung von Länden in der Bundesrepublik Deutschland in den Jahren 1960–1966* (Wiesbaden: Franz Steiner Verlag, 1968). Evoking both the ideal of the American liberal arts college, as well as the unified design of early twentieth-century town-planning, Hans Werner Rothe's vision for a new campus at Bremen was typical of blueprints for the new universities of the era.

Figure 2 Students astride a walkway over the 'moat' at the University of Sussex, *c.* 1963 (© Yves Fedida). One of the first 1960s buildings to be given Grade I listing, the Falmer campus of the University of Sussex, designed by Sir Basil Spence (1907–76), caught the media spotlight straightaway. 'Bricks and mortar can provide a background that is sympathetic to young and energetic minds which are growing and developing apace' enthused Spence.

Utopian Universities 241

Figure 3 James Birrell's James Cook University Library (James Birrell Archive, James Cook University Library Special Collections). 'Brutalist' modernism travelled far and wide, as new campuses appeared around the world. At Townsville in Queensland, Australia, James Birrell (b. 1928) came up with this unique building for the library at the James Cook University (renamed in 2008 as the Eddie Koiki Mabo Library, in honour of the Torres Strait islander who campaigned over indigenous land rights).

Figure 4 Julian Elliott and Anthony Chitty's University of Zambia campus (© Dr Ruth Craggs). British architects were responsible for many of the new campuses of the Commonwealth. Anthony Chitty (1907–76) designed university buildings in Kenya and Ghana, but his best work came at Lusaka, at the University of Zambia, where he sought out local African expertise to collaborate on the project.

242   *Utopian Universities*

Figure 5 Arthur Erickson's University Hall, University of Lethbridge, Canada (Courtesy of the University of Lethbridge Archives). Arthur Erickson (1924-2009) was the architect for Vancouver's new Simon Fraser University in 1965, then a few years later he designed another campus project: University Hall, at Lethbridge on the prairies of Alberta. The result was 'a dialogue between history and utopia' purred one reviewer.

Figure 6 Sir Keith Murray (on the right of the photograph) receiving an honorary degree from the University of Western Australia, 1963 (National Archives of Australia). A former Rector of Lincoln College, Oxford, Sir Keith Murray (1903-93), chaired the UK Universities Grants Committee during the crucial years when the new campuses were planned. He also advised on new universities in Australia (where this photograph was taken). He was made a life peer in 1964.

Utopian Universities 243

Figure 7 University of East Anglia, Norwich: an exhibition display showing a model of the proposed university being scrutinised by visitors (Lasdun Archive, RIBA Collections). Although publicly-funded, the new universities of the 1960s needed benefactors. In the UK, taxpayers' money could not be used for halls of residence. Here potential local supporters view the campus model of the University of East Anglia, including Denys Lasdun's unusual ziggurat design.

244   *Utopian Universities*

Figure 8 'Some people who matter, and some who think they do' by 'Tom', Essex, Winter 1968 (by permission of Tom Brass). The University of Essex epitomised the 1960s campus – garish architecture, vogue new social sciences, and a vibrant student counter-culture – all captured in this contemporary cartoon. Men did not dominate entirely. In February 1969 one of the first ever 'Women's Liberation' meetings was held at the university's 'Revolutionary Festival'.

Figure 9 Pradip Nayak, Founding President of the Students' Representative Council, University of York, 1964 (© *Yorkshire Post*). Student radicalism on the new campuses predated the famous protests of 1968. Pradip Nayak (b. 1944) from Nairobi in Kenya was elected the University of York's first President of the Students' Representative Council (the forerunner of the Students' Union). Active in opposition to UDI in Rhodesia, he also led a local boycott of South African goods.

Figure 10 John Fulton receiving an honorary degree from the Chinese University of Hong Kong, 1964 (Courtesy of the Chinese University of Hong Kong). Sir John Fulton (1902-86) taught at Balliol College, Oxford, alongside Alec Lindsay, before going on to become the founding Vice-Chancellor at Sussex. From there he became the doyen of university planners of the era, advising on the new Chinese University of Hong Kong (where this photograph was taken), the Open University, and around the world as Chair of the Inter-University Council for Higher Education Overseas from 1964-8. He was made a life peer in 1966.

Figure 11 Demonstration against Ronald Reagan's sacking of Clark Kerr, UC Irvine, 23 January 1967 (University Communications photographs. AS-061 (Box 37, A67-002) Special Collections and Archives, University of California Irvine Libraries). Clark Kerr (1911-2003) masterminded the new public university system of California in the 1950s and 1960s. Seen by his opponents as too lenient towards students, he was fired by the new Republican Governor, Ronald Reagan, at the beginning of 1967. Here, Dean McHenry (1910-98), Kerr's long-time colleague, addresses a meeting of protest against the dismissal, held at the University of California Irvine campus.

Figure 12 'University for Derry' protest march, 18 February 1965 (© *Derry Journal*). On 18 February 1965, 25,000 people from Derry marched on Stormont, seat of the Northern Irish government, protesting against the decision to locate the new university of Ulster at Coleraine. In a display of unity, Albert Anderson, Ulster Unionist Mayor of the city, led the march, alongside John Hume, then a young teacher, and Eddie McAteer, the nationalist politician.

Figure 13 Siné (Maurice Sinet), 'Debout les damnés de Nanterre', *Action*, 7 mai 1968 (© Éditions du Crayon). The Nanterre campus was at the epicentre of the student movement in Paris in 1968, not least because it was built alongside a neigbourhood of North African immigrants. Here 'Siné', the political cartoonist, satirises Franz Fanon's famous critique of colonialism, *The Wretched of the Earth* (*Les Damnés de Terre*).

248　　　　　　　　　　　　*Utopian Universities*

Figure 14 Sitaram Yechury, President of the Jawaharlal Nehru University Students' Union, reading a memorandum to Indira Gandhi, demanding her resignation as chancellor of the university, 5 September 1977 (© *Hindustan Times*). Perhaps the most powerful student organisation in the world in the 1970s, the Jawaharlal Nehru University Students' Union sacked its Chancellor, Indira Gandhi in 1977, shortly after she was ousted as Indian Prime Minister.

Figure 15 The Queen's visit to the University of Stirling, 12 October 1972 (© PA Images). Her Majesty Queen Elizabeth II appears unfazed during the student protest that disrupted her visit to the University of Stirling in 1972.

Part Two

14

California dreaming: Clark Kerr and the University of California

by Christopher Newfield

Climbing up that hill to Clark Kerr's hillside house in overlooking the city of Berkeley and the San Francisco Bay beyond, I told myself I wouldn't ask him about the Free Speech Movement, which among other things made him seem weak to the state's right wing business and political heavyweights, or about the disastrous aftermath that included his termination by a Board of Regents controlled by the new governor Ronald Reagan, who had launched his triumphant political career by running in 1966 against Kerr and his University of California.[1]

I had been wondering whether there was a direct line between Kerr's failed mediations and the age of Reagan. Certainly the forces at work were larger than Kerr or Reagan or the student movements themselves, which already by 1964 were as inspired by the black civil rights movement in the South as they were by First Amendment free speech issues. In that same year, the landslide defeat of right-wing Republican Barry Goldwater seemed to have confirmed the fringe nature of his opposition to the Civil Rights Act and more generally to the movement towards a post-McCarthyite society, something for which students and universities already stood. But only two years later, Reagan defeated the governor who was in effect the Lyndon Johnson of California – Edmund G. 'Pat' Brown – and set the Republican Right on a path to victory in his presidential race in 1980.[2] That in turn changed the country's political culture, perhaps permanently. In any case, I thought as I rounded the last curve, Kerr's inability to satisfy either the students or right-wing regents and his subsequent firing by UC Regents steered by Ronald Reagan in his third week after taking office were all too well studied to dwell upon. I had many questions but I wasn't sure where we would start.

The date was 13 August 1997. That was already thirty years after Kerr's firing. The general consensus in 1997 was the same as it is today: Kerr was probably the best president that the University of California had ever had, and the system's last real builder. In firing him, the Regents weakened the university's confidence and vision. In my judgement as a long-time employee, it never recovered. Or so it seemed to me then, just barely tenured in my job at the Santa Barbara campus. And so it still seems now.

I had at that point read Kerr's classic *The Uses of the University* (1963). It was heralded at the time as arguably the most important modelling of the post-war US university, and defined its missions on the threshold of the 1960s. It remains a central text today. Kerr coined the term 'multiversity' to describe the proliferation of disciplines, missions and relationships with the outside world. The potential institutional confusion was largely redeemed in Kerr's telling by the theory of the 'knowledge industry', in which knowledge had become central to economic growth and thus moving the university the centre of society. I was struck not only that Kerr was describing the explosion of universities with a fairly unitary (and mostly undergraduate) educational mission into decentred networks of research, professional and doctoral training, and technological service to the corporate and military establishments, but that he was also describing the rise of corporate management on campus – still stubbornly called 'administration' – as the inevitable result of this mission proliferation, nearly forty years before 'academic capitalism' became an identified feature of the modern university in the 1990s. A Harvard student who attended Kerr's first Godkin lecture in 1963 (which formed part of *Uses*) caught the moment: 'Clark Kerr, president of the University of California, declared last night I "The general rule is that the administration everywhere becomes, by force of circumstances, if not by choice, a more prominent feature of the university." The influence of the students and the collective faculty has declined. "The managerial revolution has been going on also in the university."'[3] I was fascinated by Kerr's combination of confidence in the university's mission and criticism of its self-understanding.

Importantly, Kerr did not associate the multiversity's evolving managerialism with a newfound power in society. To the contrary, he wrote, the university's 'directions have not been set as much by the university's visions of its destiny as by the external environment, including the federal government, the foundations, the surrounding and sometimes engulfing industry. The university has been embraced and led down the garden path by its environmental suitors.'[4] By the start of the 1960s, the public research university had already been partially 'unbundled' as we would now say in order to serve multiple interests, funders and powers that be. I noted that Kerr saw the multiversity's managerialism as both inevitable and unfortunate.[5]

I also knew Kerr as one of the architects of the California Master Plan for Higher Education. This scheme organized a number of regional colleges and several offshoots of 'Cal' – the Berkeley campus – into the two centrally administered systems of the University of California and the California State University. Two-year colleges were organized into what became the California Community College system. The Master Plan had several influential features. The structure regulated the type of degrees each campus could provide by allocating doctoral and most professional certification to UC, four-year degrees and masters programmes to Cal State, and two-year degrees to the community colleges. Next, it reduced political competition among the various campuses and improved their coordination through central administrative offices. Thirdly, it combined universality with stratification, meaning that anyone could go to a community college, while the top third and the top eighth of California high school students would be assured of a place at a Cal State or UC campus respectively. In this way it formalized and conformed to the pattern of US higher education as a whole, which was to offer easy access for all while reserving costly quality to a small elite. Fourthly, it set the 'educational fee' – what universities could charge for instruction – at zero. 'Free college' was a reality in 1950s and 1960s California.

Finally, Kerr envisioned UC as a network of general research campuses, meaning the state would have the country's most intensive research capability. This reflected a common latent but still rather original idea that research funding should not be carefully limited to a small number of superior minds (mostly at Berkeley) but instead be widely distributed across the state. Kerr was an early convert to the theory of the knowledge industry, in which the more knowledge the merrier the economy would be. The Master Plan could thus appear to be a kind of golden mean between egalitarian democracy and excellence.

Some combination of these features became a regulative ideal for many growing public university systems around the world, though honoured mostly in the breach. It is worth noting too that none of the UC campuses were selective in the modern sense. Even Berkeley was essentially open admission for students who had met the basic grade and subject requirements. Non-resident tuition was very low, offering few financial barriers to high school graduates around the country. At the same time, the campus student body was overwhelmingly white; the Master Plan subjected racial disparities to a not particularly benign neglect that left the system vulnerable to racial backlash in the 1960s and 1970s.

Preparing for the interview, I had also recently read *The Great Transformation in Higher Education* (1991), which offered insights on a range of subjects whose lack of coherence suggested that the 1960s multiversity was itself in transition.[6] I was also preoccupied with organizational psychology, having published a book on Ralph Waldo Emerson's implicit theories of corporate subjectivity while living my assistant professor years under the first round of higher education budget cuts in California, and was focusing unexpectedly on institutional budget politics. In this chapter, I will reproduce selections from my two-hour interview with Kerr and then go on to comment on the strengths and weaknesses of his legacy as it was forged in the 1950s and 1960s Cold War boom.[7] I will focus on his model of academic management, tacit as much as explicit. As disappointed as Kerr was with the university's loss of control over its fate, he shared the limited utopian hope of other builders studied in this volume in which careful planning could create an institution that protected intellectual novelty and heterodox thought from social normalization, while building a better world for which society would thank it – eventually.[8]

I

We started with Kerr's work as a labour arbitrator and in a sense never left the question of labour relations. We dwelt extensively as we did on the faculty's organizational behaviour. This is partly because, as I mentioned, I intentionally avoided asking Kerr about topics that were already burned-over districts in Kerr's biography: his role in the loyalty-oath controversy of 1949–52, his relationships with 1960s student movements and leaders like Mario Savio, his conflict with Reagan and right-wing regents, and the state politics of the Master Plan for Higher Education. I did not yet know about the extent to which Kerr had been under surveillance by the FBI from the early 1950s on: the first newspaper reporting appeared in June 2002, about fifty years after the fact.[9] This also meant I didn't know the extent to which the liberal Kerr was always surrounded by enemies. But we did spend quite a bit of time on the university's internal dynamics. Perhaps this was an area in which these two UC people, in their only meeting, with careers that began a half-century apart, found tacit agreement on one deep issue that shaped the public university's social authority in that period and ours.

Although Kerr, after his 1967 firing, never again had a leadership role at UC, his home symbolized leadership in general. It was a 1950s-style ranch house, but supersized, with hallway carpets as plush and muted as I imagined the Vatican's to be, and a living room that I thought could encompass 100 people for a formal reception. This room looked out over a sweeping lawn that formed the foreground of a two-sided historic view: ahead, to the west, was the entrance to San Francisco Bay and the Golden Gate, with a freighter just then entering on the gleaming silver sea beneath the bridge; to the left, sideways, to the south, was the green flank of the Berkeley hills and the white Campanile marking the centre of the main campus where Kerr had first become chancellor in 1952 at the age of forty-one, and then UC president in 1958, about forty years before our meeting. I imagined that to Kerr both views suggested that not much had changed.

Kerr looked great at eighty-six, as mild-mannered as I expected him to be, both comfortable and reservedly friendly. He sat me down in an armchair with the kind of big glass of water you'd give a teenage boy. At that time I was studying corporate culture and was in the middle of a series of interviews with management consultants in Silicon Valley. That may have been why I started with a softball question I'd probably used before: 'In a sector that is often ruled by conventional wisdom, how did you keep your intellectual independence?'

In response, Kerr mentioned a commencement address he'd given at his alma mater, Swarthmore College, in 1952, which got picked up by *Fortune*, in which he refers to 'this smothering of the individual'.[10] After mentioning this, Kerr skipped directly back to his work in the 1940s as an arbitrator between unions and owners, where he tried to protect individuals from both employers and the machinery of their unions.[11]

> I'd been impartial chairman on the West Coast Waterfront and the Longshore Industry between the Longshoremen's Union and the Waterfront Employers Association and there I had a rather difficult case. On the waterfront, you couldn't be employed unless you were a union member in good standing. If the union didn't like you, they would declare you not in good standing and you lost your job. In other words, they could fire people. I had a case where a man, who was black, who was a follower of Father Divine. Father Divine was a black evangelist and when you joined his sect, you changed your name because you were reborn. In fact, the members of it wouldn't even keep their own social security numbers. There was a black longshoreman who took as his name True Knowledge and gave his wife the name of True Love. During a strike they had on the waterfront, he wouldn't go on the picket line because the picket line involved the threat of force and he was against the use of force. I was convinced that he was a very, very sincere person and here the union had declared him not a member in good standing and he had lost his livelihood. Under the contract, I really had no choice but to say that

> he could no longer work on the waterfront. But I didn't like what was going on so I did something which is really outside my contract. I would have to rule on behalf of the union but I wanted to tell the union in advance that though the contract would require me to make this decision, I didn't agree with it. I would write an opinion saying why I disagreed with the all that the union had done. I would like them to give some thought as to whether or not they would wish to stand by their action considering the fact that I would be attacking their policy. That is a pretty strong thing for an arbitrator to do. Well, I wanted to warn them to see what they would do about it. They decided then to withdraw the case, so they let Mr. True Knowledge go back on the waterfront.
>
> I had other cases somewhat similar where I did a lot of arbitration attacking the industry and those kinds of places. I saw what bothered me was the treatment of the individual. There was a period in which all these personnel administrators were putting in what I thought was a heavy web of rules being placed on the dock worker, by government and by unions, and by employers.

Kerr described other similar experiences in which rules that were reasonable and that he was charged to enforce would cause problems for individuals who deserved a better outcome than the rules afforded.

After a while, I asked him whether this concern for individuals helped him in the individualistic faculty culture he faced when he became UC Berkeley chancellor in 1952.

> I had become the head of some of the labour boards in the meat packing commission. There were strikes in the meat packing industry and I had to keep the meat flowing to the troops and so forth. I was involved in that industry for over 30 years. When I stopped being an arbitrator and became an administrator, people would say to me, 'You must be relieved, moving out of the jungle in industrial relations into the moral world of academe.' I said, 'I haven't moved out of the jungle – I've moved into the jungle'.

That was really my view. The philosopher Michael Walzer argued that different types of situations define justice in different ways and I saw this moving between industrial relations and the academic world.

> First of all, in the industrial relations world, if you make a contract you are actually bound by it and in the academic world, the community can tell you one thing one day and you go ahead and follow what they've told you; then come around the next day and condemn you for it, for they have changed their minds. In the industrial relations world, if you make an agreement, that was it. On the waterfront one time, a local longshoreman leader had agreed to something and told the employers this. But Harry Bridges had trouble getting his own men to accept this. So first he said he wasn't going to be bound by what the other guy had said, since they hadn't signed the agreement. I told the longshoreman's union that when a leader had said

he would do certain things and shook hands on it, the union was bound by it. Put that way, they agreed, and the union lived under the verbal agreement for a whole year. Even that tough union would go along with this idea that they were bound by a verbal contract even though there was no written contract that they could take into a court of law.

In the academic world, I got caught a number of times where I thought I had an understanding about something and then faculty members or department chairmen would go back on it and say, 'I got new information, or I hadn't checked carefully, or something like that.' They had no feeling of obligation to keep an agreement. Very few times in my life have I lost my temper, but a couple of times I thought people were going back on their commitments so I would get pretty upset about it. That's one illustration of it.

In the industrial relations world, if somebody doesn't keep an agreement, they are just shunned. The attitude is that if a person won't keep an agreement on one thing, what's the point in negotiating with him for the rest of it? Therefore I won't bargain with him. In the faculty world, it's more a question of always having an open mind and being prepared to change your mind. The moral thing to do is always keep everything open and not be bound by any prior commitments.

The second thing I found different was that in the industrial relations world, people are very careful to try to keep personalities out of affecting the relationship. I have seen so many situations with unions where people really go out of their way to be gentle toward their opponent, to not make his job harder, because by making his job harder it is harder then to get an agreement. In academic life, people will become personal enemies over little things. There is awful lot of back biting and so forth. Can you enjoy your personal antagonisms without it costing anybody anything? In the industrial world, you didn't want to mess up things that might involve millions of dollars by having personal antagonisms. You can't afford to do that. In academic life, somebody would just think it's fun by back biting, trying to distort people's reputation and so forth and so on.

Then another thing which I thought was different in industrial relations was that you had a right to consider any factor that would affect an agreement. In academic life, faculty would call into question an administrator who would ask what impact would this outcome have on public opinion, or on some group's religious beliefs. It would be considered to be absolutely improper to raise any non-academic considerations. In other words, you couldn't look at the total situation. You had to confine yourself to the academic scheme of it.

Those are three pretty important things and it's a different kind of world to live in. The moral standards in industrial relations are higher than those of academic life. That's not a very popular thing to say about academic circles but that's the way I feel about it.

I couldn't argue with Kerr's description of faculty culture. But the 1990s and 1950s seemed so similar, which prompted me to say, 'You must have thought a number of times and about how to improve academic negotiations. What can be done to make the university world better at negotiating?'

Kerr instantly replied, 'I never figured out how to change it. It's just so basic that when a faculty member says I have new information, he feels entitled to change his mind. There is a cost – it makes it harder to be an administrator.' I could feel his resignation.

I asked whether he thought the additional security that tenured faculty had would make them easier to work with – more stable or something.

CK: I really didn't. I mostly just lived with it. I just thought that was the nature of the beast.
CN: Do you have the sense that any of that has changed in the intervening time?
CK: No, I don't.
CN: Do you see it as a permanent feature of the university landscape?
CK: Yes I do.
CN: The faculty are not going to change their fickle ways?
CK: No.

I still remember Kerr's grin as he said that.

Trying to shake something loose, I asked if he was aware of any of the group psychology work that I had recently been reading – Kurt Lewin, who had started the National Training Laboratories around the time Kerr had been arbitrating with the Longshoremen, or the Tavistock group, or others who focused on group-psychological bottlenecks in organizational life. Did anyone at UC ever try that kind of thing?

Kerr replied,

Not to my knowledge. There were some situations where as Chancellor I did intervene to try to change behavior in the psychology department of Berkeley. Psychologists are very diffuse, you know, it's a little bit of everything. Some of it is biology and some social science – many, many different groups and identifications. At Berkeley, the department almost destroyed itself. They met once a week – the whole department – and began fighting out their battles and antagonisms. They would go into a meeting and begin tearing each other apart, which nobody can do better than psychologists can do it. They had meetings Thursday afternoons and it would take Friday, Saturday and Sunday to recover. Monday they were starting to plan their attacks for the next Thursday. Then the people were complaining that they didn't have time to do their research. I got them a new department chairman and I gave him only one piece of advice, and that was to never have another department meeting. Settle things in the hallways, the men's rooms, in informal ways rather than create these scenes. He did it and he did it quite well as a matter of fact.

So the famous mediator's solution to bad group psychology was to have no group psychology at all? Did you ever hear of a success story, I implored?

> Well, one thing I discovered when I was a young faculty member, I knew that my own department had a lot of internal personnel problems and that all of the departments that had good relationships were fine. But the more departments I got to know, the more I discovered that everyone I got to know had its internal feuds going on. The only happy departments were the ones I didn't know.

> I got to know enough departments so I came to the conclusion that all departments had their own unnecessary internal strife and antagonisms which were counter-productive. But I never figured out how to do anything about it except the case with the psychology department at Berkeley.

I took that as a no.

Kerr mentioned that at one point in the mid-sixties he was serving as the chair of a relatively progressive industrial relations society that had employer, union and employee members. Called 'Work in America', the group asked him as chair to write to some universities to ask if they would like to join it.[12] 'But their faculties wouldn't accept it,' he said.

'Perhaps in a crisis they would accept some kind of guidelines for better negotiations?' I asked.

'My own experience,' Kerr replied, 'is that faculties tend to behave rather badly in crisis situations. Rather than uniting, crisis divides them and creates controversy. It is a peculiar type of institution.' I mentioned that this was also my experience inside of two English departments. 'We are under all sorts of political and budgetary pressures,' I said. 'I worry that if there isn't some mechanism for allowing groups to work out the inevitable conflicts that are part of the change process, that change will never happen and the field will die.'

Kerr replied, 'My experience is that change comes awfully hard in academic life. I've seen foundations, corporations, trade unions, and government agencies at work. Higher education is the most resistant to change of any organization that I've seen. First of all, it's not a democracy. The senior and better known people in the department tend to have an acknowledged "veto" right on what goes on. They couldn't win in a democratic vote. For a person to raise a new and controversial possibility is just something that isn't supposed to be done. It would take a pretty adventuresome and younger person to face the antagonism. But to raise an issue that you know is going to cause trouble just isn't done.' 'That's true,' I laughed. 'I've found out the hard way.' Kerr nodded.

> It is a disturbing thing so many people are very hesitant to raise new ideas to begin with and then there is the view that you shouldn't push anything through unless it has unanimous consent. There are penalties for raising new ideas and great difficulty getting them through if anybody, particularly if any important person, is opposed to it. So that means you don't change very much.

I remember feeling that Kerr, when he said that, had gotten to the bottom of UC's problem. An institution devoted to creating and disseminating new knowledge had developed an aversion to applying new knowledge in its own operations. In his view, the main problem was the tenured professors, many or most of whom pursued a narrow self-interest without basic self-awareness or concern for the university as a whole, or with any regard for rules of common political life. In spite of my stock suspicion of managers who blame employees, I was unable to refute him.

II

From the failure of faculty governance, Kerr jumped directly to a discussion of the trouble he had founding the Santa Cruz campus as a place where faculty could take teaching seriously. Swarthmore College had been the model, and the campus would hire research faculty who would nonetheless focus on undergraduate development.[13] There must be plenty of people out there, he reasoned, who thought, as he did, that teaching helps research rather than detracting from it – students help you explain things more clearly, and you read more widely than you would otherwise, and are better at seeing the big picture, which in turn feeds back to research. But, he continued,

> Santa Cruz got caught with this strong counter-culture pull, and then what really broke it down was that faculty members, when they saw at Berkeley that faculty members could withdraw within their laboratory or their cubicle at the library and not see any students at all and just turn out their pieces of work, they felt they could get away with it. So what really broke down the Santa Cruz plan was the big rewards in a university are mostly for research and not really for teaching. It was very difficult for people to take teaching seriously.

> Of course, at most places there is no such thing as educational policy anymore. Faculty members do teach whatever they want to teach and do, to some extent, even determine what their teaching hours are without negotiating with anybody else. Santa Cruz began to show up reasonably well on the research ratings and I understand that Santa Barbara has done fantastically well. When people first came in, we would ask them if they were willing to go along with this idea of paying more attention to undergraduate teaching and they say that it was one of the reasons they came here. But then they get on the job and find that the rewards are for research and that on another campus you can get away without paying much or any attention to undergraduate teaching and they move in that direction.

After discussing building other UC campuses (San Diego was the flagship project in this period), I asked him if he thought the 'corporate university' had become a model that weakened the public good.[14] He said yes: 'A corporate model is where you look upon the rewards mostly in economic terms.' He discussed some already-inflated administrative salaries, which, he noted, his predecessor as President, Gordon Sproul, had systematically prevented. But, he added,

I don't like calling this the corporate model, I like to call it the market model or the individualist model where they can spend their time as they wish, including making as much money as they possibly can. Some people do extremely well. It's not corporate in the sense of all decisions being made at the top. I guess when I say that rather than corporate model, I would say either the market model or anything that is based on the cash nexus. It's more than it used to be. There was a time when being a professor was a little like being in the ministry, it was sort of a calling, and there were moral obligations. It's good it's not that way anymore – it was a career mostly for gentlemen who had family incomes.

But what bothers me is that the university has been losing its sense of being a community and the university is much less important than the life of the individual faculty member than it used to be. That was your life and that is where you had obligations and responsibilities and there is much less of a feeling of responsibility. Faculty members now still have a sense of responsibility to their departments, but I think there is much less a sense of responsibility to the campus as a whole such as in a big system as the University of California. There are good reasons for this. There are so many opportunities outside the campus. There are new means of communication where you could have a friend at Harvard who is closer to you than the guy next to you. You may not even know the guy in the office next to you. The sense of community has gone down in many, many ways.

Kerr returned repeatedly to the theme of faculty individualism.

I know how hard I worked when I was Chancellor at Berkeley. We had gone through the loyalty oath controversy and people thought we were a disaster and we were going to go way down. A lot of people were predicting that. The campus rallied around the idea that we were not going to go down and the alternative was to move up. We moved ahead of Harvard and we improved a lot of the departments. In a 1964 study, every rated department which Berkeley had was in the top 6 in the nation – every single department. Berkeley has never made that again – every single department rated in the top 6 – nor has any other university ever done that.

It was really hard work changing those departments, you know. A lot of the departments didn't want to be changed any. One department chairman who was working hard on his department to improve it said that every department has an inalienable right not to be any better than it wants to be.

Another department chairman said that you made us climb a mountain that we didn't want to climb. So, I know how hard it is to move up.

If we wanted to move up a department, I would decide which departments which were to be worked on in conversation with the budget committee. We only took on several departments at a time. Another thing I did was I turned down an awful lot of departmental recommendations. When I was Chancellor I told them they would have to come up with a stronger person …. The Dean at Berkeley, Davis, told me that between himself and myself we were turning down 20% of all the faculty appointments for tenure that had been approved at the departmental level and at the level of the budget committee. So we just said you gotta come up with really high quality appointments here or you aren't going to get anything at all. Nobody there was able to get away with what I was able to get away with at that time because everyone thought Berkeley was going down the drain. We were building the size of the student body and consequently had a lot of new appointments we could spread around. So we took some departments which increased 5 times over in size. These were some of the sciences, engineering, and the social sciences. We had a lot of new appointments to make. We didn't push them across the board. We concentrated on department by department. I also worked very closely with the Academic Senate. When I was turning down people, I would always tell them why I had done it. I was always careful to tell them why.

Kerr stressed his high standards and careful targeting of upgrades. But he also mentioned the massive growth in enrolments and funding during this period. When some departments are growing by a factor of five, and many are doubling in size in a few years, conflicts and errors are easier to wash away. This growth was over by the mid-1970s. Certainly many factors shaped the multiversity in the 1960s: social movements were crucial, as were, for Kerr, the quality of the governing board – the Board of Regents – which in the 1960s had not yet become a site of full-time political patronage. The 1960s may have shaped the university more by sheer growth than by a progressive philosophy of education or the public good.

Kerr felt that 'the university in the 1960s had a sense of great purpose and I don't think there is that same feeling currently … Society doesn't believe in us to the same extent that we don't believe in ourselves anymore. I think that is another very major change'.

Yes, I agreed. 'Do you feel like there are things we can do to get some of that momentum back or that sense of purpose?'

Clark Kerr replied, 'I don't know.'

Although Kerr and I continued to talk for a while, that is how the formal interview wound up.

III

For Kerr, management remained the University of California's central problem. Faculty individualism was one key issue, and another was intervention from above. His administrative career was forged in response to the crisis initiated in 1949, when state politicians, in concert with activist regents, had imposed a loyalty oath as a condition of employment on all faculty and staff. Many leading researchers departed, and a group of faculty who refused to sign on principle were dismissed.[15] The loyalty oath was imposed against the wishes of both faculty and administration. It set up a political test for employment, and overrode faculty professional purview over teaching competence when it declared that membership in the Communist Party was tantamount to a disqualifying bias.

At moments like this, management itself was suspended and replaced by the kind of top-down or sovereign power that management was set up to replace. Following Michel Foucault's famous formulation, management can be understood as a practice of 'governmentality' within bureaucratic structures. Its key features are: (1) power is sufficiently distributed and decentralized such that it cannot be set aside by a sovereign centre; (2) knowledge is also distributed and circulates laterally between units rather than only hierarchically.[16]

From the start of his tenure as chancellor, Kerr stood for modern management. He had signed the loyalty oath himself, but then worked hard to mitigate the institutional damage – including the departures of many important researchers – and to get the nonsigners reinstated. His appointment as chancellor emerged from his success at restoring some semblance of campus authority in relation to the Board of Regents and the state. He developed a method of orchestration through a system of placing political moderates at the institution's control points as he'd outlined in *Uses*.[17] The method was designed to avoid opinions that would call into question the everyday practices that constituted management as such. At the University of California, the system is called 'shared governance', which refers to regularized consultation between administration and the Academic Senate leadership (itself moderate), along with parallel modes of formalized communication.

While Kerr's model of management did mean non-sovereignty, it did not mean institutional democracy or even what ordinary language means by 'shared' governance – that we all decide together as relative equals. When senior officials consulted with faculty, this meant addressing other officers – deans and their professional staff, for example, along with chairs of major Senate committees or departments. There was no regular contact with the rank-and-file. The latter could communicate up the ladder by getting items on a Senate meeting agenda, though without support from Senate leadership these would go nowhere. Special town halls could be called, but these appeared only in moments of crisis: they were reactive rather than constitutive of everyday governance.

Administrative consultation was also non-binding: senior managers ordinarily made unilateral decisions about campus policies, administrative personnel, public communications, purchasing contracts, government relations, philanthropic relations, legal matters – everything involving the business side. Kerr's memory of fickle faculty was relevant to teaching and departmental hiring. But with broader campus policy, particularly budgeting and administrative appointments, fickle faculty were not in the room.

Perhaps most importantly, budgetary and related policy information did not circulate among the faculty as a whole. Senior managers did not act like sovereigns, and yet they had sovereignty over the data on which management decisions were based, which they readily withheld. Although UC management rejected sovereignty, it ruled in a parallel mode governed by secrecy. As a result, the University could not have a coherent multilateral policy discussion, since only one side possessed the relevant facts. Commentators were forced into the position of outsider dissidents, even if they were tenured faculty. Senior managers could marginalize opposition before the open debate had begun, which therefore never took place. When this strategy did not actually forestall or end opposition – as with the Free Speech Movement on the one hand, or Ronald Reagan and his allies on the other – management had no functional Plan B to fall back on. In a tragic irony, senior managers had not brought faculty, staff and students into their political community through honest intellectual exchange and thus could not draw on them for political support. In this way, managers diminished the University's public power, and their own leverage with state politicians and business leaders.

In the 1960s, University of California management was a somewhat customized industrial bureaucracy. Tenured faculty had special privileges that blocked straightforward corporate discipline. And yet other features of the industrial corporation were routine. In particular, communication was vertical and only rarely horizontal. For example, deans convened meetings of their department chairs, but the chairs did not set up meetings amongst themselves to discuss policy issues, and develop views that were independent of the dean's framing and information. Faculty used 'the bureaucratic apparatus' as a 'buffer which protected the isolation of the individuals and the small factions on each campus'. In Laurence Veysey's description, forged in Kerr's 1960s, faculty sought safety for their professional lives in non-participation in the very system that created their professional conditions.[18]

As a result of 'patterned isolation', the university's main constituencies lacked an intellectual life in relation to their institution. UC bureaucracy was information poor and knowledge blocked. Management structurally undermined the university's ability to deliberate collectively and to bring the results of deliberation into an open discourse. This was true even when the system president was a humanist like Clark Kerr: it was a structural effect and not the decision of an individual chancellor or president.

Kerr saw a big piece of the management puzzle, but not the parts that involved his institutional level. In 1963, he observed a paradox,

> Few institutions are so conservative as the universities about their own affairs while their members are so liberal about the affairs of others; and sometimes the most liberal faculty member in one context is the most conservative in another.[19]

Kerr perfectly summarized a core problem with university faculty, which is the conjunction of elitism, conservatism and conformity. This is a problem within the liberal humanist understanding of autonomous individual agency, and not only within the political left. This fit with the description of faculty unreliability Kerr recounted to me many years later.

But his model took administration out of the picture. If we put it back in, and see the faculty's impaired and divided agency as largely a response to managerial academia, a different picture emerges. As for faculty socialism – its interest in democracy and political nonconformity – UC's management appropriated public communication to itself. Faculty focused on grant income and laboratory work in part because that was a core professional activity and in part because senior management did not allow them to design public engagement programmes, present university strategy to legislators or approach potential sponsors or outside groups on their own. Rather than being torn between incompatible guild and socialist visions, Kerr's model of management denied faculty liberals the strong versions of both. It was undesirable but also rational for them to default to bureaucratic seclusion (Veysey's 'patterned isolation') from both institutional governance and public life.

A further sorrow appears when we recollect that Veysey was describing the research university of 1910. Fifty years later, faced with different publics and difference social requirements for knowledge creation, universities like UC still had the same institutional structure. And even UC's best president, Clark Kerr, showed no interest in changing that. The result was not shared but divided governance, with faculty demobilized rather than positioned to help the university confront its dire challenges.

IV

I'll close by summarizing some of the features that came baked into Kerr's model of the public university as it circulated in other parts of the United States and in other countries. This is not to detract from Kerr's intelligence and importance, but to suggest the limits of the period's academic managerialism as he influenced it – and handed it down to us.

A first limit came through the structure of extramural research funding. Kerr's denunciations of the faculty grant seeker as a 'euphoric schizophrenic' were among the most memorable passages in *Uses*.[20] He saw sponsored research setting up an indirect form of 'federal influence' that meant that 'the university really is less than a free agent'. Its faculty researchers were setting up 'a kind of "putting-out" system' that historically had converted workplaces in other industries into 'sweat shops'. Kerr claimed that faculty had allowed federal support to evolve into a shadow government that could, in a somewhat prophetic moment, endanger the status of academic labour.[21]

But Kerr's own behaviour guaranteed that faculty would act this way, for two reasons. One was that he tied Berkeley's increasing prominence to making tenure and promotion more difficult, which in turn increased pressure to get major extramural grants: faculty who wanted to succeed had no choice but to put grant-getting ahead of university service. The second was budgetary secrecy: faculty were not told the truth – that extramural grants were a net cost to the university that required subsidy from the state. Kerr and his successors withheld exactly the kind of information that could have produced an intelligent policy discussion about budgetary solvency, the university's multiple missions and governance. Managerial secrecy has prevented university-wide discussion to this day.[22]

A second was Kerr's attempt to define the university as a 'City of Intellect' that served the entire society and not just business and defence. The 'City of Intellect' 'may be viewed in a broader context, encompassing all the intellectual resources of a society, and the even broader perspective of the force of intellect as the central force of a society – its soul.' Kerr asked whether the university as a city of intellect could be 'the salvation of our society?' 'The social sciences and humanities,' he wrote, 'may find their particular roles in helping to define the good as well as the true and to add wisdom to truth.'[23] The university needed to address *all* of society's concerns and not just its economic ones, and generate socio-cultural knowledge as well as science. And yet Kerr's industrial-style management removed the faculty and students from public engagement, allowing political and business leaders to dominate the definition of the university's purposes. Had Kerr wanted to block the instrumentalizing of the university, he would have had to open up the full spectrum of university voices and activities to the outside world – including the non-market capabilities or the radical research and teaching that disturbed the mainstream. His form of management could not. As a result, the university has had to apologize for its liberal arts dimensions rather than existing in a society that recognizes their symbiosis with the practical arts desired by business.

A third cost was the trouble UC had embracing the state's increasingly multiracial population as something to be embraced rather than contained. The embrace would have required stating clearly and often that the increased presence of African-American, Native and Latino students not only did *not* endanger quality but enhanced it. When Kerr had become UC Berkeley's chancellor in 1952, the student population was still over 90 per cent white. Fifteen years later, when he was fired, California was well on its way to having a minority–majority youth population. The university's primary mechanism of racial integration was affirmative action. Though well intended, UC managerialism translated Harvard-style discretion and 'holistic' review into a formula-driven methodology that lacked a discourse of mutual comprehension and also of racial inclusion's dynamic, exciting, public benefits.[24] The university tended to hide affirmative action's relation to social justice. It clung to an admissions formula based on standardized test scores and grades, which it then weighted with racial categories. It did not come down firmly on the side of qualitative assessment of the applicant's whole record in its social context, even though the Supreme Court had validated this practice in its famous decision, *Regents of the University of California v. Bakke* (1978).

By 1997, when I was speaking with Kerr, regental policy had become a dysfunctional colour-blindness that was drastically reducing the share of African-American and Latino students at the UC flagships. Some of this retrenchment derived from white racial backlash against the practices that were moving education towards racial equality, however slowly. But much of it derived from the university's insistence on the closed bureaucratizing of everything.

In spite of his personal values, which ran towards the egalitarian and humanistic, his firmly managerial approach bound his heirs to practices that cannot make the university a popular social institution or generate broad political support. The range of core academic activities remain shrouded in bureaucratic mystery. What does the public know of the non-market, indirect and social value of a degree; the losses universities run on sponsored research; the reasons for the high rate of increase of university costs; why so many students borrow; why universities keep touting technology that never quite produces savings; what might fix racial inequality of outcomes; why the country must address the gross unevenness of institutional quality; what to do about the weakening link between university graduation and 'middle-class' security; how to address global challenges whose dimensions were more cultural and psychological than they were technological? Confusion here is a major legacy of 1960s UC.

Can any of these issues be addressed within the framework of Kerr's managerial multiversity? I think not. And I also think, in the end, that Kerr would agree.

Notes

1. For an overview, see Michelle Reeves, '"Obey the Rules or Get Out": Ronald Reagan's 1966 Gubernatorial Campaign and the "Trouble in Berkeley"', *Southern California Quarterly* 92 (2010), 275–305.
2. Rick Perlstein, *The Invisible Bridge: The Fall of Nixon and the Rise of Reagan* (New York: Simon & Schuster, 2014).
3. *Harvard Crimson*, 24 April 1963, https://www.thecrimson.com/article/1963/4/24/kerr-says-multiversity-head-must-be/ (accessed 6 August 2019).
4. Clark Kerr, *The Uses of the University* (Cambridge, MA: Harvard University Press, 1963), 122.
5. For an analysis of the 'multiversity' and related aspects of Kerr's thought, see Simon Marginson, *The Dream Is Over* (Oakland: University of California Press, 2016).
6. Clark Kerr, *The Great Transformation in Higher Education, 1960–1980* (Albany: State University of New York Press, 1991).
7. Kerr was an accomplished interviewee. For a listing, see Clark Kerr Personal and Professional Papers, Bancroft Library, University of California, Berkeley, cartons 38 and 39, https://oac.cdlib.org/findaid/ark:/13030/kt4x0nd267/entire_text/ (accessed 6 August 2019) and two published interviews with Nancy M. Rockafellar: *Interviews with Clark Kerr, PhD: A UC President's View of the Expanding Research University* (San Francisco: Regents of the University of California, 1996) and *Interviews with Clark Kerr, PhD, and Morton Meyer, M.D.: Eyewitnesses to UC Campus Turmoil in the Mid-1960's* (San Francisco: Regents of the University of California, 1996).
8. This concern overlaps with Cristina Gonzalez's detailed study of Kerr's ideas about leadership in *Clark Kerr's University of California: Leadership, Diversity, and Planning in Higher Education* (New York: Routledge, 2010).
9. The reporting was by Seth Rosenfeld, who later published *Subversives: The FBI's War on Student Radicals and Reagan's Rise to Power* (New York: Farrar, Straus, and Giroux, 2012).
10. William H. Whyte, Jr., 'Groupthink' *Fortune* (March 1952), http://fortune.com/2012/07/22/groupthink-fortune-1952/?iid=sr-link1 (accessed 15 December 2017). Whyte later authored *The Organization Man* (New York: Simon & Schuster, 1956).
11. Kerr, *The Gold and the Blue. A Personal Memoir of the University of California, 1949-67*, 2 vols. (Berkeley: University of California Press, 2001), i, 137–8.
12. With the founder of Work in America, Jerome Rosow, Kerr later co-edited *Work In America: The Decade Ahead* (New York: Facts on File, 1985).
13. For Santa Cruz, see Kerr, *Gold and the Blue*, i, ch. 17.
14. For contemporary uses and analysis of this term, see Henry A. Giroux et al., *Beyond the Corporate University: Culture and Pedagogy in the New Millennium* (Lanham: Rowman & Littlefield Publishers, 2001).

15 For a contemporary account, see Max Radin, 'The Loyalty Oath at the University of California', *Bulletin of the American Association of University Professors*, 36 (1950), 237–45. See also Nancy K. Innis, 'Lessons from the loyalty oath at the University of California', *Minerva*, 30 (1992), 337–65, and Kerr's own analysis in *Gold and the Blue*, i, 8–9.
16 Michel Foucault, 'Governmentality', in *The Foucault Effect: Studies in Governmentality*, ed. Graham Burchell et al. (Chicago: University of Chicago Press, 1991), 93.
17 Kerr, *Uses of the University*, 39.
18 Laurence Veysey, *The Emergence of the American University* (Chicago: University of Chicago Press, 1970), 311.
19 Kerr cited in Christopher Newfield, *Ivy and Industry: Business and the Making of the American University, 1880–1980* (Durham, NC: Duke University Press, 2013), 83–4.
20 Kerr, *Uses of the University*, 59.
21 Kerr cited in Newfield, *Ivy and Industry*, 73.
22 Christopher Newfield: *The Great Mistake: How We Wrecked Public Universities and How We Can Fix Them* (Baltimore, MD: Johns Hopkins University Press, 2016), Stage 2.
23 Kerr, *Uses of the University*, 123.
24 Christopher Newfield, *Unmaking the Public University: The Forty-Year Assault on the Middle Class* (Cambridge, MA: Harvard University Press, 2008), Part II.

15

Utopian universities of the British Commonwealth

by Miles Taylor

Founded in 1942, seven years before the University College of North Staffordshire, the University of Ceylon may fairly lay claim to be the first new 'British' campus of the twentieth century. Set on a lush tropical mountainside at the centre of the island, its layout and buildings designed by Patrick Abercrombie, the progressive English town-planner, and its curriculum and governance drawn-up, almost singlehandedly, by its Principal, Ivor Jennings, the university seemed idealistic in spirit and practice from the start. In fact, with the island under threat of Japanese invasion, staffing shortages because of the war effort, and fears about examination scripts being lost or destroyed on the long sea-passage back to Britain, progress was slow, and the new campus did not open until 1952.[1] However, the early years of the university, now known as the University of Peradeniya, are routinely invoked as a 'utopia' in the public memory of contemporary Sri Lanka.[2]

The University of Ceylon was no one-off, a last throw of the imperial dice before the era of decolonization and independence. On the contrary, between 1942 and 1970 Britain was directly involved in the establishment of some twenty-five new universities in Commonwealth territories around the world. Before 1942, there were just a handful of tertiary colleges in the British Empire outside of India and the white dominion countries of Australasia, Canada and South Africa. Only three of them had independent university status: Malta (a former Jesuit college founded in the sixteenth century), Hong Kong (set up in 1911) and Jerusalem (which severed its links with Britain on the creation of the state of Israel in 1948). All the others came under the parental watch of the University of London.[3] Starting with a trickle after the end of the Second World War, then in fuller flow during the 1960s, the expansion of higher education in the colonies and ex-colonies of the British Empire became a significant part of overseas development policy. Three times as many new campuses were established with British funding outside the UK as were started up at home. No other former colonial power tried as hard as Britain to influence the future of its ex-colonies through exporting its own system of higher education, although parallel moves were made by the Belgians in the Congo: Lovanium (1954) in Leopoldville (modern-day Kinshasa in the Democratic Republic of the Congo) and the Université officielle du Congo et du Ruanda-Urundi (1955) in Elisabethville (now Lubumbashi), by the French in Senegal (the University of Dakar, established in 1957), and by the Portuguese in Angola (Estudos Gerais Universitários de Angola, founded in 1962) and in Mozambique (Estudos Gerais Universitários de Moçambique, also started in 1962).

Whilst the story of Britain's role in these new universities is relatively well-documented in institutional histories of higher education in the Commonwealth, to date there has been no study which locates these developments within the context of the new universities movement of the post-war era, and particularly of the 1960s.[4] Yet there are obvious links with many of the other case-studies which feature in this volume. The same mandarin culture that dreamed up the new campuses at home was at work in the colonies too, its inspiration coming from the 1946 report produced by the Commission on Colonial Higher Education led by Cyril Asquith, a High Court judge. Leading luminaries of the British universities played their part overseas too. Ivor Jennings is a case in point. Having master-minded the early years of the University of Ceylon, he served on the commissions that led eventually to new universities in Malaya (1949), Hong Kong (1963) and Kuwait (1966).[5] Another prominent example was Eric Ashby, vice-chancellor of Queen's University, Belfast, and later Master of Clare College, Cambridge, who turned from reforming higher education at home to drafting blueprints for the colonies.[6] The new universities of the Commonwealth also bore the imprimatur of the British template in terms of their architecture and their curricula, taking up ideas and innovations that were trending at home. Most significant of all was the interchange of teaching staff and university administrators between the metropole and the colonial and Commonwealth periphery, with careers being played out in new campuses at home and overseas.

At the same time, all these new universities were neo-colonial creations. British planners extended a model – one that had stood the test of time in Britain and also been exported to India – to create institutions that often cut across national borders, for example, the University of the West Indies and the University of East Africa. However, by the 1960s, the federated university ran counter to the separate nationalist agendas of newly independent countries. British blueprints also came freighted with occidental assumptions about what a liberal arts higher education should comprise. Often, the new curricula being proposed failed to recognize more indigenous forms of knowledge and culture and, in some cases, denied it even existed.[7] In these ways, the new overseas campuses were not so much utopias as the products of what Eric Ashby termed 'transplantation' – with all the connotations of control that this implied.[8] Moreover, this phase of new universities overseas coincided with the cultural and economic diplomacy of the Cold War. In the Commonwealth territories of the Caribbean and Africa, Britain (with American and Canadian assistance) used financial aid for higher education as a weapon to stifle Soviet and Chinese influence, and also to temper the explicit socialism of new post-colonial regimes.

The 'Asquith colleges'

An integral part of the rolling out of a welfare state in miniature for the colonies after the Second World War, the Asquith Commission saw new universities as preparing nations for self-government. In the words of the report, 'it is the university which should offer the best means of counteracting the influence of racial differences and sectional rivalries which impede the formation of political institutions on a national basis'. Ideally, this meant that the two new colonial universities envisaged by the commission – Makerere (in East Africa) and Malaya – should be residential to encourage mixing. Similarly, to overcome problems of distance and isolation, they should be unitary institutions, based at a single site, not necessarily the federal model at work in London and in India. Above all, the Asquith report valued a liberal arts education over a narrower scientific and technological curriculum. Reporting at the same time as the Asquith Commission, two other committees, chaired respectively by the former Secretary of State for Scotland, Walter Elliott, and Principal of the University of St Andrews, James Irvine, made similar recommendations for three new universities in West Africa and one in the West Indies.[9]

Asquith and his colleagues also insisted on another Western convention, the autonomy of the university, that is to say, its independence from the state. But this did not mean freedom from the mother country. A new body, the 'Inter-University Council for Higher Education Overseas' (hereafter IUC), was set up to manage the colonial universities, its membership comprising representatives from all existing UK universities. For the next thirty years, the IUC controlled the setting up of new campuses in the colonies, arranging preparatory reports and dispatching troubleshooters when problems arose, negotiating funding from the UK government as well as from US and Canadian foundations, and selecting international staff.[10] Initially, the mission of the IUC was led by men and women drawn from the colleges of London and Oxford. Its first chairman was Alexander Carr-Saunders, director of the LSE. He was joined by Christopher Cox, a Balliol philosophy graduate turned fellow of New College, who was appointed special educational adviser to the Colonial Office in 1939. Charles Morris, another Balliol philosopher (a tutorial fellow there for twenty years), now vice-chancellor of the University of Leeds, played a part, as did two women: Margery Perham, expert on Africa and fellow of Nuffield College, Oxford, and Lillian Penson, vice-chancellor of the University of London. The New Zealander, Norman Alexander, who proved the most amphibian of all of this group of university planners, broke the Oxford–London cast, having started his career in the Cambridge laboratories of Ernest Rutherford, a fellow Kiwi.[11] In the 1960s this older guard was supplemented by some of the leaders of the new English campuses, notably John Fulton, the vice-chancellor of Sussex (and another Balliol philosopher), who chaired the IUC from 1964 to 1968, and Jack Butterworth, vice-chancellor of Warwick, his successor.

The IUC leadership took to their task with a barely disguised missionary zeal. Reflecting on the new universities of Africa and the West Indies, Penson described the role of the IUC as allowing colonies to 'draw upon the experience of old-established civilisations'. Others were more explicit, lamenting the 'backwardness' of the state of education in the colonies, and asserting the necessity of the Western ideal. 'What alternative' is there, asked James Duff, vice-chancellor of the University of Durham and member of both the Asquith and Irvine Commissions, 'to a western type of higher education?' H. Jac Rousseau, the inaugural professor of Education in the University College of Rhodesia and Nyasaland, declared that we cannot 'invent a new type of university. Western civilisation and universities are the only ones we know and can promote ... [M]oreover a university that deviates far from Western traditions forfeits academic respectability'.[12] Behind the rhetoric lay two assumptions. Firstly, that the British model of the university was fit for reproduction overseas. Eric Ashby later likened this to the export of British cars, defending a process whereby 'a university system appropriate for Europeans brought up in London and Manchester and Hull was also appropriate for Africans brought up in Lagos and Kumasi and Kampala'.[13] By implication this meant that the local culture was not fit material for institutions of higher education; it was a *tabula rasa*. In Ceylon in 1942, Ivor Jennings stated this candidly: 'Though it is commonplace in political speeches to speak of the cultural traditions of Ceylon, there are in truth hardly any.'[14] The second prevailing notion of the IUC was that although the winds of change might be blowing through the British Empire, higher education would maintain the bonds of belonging. Perham expressed this aspiration thus in 1946, 'They [the colonies] are rejecting our political control, with the other hand they are reaching out for everything we can give them of our culture and social advice. So that, however difficult our political relations may become, we can meet them on the university plane in almost complete understanding.'[15] The IUC was 'training the heirs to the Empire' and adding to the 'family of British universities', its 1955 report declared.[16] The new colonial universities were, in other words, a tool of cultural diplomacy, as Britain tried to shape from afar the new elites that would take their place with the advent of independent nationhood.

The first wave of Asquith colleges prided themselves on their British parentage, most apparent in their look, in their curriculum and in composition of their staff. At Accra in Ghana, the architect Austen Harrison, whose portfolio included Nuffield College in Oxford (1938), reproduced a 'Lilliputian Oxbridge' for the new University of the Gold Coast, complete with a porters' lodge, chapel, and junior and senior common rooms. The first Principal, David Balme, formerly Senior Tutor at Jesus College, Cambridge, rolled out a Western curriculum, defending the teaching of classics to Africans, on the grounds that there was a unity to civilization, otherwise why teach Greek to Englishmen?[17] Similarly at Ibadan in Nigeria, the new campus closely followed metropolitan norms. Although the architects there – Maxwell Fry and Jane Drew – famously developed a 'tropical' style, in substance it mirrored the residential campuses being developed in Britain: halls of single-occupancy rooms around a quadrangle with a large dining hall (its waved roof repeating Fry and Drew's design for the Royal Festival Hall in London).[18] Students followed a London programme of subjects. Some concessions were made, for example, in biology, common Nigerian animals and plants were substituted for temperate species as examples. As with the Gold Coast, classics received special treatment at Ibadan, becoming the first special honours course.[19] And so to eastern Africa. At Makerere in Uganda, originally established as a technical school in 1922, now given status as a university college in 1949, 'undergraduates may well be forgiven for thinking that Makerere is a modern version of Oxford or Cambridge!', declaimed Eric Lucas, the Professor of Education.[20] Carr-Saunders resisted a radical programme of social studies and extra-mural courses devised by Reg Batten for the new university, and insisted on London rules on matriculation and examination.[21]

In other parts of the empire, the same forces of transplantation were at work. The University of West Indies began life at Mona in Jamaica, its campus built on the site of a former sugar plantation (converted to a camp for evacuated Gibraltarians during the Second World War). The architect, Graham Dawbarn, had colonial form, having designed the neo-classical Raffles College in Singapore. This time his 'ultimate layout' included a central circle of low-rise university buildings (the Senate and Convocation Hall, the Library, Arts Faculty and Administration), ringed by halls of residence and the teaching hospital, with a students' union and staff housing on opposing sides of site, the whole campus approached by a wide drive, one kilometre long, signifying the journey students would be making from the outside world into academe. The first Vice-Chancellor, Thomas Taylor, an Oxford chemist who had been scientific adviser to Southeast Asian Command during the war, arrived determined to take students 'out of their immediate surroundings so that they acquire some idea of other, and possibly higher, standards in life, thought and technical accomplishment'.[22] In Malaya, political considerations accelerated the decision to give the new university, comprising Raffles College and the King Edward VII Medical College, university status as soon as possible. Carr-Saunders recommended this in his report of 1948, in order to bring unity across the Chinese, Malay and Indian communities. The first two vice-chancellors, Sir George Allen (to 1952) and Sir Sidney Caine, presided over traditional Western rituals: high table, teaching staff in academic dress, and an Anglophone curriculum. By the mid-1950s the university under Caine was deemed to be failing the aspirations of Malay Chinese, who in 1956 started their own university at Nanyang in Singapore, which the colonial government initially refused to recognize, believing it to be a communist hotbed. Caine left to succeed Carr-Saunders as director of the LSE. Under pressure from students, a commission of enquiry looked into the problems at the University of Malaya, uncovering widespread criticism of the imported curriculum.[23]

Having dealt with British higher education interests in Southeast Asia, Carr-Saunders then turned his attention to southern Africa. He proposed a federal university in Salisbury in Rhodesia, which became the University of Rhodesia and Nyasaland (UCRN). The UCRN sought to be a 'multi-racial' university in its student body, admitting young Africans from neighbouring countries such as South Africa, Zambia and Malawi. Its faculty and curriculum were anything but multi-racial, and at first there was limited African enrolment. The dogma of keeping up London entry standards to ensure that white Europeans chose the new university instead of going abroad caused a rare disagreement between Carr-Saunders and Christopher Cox.[24] At the opposite end of the African continent, geopolitical tensions also hurried through university status at the Gordon Memorial College at Khartoum in the Sudan. Wary of Arab nationalism to the north in Egypt, Britain injected funds into upgrading the college, by increasing enrolment and adding to the expatriate staff. Lewis Wilcher, principal since 1947 (and another Balliol graduate), defended an Arab university principally run by Westerners, noting in 1955 that 'a great deal of woolly nonsense can be talked in this connection ... the basis of a healthy university is a coherent and well understood academic discipline. It is not for an Australian to say that British discipline is best, but it is clearly the kind which the Sudan knows best'. Wilcher tried to curb the influence of the students' union, politicized by Nasser's seizure of power in Egypt.

He also deflected complaints that exam results were poor, saying that London exam standards had to be kept high.[25] However, tensions grew. In 1956, students protested against the selection of another non-Sudanese as the new Vice-Chancellor, Michael Grant, a best-selling classicist seconded from the University of Edinburgh. The college was closed temporarily and Christopher Cox was sent off to investigate. When Grant did finally arrive, he abandoned the new university's inauguration ceremony, as the Suez crisis meant most on the guest list were unavailable.[26]

The protests at Khartoum and Malaya pointed up many of the problems of British policy on colonial higher education in the decade after the Second World War. Applying external University of London standards meant setting the academic bar high, unrealistically so, and maintaining an Anglophone liberal arts curriculum not always attuned to the manpower needs of local economies, or the levels of attainment, especially in the English language, amongst school-leavers. As early as 1951, S. W. R. D. Bandaranaike resigned from the Ceylonese colonial government, citing the lack of Singhalese and Tamil language teaching at Jennings' University of Ceylon as one of the reasons behind his decision to quit and form the Sri Lanka Freedom Party.[27] In turn, Western curricula and examinations relied heavily on expatriate staff for their delivery. The pace of native academic appointments was slow. In its 1955 report, the IUC noted that of the new universities, only the University of Malaya was making steady progress in achieving parity, with 37 per cent of its staff drawn from the Asian communities. The University of the West Indies came next with 26 per cent, whilst in Africa, the share dwindled: at Ibadan the proportion was 15 per cent, 9 per cent at the Gold Coast, and only 4 per cent at Makerere.[28] Even the ideal of a residential university, a relatively new concept in Britain outside of Oxbridge, did not acclimatize easily. Not only did it impose an unfamiliar style of segregated living, it also treated university students as an elite, cut off from the rest of the population.

Decolonizing the new universities

From the outset the occidental conceit that only the Western export model was fit for new universities overseas was challenged. Perhaps the most comprehensive assault came in March 1945, from the Trinidadian historian Eric Williams, then a member of the Anglo-American Caribbean Commission advising on wartime needs in the West Indies. Writing from Howard University in Washington, DC, Williams offered a careful assessment, amounting to eighty pages, of the functions and purpose of a new university for the West Indies. It should be non-residential, public and free, unitary (not federal), and authorize its own degrees, that is to say, not be an outpost of the University of London. Nor should it replicate a British curriculum, but instead be reflective of Caribbean culture. It must offer adult education and extra-mural programmes. Above all, a new university must fulfil its role as 'the developmental arm of the state'. Williams sent his manifesto for a 'popular state' university into the Irvine Commission, but it was ignored.[29] The British disliked Williams, later denigrating him as 'a disappointed and frustrated person, with a marked "chip on his shoulder" about "imperialism" and colour'.[30] But there was force to his criticism. Elements of his alternative vision for a post-colonial university can be found in the arguments of African nationalists too.

Like Williams, many of the pan-Africanists who led their countries into independence had attended university in Britain, mainly in London. Whilst benefitting from the experience, it also made them aware of the potential isolation in their own country of an elite educated overseas. They were reluctant to impose on their own countries the 'one size fits all' metropolitan system. Joseph Danquah, the first West African to gain a PhD from a British university (UCL, 1927), led the campaign for Ghana's first university, objecting to the idea that only one university was needed for West Africa, at Ibadan in Nigeria. As he told the Gold Coast Legislative Council in July 1946, 'There are nations in West Africa as there are nations in Europe. There are people among black Africans as there are peoples among white Europeans. You cannot expect to build a successful University in Anadalusia in Spain with its Moslem foundation for the education of the people of Oxfordshire … for the English are not Spanish, nor the Spanish English.'[31] One of the first African leaders to move away from the Asquith model of the neo-colonial university was the Nigerian politician Nnamde Azikiwe (known popularly as 'Zik'). Zik had taken his first degree at Howard University in Washington, DC. Now, with direct financial aid from Michigan State University in the United States, as well as support from the IUC, Azikiwe became the formative influence behind the establishment of a second campus in Nigeria in 1955, at Nsukku in the eastern region of the country, designed by the London firm of the James Cubitt, which had an agency in Accra. Azikiwe wanted the Nigerian university to break with the 'British traditional concept': not be exclusively residential, reduce reliance on expatriate staff, bring in more courses and subjects related to Nigerian culture, and offer much more vocational and teacher training. He used the word 'revolution' in relation to higher education, not to signify ambitious future plans, but rather to emphasize how in Nigeria universities could not be autonomous and independent as the 'Asquith college' vision required. In an independent Nigeria, government control was necessary as the university was there to serve national needs.[32]

Variations on Azikiwe's recipe for the new universities spread across the African continent in the late 1950s and early 1960s. The fullest critique of transplantation came from Kwame Nkrumah in Ghana. Nkrumah had studied in the United States as well as in London. He denied that African history and culture furnished no examples of advanced culture, citing Walata (Oualata), Djenné, Timbuktu and Sankore as evidence from the pre-colonial era. Nkrumah called for placing African studies at the centre of the new university curricula, to which expatriates might contribute, but they must 'adjust and reorientate their attitudes and thought to our African conditions and aspirations. They must not try simply to reproduce their own diverse patterns of education and culture'. He also took up Azikiwe's rejection of the universities' independence from the state. Nkrumah distinguished between on the one hand 'academic freedom' as scholarly enquiry, which was sacrosanct, and, on the other, the separation of the university from the community and the 'national life' of which it was key part, something he disparaged.[33] In Tanzania, Julius Nyerere deployed the language of 'revolution' as well, caricaturing the idealism of the colonial project as Don Quixote-like, and calling instead for the new university to drive forward scientific and technological development, and provide leadership for the new republic.[34]

In these ways, the Western version of the university lay at the heart of the pan-Africanist assault on 'neo-colonialism'. Higher education was the sting in the tail of dying empires. Kenneth Kaunda, Zambia's first president, included in his definition of 'neo-colonialism', the process of 'inflicting upon an African state in the closing days of colonial domination such elaborate and sophisticated governmental and administrative institutions that dependence on expatriate skills becomes inevitable after independence'.[35] Embedding financial assistance for new universities within aid budgets, making that aid dependent on leaving the universities relatively free from the state and controlling expatriate appointments all chained the new world to the old. Another aspect of the 'neo-colonial' critique was the suspicion that the old imperial powers wanted to shape the elites of new nations in their own interests. Study overseas had done this in previous decades. Kaunda lamented how African intellectuals who went abroad were no longer 'along the wavelength of the masses when they returned'.[36] Making new national universities dependent on expatriate staff created the same problem, supplying graduates disconnected from their own national culture. In this respect, Edward Shils's influential sociological hypothesis, about the over-educated intellectual alienated from his own community, resurfaced in the African critique of Western higher education. For example, Wilbert Chagula, principal of the University College of Dar es Salaam from 1965, cited Shils as he criticized the ways in which the Western style of university led African intellectuals to 'despise' their own culture. Chagula went on, insisting that universities could not be treated as a different estate; effectively they must become a branch of the 'civil service', their curricula chosen according to national needs not abstract ideals. Chagula pointed out that the Western model of two-tier university governance – a lay Council and an academic Senate – gave more control to the Senate, inevitably dominated by expatriate staff.[37]

Some British advisers only partially grasped these new realities, notably Eric Ashby, who in 1959 led a commission funded by the Carnegie Foundation to advise on new universities in Nigeria. Although he was supportive of more new campuses, as well as a more Africa-centred approach, Ashby still emphasized the 'filial' dependency of the new universities, not just in Africa but elsewhere. Writing in 1966 he called Africa a 'province of Europe' in this respect and warned of the dangers of 'premature intellectual independence'. Expatriate guidance was still needed, lest the new universities – 'nurseries of nationalism' (a favourite phrase) – practised 'racialism', whereby an Africanist agenda dictated everything they did. In frowning upon the Africanization of the university in this way, Ashby made a contrast with the University of the Witwatersrand (Wits) in South Africa, which, in defiance of the apartheid regime there, had an open admissions policy.[38] Others observed the same analogy. Robert Birley, former headmaster of Eton and visiting professor of Education at Wits, set out a classic liberal vision of the non-racial university in his lecture on 'Universities and utopia' in 1965.[39] At the same time, there were Western reformers who did heed the winds of change, and it was to some of the utopian universities that they looked for ideas and inspiration.

Tropical utopias

Following a brief lull in the late 1950s, a second wave of new universities came in the years 1960–5, when a dozen new campuses were established in British colonies. The political climate was now very different. The process of decolonization had accelerated, with most of these new universities being set up at the moment of independence, or shortly after. Moreover, British aid, channelled via the IUC, was not the only source of support for these new universities. The Americans and Canadians were involved too, and in a few cases Soviet money was on the table as well. In these circumstances the export model of Asquith colleges was not repeated. Indeed, in certain aspects, it was completely rejected. The federal concept struggled to extend in the West Indies, and in central and eastern Africa too, as new nations rejected regional sharing. And tensions over expatriate staff and university autonomy dominated the formative years of many of these campuses. At the same time, there was much evidence of new ideas about campus layout and curriculum, much of it derived from the utopian movement elsewhere around the world.

The first part of the export model that came under attack was the regional structure, nominally unitary but in effect federal. At the University of the West Indies, different Caribbean countries aspired to have their own universities, reluctant to contribute financially to the campus at Mona in Jamaica, when comparatively fewer of their students attended. In Trinidad and Tobago, Eric Williams, now prime minister, pushed for an independent campus at St Augustine. After a series of negotiations, during which the Colonial Office threatened to withdraw funding, the new campus did remain within the federal university, but on its own terms.[40] Guyana did manage to break away, setting up its own campus at Georgetown, in 1963. There Britain also struggled to maintain control. When the first Vice-Chancellor, the Welsh biologist, Lancelot Hogben, left after only one year, Harold Drayton, his deputy and a close ally of the Guyanese Prime Minister Cheddi Jagan, stepped into the role. The British objected, having been at loggerheads with Jegun's Progressive People's Party for years. A 'sinister and conspiring' Drayton was accused by Christopher Cox, who visited Guyana in 1964, of wanting to transform the university into a 'socialist Sunday school'. Drayton was quickly demoted, and a Canadian, Alan Earp, put in charge.[41]

Hybrid, supra-national universities were also challenged in Africa, as the wider colonial schemes of federation unravelled. In the summer of 1963 the Rockefeller Foundation sponsored a meeting at Lake Como in Italy of American organizations and the IUC to draw up plans for an expansion of the University of East Africa, grafting new campuses at Zambia and Tanzania onto the existing university college of Makerere in Uganda and Nairobi in Kenya.[42] The university leadership at Makarere was reluctant to be so incorporated. Likewise, the new institutions at Lusaka and Dar es Salaam paid lip service to the inter-collegiate set-up.[43] Another new campus in the region, at Malawi, which began life at Blantyre in 1964, was defiantly national, effectively a vanity project of the leader Hastings Banda: 'my university' British officials reported him saying in 1961, looking on with bemusement as he moved the university to a new campus at Zamba, directly opposite Government House, his headquarters.[44] By the end of the 1960s there was only one UEA, and that was back in the English Fens. And so the pattern repeated itself elsewhere. In Hong Kong the new university of 1963 was federal in name, but this merely reflected its early years spread across three different sites until the new campus at Shatin was fully operational in the early 1970s. In 1976 the university went over to a unitary structure.[45] In Malaya, the university split into two in 1961, with the separate campus at Singapore becoming the University of Singapore four years later when independence was declared.[46] Only in Basutoland (now Lesotho) did the federal model remain in place, as the tiny land-locked country became home to a university at Maseru not only for its own people, but for those of Swaziland (the kingdom of Eswatini) and Bechuanaland (Botswana), hundreds of miles away to the north and east, as well. In a world of emancipated nation states, national not federal universities were the new normal.

The second wave of new universities overseas looked different too. Most became modernist playgrounds in much the same way as other new campuses around the globe. Several looked to the new English campuses for inspiration. Delegations from the Chinese University of Hong Kong, from Guyana and from the University of West Indies at Barbados all came over to the UK to sample English design and to seek ideas about cheap and fast construction, principally taking in Essex, Keele, Sussex and York as part of their itineraries, whilst the University of East Africa at Dar es Salaam sought advice about halls of residence from its namesake East Anglia.[47] British architects continued to dominate commissions, but they now came up with designs that were far more bespoke. Compare, for example, Norman Dawbarn's UWI campus at Mona (1948), with the same firm's later work at Dar es Salaam and Basutoland, where their brief was fully based on the academic blueprints.[48] Or consider how the practice of Frank Rutter and Ken Draper went from generic public buildings at Fourah Bay College in Freetown, Sierra Leone (late 1950s) to much more integrated plans at Georgetown in Guyana, and also at Suva in Fiji, where the University of the South Pacific rose up from site previously occupied by the New Zealand Air Force.[49] The most striking export of 1960s brutalism was Anthony Chitty and Julian Elliott's design for the University of Zambia at Lusaka (1965–8). Chitty teamed up with local Zambian architects to complete a holistic layout that mirrored the academic structure and student culture envisioned in the preparatory reports.[50]

The new universities also proved more experimental in their curriculum and faculty structure than the Asquith colleges. At the new Chinese University of Hong Kong, which brought together three colleges in the colony (New Asia, Chung Chi and United), the famed Sussex 'map of learning' was recommended as the basis for study. John Fulton, the Sussex vice-chancellor, chaired the commission that paved the way for the new university, preliminary work having been undertaken by Cox on a visit in 1957. Determined not to make the same mistakes as in Malaya, Chinese studies were foregrounded in early planning, with the New York-based architect I. M. Pei designing an art gallery that showcased the region's cultural heritage. The new university also opted for a four-year degree, seminar teaching and reducing reliance on examination as the mode of assessment. The Sussex connection was maintained throughout the decade with David Daiches, professor of English, joining a visiting delegation from the Hong Kong Universities Grants Committee in 1969.[51]

Adaptation to local needs was also evident across the universities of Africa in the 1960s. For the University of Zambia at Luaska, Sir John Lockwood, Master of Birkbeck College in London (who in 1965 also authored the controversial report that led to the setting up of the New University of Ulster), broke rank emphatically with the Asquith model, still being served up by Ashby and others elsewhere. Lockwood backed 'O' (ordinary) not 'A' (advanced) level school grades as the standard of entry for university admission in Zambia. He also proposed a four-year degree, with the first year turned over to general studies, specialization only coming later, more vocational and technical subjects being offered, and more courses being offered as continuing part-time basis.[52] Innovation at Zambia was also helped by the fact that the university's first Vice-Chancellor, the Canadian Douglas Anglin, came from a relatively new university (Carleton in Canada, established in 1942), was an African specialist and sympathetic to nationalism. Both as president of the Republic and newly installed chancellor, Kaunda welcomed Lockwood's recommendations, although he denounced the begrudging spirit in which Britain had provided £1m in aid to the new campus.[53] At Dar es Salaam in Tanzania, another Canadian reformer and Africanist, Cranford Pratt, questioned the suitability of British methods. Pratt concluded his preparatory report of 1962 for the new university with a section entitled 'On becoming an African university' in which he argued that the elite, research-intensive university was a costly option for a country such as Tanzania. Much better to lower entry standards and focus on the manpower needs of the national economy. Educating the educators was facilitated by embedding teacher training in with standard degree programmes, as had been pioneered at Keele in the UK. Not that Pratt eschewed specialization. At Dar es Salaam there was a Faculty of Social and African Studies from the start, and he also brought in Swahili studies.[54] Further south, at the University of Basutoland, John Blake, formerly of the University of Keele in the UK, presided over a curriculum partially exported from the famous Keele experiment. Taking over a former college run by the Italian Catholic church, Blake brought in a four-year degree, training courses in public administration for local chiefs and civil servants, including a focus on sociology and psychology, and, as in Dar es Salaam, integrated teacher training within the main honours programmes.[55] By the end of the decade, little was left of the export model. For the University of South Pacific in Fiji, which started in 1968, Charles Morris and Norman Alexander recommended that Pacific Studies be taught across all departments.[56]

Alongside these innovations and adaptations, one feature of the Asquith colleges remained contested: the independence of the university. Institutional autonomy was put to the test at the new University of Lagos in 1965, and then a year later at the UCRN in Rhodesia. Lagos, partially funded by the British as well as by the Ford Foundation, the United States Agency for International Development (USAID) and New York University, appointed as its first Vice-Chancellor, Eni Njoku, an Ibo Nigerian, botanist and Dean at Ibadan, in 1962. Early in 1965, the University of Lagos Senate approved his re-election for a second term. However, another candidate, the historian Saburi Biobaku, a Yoruba, was preferred by the Provisional Council of the University. Led by Lawrence Gower, the professor of Law, Senate stood firm, backing Njoku. Gower accused the Council of favouritism in trying to impose a government nominee on the university, which was 'not to be regarded as just any other public corporation to be used for the purpose of political and tribal influence and patronage', throwing in the charge of 'racialism' as well.[57] Student protests broke out, some of it communal, with many Ibo men and women being sent home for safety. Teaching was suspended and the campus closed for three months. The Nigerian government turned on the expatriate community at Lagos, singlling out Gower for stirring up trouble, especially amongst the students. He was denounced in the Nigerian Parliament as an 'evil genius', as a 'headstrong, tactless and cantankerous' individual who has 'fouled the fountain'.[58] Both sides published their own version of events, the pamphlet produced by the pro-Biobaku camp illustrated with stunning images of James Cubitt's pristine buildings then going up on campus. The Nigerian government won out. Biobaku took over as vice-chancellor in June 1965, and promptly sacked five Deans, all of them European or American, with a further fifty staff resigning. Gower and his senior colleagues returned home, their reputations tarnished by accusations of being 'birds of passage' who had sparked off an 'insurrection'.[59]

A year later, in a rather different context, another crisis erupted at the UCRN in Salisbury (modern-day Harare). In November 1965, Ian Smith's government made its Universal Declaration of Independence from Britain, placing the university in an invidious position, subject to the laws of the country, but still partially funded by Britain, and with a large expatriate teaching staff. Walter Adams, vice-chancellor of the university since 1955, and before that former secretary to the IUC, insisted that the university retain its neutrality, and hence safeguard its autonomy, a 'theology' he feared Smith's government did not understand.[60] However, many students and staff wanted to go further, and condemn UDI. Overseas staff sent a letter of protest to the *Times*, saying academic freedom was under threat.[61] African students led demonstrations on campus and used their own press, such as *Unicorn* and *Black and White*, to criticize Smith's regime and its links to apartheid in South Africa. They also denounced the supine policy of the university authorities. Adams struggled to keep control. Lectures were boycotted. Police land-rovers and dogs appeared on campus, student discipline was tightened, with Adams using the Students' Union as the instrument of control. But Adams found no unanimity amongst his own staff as to whether to resist the Rhodesian government and risk the college's future, or carry on as normal. Appointed by the University Council, Robert Birley came over from Wits to report on the crisis, chiding Adams for not giving more legal support to the students, but to the disappointment of some staff, not putting his finger on the race issues. Matters came to a head in April when one of the African student protestors was found to be a non-registered student and was taken into custody. Adams finally broke cover, authorized the student's registration, thereby giving him some protection from the police. The university council overruled Adams, who offered his resignation, which was not accepted. Firefighters were sent out from London: Morris and Cox in August 1966, and a deputation from the University of London in November.[62] A series of student and staff expulsions eventually cleared the air, and by 1967 the university had returned to near normal. But the reputation of the campus as a multi-racial institution was destroyed, and, as at Lagos, in the stand-off between university autonomy and the national government, the state had won.

Effects of the crises in Nigeria and in Rhodesia rippled back to Britain. Deported staff found their way back into home universities. Opponents of UDI, led by Colin

Leys and Anthony Low at Sussex, turned their ire on Adams, calling for Britain to end finance for the university. By then, Adams was being courted by the LSE as its next director (to replace Sidney Caine), a move which fanned flames even more. Students at the LSE condemned the appointment, citing Adams's poor record on race.[63] Although Adams did become the new director, the incident fuelled British student protests against the white regimes of Rhodesia and South Africa. New universities – for example, York and Lancaster – saw significant anti-UDI, anti-apartheid movements emerge in 1966–7, precursors of the campus radicalism that swept the country from 1968 onwards.[64] Nowhere felt more keenly the failure of the utopian experiment in Africa, quite as much as the utopian universities at home.

Indeed, by the end of the 1960s, many of the new African universities faced an uncertain future. The older federated systems – East Africa, Rhodesia and Nyasaland, and Basutoland – reverted to their constituent parts. In Nigeria, the Yoruba-Ibo and other regional rivalries that beset the new universities there exploded into civil war in 1966, closing down many of them in the process. Elsewhere, universities maintained uneasy relations with national governments. In Zambia, Kaunda shut the university in 1971. Despite all this, the utopianism of era had worked its way into the new universities in Africa, Southeast Asia, the Caribbean and elsewhere, moderating the neo-colonialism of the IUC and creating campuses more suited to local conditions. The new universities overseas had come full circle.

Notes

1   Jennings to Christopher Cox, 21 June 1942, TNA CO1045/64 (staffing); William Ivor Jennings, *The Road to Peradeniya: An Autobiography*, ed. H. A. I. Goonetileke (Colombo: Lake House Investments, 2005), 113 (exam scripts); Anoma Pieris, *Architecture and Nationalism in Sri Lanka: The Trouser Under the Cloth* (Abingdon: Routledge, 2013), 114–16.
2   *Peradeniya: Memories of a University*, ed. K. M. de Silva and Tissa Jayatilaka (Kandy: International Centre for Ethnic Studies, 1997); K. M. de Silva, *The Making of a Historian: A Memoir* (Kandy: ICES, 2017), 47–55; Leelanada De Silva, 'Utopia Lost: The University at Peradeniya' (6 January 2018), http://ceylon-ananda.co/utopia-lost-the-univesrity-at-perdeniya (accessed 1 August 2019).
3   Bruce Pattison, *Special Relations: The University of London and New Universities Overseas, 1947–70* (London: University of London, 1984); Graeme Davies and Svava Bjarnason, 'After Empire: The "London Model" Transformed since the Second World War', in *Universities for a New World: Making a Global Network in International Higher Education, 1913–2013*, ed. Deryck M. Schreuder (New Delhi: Sage, 2013), 27–40.
4   *Universities for a New World*, ed. Schreuder, passim; A. M. Carr-Saunders, *New Universities Overseas* (London: George Allen and Unwin, 1961).
5   For his career, see 'Jennings, Sir (William) Ivor (1903–1965), jurist', *Oxford Dictionary of National Biography* (ODNB), https://www-oxforddnb-com.libproxy.york.ac.uk/view/10.1093/ref:odnb/9780198614128.001.0001/odnb-9780198614128-e-34181 (accessed 1 August 2019); *Constitution-Maker. Selected Writings of Sir Ivor Jennings*, ed. Harshan Kumarasingham (Cambridge: Cambridge University Press, 2014), 1–18.

6   Alan Burges and Richard J. Eden, 'Ashby, Eric, Baron Ashby (1904–1992), Botanist and University Administrator', *ODNB*, https://www-oxforddnb-com.libproxy.york.ac.uk/view/10.1093/ref:odnb/9780198614128.001.0001/odnb-9780198614128-e-50791 (accessed 1 August 2019).
7   For Africa, see Toyin Falola, *Nationalism and African Intellectuals* (Rochester, NY: University of Rochester Press, 2001), ch. 5; Mahmood Mamdani, 'The African University', *London Review of Books*, 40 (19 July 2018), 29–32. For the Caribbean, see Harry Goulbourne, 'The Institutional Contribution of the University of the West Indies to the Intellectual Life of the Anglophone Caribbean', in *Intellectuals in the Twentieth Century Caribbean. Vol. 1, Spectre of the New Class: The Commonwealth Caribbean*, ed. Alistair Hennessy (London: Macmillan, 1992), 21–49.
8   Eric Ashby (with Mary Anderson), *Universities: British, Indian, African. A Study in the Ecology of Higher Education* (London: Weidenfeld and Nicolson, 1966), ch. 8.
9   'Report of the Commission on Higher Education in the Colonies', *Parl Papers* (1944–5), vol. iv, Cd 6647, 10, 13–15, 104. For a recent study of the Elliot Commission, see Timothy Livsey, 'Imagining an Imperial Modernity: Universities and the West African Roots of Colonial Development', *Journal of Imperial and Commonwealth History*, 44 (2016), 952–75.
10 For the IUC, see I. C. M. Maxwell, *Universities in Partnership: The Inter-University Council and the Growth of Higher Education in Developing Countries* (Edinburgh: Scottish Academic Press, 1980); Martin Kolinsky, 'The Demise of the Inter-University Council for Higher Education Overseas: A Chapter in the History of the Idea of the University', *Minerva*, 21 (1983), 37–80.
11 See 'Saunders, Sir Alexander Morris Carr- (1886–1966), sociologist and academic administrator', *ODNB*, https://www-oxforddnb-com.libproxy.york.ac.uk/view/10.1093/ref:odnb/9780198614128.001.0001/odnb-9780198614128-e-32308 (accessed 1 August 2019); Clive Whitehead, 'Sir Christopher Cox: An Imperial Patrician of a Different Kind', *Journal of Educational Administration and History*, 21 (1989), 28–42; 'Morris, Charles Richard, 'Baron Morris of Grasmere (1898–1990), Writer on Philosophy and University Administrator', *ODNB*, https://www-oxforddnb-com.libproxy.york.ac.uk/view/10.1093/ref:odnb/9780198614128.001.0001/odnb-9780198614128-e-39813 (accessed 1 August 2019); C. Brad Faught, *Into Africa: The Imperial Life of Margery Perham* (London: I.B. Tauris, 2012); 'Penson Dame Lillian Margery (1896–1963), 'Historian', *ODNB*, https://www-oxforddnb-com.libproxy.york.ac.uk/view/10.1093/ref:odnb/9780198614128.001.0001/odnb-9780198614128-e-35468 (accessed 1 August 2019). For Alexander's career, see Mary Harris' obituary, *Independent*, 5 April 1997, https://www.independent.co.uk/news/obituaries/obituarysir-norman-alexander-1265280.html (accessed 1 August 2019).
12 Lillian Penson, *Educational Partnership in Africa and the West Indies* (Glasgow: Jackson, Son and Company, 1955), 21; James Duff, *Foundations of Freedom: The New Universities Overseas* (London: Longmans, 1955), 15; H. Jac. Rousseau, *The University in Africa* (London: Oxford University Press, 1957), 6.
13 Eric Ashby, *African Universities and Western Tradition* (London: Oxford University Press, 1964), 11, 19.
14 Jennings, 'The Task of the University of Ceylon' (enclosed in a letter to Christopher Cox, 25 May 1945), 1, TNA CO145/164.
15 Margery Perham, 'Relation of Home Universities to Colonial Universities' (1946)', in Perham, *Colonial Sequence 1930 to 1949. A Chronological Commentary upon Colonial Policy Especially in Africa* (London: Methuen, 1967), 288.

16 'Inter University Council for Higher Education Overseas, 1946-54', *Parl. Papers* vol. xiv (1955-6), Cd 9515, 4.
17 Katya Leney, *Decolonisation, Independence, and the Politics of Higher Education in West Africa* (Lewiston: Edwin Mellen Press, 2003), 160; cf. Barbara Goff, *'Your Secret Language': Classics in the Colonies of British West Africa* (London: Bloomsbury, 2013), 188.
18 Timothy Livsey, '"Suitable Lodgings for Students": Modern Space, Colonial Development and Decolonization in Nigeria', *Urban History*, 41 (2014), 664-85; Rhodri Liscombe, 'Modernism in Late Imperial British West Africa: The Work of Maxwell Fry and Jane Drew, 1946-56', *Journal of the Society of Architectural Historians*, 65 (2006), 188-215.
19 Kenneth Mellanby, 'Establishing a New University in Africa', *Minerva*, 1 (1963), 149-58.
20 Eric Lucas, *English Traditions in East African Education* (London: Oxford University Press, 1959), 16. For the earlier history of the university, see Margaret Macpherson, *They Built for the Future: A Chronicle of Makerere University College, 1922-1962* (Cambridge: Cambridge University Press, 1964).
21 T. R. Batten, 'Proposals for the Further Development of Social Studies Training at Makerere' enclosed in Batten to W. D. Lamont 9 December 1947, TNA BW90/1306; Carr-Saunders to Lamont, 9 June 1948, ibid.
22 Taylor to Hammond, 8 September 1946, TNA CO1042/176. For Dawbarn's 'ultimate layout' see TNA CO1042/177.
23 A. J. Stockwell, '"The Crucible of the Malaya Nation": The University and the Making of a New Malaya, 1938-62', *Modern Asian Studies*, 43 (2009), 1149-87.
24 Cox to Carr-Saunders, 2 March 1954, TNA BW90/1304; cf. B. A. Fletcher, *The Building of a University in Central Africa* (Leeds: Leeds University Press, 1962), 7-10. For the early years of the university, see Michael Gefland, *A Non-Racial Island of Learning. A History of the University College of Rhodesia from Its Inception to 1966* (Gwelo: Mambo Press, 1978), ch. 4.
25 'Principal's Report 1954-55', 1, 3, C. M. Cox Papers, Durham University Library, 667/8/24-6; cf. Wilcher, 'Some Problems of University Education in the Sudan', *Sudan Notes and Records*, 34 (1953), 62-72.
26 Carr-Saunders to Cox, 29 December 1955, C. M. Cox Papers, Durham University Library, 670/5/13; 'Report by Professor Michael Grant' (March 1958), 7, ibid., 667/8/26-37. For the background, see El Subki Mohamed El Gizouli, *Higher Education in the Sudan, 1898-1966* (Khartoum: Khartoum University Press, 1999), 200-28.
27 'Speech on the Appropriation Bill' (12 July 1951) in *Towards a New Era. Selected Speeches of S. W. R. D. Bandaranaike made in the Legislature of Ceylon, 1931-1959*, comp. G. E. P. de S. Wickramaratne (Colombo: Government of Ceylon, 1961), 692-3.
28 'Inter University Council for Higher Education Overseas, 1946-54', 7-8.
29 Williams to Irvine, 7 March 1945, enclosing 'Memorandum on the British West Indian University', TNA CO1042/176. Williams's paper does not feature amongst the forty-eight memoranda listed in the Appendix: 'Report of The West Indies Committee of the Commission on Higher Education in the Colonies', *Parl. Papers* (1944-5), vol. v, Cd 6654, Appdx 1. It was reprinted in 1951 in his *Education in the British West Indies* (New York: Teachers Economic and Cultural Association, 1968). For Williams, see Maurice St. Pierre, *Eric Williams and the Anticolonial Tradition. The Making of a Diasporan Intellectual* (Charlottesville: University of Virginia Press, 2015), 49-51.

30. 'Trinidad and Tobago Elections: CO Note on the General Election and the New Government of Dr Williams' (November 1956), reprinted in *British Documents on the End of Empire, Series B Volume 6, The West Indies*, ed. S. R. Ashton and David Killingray (London: The Stationery Office, 1999), 256.
31. *Dr Joseph Boakye Danquah Historic Speeches and Writings on Ghana*, comp. H. K. Akyeampong (Accra: George Boakie, 1966), 44.
32. Speech at the Inaugural Meeting of the Provisional Council of the University of Nigeria (3 March 1960), reprinted in *Zik. A Selection of the Speeches of Nnamdi Azikiwe* (Cambridge: Cambridge University Press, 1961), 293–6. For the background, see B. I. C. Ijomah, 'The Origin and Philosophy of the University', in *The University of Nigeria, 1960-1985: An Experiment in Higher Education*, ed. Emmanuel Obiechia et al. (Nsukku: University of Nigeria Press, 1986), 1–11; Egerton O. Osunde, 'The American Higher Educational System and the University Education System in Nigeria: An Analysis of Reform Influences and Processes', *International Education*, 17 (1987), 32–42.
33. 'Speech at his Installation as First Chancellor of the University of Ghana' (25 November 1961), reprinted in *Selected Speeches Kwame Nkrumah, vol. 2*, comp. Samuel Obeng (Accra: Afram Publications, 1979), 139 and cf. Nkrumah, *The Role of Our Universities* (Accra: Ministry of Information and Broadcasting, 1963), 1–3.
34. Nyerere, 'University, an Investment of the Poor in Their Own Future' (21 August 1964), reprinted in *Nyerere on Education. Selected Essays and Speeches, 1954–1998*, ed. Elieshi Lemba et al. (Dar es Salaam: HakiElimu, 2004) 21.
35. *A Humanist in Africa: Letters to Colin M. Morris from Kenneth D. Kaunda, President of Zambia* (London: Longmans, 1966), 116.
36. Ibid., 97.
37. W. K. Chagula, *Academic Freedom and University Autonomy in the Economic, Social and Political Context of East Africa* (Nairobi: East African Academy, 1967), 7–8, 23.
38. Ashby, *Universities: British, Indian, African*, xii.
39. Robert Birley, *Universities and Utopia* (Johannesburg: Witwatersrand University Press, 1965).
40. The negotiations through 1962 can be followed in TNA OD17/129. For some of the background, see Bridget Brereton, *From Imperial College to University of the West Indies: A History of the St. Augustine Campus, Trinidad & Tobago* (Kingston, Jamaica: Ian Randle, 2011), ch. 4; Anthony Payne, 'One University, Many Governments: Regional Integration, Politics and the University of the West Indies', *Jamaican Historical Review*, 16 (1988), 33–53.
41. 'Summary of Sir Christopher Cox's Report on the University of Guyana', 5, TNA OD17/223; Drayton recounted the episode (without Cox's invective) in his *An Accidental Life* (Hertford: Hansib, 2017), 652–5.
42. Edward H. Berman, *The Influence of the Carnegie, Ford, and Rockefeller Foundations on American Foreign Policy: The Ideology of Philanthropy* (Albany: State University of New York Press, 1983), 77–8.
43. 'The Future of the University of East Africa' (Memorandum), TNA OD17/235. For the background, see Michael Mwenda Kithinji, 'An Imperial Enterprise: The Making and Breaking of the University of East Africa, 1949–1969', *Canadian Journal of African Studies/La Revue canadienne des études africaines*, 46 (2012), 195–214; Carol Sicherman, *Becoming an African University: Makarere, 1922-2000* (Trenton: Africa World Press, 2005), ch. 4.

44 J. D. Hennings to K. J. Neale, 20 October 1961, reprinted in *British Documents on the End of Empire, Series B Volume 9, Central Africa, Part II*, ed. Philip Murphy (London: The Stationery Office, 2005), 277. For the background, see Gordon Hunnings, *The Realisation of a Dream: The University of Malawi, 1964–74* (Zomba: The University, 1974).

45 Tak Sing Cheung, 'Institutional Changes', in Alice N. H. Lun Ng (ed.), *The Quest for Excellence. A History of the Chinese University of Hong Kong from 1963 to 1993* (Hong Kong: Chinese University Press, 1994), 81–124.

46 Stockwell, 'The Crucible of the Malaya Nation'.

47 'Memorandum', Executive Council, Hong Kong (7 March 1961), Chinese University Preparatory Committee, Secretariat papers, Hong Kong Public Records, HKRS457-3-14; 'Programme for Dr Earp' (November 1965), TNA BW 90/537 (Guyana); *Report of a Visit to Some Universities in the United Kingdom by the Pro-Chancellor, the Vice-Chancellor and the Pro-Vice-Chancellor of the University's College Situated in Barbados* (Mona: University of the West Indies, 1965); Frank Thistlethwaite to Ian Maxwell, 17 December 1963, TNA BW90/136 (Dar es Salaam); 'Visits in Britain, USA and Canada Made by the Vice-Chancellor, etc., 13 February–7 April 1966', Ian Michael Papers, UCL Institute of Education, DC/IM (Malawi).

48 'Development Report of Messrs Norman and Dawbarn' (October 1964), TNA OD17/218.

49 For Guyana: *UG Newsletter*, 2 (June 1966), TNA BW90/537; and for Fiji: 'The University and the Laucala Bay Site' (14 August 1969), TNA OD40/15.

50 Chitty's two project development reports for 1964 and 1965 are in TNA BW90/755. See also *Design for Zambia's University* (Lusaka: Government of Zambia, 1966), a brochure printed for the opening ceremony, Ian Michael Papers, DC/IM; cf. *African Modernism. The Architecture of Independence*, ed. Manuel Herz et al. (Zurich: Park Books, 2015), 546–53.

51 Alice N. H. Lun Ng, 'The Founding', in Ng (ed.) *The Quest for Excellence*, 22–3; *Report of the Fulton Commission* (Hong Kong: Government Press, 1963), 24; *The First Six Years, 1963–1969. The Vice-Chancellor's Report* (Hong Kong: Chinese University of Hong Kong, 1969), 33; *University Bulletin*, 5 (March, 1969), 1–2.

52 *Report on the Development of a University of Northern Rhodesia* (Lusaka: Government of Zambia, 1964).

53 *Addresses at the Installation of His Excellency the President as First Chancellor of the University of Zambia* (Lusaka: Government Information Services, 1966), 4–7; John B. Stabler, 'The University of Zambia: Its Origin and First Year', *Journal of Higher Education*, 39 (1968), 32–8. Anglin co-edited with Millar McClure a collection of essays by the Nigerian pan-Africanist, Jaja Wachuku and others: *Africa. The Political Pattern* (Toronto: University of Toronto Press, 1961). See also his 'A University for Zambia: Problems and Prosepcts', *Mindolo Bulletin*, 1 (1963), 10–26.

54 Pratt, 'The University College Dar es Salaam' (July 1962), TNA BW90/136; *Guardian*, 25 August 1964, 7. With Colin Leys, Pratt edited *A New Deal in Central Africa* (London: Heinemann, 1960). For his career, see Robert O. Matthews, 'A Tribute to Cranford Pratt', *International Journal*, 57 (2002), 167–74, and for the background, see David Court, 'The Experience of Higher Education in East Africa: The University of Dar es Salaam as a New Model?', *Comparative Education*, 11 (1975), 193–218.

55 'University of Basutoland, Bechuanaland Protectorate and Swaziland. Development Plan' (October 1964), TNA OD17/218; *Times Educational Supplement*, 9 April 1965, 7.

56 'Report of the Higher Education Mission to the South Pacific' (Ministry of Overseas Development 1965), 47, TNA T317/2171. Morris led this mission. Norman Alexander, 'University of the South Pacific. Interim Report' (29 July 1966), TNA BW90/670; cf. Colin M. Aikman, 'Establishment: 1968-74', in *Pacific Universities. Achievements, Prospects, Problems*, ed. Ron Crcombe and Malama Meleisea (Suva: University of the South Pacific, 1988), 35-54.

57 Gower to the chairman of the Provisional Council, 31 March 1965, 'Higher Education in Anglophone Tropical Africa', Bodleian Library, University of Oxford, MSS. Afr. s. 1825, Box XXI ('public corporation'); 'An Open Letter to the Nigerian People' ('racialism'), Box XXII (2). For the background, see A. B. Aderibigbe, 'Emergence of the University, 1962-1967', in *A History of the University of Lagos, 1962-1987*, ed. A. B. Aderibigbe and T. G. O. Gbadamosi (Lagos: University of Lagos Press, 1987), 11-15.

58 'Government Statement' (9 April 1965), 'Higher Education in Anglophone Tropical Africa', Box XXII (1).

59 *Change in Vice-Chancellorship* (University of Lagos, 1965), ibid., Box XXII (2).

60 For a contemporary account, see the article by Paul Nursey-Bray (a lecturer in government at UCRN), 'Rhodesia: A University in Crisis', *Mawazo* (December 1967), TNA BW90/772; cf. Gefland, *A Non-Racial Island of Learning*, ch. 17. Adams to Paul Yowles, 12 October 1964, Adams Papers, LSE Library, W/5/A ('theology').

61 *Times*, 30 December 1965, 9.

62 *University College of Rhodesia. Report by Robert Birley* (April 1966), TNA BW90/771; 'Draft Report of Visit of Professor M. Guthrie, etc' (November 1966), TNA BW90/772.

63 Leys and Low to Barbara Castle, 2 September 1966, TNA BW90/772 (Castle had been Minister for Overseas Development until the end of 1965); *Observer*, 16 October 1966, 1; 'Pamphlet Issued by Students of L.S.E.' (*c.* October 1966), TNA BW90/806.

64 *Guardian*, 28 November 1966, 1 (York); ibid., 11 February 1967, 3 (Lancaster).

16

The other 1960s: University leaders as agents of change in Canadian higher education

by Paul Axelrod

Students dominate the literature on universities in the 1960s. Their idealism, activism and rebellion drew the attention of sociologists and psychologists in the immediate wake of that turbulent era. In a growing body of published work, historians now seek to make meaning of those times and students and youth remain a central focus.[1] Given their headline-making activities, which included demonstrations, occupations, strikes and periodic violence, this is understandable. But it is far from the whole story of university life in that famous decade.

Largely neglected in the historical narrative is the crucial role of university planners and administrators in shaping campus culture, particularly in new universities. They literally built the platforms on which student theatrics were staged. Many saw themselves as educational innovators, even rebels, determined to reinvent, revive and rescue higher learning from its post-war stasis. They had utopian visions of their own and were taken aback by the uprisings of the 1960s, in which they were deemed by radical students to be enemies rather than agents of change.

This chapter explores this dynamic primarily through the lenses of three new Canadian universities established in the 1960s: Simon Fraser University in British Columbia; York University and Trent University, both in Ontario. I will describe the visions of university planners and leaders through three aspects of university development: academic programming, design and architecture, and university governance. The reform of higher education, promoted by liberal faculty and administrators, was in the wind well before the student movement erupted, and their impact was deeper and more enduring than that of their radical student critics.

I

Prior to the 1960s, Canadian universities served a small but socially significant segment of the population. Some 6 per cent of the eighteen to twenty-four age group attended university in 1951, far lower than that in the United States, but higher than that in Britain.[2] Typically, students were the male children of professional and business elites, though women and lower-income youth comprised a meaningful minority of the student population, and had done so throughout the twentieth century.[3] The flood of post-war veterans doubled Canada's university population between 1944 and 1948, and heightened the profile and perceived importance of higher education. This cohort included thousands of lower income youth who would never have attended university without the opportunity offered by a funding programme that included free tuition and a living allowance (similar to the American GI Bill), an initiative that helped embed the idea that university education could be and should be a vehicle for social mobility.[4]

This idea gained traction in the 1960s with the arrival of the post-war baby boom generation. Notwithstanding the enormous expansion of the university student population, from 90,000 in 1958 to almost 300,000 in 1973 (nearly 20 per cent of the eighteen to twenty-four age group), working class and poor Canadians were still under-represented in advanced education, something liberals, socialists and other activists, on and off-campus, sought to change. Indeed, for a variety of reasons, higher education generated unprecedented support from business leaders, economists, politicians and voters. Public and private investment in higher education, they argued, would provide the 'human capital' required to improve the country's standard of living, make it more competitive and help Western powers confront the stunning technological challenges of the Soviet Union, illustrated by the launch of the Sputnik satellite in 1957.[5] Politicians frequently referenced the impressive, and ominous, educational progress of the Soviets, and the urgency for the West to address it. The president of the University of Toronto wrote an article for *Maclean's Magazine* entitled, 'Universities must answer sputnik with higher standards' and in a subsequent speech, 'Sputnik and the humanities', he contended that universities played a key role in the Cold War 'struggle for men's minds'.[6] Population growth alone required substantial spending on higher education; raising participation rates, considered essential by educational advocates, would cost even more. Existing institutions would need to expand, and brand new universities had to be built. Canada's future prosperity and its identity as an enlightened country, proffering equality of opportunity, evidently, demanded it. The expansionist liberal state facilitated it.[7]

What form would higher education take in this extraordinary growth phase? In the four decades prior to the 1960s, university life in both French- and English-speaking Canada was relatively uniform and unchanging. In the former, within a decade, higher education in Quebec and New Brunswick was reshaped.[8] In the latter, students in arts and science enrolled in heavily prescribed bachelors programmes with few course options.[9] Arts courses were steeped in the intellectual traditions of the Western world. English students studied the works of Milton, Johnson, Shaw and Shakespeare. History, philosophy and political science classes traced the roots and development of European civilization, in which the study of Canada played a minor role, though the country's emergence within the British Commonwealth was duly noted. Students were encouraged to cultivate an appreciation of their 'cultural inheritance'. An overwhelmingly male faculty taught in a paternalistic and hierarchical atmosphere, which some students found intimidating, but to which almost all acquiesced. As a future president of the University of Toronto recalled, 'In the classrooms students expected and welcomed the [professors'] grand manner and authoritative stance; and if they mimicked or ridiculed their instructors, it was a form of flattery.'[10]

Some professors and academic administrators were dissatisfied with the traditional orientation of Canadian undergraduate education and used the opportunity to build new campuses as a vehicle for programmatic change. Murray G. Ross, the founding president of York University (1959–70) in Toronto, Ontario, held an EdD from Columbia University in New York, and served as the University of Toronto's academic vice-president from 1957 to 1959. A sociologist with interests in community development and leadership, he later published several works on the development and character of higher education.[11]

By the late 1950s Ross had grown impatient with the 'conservatism' and 'rigidness' of the University of Toronto, and was irked by the growing culture of specialization generally in North American universities. He believed students were not ready to specialize in first year,[12] and he was heavily influenced by the philosophy of general education, expressed most fully in the 1945 Harvard University Report, General Education in a Free Society; the Columbia University Contemporary Civilization programme, to which Ross had undoubtedly been exposed during his graduate school days in the late 1940s; and the 'New Plan', designed by Robert Hutchins for the University of Chicago in 1930 (where Ross also did graduate work). These programmes 'wrestled with the problem of how to sustain and nurture the value and content of Western learning in an age of fragmented and specialized knowledge', and to counteract students' 'early induction into a disciplinary major'.[13]

Ross also sought more innovative forms of pedagogy than he encountered at the University of Toronto. 'I was never devoted to the lecture method and always sought to find some way to encourage dialogue in the classroom.' In his own classes, he fostered student engagement and informal interaction, but found that in 'tradition-bound Toronto such methods smacked of progressivism.' His commitment to establishing, from the outset, a seminar-tutorial system at York arose, in part, from his critique of University of Toronto teaching practices.[14]

When York opened in 1959, it was obliged to offer the University of Toronto undergraduate curriculum for its first four years, after which it would become fully independent and free to forge its own academic direction. To prepare for this change, Ross initiated a curriculum design exercise that, ideally, would reflect his general education vision. He wanted to ensure that graduating students, whatever their areas of concentration or specialization, had the breadth of knowledge that would contribute to the education of the 'whole man' and equip them to address 'the modern problems of the world'.[15]

The original plan envisioned two years of general education before specialization, as well as an 'alternative' four-year general education programme which would include, for a full year, the study of an 'alien culture'.[16] Students would be required to take a common course in each of the interdisciplinary 'divisions' (vs 'departments') of humanities, social science and natural science. Owing to major disagreements among the faculty about the content of these core courses, and to concerns that students with a 'general education' degree would have difficulty being admitted to graduate school, the programme that actually emerged took a different form, one that preserved the principle of general education. While this outcome proved 'conventional' in most respects, York's interdisciplinary 'divisions', and its emphasis on the value of a general, liberal education earned it a degree of distinctiveness on the Canadian educational landscape.[17]

Ross's vision of the 'new' university included an Oxford-like college system, which he hoped would be an antidote to the burgeoning 'multiversity', described so vividly in 1963 by Clark Kerr, the president of the University of California.[18] Ross believed that York could reap the benefits of a large campus – diversity, prestige, public support – but serve individual students well if they were placed in colleges with no more than 1,000 members. There they would eat, sleep, socialize, meet faculty and ideally develop a 'sense of identity … even as the university grew'.[19] The college system that emerged (each with its own master) achieved only partial success. It worked best for residence students, who spent virtually all their time on campus, but much less well for commuter students, who formed the vast majority of the rapidly growing student population. Because classes and extra-curricular activities could be held anywhere on campus, these students' college identity was generally little more than an administrative designation. Over the following decades, successive university administrations tried to radically modify, and even eliminate, the costly college system, and always failed because its proponents (the masters, faculty fellows and residence students) mounted effective campaigns claiming that York would undermine its founding mission to create an alternative to the multiversity if the college system was abolished. Ross's plan, for better or worse, survived.[20]

Like Ross, Tom Symons, the founding president (1964–72) of Trent University in Peterborough, Ontario, revered small colleges, an attitude shaped by his experiences pursuing a BA at Trinity College, University of Toronto and an MA at Oxford's Oriel College (1951–53). In these environments, he recalled, 'Every student knew every other student. Every professor knew every one of those students. You could see your professors in their offices. You could stop them in the hall or have a coffee with them in the cafeteria. It was profoundly valuable.'[21] Back in Canada, where higher education enrolments were ballooning (and now working as the dean of Devonshire House at the University of Toronto), Symons grew increasingly concerned about the quality of the student experience. Universities, he contended, were suffering from 'academic elephantitis', and faculty and students were 'drifting apart'.[22] As president of Trent, he was determined to avoid this. He set out to build 'a small, residential national university'[23] where each student was expected to 'find a close direct contact with his teachers through the tutorial and seminar approach to instruction'.[24] To advise in course selection 'and other matters', faculty supervisors were assigned to every student, an unusual practice in Canada, which continued unchanged at Trent until 1987.[25] Located in the small urban centre of Peterborough (built on land provided by Canadian General Electric, one of the community's largest employers), and resolving (with provincial government support) to remain primarily a modestly sized residential college, Trent was better positioned than York to realize its president's ambitions, and students were drawn to the new campus by its intimacy and clear focus on undergraduate teaching. Financial and political challenges would soon tarnish elements of the founders' dream, but Trent did offer an alternative university experience for those looking to learn and live (and able to afford the residence fees) in a small academic community.

As a new university in western Canada, Simon Fraser, which opened in 1965 in Burnaby, British Columbia, followed a route to its founding that differed from that of York and Trent. The latter two institutions arose from the lobbying efforts of local educators, community groups, political advocates and business leaders. Ontario's government made clear its interest in expanding higher education capacity but had no elaborate plan to achieve this. Instead, it expected existing institutions to increase enrolments, and it responded favourably (and sometimes unfavourably) to local proposals for new universities. Simon Fraser, on the other hand, was largely a top-down initiative, mandated into existence by W. A. C. Bennett, the premier of British Columbia, and initially governed by his hand-picked designate, Chancellor Gordon Shrum, a former University of British Columbia (UBC) physicist, dean of graduate studies and subsequent co-chair of B. C. Hydro. John B. Macdonald, the president of UBC, had prepared a report that recognized the need for additional higher education facilities in the province, but recommended that the new institutions, the University of Victoria and Simon Fraser, have limited autonomy and be confined to teaching undergraduate students. However, advocates for the University of Victoria, which opened in 1963, objected to this restricted academic role and persuaded Bennett, on the eve of a provincial election, to award the new institution full university status, including the establishment of graduate studies, a standing also granted to Simon Fraser.[26]

At Bennett's insistence, Simon Fraser, which was authorized in 1963, opened in 1965 with an initial enrolment of 2,500 students, a remarkably compressed time span. Building the physical plant, as we shall see, was challenging enough. The administration – which consisted essentially of Shrum, president Patrick McTaggart-Cowan (Shrum's choice, approved by Bennett), and academic planner Ron Baker – had to quickly map a curriculum plan and recruit students. Shrum was hardly an educational visionary, but he had some definite ideas about what academic shape the new university should take, and he had the authority to implement them. Chief among them was a 'year-round university'. While there were examples of quarter or trimester systems in the United States, this approach was untried in Canada. Shrum and Baker believed that if established efficiently, a trimester programme would be economical, speed the time to graduation and attract students with diverse needs. Critics claimed that academic courses would be too compressed, that quality would suffer and that the credit accumulation system would lack the integrity of the traditional degree programme. It was subsequently discovered that because students needed to find jobs, they were disinclined to study during the summer semester. Still, the programme provided flexibility for those who did enrol, including mature students whom Simon Fraser, from the outset, deliberately sought to attract. Baker himself who had left school at fifteen attended university as a post-war veteran and cherished the extended educational opportunity that it afforded him. York University similarly introduced a programme for part time, adult students at Atkinson College in 1961, and, like Simon Fraser, was credited with broadening accessibility to post-secondary education beyond the traditional university-aged demographic.[27]

Simon Fraser's academic structure also departed from the Canadian norm. Shrum chose to govern the academy without deans or traditional faculties such as Arts and

Science. His experience at UBC persuaded him that 'deans were people who got in your way' because they were academic empire builders. He favoured 'government by department heads' who would report 'directly to the president'.[28] The one exception to this model was the establishment of a Faculty of Education, which from the outset had its own dean. A large proportion of Simon Fraser's students were expected to pursue teaching careers, and Shrum was determined to create a form of teacher education different from that at UBC which he believed was preoccupied with teaching methods and academic psychology, and which devoted too little time to subject content. He wanted a strong dean to drive a sophisticated, alternative programme that, among other innovations, included more practice teaching time in the student's schedule than that at UBC. For several years, the Faculty of Education thrived under the deanship of Archie MacKinnon, a proponent of integrated learning who attracted an eclectic group of professors, helping make Education a 'happening place'.[29]

Committed to a curriculum with strong liberal arts and science foundation, Simon Fraser's planners were drawn as well to interdisciplinarity. Some departments, notably (and infamously), political science, sociology and anthropology (PSA), were constituted as a single unit, and professors were encouraged to teach outside their disciplinary specialties. The university stressed the value of tutorials and term work over examinations, and change was stimulated by a youthful faculty cohort which relished the opportunity to build new university departments from the ground up. Some shared Cardinal John Henry Newman's mid-nineteenth-century faith in a broad undergraduate liberal education as the core of higher education and his aversion to university-based professional training. Others, who had taught in large American universities, soured on the emerging multiversity, with its large classes, major research centres and heavy corporate funding. They were drawn to the promise of greater intimacy and inventiveness at Simon Fraser which had, early on, cultivated an image of 'innovation, academic excellence and openness'.[30]

As we shall see, all three universities – York, Trent and Simon Fraser – were unable to sustain the idealism and experimentation of their early days. Demographic and financial pressures, student unrest, faculty-administration conflict and public criticism pervaded these and other campuses in the latter half of the 1960s. University administrators who considered themselves educational reformers were suddenly on the defensive – attacked by radical students for being too timid and by bemused citizens for apparently losing control of their campuses. Still, the academic changes they did introduce in the early 1960s, prior to the student movement, should not be overlooked in the wake of the confrontations and cynicism that followed. This included the design and architecture of the new institutions, propelled, in interesting ways, by utopian academic concepts.

II

New universities required new buildings and the construction of the campuses preoccupied the institutions' founders. They engaged this task at a time when modern architecture increasingly dominated the landscape. Walter Gropius, Ludwig Mies van der Rohe, Frank Lloyd Wright and Le Corbusier (born Charles-Édouard Jeanneret) captivated observers in Europe and North America in the first half of the twentieth century with their inventive use of steel, glass and reinforced concrete, building materials deemed most efficient and adaptable to newly industrialized, heavily populated urban centres. Purveyors of what became known as 'brutalism', these architects and their followers turned away from the 'traditional hierarchy' and 'formalism' of the past which featured grandiose churches and palaces catering to elite tastes and inconsistent with the modern spirit of an 'open and dynamic world'.[31]

The post-war construction boom in schools and universities, in addition to that in housing, shopping malls and office buildings, offered unprecedented opportunities for designers and architects to shape living and learning spaces. Inspired by the trailblazers and groundbreakers of modern architecture, some were intensely idealistic about the transformative possibilities of their work. 'The buildings emphasized material simplicity and secular anonymity. They spoke of the ethos of managerial grandness and implied a transnational utopianism.' Because universities required multi-purpose, 'large scale' facilities, brutalism was 'particularly popular' on campuses, both old and new. In the United States, Kane Hall at the University of Washington in Seattle, the Stratton Center at the Massachusetts Institute of Technology and the Modern Art and Architecture Building at Yale University all embraced this style.[32] Canadian universities were equally affected by these architectural trends and the idealism that infused them, though the material results were far from uniform, or universally acclaimed. York, Simon Fraser and Trent all began with blank slates – empty landscapes – upon which new educational forms would be forged, preferably in ways that reflected the visions of the institutions' founders and leaders. Like the ideas they espoused, the buildings they constructed were intended to be agents of change.

Arguably, the most direct descendants of the brutalist style were York's new buildings, the first of which arose on the site of what later became Glendon College. York Hall, a three-story complex, completed 'quickly and inexpensively' in 1961, housed classrooms, offices, laboratories and eating facilities in a 'squat, functional, and architecturally undistinguished' manner.[33] More elaborate thinking preceded the building of the university's expanded site on the northern edge of suburban Toronto. A consortium of three Toronto architectural firms sought to embody York's academic principles in 'forms appropriate to modern conditions and techniques'.[34] Murray Ross revered the collegiate atmosphere of European universities where students from various disciplines would intermingle in accessible open spaces, undisturbed by cars. A ring road circling the York site kept automobiles and parking lots well outside the campus core, requiring staff and students (on an overwhelmingly commuter campus), to trek long distances in cold wintery weather.

The main university-wide gathering site, designed (and ultimately spurned) for convocation ceremonies and other celebrations, was a paved outdoor square adjacent to the Humanities and Social Sciences Building (later named the Ross Building) completed in 1968. The nine-story grey, concrete structure dominated and defined the campus visually, and for many, in an entirely negative way. Appalled by its barren, institutional form, the student newspaper denounced it as Orwellian, and (recalling the novel, *1984*) dubbed it, sarcastically, 'The Ministry of Love'.[35] Harry Arthurs,

York president from 1985 to 1992, described the campus's physical state, prior to the completion of new building projects in the 1990s, as 'isolated, ugly, windswept and inefficient', dominated 'unpleasantly [by] a series of giant parking lots'.[36]

New brutalist architecture was not confined to York. Campuses across the country constructed immense, functional facilities that, at best, elicited mixed responses and increasingly critical ones as time passed. The University of Toronto, for example, built Canada's best-equipped research library, the design of which aroused fierce censure from architects and users alike. Considered 'out of proportion to the neighbourhood' and a reflection of the 'sterility of our restless, anomic society', the Robarts Library was christened 'Fort Book' by student critics who deplored its bulk, greyish hue and seeming impenetrability.[37]

Such bleakness, however, was not the entire story of the concrete building revolution. Some architects, including those hired by Simon Fraser and Trent Universities, designed facilities that were universally applauded for their functionally imaginative and environmentally sensitive qualities. Chancellor Gordon Shrum, who considered himself something of an architecture authority, was tasked with creating a new facility that could be completed in less than two years.[38] The provincial government agreed with Shrum that Simon Fraser should be located on Burnaby Mountain, a stunning venue east of Vancouver, overlooking the Burrard Inlet, the Fraser River and Vancouver Harbour. The winner of the building contract competition was the Vancouver firm of Erickson/Massey, headed by Arthur Erickson, an emerging presence on the Canadian architectural scene. Erickson, who was influenced by the modernists – including Gropius, the Bauhaus school, van der Rohe and Le Corbusier – shared Shrum's view that 'architecture determines the nature, the inner philosophy of the university'.[39] That philosophy opposed intellectual silos and favoured interdisciplinarity. It required 'open-air spaces' (inspired by the magnificent Plaza Mayor in Salamanca, Spain) that would encourage free expression and campus unity. In Erickson's words, Simon Fraser 'casts the contemporary university as an appropriate Acropolis for our time'.[40] And given the spectacular location of the university, this intellectual marketplace should cultivate harmony between humanity and nature – the university would be built along the mountain ridge and extend down the slopes, allowing the 'surrounding landscape to be left undisturbed'.[41] Completed largely on time, the new facility evoked international acclaim and earned Erickson 'a degree of fame unparalleled by a Canadian architect'.[42]

Erickson was subsequently hired to design the University of Lethbridge in Alberta, opened in 1971, which he deemed 'his supreme achievement', a sentiment shared by Venetian architect Giuseppe Mazzariol who praised Erickson's success at fostering a 'dialogue between history and utopia'.[43] Notwithstanding such plaudits, Erickson's campus creations were not unqualified successes. Both Simon Fraser and Lethbridge had construction flaws leading to serious water damage in the former case and costly structural reinforcements in the latter, and both suffered significant cost overruns. Lethbridge, indeed, sued Erickson (unsuccessfully) and denied him the contract for the second phase of the building. He was 'sickened' by his successor's work, which he believed undermined his naturalist vision.[44]

While Erickson was making his mark in western Canada, Ron Thom, a Vancouver architect, was enhancing his reputation through his work in eastern

Canada at Trent University. As Trent's master architectural planner, he conceived of structures that, ideally, would fulfil Tom Symons's vision of an intimate, residential-based liberal-arts college offering the type of 'communal' experience largely 'lost or ignored in the bustle of modern education'.[45] Thom shared Symons's enthusiasm for the Oxford–Cambridge aura and returned from a visit persuaded that the two 'venerable institutions should serve as a template for Trent'. He sought to 'create a ... live community on a human scale, with pedestrian paths, cloisters, quadrangles, clusters of buildings ... and a riverfront that [could] be used for sports or walking'.[46] Perched on the banks of the Otonabee River, Champlain College was the most 'fully integrated' of the new buildings, featuring a 'dominating fireplace and a low wooden trellis ... like many of [Frank Lloyd] Wright's cozy, intimate interiors'.[47] Thom used 'rubble-aggregate concrete' in a way that was designed to 'offset the trend in North America during the 1960s for dull, prosaic, institutional and commercial buildings made of cheap materials'.[48] According to one architectural reviewer, Thom's creation would 'remain a watershed in the history of Canadian modern architecture'.[49] Arthur Erickson agreed. He viewed it as 'an inspired plan' which 'gained that elusive of all qualities – an ambience towards which everyone responds'.[50]

Unfortunately, this inspiration would not survive the next phase of building. While requiring additional space, Trent faced serious financial challenges and chose to meet its needs as inexpensively as possible. Otonabee College (designed by a different architect and opened in 1973) was a more conventional and less creative brutalist-type structure. Constructed with grey cinder blocks, and topped by a red metal roof that resembled the style of rural Ontario barns, it was less welcoming than the first series of buildings. Impudent students called it the 'college that ought not to be'.[51]

Like the activist students in the late 1960s who were intent on changing the world, the designers and builders of new university facilities in Canada had ideals. They were obliged to meet the utilitarian expectations of state funders – burgeoning enrolment demands had to be satisfied – but they were determined to reimagine both the content of higher learning and the venues in which it would be offered. Among the new buildings were some inventive and enduring legacy projects which co-existed, sometimes side by side, with unloved structures that merely endured.

III

Liberal administrators, who viewed themselves on the leading edge of higher educational reform, did not anticipate the student movement of the 1960s. In Canada, pranks and initiation rituals gone awry had long been part of extra-curricular rites of passage, but as classes got underway most students behaved compliantly and only occasionally challenged authority. Even political activists in the 1930s, who campaigned for student aid and world peace, did so largely in respectful and orderly ways.[52]

In 1964, the free speech movement at the Berkeley campus of the University of California signalled the changed politics and tactics of the times. The right of students to speak and write freely, protest, assemble and control their own organizations fuelled many subsequent campus battles. Militant politics and revolutionary rhetoric also filled the airwaves and television screens. For these activists, reforming campus governance and ending the war in Vietnam were insufficient political objectives. They opposed capitalism, and the role universities played in sustaining it, in part through the institutions' corporate-dominated boards of governors. Inspired by neo-Marxist (New Left) thought, they sought alliances with working-class movements at home, and anti-imperialist struggles abroad.[53] The student movement, then, had both liberal-reformist and radical-revolutionary propensities. Besieged university administrators sought to defuse unrest and maintain order by responding to and sometimes directly negotiating with more moderate activists, and in the long run, this strategy worked.[54] Though in ways that went beyond what presidents like Murray Ross and Tom Symons had in mind, Canadian university governance, in the wake of the 1960s, was indeed reformed.

In pursuing his vision of the new university, for example, Ross did not take the prospect of student unrest into account. He knew that he needed to persuade board members and faculty of the wisdom of his plan, but students could be expected to bend to the will of university authorities as they always had. The university, he wrote, was not a

> full democracy, and indeed, I think it foolish to conceive of it as such ... I have always favoured listening to the student voice, indeed for some degree of student participation in the government of the university, but I agree with Richard Lyman, a president of Stanford University, when he said, '[t]he university is not a democratically organized institution and cannot become one without destroying itself as a university.'[55]

That said, change was possible in the liberal university. York student activists demanded and ultimately achieved a number of reforms. Following sustained protest, Modes of Reasoning, a compulsory first-year logic course taught on closed-circuit television to the vast majority of students, was turned into an elective. Parking lots, washrooms and high table dining facilities, once reserved exclusively for faculty, were made available to all York members. Students obtained ten seats on the university senate and a significant presence on the search committee for a new university president. Paternalistic residence rules at York and elsewhere were relaxed. On the other hand, more radical proposals to ban corporate recruitment on campus, abolish the use of grades and give students 'parity' on faculty hiring committees were rebuffed; indeed, there was little evidence of significant student support campus-wide for these demands.[56]

York's student politics were positively sanguine compared to those at Simon Fraser, dubbed by later observers as 'Berkeley north'.[57] Following its accelerated building schedule, Simon Fraser opened in 1965 with great expectations and an unfinished infrastructure. Having now found their political voices, students complained vehemently about inadequate cafeteria space, the quality of food, unfinished residences and unpaved parking lots. Of particular concern was the location of a new gas station, which Shell built after promising to help fund the construction of campus residence.

Anti-corporate activists regarded the arrangement as a nefarious example of business influence over academic life.[58]

More significant were conflicts involving the PSA department, many of whose youthful faculty were themselves supportive of New Left political objectives. When five graduate teaching assistants (four from PSA) were dismissed from their positions in 1967 for organizing in support of a local high school student who had been disciplined for mocking a teacher, PSA students threatened to strike, rousing strong campus support. The administration retreated and rehired the teaching assistants.[59] The PSA was at the centre of continuing confrontations over the issue of university and departmental governance. Radical students occupied administrative offices in November 1968, leading to their eviction by police, the arrest of 114 occupiers and the placement of the PSA department under trusteeship. With the status of the department unresolved, militant PSA students, supported by nine faculty members, went on strike in September 1969. The professors, whose actions were illegal, were suspended and five ultimately lost their jobs. After nearly six weeks, the strike 'collapsed', though not before the Board of Governors had dismissed President McTaggart-Cowan and the Canadian Association of University Teachers censured the university for its 'undemocratic' practices.[60]

By the early 1970s, a sense of academic normality and political stability had been restored to Simon Fraser, though it retained its reputation for unconventionality. Historians attribute the period of turmoil, which, in its intensity, was unusual in Canadian universities, to some unique aspects of the institution's early history. Like York, it attracted faculty and students who were looking for alternatives to the perceived traditionalism and elitism of the University of British Columbia and the University of Toronto. Coinciding with the (post-Berkeley) eruption of counter-cultural and political activism on campuses across the continent, and governed initially by a paternalistic chancellor and board anxious to meet government-imposed deadlines, the newly founded university was unprepared to manage the internal criticism and active resistance which were buoyed by New Left faculty like those in PSA. They, in turn, attracted radical students to the campus, though even at the height of the tensions, 'the politically active were a tiny minority' of the student body.[61] Activists were most effective when they campaigned for moderate reforms and unsuccessful in their efforts to shut the university down, overturn the governing structure and forge enduring political links with non-university workers and unions. Enhancing the student voice in university governance, which was endorsed by a national administrator/faculty report in 1966 on this subject,[62] was achievable. Simon Fraser was the first university in Canada to admit students to the senate. The student council created the office of ombudsman to represent students with grievances against the administration, an innovation which became permanent at Simon Fraser. In loco parentis practices, including strict conduct rules in female residences, ended. In 1969, a moderate slate of student council candidates prevailed over a radical group as 'many bestirred themselves to vote because they had enough of campus disruption'.[63] Incremental academic reform encouraged, or at least abided, by liberal university leaders in the 1960s, especially at newly established institutions, ultimately became the norm at Simon Fraser and elsewhere.

IV

The tumultuous student movement, in new and existing universities alike, blew through North American higher education between 1965 and 1970 like an unforeseen tempest. While student activism was by no means invented in that period, unique circumstances conspired to intensify its impact and heighten its profile. The unprecedented number of students in the wake of the post-war baby boom surge meant that political activists, while comprising a minority of the campus population, could, by filling a meeting hall or occupying an administrator's office, exercise power, even more so when radicals and moderates united. Television, which revelled in images of conflict, instantly spread the news of unrest from one university to the other. The anti-Vietnam War Movement pervaded American campus life and spilled into Canada in the late 1960s. Young people, in general, were less inclined than in the past to accept unquestioningly the authority of parents, teachers and university administrators. They insisted on greater autonomy, sexual freedom, the right to freely assemble for social or political reasons and greater influence over educational policies that affected them. University administrations were clearly being tested in these challenging times.

But taking a long historical view, the disorder on campus was episodic and temporary. Change had been in the air before the student movement erupted, particularly in the case of new universities, and it would continue in its wake. Restless administrators like Murray Ross, Tom Symons and even Gordon Shrum promoted undergraduate curricular reform, including general education, interdisciplinarity, adult learning and trimester programming. They approved campus designs that were intended to be student friendly, intimate and humane – though more than once, these lofty plans foundered in the face of academic politics, architectural romanticism and financial exigencies.[64]

Interdisciplinarity, for example, a foundational academic value at York and Simon Fraser, failed to realize its original conception, though it remained part of these institutions' discourse and mythologies. Not only did the PSA department at Simon Fraser break down into the separate departments of Political Science and Sociology/Anthropology, but Economics and Commerce also divorced: Economics remained in Arts, and Commerce became a core component of a new professionally oriented Faculty of Business Administration. There were deep intellectual and pedagogical divisions within the Department of Modern Languages (which changed its name to the Department of Languages, Literature and Linguistics (DLLL) in 1978), among language specialists, linguistic theorists and literature scholars. DLLL disbanded in 1989, morphing into the three departments of Linguistics, French, and Spanish and Latin American Studies.[65]

Simon Fraser clung to the idea of interdisciplinary by establishing a Faculty of Interdisciplinary Studies in the early 1970s, a catchment which included computing science, kinesiology, communications, fine and performing arts, and criminology. Rather than countering specialized study, 'these [and other] new disciplines took Simon Fraser further along the path of specialization and professional education. It meant more departments within the university, not fewer, more divisions of the curriculum and more programmes with an employment focus'.[66]

York University followed a similar trajectory. The humanities and social science 'divisions' became 'departments' in 2008.[67] The original nomenclature, although central to Murray Ross's vision of interdisciplinary general education (he associated 'departments' with specialized disciplines), had always been a source of confusion, particularly as the divisions eventually developed their own interdisciplinary specializations such as communications studies in social science and religious studies in humanities. York also created Canada's first faculty of environmental studies in 1968 and pioneered the teaching of women's studies in the early 1970s.[68] Interdisciplinarity remained an important part of York's identity and branding – a marketing campaign in the early 2000s promoted the institution as the 'interdisciplinary university'[69] – but the concept, in practice, at York and elsewhere, including Trent,[70] had evolved into more of a 'multidisciplinary' model.[71] Thematic studies attracted students, stimulated new research, scholarly journals and graduate streams, and became specialized programmes in all of the conventional university ways. They also attracted strong criticism from traditional disciplinary scholars who condemned the 'trendy', 'shallow', and politicized nature of thematic programmes.[72] How, they argued, could academics, particularly undergraduate students, sufficiently master multiple disciplines? Acquiring deep knowledge of a single discipline was challenging enough. Notwithstanding such objections, these new teaching and research currents flowed widely, representing a significant diversification of academic study. It was driven by social, political and cultural interests that had been spawned in the universities of the 1960s.

There were limits to academic reform. Boards of governors retained strong business connections, much to the dismay of radical students. Neither faculty nor administrations supported student parity on hiring and important curriculum committees, nor were tuition fees abolished. Nevertheless, piecemeal change became the norm in the decades that followed. Women's issues, which had been on the margins of the 1960s student movement, assumed major significance in university policy discourse and practice by the 1980s.[73] Curricula, as noted above, became far more diverse and academic options multiplied. Student service initiatives, including programmes for those with learning disabilities, expanded.[74] On many campuses, relations were forged with local neighbourhoods, including Aboriginal communities. Bridging programmes, facilitating access for students from lower income families, emerged. To the dismay of militant student activists, the revolution they envisioned in the 1960s was thwarted, but campus life in Canada was significantly altered, and administrators with ideas had played a significant role in this evolution.

Notes

1   Canadian examples of this genre include: Cyril Levitt, *Children of Privilege: Student Revolt in the Sixties* (Toronto: University of Toronto Press, 1984); Bryan D. Palmer, *Canada's 1960's: The Ironies of Identity in a Rebellious Era* (Toronto: University of Toronto Press, 2009); *The Sixties in Canada: A Turbulent and Creative Decade*, ed. M. Athena Palaeologu (Montreal: Black Rose Books, 2009); *Debating Dissent: Canada and the Sixties*, ed. Lara Campbell, Dominique Marshall and Greg S. Kealey (Toronto:

University of Toronto Press, 2012); Ian Milligan, *Rebel Youth: 1960s Labour Unrest, Young Workers and New Leftists in English Canada* (Vancouver: University of British Columbia Press, 2014); J. Paul Grayson, 'The "Feminine Mystique" and Problems of a Cohort of Female Canadian University Students in the Early 1960s', *The Sixties*, 8 (2015), 50–74.

2 W. B. W. Martin and A. J. Macdonell, *Canadian Education: A Sociological Analysis* (Scarborough, Ontario: Prentice-Hall, 1978), 22.

3 Paul Axelrod, *Making a Middle Class: Student Life in English Canada during the Thirties* (Montreal and Kingston: McGill-Queen's University Press, 1990), ch. 2.

4 Peter Neary, 'Canadian Universities and Canadian Veterans of World War II', in *The Veterans' Charter and Post-World War II Canada*, ed. Peter Neary and J. L. Granatstein (Montreal and Kingston: McGill-Queen's University Press, 1998), 110–48.

5 Paul Axelrod, *Scholars and Dollars: Politics, Economics and the Universities of Ontario, 1945–1980* (Toronto: University of Toronto Press, 1982), ch. 1.

6 Cited in Claude Bissell, *Halfway up Parnassus: A Personal Account of the University of Toronto, 1932–1971* (Toronto: University of Toronto Press, 1971), 49.

7 James J. Rice and Michael J. Prince, *Changing Politics of Canadian Social Policy*, 2nd edn. (Toronto: University of Toronto Press, 2013), ch. 3.

8 The Université du Québec, with nine branches throughout the province, opened in 1968, whilst in 1963, the province of New Brunswick established the francophone institution, the Université de Moncton. See Lucia Ferretti, *L'Université en Reseau* (Sainte-Foy: Presses de l'Université du Quebec, 1994); Joel Belliveau, 'Acadian New Brunswick's Ambivalent Leap into the Canadian Liberal Order', in *Creating Post-War Canada: Community, Diversity and Dissent, 1945–75*, ed. Magda Fahrni and Robert Rutherdale (Vancouver: UBC Press, 2008), 61–88.

9 Robin Harris, *A History of Higher Education in Canada, 1663–1960* (Toronto: University of Toronto Press, 1976), 502–3.

10 Bissell, *Halfway up Parnassus*, 15; Axelrod, *Making a Middle Class*, 93–7.

11 These included: *The New University* (Toronto: University of Toronto Press, 1961); *New Universities in the Modern World* (London: Macmillan, 1966); *The University: the Anatomy of Academe* (New York: McGraw Hill, 1976).

12 Murray G. Ross, *The Way Must Be Tried: Memoirs of a University Man* (Toronto: Stoddart Publishers, 1992), 75, 87.

13 Michael Bisesi, 'Historical Developments in American Undergraduate Education: General Education and the Core Curriculum', *British Journal of Educational Studies* 30 (1982), 202–3.

14 Ross, *The Way Must Be Tried*, 58–9.

15 Cited in Michiel Horn, *York University: The Way Must Be Tried* (Montreal and Kingston: McGill-Queen's University Press, 2008), 30.

16 Murray G. Ross, 'York University', in *New Universities in the Modern World*, ed. Ross, 72.

17 Horn, *York University: The Way Must Be Tried*, 52–3.

18 Clark Kerr, *The Uses of the University* (London: Oxford University Press, 1963), esp. ch. 1.

19 Horn, *York University: The Way Must Be Tried*, 64; Douglas V. Verney, 'The Government and Politics of a Developing University: A Canadian Experience', *The Review of Politics*, 31 (1969), 300.

20 In 2000, a small minority – some 5% of York students – lived in college residences. As President Lorna Marsden (1997–2007) noted, 'The college model created by the founding president for a much smaller institution was a poor fit with a very large commuter university': Lorna R. Marsden, 'Years of Transition, 1997–2007', in *Leading the Modern University: York University's Presidents on Continuity and Change, 1974–2014*, ed. Lorna R. Marsden (Toronto: University of Toronto Press, 2016), 125.
21 D'arcy Jenish, *Trent University: Celebrating 50 Years of Excellence* (Toronto: ECW Press, 2014), 36.
22 Ibid., 36.
23 Denis Smith, 'Tom Symons and the Founding of Trent University', in *Tom Symons, a Canadian Life*, ed. Ralph Heintzman (Ottawa: University of Ottawa Press, 2011), 45.
24 Jamie Benidickson, 'Classroom, Conversations, and Communities: Tom Symon's Perspectives on Learning', in ibid., 66.
25 Jenish, *Trent University*, 40, 78.
26 Hugh Johnston, *Radical Campus: Making Simon Fraser University* (Vancouver and Toronto: Douglas and McIntyre, 2005), 29–30. In Canadian universities, the chancellorship is normally an honorific, ceremonial leadership position. Premier Bennett evidently was unaware of this, which suited Shrum. As well as picking the president, Shrum selected the first members of the Board of Governors. As Johnston notes, 'The freedom that Shrum enjoyed at SFU was without precedent in Canadian higher education', 12.
27 Johnston, *Radical Campus*, 69–74, 110; Horn, *York University: The Way Must Be Tried*, 54–7; *Toronto Daily Star*, 6 May 1967, 2; James Laxer, 'Farewell to Atkinson College', rabble.ca, 30 June 2009, http://rabble.ca/blogs/bloggers/james-laxer/2009/06/farewell-atkinson-college (accessed 4 January 2017).
28 Johnston, *Radical Campus*, 82–3.
29 Ibid.,101–5.
30 Ibid., 120.
31 Christian Norberg-Schulz, *Principles of Modern Architecture* (London: Andreas Papadakis Publisher, 2000), 18.
32 Francis D. K. Ching, Mark Jarzombek and Vikramadiya Prakash, *A Global History of Architecture*, 2nd edn. (Hoboken, New Jersey: John Wiley & Sons, 2011), 777; Beverly Russell, *Architecture and Design, 1970–1990s: New Ideas in America* (New York: Harry N. Abrams Publishers, 1989), 26.
33 Michiel Horn, *York University: The Way Must Be Tried*, 30.
34 Cited in John Sewell, *The Shape of the City: Toronto Struggles with Modern Planning* (Toronto: University of Toronto Press, 1993), 130.
35 Horn, *York University: The Way Must Be Tried*, 115. See also Christopher Armstrong, *Making Modern Toronto: Architecture and Design, 1895–1975* (Montreal and Kingston: McGill-Queen's University Press, 2014), 323.
36 Harry Arthurs, '"The Economy Is the Secret Police of Our Desires": York University, 1985–92', in *Leading the Modern University*, ed. Marsden, 70.
37 Armstrong, *Making Modern Toronto*, 331.
38 Johnston, *Radical Campus: Simon Fraser University*, 46.
39 Ibid., 43.
40 Cited in Harold Kalman, *A History of Canadian Architecture*, 2 vols. (Toronto: University of Toronto Press, 1994), ii, 793. See also Arthur Erickson, *The Architecture of Arthur Erickson* (Vancouver and Toronto: Douglas & McIntyre, 1988), 34.

41 Jonathan Coulson, Paul Roberts and Isabelle Taylor, *University Planning and Architecture: The Search for Perfection* (New York: Routledge, 2011), 102.
42 Ibid., 104.
43 David Stouck, *Arthur Erickson: an architect's life* (Madeira Park, British Columbia: Douglas & McIntyre, 2012), 239–40.
44 Ibid., 240.
45 Jenish, *Trent University*, 40.
46 Ibid., 41.
47 *Globe and Mail*, 20 May 2005, G15.
48 Lisa Rochon, 'Trent University', in *Concrete Toronto: A Guide to Concrete Architecture From the Fifties to the Seventies*, ed. Michael McClelland and Graeme Stewart (Toronto: Coach House Books, 2013), 274.
49 *Globe and Mail*, 20 May 2005, G15.
50 Jenish, *Trent University*, 47. Thom's other major contribution to university architecture was Massey College at the University of Toronto (1963). Though considered by some early critics to be ostentatious, it was eventually 'much loved' and 'acclaimed'. See Armstrong, *Making Modern Toronto*, 245, and Martin L. Friedland, *The University of Toronto: a History* (Toronto: University of Toronto Press, 2004), 463.
51 Jenish, *Trent University*, 138.
52 Keith Walden, 'Hazes, Hustles, Scraps and Stunts: Student Initiations at the University of Toronto, 1880–1925', in *Youth, University, and Canadian Society: Essays in the Social History of Higher Education*, ed. Paul Axelrod and John G. Reid (Montreal and Kingston: McGill-Queen's University Press, 1989), 94–125; Axelrod, *Making a Middle Class*, chs 5–6.
53 Milligan, *Rebel Youth*, 1–7; Palmer, *Canada's 1960s*, 253–6.
54 The president of the University of Toronto in the 1960s employed this approach. See Bissell, *Halfway Up Parnassus*, 130–1.
55 Ross, *The Way Must Be Tried*, 74.
56 Horn, *York University*, 110–11; Ross, *The Way Must Be Tried*, 100–1.
57 Johnston, *Radical Campus*, 114.
58 Doug Owram, *Born at the Right Time: A History of the Baby Boom Generation* (Toronto: University of Toronto Press, 1996), 44; Johnston, *Radical Campus*, 257–60.
59 Johnston, *Radical Campus*, 264; John Cleveland, '"Berkeley North": Why Simon Fraser Had the Strongest 1960s Student Power Movement', in *The Sixties in Canada*, ed. Palaeologu, 213–16.
60 Johnston, *Radical Campus*, 267, 283, 293–4.
61 Ibid., 152.
62 Sir James Duff and Robert O. Berdahl, *University Government in Canada: Report of a Commission Sponsored by the Canadian Association of University Teachers and the Association of Universities and Colleges of Canada* (Toronto: University of Toronto Press, 1966).
63 Ibid., 158. On the end of *in loco parentis*, see Catherine Gidney, *Tending the Student Body: Youth, Health and the Modern University* (Toronto: University of Toronto Press, 2015).
64 In the wake of economic disruptions caused by the 1973 international oil crisis, public spending was restrained in Canada, and universities struggled throughout the decade to cover their growing costs. See Paul Axelrod, *Scholars and Dollars*, ch. 7.
65 Johnston, *Radical Campus*, 225–33.
66 Ibid., 241.

67 Senate of York University (Toronto), Minutes of the Meeting of 24 April 2008, 2 (MS 2017-023/002, York University Archives).
68 Horn, *York University: The Way Must Be Tried*, 90, 197. See also Darryl Reed and Richard Wellen (ed.), *Social Science@50: A Half-Century of Critical Interdisciplinary Teaching and Research at York University* (Toronto: Department of Social Science, York University, 2013).
69 Lindsay Carrocci Bolan and Daniel J. Robinson, '"Part of the University Lexicon": Marketing and Ontario Universities, 1990–2013', *Canadian Journal of Communications* 38 (2013), 573.
70 Trent shared the enthusiasm for interdisciplinary studies. Its courses in Indian-Eskimo studies (1969) led to a programme in native studies, renamed indigenous studies, while comparative development studies (1976) became international development studies. Tom Symons strongly promoted Canadian studies (1972). Other initiatives included environmental and resource studies (1975), and Gender and women's studies (1987): Jenesh, *Trent University*, 103–21.
71 In his discussion of recent academic developments, Mamdouh Shoukri, York's current president (2007–2017), uses the terms 'interdisciplinary' and 'multidisciplinary' interchangeably. Mamdouh Shoukri, 'This Is Our Time: Into a New Era at York University, 2007–14', in *Leading the Modern University*, ed. Marsden, 188.
72 David J. Bercuson, Robert Bothwell, J. L. Granatstein, *The Great Brain Robbery: Canada's Universities on the Road to Ruin* (Toronto: McClelland and Stewart, 1984), 136–7. Famously, Alan Bloom's *The Closing of the American Mind* (New York: Simon and Schuster, 1987) critically explored curriculum reform and cultural change in the United States.
73 See *Minds of Our Own: Inventing Women's Scholarship in Canada and Quebec, 1966–1976*, ed. Wendy Robbins et al. (Waterloo, Ontario: Wilfrid Laurier University Press, 2008).
74 *Achieving Student Success: Effective Student Services in Canadian Higher Education*, ed. Donna Hardy Cox and C. Carney Strange (Montreal and Kingston: McGill-Queen's University Press, 2010).

17

From progressive pedagogy to 'capitalist fodder': The new universities in Australia

by Hannah Forsyth

In the 1960s, 'the cultured confidence of the ivory tower' gave way to 'the agony of introspection', according to Peter Karmel, vice-chancellor of Adelaide's brand-new Flinders University in 1969. Higher education in Australia was subject to a radical re-evaluation against, as Karmel labelled it, 'the permissive society'.[1] Peter Karmel, economist and chair of the Commonwealth Tertiary Education Commission, was respected across the political spectrum as a guru of Australian higher education policy. He wrote submissions for every Australian review of higher education from the first (the Murray Review of 1957) to the most recent (the Bradley Review of 2008), just prior to his death.[2] Bridging establishment and progressive aims, in 1969, Karmel was optimistic about the effect of change on higher education. He hoped that including new ideas in new universities would prevent what some feared – or hoped – was a revolution in higher education. He believed that by welcoming some of the progressive pedagogies that the growing cohort of radical staff and students advocated, they would bolster a university focused on producing graduates for the labour market, that was at that time experiencing important shifts. This was at odds with the sentiments and politics underpinning the more radical reforms that advocates of progressive pedagogies hoped would turn the values of the capitalist university upside-down. Ironically, by embracing some of the changes radicals proposed, I argue, universities helped reinforce the role of Australian higher education in producing what post-war reconstruction planner H. C. ('Nugget') Coombs called 'capitalist fodder'.[3]

The 1960s reforms were in part a result of the massive growth of the universities. As Karmel pointed out, student enrolments in higher education in Australia in 1951 were 32,000; by 1969, this now numbered 105,000. In that same timeframe, around 5,000 new lecturers were appointed, almost 2.5 times the number employed in 1951. Twelve per cent of young Australians were now seeking entry to university, double that of the already-growing post-war reconstruction phase of student enrolment. The system was booming.[4]

By the mid-1960s, student demand placed new pressure on institutions. At that time, these consisted of nine universities. The original six were built in the state capitals of Sydney, Melbourne, Adelaide, Perth, Brisbane and Hobart. After the Second World War, two technology-focused universities were planned for Sydney and Melbourne, with a truly utopian vision to be found at the federally funded Australian National University (ANU) focused on research, established in the national capital, Canberra. These universities, which in 1957 were intended to hold no more than 2,000 students each, were bursting at the seams within a decade. In Victoria in 1966, 2,000 students who successfully passed the matriculation exam were nevertheless denied admission to any university. Community pressure for increased enrolment intakes intensified.[5]

Australian state governments built several new universities between 1958 and 1974. Suburban universities sought greenfield sites, as at Monash in Melbourne in 1958, Flinders in Adelaide in 1966, La Trobe and Macquarie in Melbourne and Sydney in 1967, Griffith in Brisbane in 1971 and in Western Australia, Murdoch University in 1974. Demand in rural Australia established James Cook University in far north Queensland in 1970 and Deakin in regional Victoria in 1974. University colleges in regional NSW became the standalone universities of Newcastle in 1964 and Wollongong in 1975. In this time too, colleges of advanced education were established and long-established schools of mines and urban technical colleges were re-invented as degree-awarding Institutes of Technology.[6]

This expansion built on the utopian vision that post-war reconstruction brought to the ANU, founded in Canberra in 1946. With their efforts focused on this institution, key leaders of post-war reconstruction had a framework in which they could reconsider the place of universities for Australian society and for its economic future.[7] Economists such as Nugget Coombs joined with educational leaders, including Eric Ashby, to identify the ways that developments in science, technology and the social sciences might offer innovative solutions to the economic, political, diplomatic and even medical problems confronting the world, believing that scholars were able to see 'from Australia' a unique 'view of the universe'.[8] Ashby, who became Sydney's professor of Botany at the age of thirty-four, returned to the UK in 1947, growing his reputation as an administrator in higher education, after an influential eight years in Australia.[9] On the basis of visionary planning, including Ashby's, the ANU was a research-only institution, structured into schools that brought research to bear on emerging problems. This model was influential for the new universities, but not even there was it possible to sustain research without teaching for long.[10]

Teaching was a central consideration, in fact, by the 1960s. As the new institutions emerged, they were, as Peter Karmel suggested, 'uncommitted to established practices', embracing pedagogical experimentation. Karmel believed new vigilance over teaching was due to the substantial growth in public investment in higher education, from less than $7 million in 1951 to around $170 million at the end of the 1960s.[11] With the 'new left' emerging, student activism, too, was putting university governance, teaching and assessment practices under the microscope. Changing social values, Karmel believed, were transforming the role of university students and graduates in the economy and the public sphere. He maintained that this scrutiny would lead to improvements in the ways universities operated.[12]

Building on two decades of Cold War growth

Pedagogical innovations addressed the severe criticisms of university teaching that the 1957 Committee on Australian Universities, chaired by head of the British Universities Grants Committee, Sir Keith Murray, levelled at Australian higher education. Student attrition, the committee argued, was unacceptably high: 'It is difficult to exaggerate,' the report argued, 'the cost in time, effort and money to students, universities and the nation of this low rate of graduation.' The committee insisted that if universities were to receive federal funding, this 'wastage' must be reduced.[13]

Indeed, Australian Commonwealth politicians were divided over whether they should provide funding for higher education after 1951, when the Commonwealth Reconstruction Training Scheme, aimed at soldier repatriation, ended. University vice-chancellors too were divided, for while centralized government funding would help secure their future – and assist them to recover from the ravages of the great depression – they were worried that it would result in increased meddling in university affairs. Ian Clunies Ross, chair of the Commonwealth Scientific and Industrial Research Organisation, worked through the 1950s to encourage both vice-chancellors and politicians to support a review of higher education. University vice-chancellors looked to Clunies Ross as a leader and his letters to Eric Ashby, maintained long after Ashby's return to the UK, reveal his commitment to a liberal idea of the university.[14]

Clunies Ross wanted Ashby to chair the committee, but was also happy with Keith Murray, who was preferred by the prime minister. Murray's concern to secure a diversity of income sources to assure the independence of UK universities may have influenced his selection as chair of the committee, though there is little evidence of this concern in the final report. Minutes of the meetings of the Murray Committee with university leaders, in fact, suggest that it was Clunies Ross who was the committee's most influential member. It was Clunies Ross who kept pushing them on student attrition rates and the quality of university teaching.[15]

Australian historians depict the Murray Review as launching a golden age for Australian higher education. Histories of universities joyfully recount the expansion of student numbers, the increase in staff, the building programmes and the general ambience of relative prosperity that followed the review. For students, a substantial Commonwealth Scholarships Scheme meant that over the following fifteen years, up to two-thirds of university students at any given time also received fee-free tuition.[16] It was an expensive undertaking for the federal government, one for which then Prime Minister Robert Menzies later claimed personal responsibility. Despite controversies in the 1950s, it was recognizably a successful and popular policy.[17]

Policy makers were conscious of the need for research to ensure Australian scientists and other scholars maintained networks across the former British empire and into the United States in the Cold War environment. In the same spirit, the Australian government also used higher education to try to shore up its place in the Asia-Pacific region. The Australian government conceptualized this as a civilizing mission, grounded in British colonialism. Such sentiments underpinned the establishment of the University of Papua and New Guinea (UPNG) in 1967, where Australia's colonial responsibility in the post-war era required both research and education to be kept on the agenda. No one was watching the new university terribly closely, however. Progressive thinkers populated the university and actively undermined British norms of scholarly 'quality' to produce a uniquely local intelligentsia – a 'decolonisation of the mind', as scholars from Africa also called it. Like the postcolonial Jawaharlal Nehru University, UPNG proved to be central to the development of national culture, providing links between traditional culture and modern Papua New Guinea.[18]

For the most part, however, Australian hegemony in the Asia Pacific was achieved through less direct action, in a policy that became known as the 'Colombo Plan'. Its overt aim was to use universities to stabilize the region, though it was in fact primarily for the purpose of securing favourable political and economic conditions that would enable Australia to dominate trade and economic relations in the region. The new universities were a part of this scheme and the geopolitical environment that shaped it. Students from across Southeast Asia were recruited to Australian universities in key areas that the planners believed would support economic stability in the student's home country. Consistent with the race-based immigration restrictions known as the 'white Australia policy', students from Asia were required to sign an agreement affirming that they would return home after they graduated. Politicians hoped that graduates would not only support economic stability by contributing their university knowledge to the local economy, but they would also foster friendly feelings towards Australia by taking (potentially influential) senior roles in government, the public service and corporate Asia.[19]

While the Cold War shaped the 1960s political environment for higher education, it was in fact the changing employment outlook that fuelled the growth that required new institutions. Post-war reconstruction was about the re-deployment of a workforce largely employed by the military; their labour would be refocused to manufacturing. The brightest would be invited to train as the technologists needed to ensure the new manufactured goods were innovative and able to be kept in good repair. By contrast, by the mid-1960s, employment growth was no longer in manufacturing, mining or agriculture. Rather, professional, white-collar work was taking over. Constituting just 3 per cent of the work force in 1901, professional work, including managers, grew to around 50 per cent of the labour market by the end of the century. The 1960s was a tipping point for this shift, which was partly enabled by the expansions in tertiary education implemented in the mid-1960s.[20]

Planning Australia's new universities

The states, and not the federal government, retained formal responsibility for education since Federation in 1901. By the early 1960s, most Australian states faced growing popular discontent about higher education. Older, established universities like Sydney, Melbourne and Adelaide were unable or unwilling to expand their enrolment. Quotas were being set in particular disciplines and soon applied to all of them. To a population accustomed to the convention that those who matriculated gained automatic entry to university, these changes were affronting. The new planning went further than that. It became evident that in response to changing international competition in key industries, the so-called technological society required more science and engineering graduates, while corporate Australia was both conglomerating in some areas and diversifying in others; and in both cases, the result was larger, more complex firms requiring more white-collar workers.[21] State politicians were increasingly persuaded that they should send as many students as possible to university, and in a wider range of fields, in order to be ready for economic change.[22]

In response, in 1960, the government of New South Wales launched a formal inquiry into higher education, with Macquarie University the result. Flinders University in South Australia had less official origins, evolving from an initial suggestion that the University of Adelaide might build another campus, as space at the inner city site was becoming tight, with no room for expansion. In Victoria, a decision to build a third university was easily made, and a committee established to form La Trobe University. In all cases, new universities were the solution to the states' urgent need for higher enrolments. This was a result of the demographic bubble known as the 'baby boom', of course, but also because of growing expectations among young people and their parents that they would be able to access tertiary education. As well as a growing discourse that suggested government services like education were a citizen's right, Australians faced a changing economy, which was progressively shepherding more young people into white collar work. This work was increasingly also associated with a university degree, raising demand for enrolment.[23]

Founders of the new universities suffered few public expectations that new institutions should be modelled on the old. In later decades, public intellectuals expressed mistrust of fusty old British traditions that many believed the first universities maintained, but in the early 1960s, the reverse criticism dominated public debate. Commentators showed a lack of faith in locally grown institutions; perhaps Australia's famous 'cultural cringe', or sense of intellectual inferiority, was at work. According to popular writer Donald Horne, Australian academics were as lazy and mediocre as the rest of the nation. Horne's *The Lucky Country* – a very widely read, polemical, book on the state of contemporary Australia, published in 1964 – depicted Australian scholars as immensely arrogant, wannabe-elites, seeing their students as nuisances who interfered with their middling, unexceptional academic thinking. It was high time, as public intellectuals like Horne frequently goaded, that youthful exuberance shook up the system.[24]

As if seeking to reflect this spirit of fresh starts, the new universities were built on previously undeveloped, suburban sites. In southern Brisbane, Griffith University, for example, is surrounded by native bush. Flinders University, which like Griffith sought crown land (that was therefore also cost free), was built on an extremely windy hill in the south of Adelaide. Often the new institutions were seeking the self-contained community of their new British counterparts, with enough space for all the required buildings, green areas and residential colleges. By selecting suburban sites, however, university planners also positioned the new institutions within reach of their target demographic. La Trobe was built in Melbourne's unfashionable north, seeking to attract students from a wider demographic than Melbourne or Monash (located in Melbourne's wealthier southern suburbs).[25]

Macquarie University was built in Sydney's north, but for opposite reasons. The majority of Sydney's unmet student demand was on the city's wealthy north shore: these students would get first dibs on new developments in higher education, just as they had on the old.[26] Macquarie made their elite catchment more politically palatable by also recruiting external and part-time students, for it was established against powerful arguments to develop instead a university in Sydney's working-class west. The University of Western Sydney was not established until the late 1980s.[27]

In their locations, layout and design, the new suburban universities were modelled on British innovations such as the University of Sussex.[28] By doing so, however, their campuses developed (and retain) a particularly Australian feel. Wide open spaces littered with eucalypts, such as at Wollongong and Macquarie Universities, recall Australian rural scenes. While the design of the new universities in Australia was primarily about the spaces between buildings, the utility of building did matter, nevertheless: by eschewing the neo-gothic sandstone designs with which the nation's first universities were built, these institutions carried with them connotations of Australian egalitarianism.

In some cases – at Deakin University, for example – such egalitarianism was an explicit aim.[29] This was not at all the case at Macquarie. Peter Karmel's Flinders University was ambivalent, wishing to open access to more students, but without seeming like a second tier institution.[30] This was a battle that Australian universities consistently waged. The myth of merit as the basis of performance had a disproportionate importance in Australia's higher education sector. It intersected, through Australia's history, with other powerful national myths of 'mateship' and a 'fair go', so that describing Australia's higher education sector in tiers has not normally been politically acceptable, despite its obvious, lived truth. This was especially the case for the universities built after the original six, though even for those it had mattered. Even when debating those first pre-Federation universities, the Australian colonies were often reluctant to support government-funded universities when schooling needed so much attention. In most instances, popular consent was only obtained by promising that anyone would be able to access higher education. In New South Wales, this meant demonstrating that sectarianism would not prevail; in Western Australia, students initially attended university free of all tuition fees; and in Queensland, special measures were instated for rural students, from the beginning. For the utopian universities of the 1960s, such concerns were all the more prominent. It was not just to assure equity, for the new universities did not in fact achieve this. Rather, the legitimacy of the system, and support for growing public investment in it, rested on diversity of access as evidence that university degrees and the class status that they afforded were well earned.[31]

As a consequence of concerns for equity, merit and legitimacy, then, the public required the new universities to perform accessibility. As a result, they often offered adult and distance education in addition to traditional, campus-based undergraduate programmes. Deakin still does.[32] New institutions also took the opportunity to re-think disciplinary boundaries. Flinders and Macquarie chose school-based structures, replacing the faculties and departments that dominated at traditional universities – they were perhaps inspired by the ANU. Founders hoped this would lead to greater interdisciplinary cooperation.[33] At Monash, scholars developed 'trans-faculty' courses, such as 'The economics of medicine'. State governments supported such 'openness to the interplay of … categories of knowledge', specializing in new areas of study, such as Asian studies.[34] With this they mollified influential scholars in older institutions and avoided replicating disciplines across institutions, which they hoped would make the sector more cost-efficient.[35] There, as at other institutions which made similar efforts, the innovators met too many barriers, caused by cultural divides and persistent turf wars. Disciplines almost always remained intact. While transcending disciplinary boundaries may have been just a little too utopian for the new batch of Australian scholars, there is evidence that in some areas at least, disciplinary norms were nevertheless re-considered. At Macquarie, R. W. Connell's sociology team refused to be linked with anthropology, repudiating the remnants of imperialism they detected in that discipline. This did not prevent innovation, however and their team-based research resulted in lasting changes to the practice of sociology in Australia.[36]

Teaching in particular was subject to a considerable re-evaluation. At Flinders, longer laboratory times were intended to allow students time to learn, rather than just follow instructions. 'Continuous assessment', Flinders founders suggested, would soften what the public increasingly saw as the 'tyranny of examination' in educational institutions, both schools and universities.[37] In the humanities, small-group work was becoming the norm. Tutorials were already in place at the older universities, but the new universities sought to enhance this innovation, drawing on techniques that were successful in the Workers' Education Association.[38] Small group teaching was designed at Macquarie and elsewhere, inspired by the British psychologist, Jane Abercrombie at University College London, who claimed that it was important that group work minimize the dominance of the teacher and stress student contributions.[39]

Not just universities

The new universities represent just a fraction of the developments in Australian tertiary education in the mid-1960s. Within just five years of the Murray Report, Menzies commissioned Leslie Martin to consider expanding tertiary education still further. Martin was tied to the Atomic Energy Commission and was a founder of the Academy of Science; he later joined the Royal Military College. He was a truly establishment man with a reputation for hard-headed pragmatism. Martin sought an alternative model for higher education that would fuel economic development: 'It would hardly seem to be practicable to continue creating new universities all patterned after the nineteenth century models,' he argued.[40] Martin talked and travelled to the United States, where he found examples of 'what not to do in Australia'. He was similarly unimpressed with the British colleges of advanced technology (CATs), a view that meant the state institutes of technology were not included in the 1965 proposals.[41]

When Martin embarked on his task, Australian political and university leaders were conscious that Lionel Robbins had recently conducted a similar review in the UK. Martin seems to have had quite different priorities, however. Whereas Robbins sought a larger number of conventional universities in an expanded sector, Martin pursued a second-tier college system. He did not want America's liberal arts colleges, however, seeking something more vocational, fulfilling the government's economic goals.[42] Menzies aspired to this too, but he was very attached to liberal ideas of the university, which he saw as central to the British culture he hoped to sustain in postcolonial Australia. The purpose of the binary system was to keep the culture that universities nurtured untainted by lesser knowledge; and yet to also reassure Australians sensitive to inegalitarian sentiments, that educating their children at colleges of advanced education was not lesser, just cheaper. This delicate balancing act meant that tertiary education was expanded, without losing its elitism.[43]

In 1972 when a newly elected Labour government led by Gough Whitlam sought further educational reform, the new prime minister saw no need to meddle with the seven-year-old colleges of advanced education. These colleges continued to educate teachers, allied health professionals, childcare workers and others until 1987, all united under the Commonwealth Tertiary Education Commission. This was the buffer body between government and higher education that Karmel led between 1971 and 1981, only to be dismantled in 1987.[44] In Australian public memory, Whitlam is often credited as the prime minister most invested in higher education. He abolished all tuition fees and welcomed the contribution of radical students to the public sphere, arguing that their efforts had alerted the Australian public to the need for reform in important areas, leading (for example) to the dismantling of the white Australia policy (though others maintained it was mostly already gone by then).[45] With the prime ministership of Gough Whitlam, the aims of government and those who sought reform within universities appeared to be aligned.[46] Unfortunately for this brief truce between radicals and the establishment, Australia's governor general dismissed Whitlam in 1975. By the time of the dismissal, however, the radical moment, urging progressive reform, had largely dissipated.

Seeking progressive pedagogy

Leaders of change, like Peter Karmel, hoped that the new institutions would lead the way with pedagogies supporting better outcomes for students and graduates – and government funders and the national economy. Advocates of progressive pedagogies on the ground, however, tended to wish, by contrast, to use innovative teaching to undermine higher education's links with the political economic aims of government. In 1967, for example, Terry Irving, left-wing lecturer in politics in Sydney, argued that Australia's universities legitimized the nation-state, capitalism, elitism and structural inequality.[47] For Irving, this made lecturers complicit in repressive systems. 'Just as the university serves the nation,' he claimed, 'so the "good" teacher serves the university by instructing his students efficiently in those skills whose acquisition the nation has already made a condition of his entry to the university.'[48]

While Irving was lecturer at Australia's oldest university, he was also co-founder of the 'Free University', an experiment with progressive pedagogies that operated outside of the university system. Irving, along with Rowan Cahill and R. W. Connell (who then moved to sociology at Macquarie University), established what was known as the 'Free U' in a terrace house near the University of Sydney.[49] Launched in 1968 with 150 students, at its peak the Free U had approximately 300 participants, of which around twenty were university staff.[50] The Free U is emblematic of the ways that those interested in university reform believed that progressive teaching was really only possible outside of the old institutions. Karmel, among other university planners, hoped that this would include the new universities.

Monash University, opened in 1958 in suburban Melbourne, was such a site. As well as serious attempts at innovation in teaching and disciplinary structures, Monash also acquired the reputation as the key centre for student dissent in Australia. Student radicals at Monash, often of a Maoist persuasion, sought a revolution within the university as well as in wider society.[51] They challenged the structure of academic authority in classrooms; they sought to undermine the division between scholars and students, which they believed reflected inequalities across society. This resulted in a series of sit-ins, occupations, violent confrontations and peaceful protests that made Monash a challenging environment for its academics, especially its first vice-chancellor, Louis Matheson.[52]

The concerns of radical students had a lasting, though more tempered, expression in the pedagogies of the new universities, and these eventually also influenced lasting changes at the older universities, too.[53] For example, an Academic Board inquiry at the University of Sydney in 1975 instigated continuous assessment at that institution.[54] Led by the new institutions (especially Monash), universities across Australia incorporated aspects of the radical critique of the old universities into their structures. The older institutions suffered long and sometimes violent struggles, but eventually followed the newer universities in restructuring professorial boards as academic boards, including junior staff and student representatives.

Departments and faculties too began to take on the egalitarian values embodied in the new university structures, acquiescing as minimally as possible, usually, to student demands that the university reflect their quest to instate participatory democracy in Australia more generally, at their local level. Only at Sydney University's philosophy department – long a haven for non-conformist intellectuals, notably producing the libertarian movement known as the 'Sydney Push', with members such as Germaine Greer, Clive James and Robert Hughes – was this demand actually met.[55] University vice-chancellors colluded on finding a minimum number of student representatives that they thought they could get past the rebels, hoping to avoid the probability that students would point to whichever institution allowed the highest number of representatives and insist on the same.[56] This ended the days of the 'god-professor', who chaired their departments ruthlessly, allowing increased freedom in teaching and research, mainly for junior academic staff.[57]

Another lasting effect of the radical embrace of the new pedagogies was in changes to assessment practices. In 1974, history students at Flinders University occupied university buildings in protest over exams, suggesting that the university's initial aspiration to instate continuous assessment had not been universally applied. A nation-wide anti-exam movement flourished among students, informed by educational research conducted by the National Union of Australian University Students.[58] The research students conducted became increasingly formalized, eventually becoming what is now the Centre for the Study of Higher Education at the University of Melbourne, supporting ongoing scholarship in the field, including that by high-profile higher education researcher, Simon Marginson.[59] At other universities, centres for teaching and learning were established from the same cohort of former student activists, seeking to connect university strategy to new pedagogical thinking.[60] Supporting pedagogical innovation across the sector, the Higher Education Research and Development Society of Australasia (HERDSA), was established in 1972. This organization sought to encourage socially situated, student-centred approaches through professional development for university teachers.[61]

A key goal of student-centred pedagogy was to increase the sense of power over one's own world, freeing students from authoritarian control: 'Most students do not feel that they are able to control their own destiny,' read the educational policy of the Australian Union of Students.[62] Such a desire, to control one's own destiny, resonated with Australian youth in this period; from 1965 until 1972, young men were entered into lotteries that decided their conscription to the Vietnam War.[63] For the following decade, young people, often through university student groups, battled for more control over their social and political environment, including lowering the voting age to eighteen.[64] University students at all universities (though especially Monash) were an integral part of the growing movement that opposed the war.[65] They were especially affronted when the universities were used in support of the war. La Trobe University students in Victoria took over the university's Careers and Appointments office that had been used for military recruitment and redirected its resources to 'the revolution', using its stationary and stencil duplicator to add to the copious mimeographed broadsheets student radicals produced.[66] By 1970 when most university students were united in opposing the war, the universities seemed a part of the establishment's apparatus of power, in need of a shake-up. To them, the Vietnam War represented a great deal that was wrong with the establishment, including the universities.[67] This was not entirely fair. Monash Vice-Chancellor Louis Matheson encountered significant pressure from both directions when he refused to use Monash student records to help the government enforce national service. On the one hand, Matheson's students felt that this did not go far enough to consider the university innocent of government wrongdoing; on the other hand, conservative commentators were concerned that Monash and other universities were beginning to display an alarming softening to student demands.[68]

Despite government fears, however, the transformations to the university system in response to radical student support for new pedagogies were quite confined. Students and junior staff worked hard to embed changes in university governance and in assessment, but these did little to disentangle the universities from their obligations to support government political and economic strategy. The vision, among radical students, of a university that actively sought to re-think the establishment and bring about revolutionary ideals was instead limited to some minor developments in teaching, assessment and committee structures.

Seeking capitalist fodder

When Whitlam was elected in 1972, he headed a Labour government that had been in opposition since 1949. It is perhaps for this reason that his reforms seemed too radical for some. Placed against the drastic changes to the economy wrought by the oil shocks and other international shifts from 1973 into the following decade, Whitlam's programme of government expenditure, including that on education, began to look, to members of government and the public, like reckless economic management. Historians continue to debate the blame for the events that followed: the Senate blocked supply and the governor general sacked the elected government in November 1975. What is indisputable, however, is that the period of rapidly escalating inflation coupled with a marked slowdown of economic growth meant that the government succeeding Whitlam would need to be attentive to economic concerns, in reality as well as in appearance.[69]

By 1975, the economic downturn was also wreaking havoc on the expectations of the now-large number of university-trained youth. Having entered higher education with the expectation of attaining well-paid, white-collar jobs, they now encountered a contracting economy with increased unemployment. Disillusionment was setting in, augmented by the critiques that radicals levelled at the university in the 1968-74 era. It was ironic: cynicism in popular discourse emerged in this period at least in part because going to university now was no guarantee of financial success. And yet, it was this exact alignment with capitalism that had also been the subject of youthful critique just a few years earlier.[70]

The mood in government also shifted, with the economic predicaments rendered by global change. No longer could politicians afford, they believed, to make higher education available as a citizens' right. Any further expansion of higher education – and indeed there were many who believed it ought to contract – must be aligned with the nation's economic needs. With the election of Liberal Prime Minister Malcolm Fraser, a new review of higher education was conducted by economist and University of Sydney Vice-Chancellor, Bruce Williams. With an academic background at Belfast, Keele and Manchester, Williams brought experience of one of the British utopian institutions to his review. By contrast to the compact Murray Report, however, the Williams document was a bulky, two-volume affair that Australian historians of education privately describe as the longest and least readable of our key primary sources. Williams did not advocate radical change, instead reinforcing many of the conclusions of the 1965 Martin Report. Nevertheless, he provided the government with the data and analysis it needed to address connections between higher education and the economy.[71] While largely forgotten, the economic connections he made mark a key shift in the public discussion about universities.[72]

The new universities were the target in this debate. Some critics suggested that the likes of Griffith, Murdoch and Deakin Universities should never have been built. This was because the most urgent issue, given the economic environment, was that higher education seemed too costly, particularly since student demand was rapidly declining. Growth of 8-9 per cent had been the norm as the new universities enabled increased enrolments, but in 1979 this growth rate dropped to just over 2 per cent. It fell even further in the early 1980s, even though higher education was free of tuition fees.[73] Williams had a different perspective. There might currently seem to be too many universities, he argued, but with projected growth in student participation in the following decades, they would soon be required, a prediction that proved correct, but not until the 1990s.[74]

In the meantime, universities faced looming financial deficits, as income failed to grow at its previous pace, decelerating in concord with the slowdown in student participation. As in the rest of the economy, university income was not able to keep pace with salary and other costs. Student numbers were clearly the key to maintaining income streams. And so, for Australia's older institutions, the new universities now posed a competitive threat. For the first time in Australian university history, the old elite institutions began to advertise. Members of other universities across the sector were outraged; it was a betrayal of principles.[75] Their outrage had no effect. In fact, once it began, advertising coerced them all.[76] Such advertising symbolized a shift in focus. University leaders were now making unprecedented compromises to shore up their shaky financial situation.

Conclusions

When he was leading Flinders University and CTEC, Peter Karmel struggled to reconcile radical investment in using education for revolutionary ends with those of government and university leaders. The latter sought a sector that would support political economic structures and bolster Australia's position internationally in the Cold War environment – a significant contrast to the revolutionaries, who (especially as the Vietnam War proceeded) sought to dismantle both. To some extent, to be sure, it appeared to be possible. Government and academic leaders like Ian Clunies Ross disparaged the traditional universities for their approach to teaching, arguing that failing students was bad for the nation. As the new universities began, unconstrained by the traditions and habits of the older institutions, they embraced new methods of teaching. Continuous assessment, working in small groups or in longer laboratories, where the emphasis was on student learning rather than the teacher's instruction and refusing to allow disciplinary boundaries to define the problems confronted by research, was all approaches advocated by both establishment leaders and revolutionaries.

If radical and establishment solutions to problems in universities resembled each other, so too did the language attached to their respective critiques of higher education. Radical student discontent with the relevance of university education was later echoed, perhaps even appropriated, by economic conservatives and underemployed university graduates after the oil shocks prompted a shift in the economy that disrupted the process of professionalizing the labour market, which had formerly propelled university growth. Here is where the similarity ends, however. The progressive pedagogies that radicals idealized, present in such experiments as the Free U, were for the purpose of undermining the university's complicity with 'establishment' systems that they saw as causing inequality and oppression. The new universities, by contrast, may have drawn on similar teaching philosophies, but for opposing purposes. Instead, the new universities placated the radical position, but became a part of the structure that bolstered government aims.

All of this was also underpinned by changes in the Australian government's approach to university education. After the Second World War, higher education was key to redeployment of ex-service women and men. By the 1950s, it was associated with ideals of 'civilization', seen as a tool to bolster democracy and offer citizens opportunities to improve their lives and livelihood. There were economic aims as the post-war labour market emerged, but both progressive and conservative planners in the 1940s and 1950s resisted reducing students to 'capitalist fodder'. The new universities entered Australia's higher education scene while this philosophy remained in place, but the last of them were built as the idea of the university was rapidly transforming. By the mid-1970s, universities were primarily tools for the government's economic management.

The Australian government's preoccupation with using the universities to this end was also reflected in universities' changing stance towards students. Not only did government planners see students as tools for propping up capitalism, but in their advertising, universities too were beginning to position students as consumers. The new universities were most vulnerable in this environment, as the older universities lured scarce students away from them. Although they were launched with the youthful exuberance that commentators like Donald Horne hoped might renew Australia, the new universities receded into an ancillary role in government economic planning.

Notes

1   *Canberra Times*, 24 April 1969, 2.
2   Hannah Forsyth, *A History of the Modern Australian University* (Sydney: New South, 2014), 201–25; John McLaren, 'Karmel Report: Schools in Australia', *Dictionary of Educational History in Australia and New Zealand (DEHANZ)* (2014), http://dehanz.net.au (accessed 25 January 2017).
3   H. C. Coombs, 'Science and Technology: For What Purpose?', in *Science and Technology: For What Purpose?*, ed. A. T. A. Healy (Canberra: Australian Academy of Science, 1979), 21–47.
4   Ibid., 21. Department of Education Training and Youth Affairs, *Time Series Report* (Canberra: Australian Government Publishing, 2000).
5   *Canberra Times*, 7 June 1976, 1.
6   Stephen Murray-Smith and Anthony John Dare, *The Tech: A Centenary History of the Royal Melbourne Institute of Technology* (Melbourne: Hyland House, 1987). Michael White, *WAIT to Curtin: A History of the Western Australian Institute of Technology* (Bentley: Paradigm Books & Curtin University, 1996). Noeline Kyle, Catherine Manathunga and Joanne Scott, *A Class of Its Own: A History of the Queensland University of Technology* (Alexandria: Hale & Iremonger, 1999).
7   Stuart Macintyre, *Australia's Boldest Experiment: War and Reconstruction in the 1940s* (Sydney: NewSouth, 2015), 214–18, 327–32, 465–7.
8   Robin Hughes, 'Interview with H. C. "Nugget" Coombs' (Canberra: Film Australia Australian Biography Series), http://www.australianbiography.gov.au/subjects/coombs/ (accessed 21 December 2011).
9   Hannah Forsyth, 'Academic Work in Australian Universities in the 1940s and 1950s', *History of Education Review*, 39 (2010), 44–52.
10 S. G. Foster and Margaret M. Varghese, *The Making of the Australian National University* (Sydney: Allen & Unwin, 1996), 197–228.
11 Peter Karmel 'The Role of the University' *Canberra Times*, 24 April 1969, 2; cf. David Hilliard, *Flinders University: The First 25 Years* (Adelaide: Flinders University, 1991), 29–30.

12 Karmel, 'The Role of the University'; Hilliard, *Flinders University*, 29.
13 Keith A. H. Murray et al., *Report of the Committee on Australian Universities* (Melbourne: Committee on Australian Universities, Commonwealth Offices, Treasury Place, 1957), 35–41.
14 Forsyth, 'Academic Work in Australian Universities in the 1940s and 1950s'; C. B. Schedvin, 'Clunies Ross, Sir William Ian (1899–1959)', *Australian Dictionary of Biography*, National Centre of Biography, Australian National University. URL?
15 Meeting held at New South Wales University of Technology, 16 July 1957 Committee on Australian Universities Documents: Draft Summary Records, NAA/CAU/A7691/PART11, National Archives of Australia, Canberra.
16 Patrick O'Farrell, *UNSW: A Portrait. The University of New South Wales, 1949–1999* (Sydney: UNSW Press, 1999), 75. Brown, *Governing Prosperity*, 236. Simon Marginson *Education and Public Policy in Australia* (Cambridge: Cambridge University Press, 1993), 180.
17 Simon Marginson, 'The Whitlam Government and Education', in *It's Time Again. Whitlam and Modern Labor*, ed. Jenny Hocking and Colleen Lewis (Melbourne: c., 2003), 251–3.
18 Ulli Beier, *Decolonising the Mind: The Impact of the University on Culture and Identity in Papua New Guinea, 1971–74* (Canberra: Pandanus Books, 2005), 43–94.
19 Alex Auletta, 'A Retrospective View of the Colombo Plan: Government Policy, Departmental Administration and Overseas Students', *Journal of Higher Education Policy and Management*, 22 (2000), 47–58; Daniel Oakman, 'Young Asians in Our Homes: Colombo Plan Students and White Australia', *Journal of Australian Studies* 26 (2002), 894–8. David Lowe, 'Percy Spender and the Colombo Plan', *Australian Journal of Politics and History*, 40 (1994), 146–76.
20 Australian Bureau of Statistics, Labour Force Historical Timeseries, Australia – Occupation by Full-time/Part-time 1966–1984 ABS/6204055001TS0011/Table1; Census of Australia, 1911, 1921, 1933, 1947, 1954, 1961, 1966, 1976. Canberra: Australian Bureau of Statistics http://www.abs.gov.au (accessed 18 May 2016); Census of Australian States 1901. Canberra: Australian Data Archive https://ada.edu.au/ (accessed 28 March 2019).
21 Grant Fleming, David Merrett and Simon Ville, *The Big End of Town: Big Business and Corporate Leadership in Twentieth Century Australia* (Cambridge: Cambridge University Press, 2004).
22 Bruce Mansfield and Mark Hutchinson, *Liberality of Opportunity: A History of Macquarie University 1964–1989* (Sydney: Hale & Iremonger, 1992), 17–34.
23 Australian Bureau of Statistics, Labour Force Historical Timeseries, Australia – Occupation by Full-time/Part-time 1966–1984 ABS/6204055001TS0011/Table1 http://www.abs.gov.au (accessed 18 May 2016); Forsyth *History of the Modern Australian University*, 20–45.
24 Donald Horne, *The Lucky Country: Australia in the 1960s* (Harmondsworth: Penguin, 1964), 205–8.
25 William J. Breen, *Building La Trobe University: Reflections on the First 25 Years, 1964–1989* (Melbourne: La Trobe University Press, 1989).
26 Mansfield and Hutchinson, *Liberality of Opportunity*, 17–56. Mark Hutchinson, *A University of the People: A History of the University of Western Sydney* (Sydney: Allen & Unwin, 2013), 33–41.
27 Hutchinson *A University of the People*, 28–32.
28 Hilliard, *Flinders University*, 1–28.

29 Mary Jane Mahony, 'Warp and Weft in Policy Analysis: Australian Distance Education Policy 1901–1989' (unpublished PhD thesis, University of Wollongong, 1994), 68–94.
30 Hilliard, *Flinders University*, 1–28.
31 Forsyth, *A History of the Modern Australian University*, 11–13; idem., 'Dreaming of Higher Education', *Southerly* 74 (2014), 119–42; 'Expanding Higher Education: Institutional Responses in Australia from the Post-War Era to the 1970s', *Paedagogica Historica: International Journal of the History of Education*, 51 (2015), 365–80.
32 Forsyth, *A History of the Modern Australian University*, 176–225.
33 *Canberra Times*, 4 March 1966, 2; Mansfield and Hutchinson, *Liberality of Opportunity*, 57–86, 185–216.
34 Noel Quirke, *Preparing for the Future: A History of Griffith University 1971–1996* (Brisbane: Boolarong Press, 1996), 8; see also Terry Hogan, *Coming of Age: Griffith University in the Unified National System* (Melbourne: Melbourne University Press, 2016).
35 Mansfield and Hutchinson, *Liberality of Opportunity*, 35–56; Hutchinson, *University of the People*, 33–41.
36 Mansfield and Hutchinson, *Liberality of Opportunity*, 187–9; Graeme Davison and Kate Murphy, *University Unlimited: The Monash Story* (Sydney: Allen & Unwin, 2012), 117–50.
37 *Canberra Times*, 4 March 1966, 2; ibid., 8 June 1966, 3.
38 See Chris Hilliard, *English as a Vocation: The* Scrutiny *Movement* (Oxford: Oxford University Press, 2012).
39 Mansfield and Hutchinson, *Liberality of Opportunity*, 61–3.
40 Susan Davies, *The Martin Committee and the Binary Policy of Higher Education in Australia* (Melbourne: Ashwood House, 1989), 23; R. W. Home, 'Martin, Sir Leslie Harold (1900–1983)', Australian Dictionary of Biography, National Centre of Biography, Australian National University, http://adb.anu.edu.au/biography/martin-sir-leslie-harold-14939/text26128 (accessed 27 March 2019).
41 Davies, *Martin Committee*, *83,* 97.
42 Ibid., 97.
43 Bob Bessant, 'Robert Gordon Menzies and Education in Australia', in *Melbourne Studies in Education 1977*, ed. Stephen Murray-Smith (Melbourne: University of Melbourne Press, 1977), 163–87; Davies, *Martin Committee*, 123–142.
44 Peter Karmel, 'Education and the Economic Paradigm: The Cunningham Lecture', in *Confusion Worse Confounded: Australian Education in the 1990s*, ed. Brian Crittenden (Canberra: Academy of the Social Sciences in Australia, 1995), 10–12.
45 Whitlam, Commonwealth of Australia Parliamentary Debates (House of Representatives), 27 May 1969, col. 2248.
46 See, for example, Don Beer, *A Serious Attempt to Change Society: The Socialist Action Movement and Student Radicalism at the University of New England, 1969–75. Transcripts of Interviews* (Armidale: Kardoorair Press, 1998).
47 Terry Irving, 'The Free University', *Honi Soit,* 14 September 1967, 8.
48 Terry Irving, 'The Mass University and the Free University as Utopia', in *Counterpoints: Critical Writings on Australian Education*, ed. S. D'Urso (Sydney: John Wiley, 1971), 21.
49 Irving, 'The Free University'; Brian Freeman and Bob Connell, 'Free University', *National U*, 4, 1968, 2.
50 W. F. Connell et al., *Australia's First: A History of the University of Sydney. Volume 2 1940–1990* (Sydney: Hale & Iremonger, 1995), 360–1.

51 Paul Francis Perry, *The Rise and Fall of Practically Everybody. An Account of Student Political Activity at Monash University 1965-1972* (Melbourne: P. F. Perry, 1973), 34.
52 Davison and Murphy, *University Unlimited*, 117-50.
53 John Docker, '"Those Halcyon Days": The Moment of the New Left', in *Intellectual Movements and Australian Society*, ed. Brian Head and James Walters (Melbourne: Oxford University Press, 1988), 296. For their legacy, see David Boud, 'Aren't We All Learner Centred Now?: The Bittersweet Flavour of Success', in *Changing Higher Education: The Development of Learning and Teaching*, ed. Paul Ashwin (London: Routledge, 2006), 19-32.
54 Report of the Committee on Examination and Assessment, Minutes of the Academic Board 30 August 1975, USYD/AB/70, University of Sydney Archives, Sydney.
55 Lewis D'Avigdor, 'Participatory Democracy and New Left Student Movements: The University of Sydney, 1973-1979', *Australian Journal of Politics and History*, 61 (2015), 233-47; Anne Coombs, *Sex and Anarchy: The Life and Death of the Sydney Push* (Melbourne: Viking, 1996).
56 Minutes of the Meeting of the Victorian Universities Committee (April 1968), 'Student Participation File', MON84/96/012, Monash University Archives.
57 Julia Horne and Geoffrey Sherington, *Sydney: The Making of a Public University*. (Melbourne: Miegunyah Press, 2012), 131-4.
58 Graham Hastings, *It Can't Happen Here. A Political History of Australian Student Activism* (Adelaide: Empire Times Press, 2002), 75-128.
59 Australian Union of Students, Examination Papers: 'Background Papers on Assessment, Prepared for Campus Activists by the Australian Union of Students' (1975), Tom Hurley Papers, UMA/1984.0001/6/2/4 University of Melbourne Archives.
60 Catherine Manathunga, 'The Field of Educational Development: Histories and Critical Questions', *Studies in Continuing Education*, 33 (2011), 347-62.
61 Robert A. Cannon, 'The Professional Development of Australian University Teachers: An Act of Faith?', *Higher Education*, 12 (1983), 19-33.
62 Australian Union of Students, 'Education Policy', *Woroni* (undated, c. July 1973), 4.
63 Peter Siminski and Simon Ville, 'I Was Only Nineteen, 45 years Ago: What Can We Learn from Australia's Conscription Lotteries?', *Economic Record*, 282 (2012), 351-71.
64 Terry Irving et al., *Youth in Australia: Policy Administration and Politics* (Melbourne: Macmillan Education, 1995), 146-53.
65 Verity Burgmann, *Power and Protest: Movements for Change in Australian Society* (Sydney: Allen & Unwin, 1993), 190; 'The Facts about the Anti-LBJ Demonstration' (pamphlet), 'Student Ephemera File, 1967-1968', MON342, Monash University Archives, 1968.
66 Barry York, *Student Revolt!: La Trobe University 1967-1973* (Campbell: Nicholas Press, 1989), 91.
67 For example: *Grass Roots*, 2 (4) (20 April 1971) (one page broadsheet of the Students for Democratic Action), AUA/SDA/M-L/GR Adelaide University Archives; *Student Guerila*, 13 June 1968 (one page broadsheet of the Society for Democratic Action University of Queensland), AUA/SDA/M-L/SG Adelaide University Archives.
68 See Peter Wetmore, 'The Strike at Sydney University', *Quadrant*, 17 (1973), 23-9.
69 Martin Shanahan, 'Economic Policy of the Hawke Years', in *The Hawke Legacy*, ed. Gerry Bloustien (Adelaide: Wakefield Press, 2009), 168.
70 Ian Spicer, 'An Employer's View', in *The Future of Higher Education in Australia*, ed. Terry Hore (Melbourne: Monash University Press, 1978), 103; Edward Gross

and John S. Western, *The End of a Golden Age: Higher Education in a Steady State* (St Lucia: University of Queensland Press, 1981), 1-9.
71 *Education, Training, and Employment: Report of the Committee of Inquiry into Education and Training* (Canberra: AGPS, 1979).
72 Bruce Williams, *Making and Breaking Universities: Memoirs of Academic Life in Australia and Britain 1936-2004* (Sydney: Macleay Press, 2005).
73 Department of Education Training and Youth Affairs, *Time Series Report* (Canberra: Australian Government Publishing, 2000).
74 *Education, Training, and Employment: Report of the Committee of Inquiry into Education and Training* (Canberra: AGPS, 1979) 190.
75 'Problems of Campus Promotion', *The Australian Higher Education Supplement*, 16 April 1980, 13; 'Rationalisation Could Be Irrational', ibid., 7 May 1980, 12.
76 'Competing for Student Advantage', *The Australian Higher Education Supplement*, 27 January 1982, 11.

18

Jawaharlal Nehru University: A University for the Nation

by Rajat Datta and Shalini Sharma

Introduction

In April 1969 the Indian Prime Minister Indira Gandhi appointed an academic body to establish a new university that would bear the name and embody the beliefs of her father, Jawaharlal Nehru. Thus began an often fraught and complicated relationship between the national government and a new national university, an institution burdened with high and often contradictory expectations, and constantly in the public eye. Before JNU was even thought of, Indian intellectuals and politicians were bemoaning the state of the nation, the rising tide of regionalism that threatened to break up the new independent republic and the lack of coherence in the federal education system. A new national university was reckoned to be a panacea for all. Upon the death of Nehru in 1964 the university took on even-greater responsibilities: to 'promote the study of the principles for which Jawaharlal Nehru worked during his lifetime, national integration, social justice, secularism, democratic way of life, scientific approach to the problems of society'.[1] And yet within its first few years criticism and scrutiny began that keeps JNU in the public domain to this day. It was seen as a vehicle for the country's left-wing intelligentsia and for a weak Prime Minister, Indira Gandhi, and also as a corrupt and pampered elite institution, one based on foreign assumptions. Above all, JNU epitomized an overly politicized university producing so-called Westernized women and unwieldy anti-national men.

JNU catered principally for graduate students, and across the five decades of its history JNU students have contributed hugely to its evolving identity and its way of life. In his recent comprehensive history of JNU Rakesh Batabyal has uncritically accepted conventional assumptions about the disruptive nature of student politics, the undue influence and insidious behaviour of the left on JNU and the 'logic of backwardness' which quota admissions systems such as the one established in JNU can generate.[2] This chapter seeks to revise these assumptions and provide a short fresh survey. It will

---

The authors wish to acknowledge the research assistance provided to them by Jeena Sarah Jacob and Preeti Gulati of the Centre for Historical Studies, Jawaharlal Nehru University.

describe the unique and utopian nature of JNU at its birth, the public denunciations of its failed mission that set in during the second half of the 1970s, and argue that ultimately the students of the university have fought for the decreed charter of the institution, by consistently pressing for an equitable system of positive discrimination and introducing innovative measures to bring about gender awareness on the campus. The chapter will also provide for the first time an assessment of the success over time of the university's distinctive policy on equity in student admissions and staff appointments. In the climate of Indira Gandhi's Emergency of 1975, the conflicts over admissions policy in the 1980s, economic liberalization and the constant threat of the privatization of education in India, JNU students have time and again upheld their interpretation of what it is to be a 'national' university. If JNU retains any of its utopian nature, it is because of them.

A utopian charter?

The period from Indian independence in 1947 through to the creation of JNU in 1969 was one of immense national upheaval. The partition of the country accompanying independence left a festering sore, and an insecurity that could imagine any regional opposition as potential secessionism. The call for the 'emotional integration' of the nation saw education as the means to stop divisive regional, communal and linguistic affinities being aired throughout the country.[3] The builders of the new Indian constitution during and after independence offered a new perspective on national unity by promising social justice for communities of people who were beyond the pale of the caste and social system observed by the majority of Indians.[4] The University Education Commission of 1948 chaired by Dr Radhakrishnan and the Indian Education Commission 1964–6 chaired by Professor D. S. Kothari both maintained that education, especially university education, was the remedy for social injustice.[5] As Prime Minister Nehru averred repeatedly these ideals were important because divisions such as caste, religion and language were holding India back and preventing the much-desired modernization of the new country.[6]

However, the consensus throughout the 1950s and 1960s agreed that existing university provision, even if expanded, would fail to deliver the desired results. Education policy came under state legislatures rather than the central government. The older presidency universities in Calcutta, Bombay and Madras were too parochial, recruiting locally and becoming seats of local political conflict. The University of Delhi was a set of affiliated colleges that worked against a united pedagogic vision, let alone a national one. Commentators also bemoaned the ill-thought expansion of university education since independence, seen as a means of satisfying state level politicians rather than a response to part national planning criteria. Indian commentators in education journals published abroad such as *Minerva* blamed this haphazard approach for producing swathes of frustrated educated and unemployed youth, typified in the Bollywood version of the 'angry young man' but more dangerously as a symptom of the 'rampant student unrest' throughout the country.[7] In this way demands were made for a new national university or set of universities, which would recruit from the whole of India and repair the damage of parochialism. The government had already created five Indian Institutes of Technology from 1951 to 1961. These were addressing some of the problems raised. However, they were primarily institutions to promote science education from which many elite students went to work abroad. And they did not have universal range – the broad church of science and humanities – encompassed in the ideal of the 'national university'. As the University Grants Commission reported in 1969, central universities needed to 'function effectively and vigorously on an all-India basis, to help build up a corporate intellectual life in the country and to further national integration'.[8] They were to provide courses that could not be provided by individual state universities and they were to remove 'imbalances from the academic life of our country' by taking action to help 'deserving students from educationally backward areas'. Most importantly, the report mandated that 'Central Universities have an all-India character to be reflected in admissions, appointments and the nature of their courses and programmes'.[9]

The most utopian ambitions were laid out in the first schedule of the original Act of 1966, stating JNU's attachment to the principles which its namesake was said to have promoted in his life: national unity, democracy, social justice, secularism and science. The founders of the university felt that these principles would be achieved through the establishment of a different way of teaching, a new type of relationship between staff and student and physical environment compared with existing Indian universities. In terms of teaching and assessment, Gopalaswami Parthasarathy, eminent polymath, diplomat, journalist and test cricketer, the first vice-chancellor of the university, expounded a set of ambitious plans in April 1970. His university would be made up of centres, modelled on the institutional experiments of the Australian National University and the University of Sussex.[10] Parthasarathy's three stated objectives of the university were to promote interdisciplinary studies, provide integrated courses in the humanities, science, technology and social science and 'strive to create in its teachers and students an awareness and understanding of the social needs of the country, and to prepare them for fulfilling such needs'.[11] The university would begin with research students and then move on to teaching postgraduate degrees while only students coming to learn a foreign language could study for an undergraduate degree. In December 1970 he reported a successful beginning, stressing especially the establishment of the School of Social Sciences bringing together sociologists, historians, international relations specialists, political scientists, economists and educationalists to work together as teams on 'selected sectors of vital importance in our national life'.[12]

The architecture of the university was equally visionary. A New Delhi architect, C. P. Kukreja, working on his first major project, won 35,000 rupees for his design.[13] For Parthasarathi, the arrangements for each university building in the new university needed to reflect the interdisciplinary character of the academic programmes. The library was the symbol of unity and each building 'should radiate from this focus and be closely linked to each other'. This would, he felt, facilitate conversation across specialisms and break down disciplinary barriers. Equally 'the residential complex should reflect unity in diversity in India and should be imbued with the spirit of democracy and social justice'.[14] It was especially important, in his view, that students and teachers should live close to each other.

However, it was primarily JNU's commitment to student participation and social inclusion that made it so utopian. Students were drawn into the governance of the university from the start. A 1969 report by Dr P. B. Gajendragadkar had already argued that students needed to be involved in the running of their universities. Amar Chand Joshi, the illustrious botanist, veteran vice-chancellor of several other Indian universities and UGC member, mooted the idea of joint teacher–student councils.[15] JNU was unique at its birth in ensuring that the findings of the Gajendragadkar report would be upheld in its governance structures. Parthasarathy agreed with the students union, newly established in 1969, that the university could not ban the union.

Parthasarathi also ensured that an obligation to recruit from all parts of the country was written into the university's founding charter. On 24 July 1973 an agreement was signed between the JNU Student Union (hereafter JNUSU) President V. C. Koshy and the teachers' representative, Yogendra Singh, confirming that the university admission policy would be worked out by a joint student–faculty committee. The JNUSU had demanded that admissions at the university must satisfy the following criteria: expansion of numbers over time, regional parity in the allocation of places, weight given to academic merit and reservation of places for 'Scheduled Castes' (SCs) and 'Scheduled Tribes' (STs) in 'accordance with the law'.[16] This agreement was followed in 1974 by the establishment of a new method of admission. Each student was to be assessed on a 100-point scale. If they were a member of the scheduled caste or tribe, they would be given nine points automatically; if they were from a low-income family, they were given four points and if they were from a designated 'backward' region, they were given four points. The whole process involved an exam and interview by faculty members with input from students serving on the Student Faculty Councils. One academic, D. Bannerjee, has reminisced that Parthasarathi observed in the Academic Council that JNU should attach greater importance to teachers who could help an underprivileged student increase her mark from 45 to 50 than to those who helped a privileged student improve a grade from 60 to 85 marks.[17] Five years later, the JNUSU went further, calling for 'a more scientific backward area classification' taking into account the existing provision of educational facilities in the locality from which the applicant originated.[18]

Such a procedure of positive discrimination was important in a country trying to tackle caste injustices through the implementation of reservations in the public sector. Article 341 of the Indian Constitution provided that certain quotas be set aside to reserve places in public employment and higher education for people of castes and tribes on a scheduled list of 'untouchable' or depressed groups (these groups are now called Scheduled Castes and Scheduled Tribes). However, the measures adopted in JNU went much further, taking into account inequalities in parental income and region of origin as well. This was unique and revolutionary for the time. Greater inclusivity was not straightforward. Students of the time later recalled fighting for each point of the admissions policy as well as demanding increased scholarships for students of families with low parental income.[19] They fought by pamphleteering, negotiating and blockading the vice-chancellor's office for hours on end. For this tactic to succeed the students needed an amenable vice-chancellor. Parthasarathi was one such vice-chancellor. His successors considerably less so.

Left-wing student leaders also characterize this early period in JNU's history as utopian. Prakash Karat, a student at JNU who was part of the Student Federation of India, the student wing of the Communist Party of India, champions the students' democratic victories, pointing to the change in admission policy and the lack of sexual harassment of women students, or 'eve teasing,' as stemming from the strength of student democracy on the JNU campus. Indeed, he felt that this was why 'healthy relations have developed among men and women students', and why JNU had also escaped other trends in student life. 'Unlike at other elitist institutions, drugs or other symbols of western degenerate culture have no attraction for the overwhelming majority of students,' he observed.[20]

Students not only had power to approve admissions. The first vice-chancellor also heeded their opinions on the establishment of ceremonies at the institution. Thus when the first JNUSU president reprimanded the university for wasting money on an extravagant convocation ceremony in 1972 asserting that the money would be better spent on additional scholarship funding for needy students, the university complied and ceased organizing such events. At this one convocation itself, instead of inviting an eminent academic to address the graduands, the JNUSU invited Balraj Sahni, a veteran communist actor, much to the bemusement of the gathered liberal intelligentsia.[21]

JNU under fire

Within the first few years of its existence, JNU had acquired its own distinct identity. Despite being funded by the government, JNU was effectively an autonomous institution. It was quickly open to criticism. Student action and non-compliance with Indira Gandhi's Emergency from 1975 to 1977 led to the university losing the protection of Congress politicians at the centre, while the government that Indians elected in 1977, the first non-Congress government since independence, formed by the Janata Party (JNP), looked upon JNU as a personal legacy of Nehru and his daughter. JNU could not escape the political earthquakes that divided the Indian polity in the 1970s and 1980s. In June 1975 Indira Gandhi imposed emergency rule. The Emergency lasted almost two years, and police rounded up 100,000 people, mostly members of the political opposition, including JNU students, keeping them in indefinite detention under the Maintenance of Internal Security Act. In JNU the ramifications of the Emergency were complicated. Students were divided between left groups whose parliamentary leaders supported Gandhi and hence the Emergency and those, on both the left and right, who opposed her and were imprisoned. These divisions also permeated the teaching ranks at the university.

Other criticism ran deeper. In a survey on Indian intellectuals, the novelist Akhileshwar Jha bemoaned the fact that JNU has 'produced not a single pioneering research work, of a kind that could justify its existence as an exclusive elite institution of higher research',[22] while the eminent journalist Usha Rai wrote in 1982 that the basic concept of multi-disciplinary research had floundered. The university was seen as a cosseted and exclusive institution on which the central government was wasting resources.[23] V. C. Koshy, the second president of the JNUSU, echoed these sentiments.

He insisted that students came last in the priorities of the JNU administration after carpets, landscape design, architecture, accommodation for the higher echelons, swimming pools and artificial lakes. He also attacked JNU's small student–staff ratio, 1:5, suggesting that it should come closer to the Indian average of 1:20. The most vehement attack on the elitism of JNU students came from veteran left-winger Amiya Rao who wrote in 1983 that 'the whole idea of producing men and women whose discoveries in the field of knowledge would benefit the nation has gone awry – only half-baked bureaucrats come out who, as students in this prestigious University, develop a kind of obnoxious superiority towards fellow students getting less expensive education in less advantaged universities'.[24]

The admissions policy was targeted by the left and right for a number of reasons. Some saw it as a ploy to gain political leverage on campus, to reward applicants from particular regions or those with a particular political bias. In the early 1980s, rumours about some applicants producing fake certificates to gain deprivation points were confirmed to be true.[25] Others saw the deprivation points as a reason for falling academic standards. Above all the admission policy was blamed for the strength of student politics at JNU. Such criticism began in the early 1970s when the conveners of the Centre of Historical Studies were held to account for recruiting too many local Delhi students. Other protests followed, demanding fairer working arrangements for *karamcharis* or mess-workers and better food for the students. There were also protests conducted against the Emergency in an underground fashion. Some commentators refused to take such politics seriously. Responding to the mess struggle in 1975 when students took over the running of the kitchens in JNU hostels, an *Economic and Political Weekly (EPW)* editorial called their protests an 'intra-elite struggle' and urged them to remember that most Indians lived in poverty and could never hope to eat the sort of food for which JNU students were fighting.[26]

Some JNU vice-chancellors did take on the students. For example, in the academic year 1981–2, Vice-Chancellor Nag Chaudhuri tried to dilute the admissions policy by cutting the threshold at which deprivation points began. The students prevented this move, even though they were divided on how best to reform the admissions policy. They dismissed pure 'merit' as the basis for admission. One students' organization, 'Samara Yuvajan Sabha' ('Organisation for Equality among the Youth'), deemed merit a 'bogey', as it 'is provided by an educational system which works with many biases. Under this system, merit is, to a large extent, determined by one's background [...] Thus merit cannot be looked at in isolation [and] due considerations have to be given to the socio-economic conditions of regional backwardness'.[27] Undeterred, Chaudhuri came down hard. Following a relatively minor dispute over student accommodation, he shut down the university for an entire year, forcing even overseas students to quit their rooms. Playing the blame game, the UGC reported that the 'false admissions' had contributed to disturbances in the university.[28]

Such developments attracted a swathe of negative commentary from press and politicians alike. Three government inquiries are known to have investigated the affairs on JNU between 1978 and 1981 alone. Although their findings were filtered through the press, they were never officially made public.[29] Whilst a number of causes were identified, foreign influence was a common denominator. Even before the opening of JNU, Indian MPs tried to control its academic centres by insinuating that intellectuals who may or may not have received money from agencies like the CIA threatened the national security of the country. The celebrated librarian, Girja Kumar, suggested that foreign experts who came to India were invariably of low academic standing, usually at the end of their careers. He wanted to ban all foreign travel for Indian academics.[30] Others were less severe. Several speculated that aspects of JNU radicalism and research were adverse developments of foreign influence. For instance, in 1971 Barun De lambasted the style of 'participative management', imported from the West as a means by which student ringleaders could be co-opted.[31] Accusations that JNU was recolonizing the Indian mind were frequent. In his diatribe against JNU scholars written in the aftermath of the Emergency, when the Janata government was targeting JNU historians, Sumanta Bannerjee lamented 'this habit of being taken in by internationalist postures and mouthing of Western liberal ideology has become almost a tradition in Marxist historiography in India'.[32] JNU scholars themselves made similar claims. For example, sociologist Satish Saberwal called for an end to 'acacolonialism' or the political dominance of the Western agenda on pedagogy and teaching in the postcolonial world.[33] K. R. Bhattacharya took it further, suggesting that Indian research was deeply affected by Western superiority: Western academic dominance was fostering class philosophies and focusing research on topics such as modernization, development and industrial economics, instead of more Indian concerns. As he stated, 'Only this explains the mistaken growth of social science and management institutes in the country (largely foreign financed) in the last decade or so.'[34] Students also showed wariness towards foreign finance and influence. As early as 1973 Koshy wrote against the teacher's 'constant foreign tours to attend international conferences and seminars', while the Communist leader Prakash Karat listed the many international, especially American connections, that top JNU academics enjoyed, keeping them away teaching JNU students. He was also convinced that pro-US bias influenced JNU programmes. For example, he questioned why the Centre for South East Asian Studies overlooked Vietnam, Cambodia, North Korea and Laos.[35] A pertinent question indeed in 1975. When asked why JNU had failed one former student summed it up: 'The foreign degree still holds more value than our own.'[36]

JNU, the government and the 'nation'

Another reason for JNU's retreat from utopia was the level of political interference by government. Since it opened in 1969, JNU was seen by many as embodying the Congress–Communist alliance. In the late 1960s and early 1970s the Congress government entrusted a number of scholars who were sympathetic to secularism and some shade of socialism to lead institutions such as the Indian Commission on Historical Research and the Indian History Congress as well as inhabit the teaching corridors of JNU. The press and disgruntled academics on the outside of this clique saw this as an exercise of patronage in which pro-Congress academics were given jobs to propagate pro-Congress views. As an *EPW* editorial read,

A good proportion of the teachers feels a little Leftist and left-out, committed as they are to the triple virtues of secularism, socialism and democracy. To play their role they need to be guaranteed the right standard of living, the proper number of trips abroad, and adequate opportunities for social climbing.[37]

This view gained ground as JNU staff were reprimanded for not being an effective voice of protest during the Emergency. Even Prime Minister Morarji Desai's 1978 inquiry into the affairs of JNU during the Emergency was far more focused on the leftist ideology of students and staff rather than with administrative compliance with the Emergency, for example, the arrest, torture and year-long detainment of a man who was mistakenly taken to be the JNUSU president. As Sumanta Bannerjee concluded in 1978, such enquiries that were meant to look into bad recruitment practice and nepotism were corrupt in themselves, as they were instructed to ignore appointments involving supporters of the Janata government. Desai was supposed to have remarked to the vice-chancellor, 'a lot of communists' were employed at the university.[38] Such criticism appeared again in 1983 in the aftermath of the university closure. Veteran educationalist Amrik Singh called it a golden opportunity to reform the university, claiming that the university, now vilified by the public, needed to wrestle control back from students. This was a propitious moment because the political supporters of the left and Congress who had previously thwarted attempts to reform the university 'are at the moment in a state of disrepute as disarray. Not to understand the dynamics of the current situation is to miss the elementary fact about the JNU'.[39]

While these claims and counter-claims are to be expected because of the legacy of Jawaharlal Nehru and his close identification with Congress politics, the fevered rhetoric underlines how the JNU experiment exposed a contradiction between the widely different expectations of the press, public, government and educational reformers. In the 1960s, the higher education press in India was dominated by columnists and contributors critical of the unplanned expansion of higher education, voicing the rising expectations of the unemployed and of the educated middle classes. Trying to explain student unrest in 1967, A. C. Joshi pointed to similar agitations in 'Korea, Vietnam, Turkey and in almost every developing country in Asia, Africa and Latin America'. For him these agitations were inevitable because three or four centuries of development were being rushed into a few decades.[40] In JNU the expectations were even more pronounced. After a year at its helm, Parthasarathi gave JNU a glowing report. In line with the high expectations of the university named after Nehru, it was going to trump British universities. For 'while the experience of UK universities has been that it takes three years after the appointment of the first VC for the first batch of students to be enrolled, we expect to make quicker progress.'[41] And indeed it did.

What kind of students was this new university enrolling, critics asked. One of the first JNUSU presidents bluntly delineated the contradictions in admission practice in 1973. How could a university committed to improving social justice in its teaching and admission practices also be an elite and exclusive centre of excellence? The only way it could be a national university in the true sense of the word was to encourage more rural and poor entrants. However, the entry requirements for JNU emphasized 'abstract academic quality'. Advertisements for JNU only appeared in the English-language dailies. While Indira Gandhi talked at JNU's inception about mass education, her Minister for Education, V. K. R. V. Rao, talked about elite education in which students would be instilled with a sense of social awareness to assist Indian development. As V. C. Koshy concluded, 'It is futile to build a "model institution" in isolation from the existing realities of our educational and social framework ... The university is being used as a progressive façade, an attempt to patchwork the rotten education system that this country is burdened with.'[42]

This continuous battle over the admissions policy illustrates the dilemmas inherent in JNU's utopian project. From the early 1980s, the university administration, along with an extremely vocal professoriate, pressed the university to change the method of admissions and get rid of the special deprivation points given to applicants from so-called educationally backward regions or poor income families. The joint secretary of the UGC encouraged them, stating in 1981 that merit be the only consideration of admission. In his view, 'our universities have been following an open door admission policy with the result that we find that a large number of students who have neither the ability nor the aptitude for higher education but are there like jardinière. Access is a free as in any film with "U" certificate'.[43] Continuing this film theme, Bipin Chandra, the JNU historian who would later be commissioned to look into the admission system of the university, publicly attacked the deteriorating standards of university education, 'which was producing unthinking people incapable of reading anything more taxing than a film magazine'.[44] On 22 January 1984, the JNU academic council announced a complete transformation in the admissions procedure. They took on board the entirety of Bipin Chandra's committee findings on admission policy. Merit would be the only criterion, while reservations for SC and STs would be the same as in other public institutions. In one fell swoop the radical admission policies of JNU were changed, just at the moment when students were returning after the closure of the campus. Although the universities easily implemented the new admission policy and even gained the compliance of the leading Communist Party student leaders in JNU, the issue never really went away. In 1990, Solidarity, a left wing collective, won the JNUSU election with the promise to overturn the 'new admission policy' and in 1993, the Marxist–Leninist student wing, 'All India Students Association', finally managed to change the admissions policy back to not only include positive discrimination for poor students coming from backward regions, but also, for the first time, gave deprivation points to women students.[45] The students had shown that their party affiliations were less important than their desire to stick to the principles of social justice embodied in JNU's founding charter.

Not surprisingly, student power came to be seen as JNU's biggest problem. Remarkably, lecturers at the university used the press to denounce their own students. As early as 1972, after the students invited the radical actor Balraj Sahni to speak at the Convocation ceremony, a letter signed by fourteen faculty protested that, 'if this trend continues it cannot be long before cabaret dancers address convocations'.[46] Nine years later Bipin Chandra complained about student agitation in the name of radicalism, so violent that vice-chancellors required police support. Students were blamed for all manner of problems on campus. Satish Saberwal, a sociologist in the Centre of Historical Studies, claimed that Indian students could not cope with conflict and impersonal normative order and that they found engaging with debate difficult. This is why they could not rethink the admissions issue.[47] With one eye on JNU, the UGC reported that Indian universities housed a large number of students who stayed on long after the normal period of a degree. These 'over stayers' were to blame for student agitation,[48] a view that gained ground as it was the removal of an 'over-stayer' student at JNU that sparked off the conflict in 1983. Amrik Singh also criticized the amount of power enjoyed by JNU students. In his view 'no university can run if the students call the tune; it is the teachers who should do that'.[49] Amiya Rao put his finger on the main issue, 'the root of the trouble is that the very system, imported from abroad on which JNU is based – the democratic liberal system believing in the concept of student participation in its functioning – is causing friction and frustration. The teachers have begun to question the wisdom of "over-democratisation" of the University'.[50]

Equity at work

Equity in the admissions process has long been battled over at JNU across the last forty years or so, but what has been the experience in real terms? JNU prepares extensive annual reports which provide detailed information on various aspects of the University's performance, over each fiscal year from 1 April to 31 March. The analysis that follows is derived from this data during the last three decades.[51]

Whilst this data shows some skewed distribution of opportunity across the different categories of entrants to JNU, there is a clear trend in the direction of social equity over

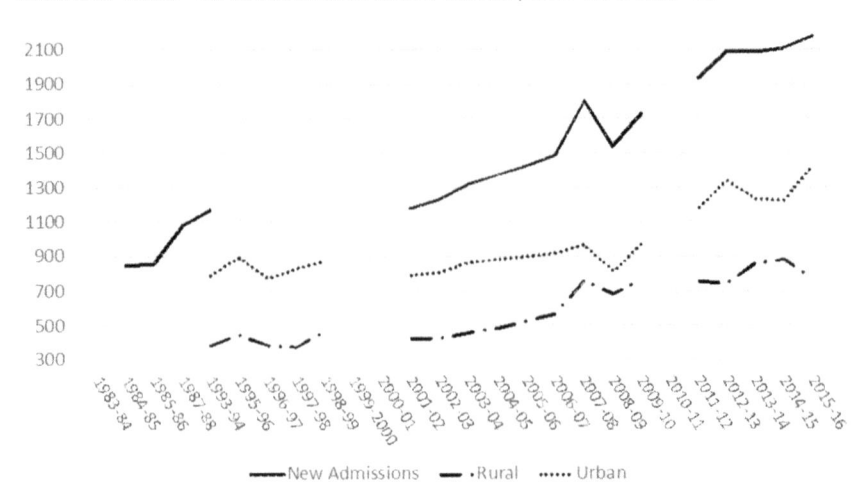

Table 18.1 Rural – urban distribution of new entrants, 1983–84 to 2015–16

Table 18.2 Gender distribution of students across all disciplines

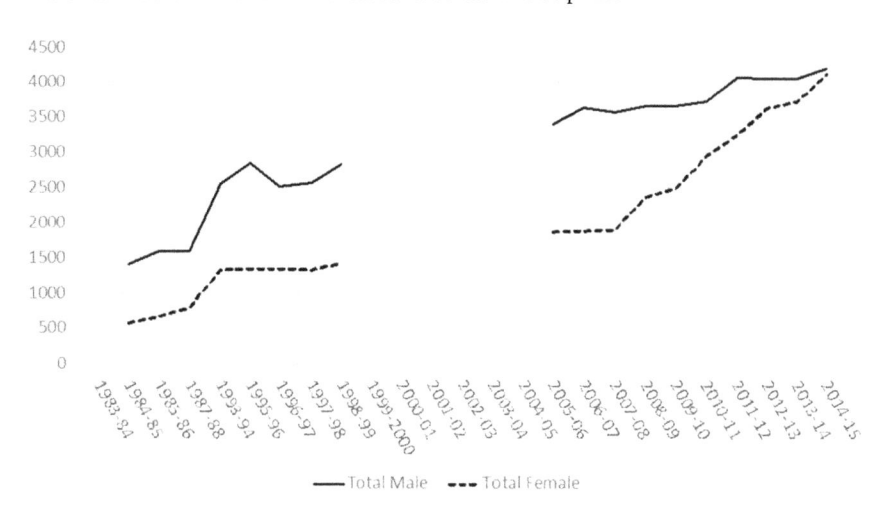

Table 18.3 Profile of social reservations for students

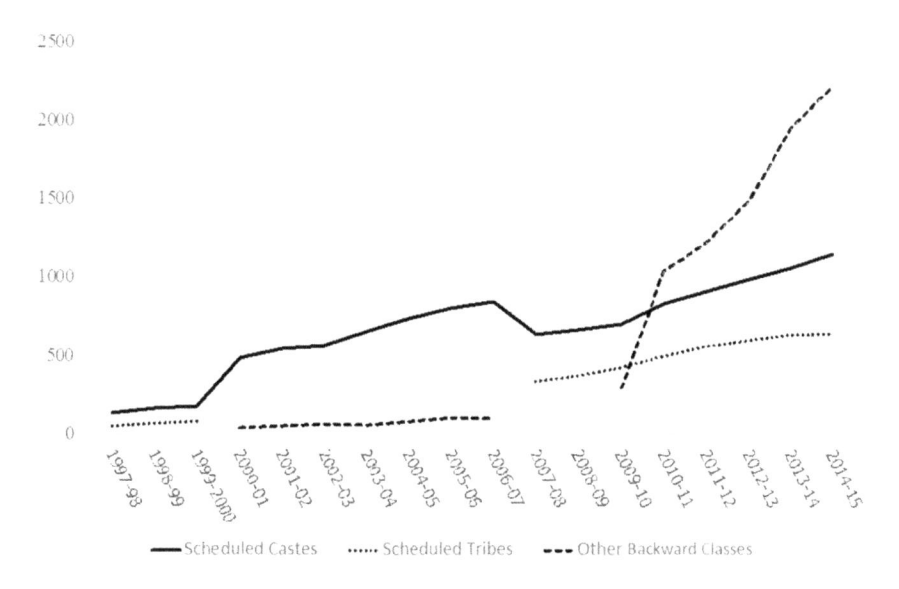

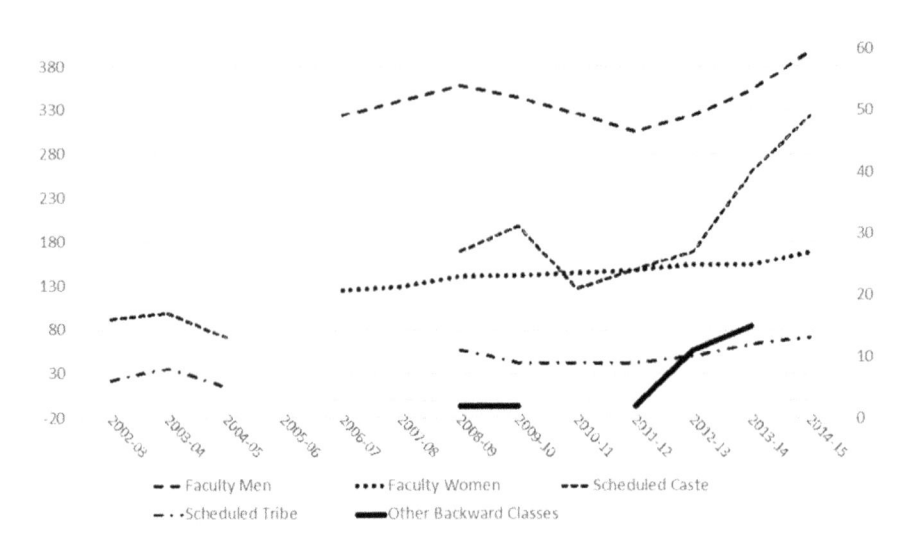

Table 18.4 Gender and social distribution of faculty

Table 18.5 Admissions at JNU, 2008–14

Academic year	Numbers admitted	General	OBC	SC	ST	Persons with disabilities
2008–9	1798	958	351	254	137	38
2009–10	2023	950	446	317	178	49
2010–11	2207	1026	530	311	194	64
2011–12	2203	900	672	320	193	58
2012–13	2347	946	722	343	209	51
2013–14	2110	847	730	323	166	44
2014–15	2184	885	737	319	187	56

the years. The largest lacuna seems to lie in the rural–urban ratio of students. For this reason, in 2016 JNU adopted a differentiated system of deprivation points, in the hope that this will equalize further opportunity. Table 18.5, which provides admissions data by programme of study over a seven-year period, offers further corroboration of greater equity.[52] The Indian Parliament remains constantly vigilant in the enforcement of the mandated reservations policy in the public education sector, particularly at JNU because of its national status. Such data as this on student admissions tells its own story. Even the Indian courts have on occasion cited the JNU admission policy as a model for an equity-linked protocol that other Indian universities might emulate.[53]

The pattern of faculty appointments, by contrast, shows less of an onward march towards equity, as Table 18.4 reveals. One in every seven staff appointed at JNU currently belongs to a reserved category. In a reply to a question raised in the Indian Parliament in July 2016, it was revealed that out of a total of 900 sanctioned faculty posts, 609 were occupied (291 positions were still lying vacant). Of those 609 posts, only 10 per cent and 3 per cent were staffed by Scheduled Caste and Scheduled Tribes categories respectively, whereas OBC constituted 7 per cent.[54] The constitutionally mandated quotas are 15 per cent (SC), 7.5 per cent (ST) and 27 per cent (Other Backward Class) of the total sanctioned posts for these three categories. At the same time, Table 18.4 also shows a gender bias towards men, despite some improvement over the years. JNU still has a long way to go to fulfil its aspirations towards equity as far as staffing is concerned.

However, here too, there has been criticism of the very reservation policies upheld by the University. For example, in 2009, when JNU advertised 149 new teaching posts with a reserved quota in all ranks and not just assistant professorships, the anti-reservation lobby managed to freeze the selection process despite students and pro-reservation teacher protest and in a note to the vice-chancellor, the former vice-chancellor and emeritus professors such as Alagh, Oomen and Bipin Chandra argued, 'If steps are taken which prevent it (JNU) from remaining one of the premier centres of excellence (which is what we fear will happen by limiting, through reservation, the scope for selecting the best faculty at the senior professor and assistant professor levels) the chief victim will be the disadvantaged sections of Indian society. If JNU declines, the well to do will move to foreign and private universities and the disadvantaged will no longer be able to get world-class education, which JNU has been so proud to offer them so far.'[55] Thus in their eyes, the problem is not inherent in the teaching system which continues to perpetuate stark inequalities in Indian society, but in the proposed solutions of quotas which the government has suggested will go some way in remedying some aspect of those problems.

Women at JNU

Nowhere have the founding ideals of JNU been more fiercely fought out than in the conditions of women at JNU. Special treatment for women was covered in JNU's founding Act (2nd Schedule, para. 35 (f)), but ultimately was something that could only have emerged as a result of the politics of female liberation on the global and Indian scene, as well as from the experience of women students at JNU. Beginning with the first student agitations on the JNU campus in 1973, women were as active as the men, despite journalists and academics venting against the docile and un-academic nature of women students in India.[56] Even friends of feminism regretted the tragic influence of the West which had reduced the modern Indian university woman into someone interested in Western fashion, miniskirts, blue films, speed, smoking and even drug addiction![57] The JNU 'type', modern, assertive and confident, certainly contrasted with conventional conservative images of Indian women. Many have categorized this as a result of the predominance of the left on campus. For example, Nina Rao, a former student who joined the campus in 1972 and became active in the Students' Federation of India, remembers that 'JNU perhaps played the biggest role in empowering girl students because of the influence of the left. There was no formal woman's or feminist movement, but there was equality of opportunity and both sexes had access to the JNU student body. In the alternative JNU culture, all marginalised and oppressed groups were made visible through affirmative action and weightage was given to ensure a level ground for all to enhance their potential.'[58] While this account is tinged with nostalgia and romance, it does spell out the importance of the progressive left. Equally biased is the converse view, presented by Batabyal, in his 2014 history of JNU, who suggests that the SFI actually used attractive women as sexual bait in order to ensnare innocent new recruits to its ranks.[59]

In terms of participation in student politics, women were arrested in equal measure during the Emergency and in the 1983 agitation. Women were also elected officials in the JNUSU from the very beginning. In 1983 Rashmi Doraiswamy became its first woman president. In 1990, the winning JNUSU movement, 'Solidarity', campaigned for less stringent security measures in women hostels, thus giving more freedom of movement for the women on campus. And in 1993 the incoming JNUSU President, Pranay Krishna, forced through deprivation points for women in order to facilitate more women admissions.[60] However, women students did not just campaign for more of a say in conventional student politics. A group of women set up 'Spruha' ('consciousness') in the 1990s which called for political intervention on women's issues as women and not by men on behalf of women. They held women-only meetings that not only discussed women's politics and theoretical concerns, but also how to intervene on practical issues of campus security, 'eve teasing' and gender awareness. In the late 1990s students began to demand formal commitments from the university to redress sexual harassment and punish sexual predators on campus. This was intensified after a landmark Supreme Court judgement in 1997 that ruled against the state of Rajasthan for not granting justice to a rape survivor. Coinciding with a number of molestation cases on campus, students and faculty organized a state convention on gender awareness and put pressure on the university authorities to act. Like most change that has happened in JNU this was not something to which the university Executive Council responded positively. Indeed, Asis Datta, the vice-chancellor at the time, claimed that any committee on gender awareness would give the university a bad reputation.[61] However, the students did not let up. The result was the Gender Sensitisation Committee against Sexual Harassment, established in February 1999 after two years of pressure. Through the work of faculty and students, this committee drew up a set of guidelines and procedures to deal with harassment. In 2001, after much debate the guidelines were added into the constitution of the University. Here was a committee that would report annually on complaints, gender grievances and discrimination, address general problems of gender inequality and awareness on campus, and punish individual perpetrators of sexual harassment. It is a formula that has been adopted by a number of Indian universities since its inception and one which would not have been possible without the persistence and resilience of JNU women.[62] The conflicts over womanhood and sexuality in JNU, and its depiction by Indian press and politicians rage to this day. In the recent agitation over Kashmir which threatened to shut down the university in 2016, political critics talked of the numbers of condoms and cigarettes being used in JNU,[63] caricaturing the sexually liberated woman of the campus as a contrast to the ideal Indian woman.

Conclusion

Has JNU failed in its professed mission both, utopian and conventional, as some claim? Not really. JNU has consistently been in the top 100 rankings of Asian universities, ahead of its national counterparts, while its students have joined the faculties of universities across India and overseas. JNU students continue to fulfil the ambitious desires of 'national integration' that were voiced in the 1960s, through the steady recruitment of JNU students, through competitive exams, into the Indian Administrative Service every year. The university has also increased its offering by introducing more interdisciplinary centres of scholarship. The School of Arts and Aesthetics and the Centre for Studies in Law and Governance are two such innovations. Artists, filmmakers, dancers, musicians and constitution scholars and cultural historians have joined the teaching ranks of the university. More recently, the university has unveiled an innovative range of fifteen 'trans-disciplinary clusters', combining the methodologies of the social, biological, environmental and physical sciences in order to stimulate new systems of knowledge.[64] Here might be seen a reply to critics: a vindication of an admissions policy that does not deter world-renowned scholars from coming to JNU.

In this way, the experience of JNU belies arguments that affirmative policy through reservations, a hallmark of the Indian constitution, is the antithesis of merit. In a society like India, where so many hierarchies based on caste, gender and community still exist, merit as an objective measure of competence seems a chimera. According to official statistics, Scheduled Castes and Scheduled Tribes are still shut out, on a very large scale, from the kind of educational opportunities enjoyed by other Indian boys and girls at primary and secondary school level.[65] The principle of reservations in university admission, extended and deepened across the decades at JNU, embodies a key principle of John Rawls' classic theory of justice, namely that social primary goods – liberty and opportunity, income and wealth, and the bases of self-respect – are to be distributed equally unless an unequal distribution of any or all of these goods is to the advantage of the least favoured.[66] The whole point of reservations is that work to the advantage of the least favoured.

At the same time the on-going battles over admission policy and the wider challenge of Indian patriarchy and competing ideas of the 'nation' are a reminder that the utopian ideals of JNU have been constantly debated, decried and developed by supporters and detractors alike. As the 2016 row over 'sedition' on campus demonstrated, JNU remains one of the most politically radicalized universities in the world. Merit has become laced with empowerment. Originally conceived as a holding station for the liberal intelligentsia, JNU evolved into a 'leftist bastion' because of its 'unique reservation policy'. Social inclusion changed the texture of the campus. As Pramod Ranjan expresses it: Marx, Lenin and Mao co-habited with subaltern radicals such as B. R. Ambedkar and Jyotirao Phule: 'The students from the deprived sections brought with them life experiences and thinking that were hitherto alien to the academic world'.[67]

As with many other post-colonial new universities, especially those in Southeast Asia and sub-Saharan Africa, where newly independent republican regimes created flagship 'national' institutions of higher education, JNU could never escape the immediate political agendas of those in power. However, it is to the credit of the students in particular, that JNU's defining mission of equity has always been kept in view. If JNU can be seen as a utopian university, it is only because each new generation of its students continues to fight afresh for its utopian ideals.

Notes

1. Report of the Visiting Committee to the JNU, 1–2 February 1972, Appendix, 46–48th Interuniversity Board Annual Meeting, 1971–3, 49.
2. Rakesh Batabyal, *JNU: The Making of a University* (Noida: HarperCollins, 2014).
3. For example, see the 'Report of the Committee on Emotional Integration' (Ministry of Education, 1962); A. B. Shah, *A National University, Papers and Proceedings of the Seminar Convened on September 28–30, 1962* (Bombay: India Committee for Cultural Freedom, 1964).
4. Rochana Bajpai, *Debating Difference: Group Rights and Liberal Democracy in India* (Delhi: Oxford University Press, 2011).
5. For some of the background, see Philip G. Altbach, 'Problems of University Reform, (1974) in *A Half-Century of Indian Higher Education. Essays by Philip G. Altbach*, ed. Pawan Agatwal (New Delhi: Sage, 2012), 24–6.
6. This is a sentiment repeated in many of his speeches, for example, 'The Bases of Indian Unity' (6 August 1954), *Selected Works of Jawaharlal Nehru*, ed. Ravinder Kumar and H. Y. Sharada Prasad, second series, 26 (1 June 1954–30 September 1954) (New Delhi: Jawaharlal Nehru Memorial Fund, 2000), 90–8.
7. Joseph E. DiBona, *Change and Conflict in the Indian University* (Durham, NC: Duke University, 1969), G. S. Mansukhani, *Crises in Indian Universities* (New Delhi: Oxford & IBH Pub. Co., 1972).
8. *Report of the Committee to Enquire into the Working of Central Universities* (New Delhi: Universities Grant Commission, 1984), 2.
9. Ibid., 5
10. Sabyasachi Bhattacharya, 'Reflections on the Cultural Personality of JNU', in *JNU: The Years: An Anthology by the Silver Memoir Committee*, ed. Kanjir Lochan (Delhi: Popular Prakasan, 1996), 129. On Parthasarathri, see H. Y. Sharada Prayad (ed.), *G. P.: The Man and His Work* (New Delhi: New Age International, 1998). See also https://scroll.in/article/884499/the-man-who-built-jnu-understood-what-a-university-stands-for-the-freedom-to-question-and-debate (accessed 20 September 2019).
11. *University News* (April 1970), 1.
12. Ibid. (December 1970), 14.

13 Ibid. (October 1970), 13; A. Kanvinde and H. J. Miller, *Campus Design in India: Experience of a Developing Nation* (Topeka, Kansas: American Yearbook Company, 1969); Batabyal, *Making of a University*, 172–9.
14 *University News* (December 1970), 17.
15 A. C. Joshi, 'On Student Unrest. II', *University News* (March 1967), 3–4.
16 'JNUSU: An Admission Policy for JNU' (21 July 1973), in *Ideology of/in Admissions Policy: Documents of a Debate at JNU* (JNU: Students Locus, 1982) (unpaginated); Durendra Kumar Pattnaik, *Student Politics and Voting Behaviour: A Case Study of Jawaharlal Nehru University* (New Delhi: Concept, 1982), 30.
17 D. Bannerjee, '25 Years of JNU: Betrayal after a Promising Start', in *JNU: the years*, ed. Lochan, 36.
18 'JNSU: Resolution on Admission Policy' (30 January 1979), in *Ideology of/in Admissions Policy*.
19 Prakash Karat, 'JNUSU: The Early Years' in *30 Years of the JNUSU* (Delhi: JNUSU, 2002), 4.
20 Prakash Karat, 'Student Movement at Jawaharlal Nehru University', *Social Scientist*, 3 (May 1975), 52.
21 *Times of India*, 18 November 1972, 6.
22 Akhileshwar Jha, 'A University Survey', *Seminar*, 222 (February 1978), 31.
23 Usha Rai, 'An Elite Level', ibid., 272 (April 1982), 36.
24 Amiya Rao, 'Jawaharlal Nehru University', *Economic and Political Weekly*, 9 July 1983, 1222.
25 'Administrative Changes over the Years' in *JNU: The Years*, ed. Lochan, 26.
26 (Anon.), 'Bread, Butter, and Jam Stir', *Economic and Political Weekly*, 26 April 1975, 689.
27 'JNUSU: An Admission Policy for JNU'.
28 *Report of the Committee to Enquire into the Working of Central Universities*, 31.
29 'JNU, An Introduction by the Committee', in *JNU: The Years*, ed. Lochan, 7.
30 Girja Kumar, 'Servitude of the Mind', *Seminar*, 112 (December 1968), 20–4. Kumar was then the Librarian of the Indian Council of World Affairs.
31 Barun De, 'The Problem', *Seminar*, 148 (December 1971), 14.
32 Sumanta Bannerjee, 'Devaluation of Marxism by Leftist Academicians', *Economic and Political Weekly*, 8 April 1978, 605–8.
33 Satish Saberwal, 'The Problem', *Seminar*, 112 (December 1968), 9.
34 K. R. Bhattacharya, 'Science Politicking', ibid., 148 (December 1971), 22.
35 Karat, 'Student Movement', 51.
36 Nina Rao, 'A View from the Past' in *JNU: The Years*, ed. Lochan, 122.
37 'Bread, Butter and Jam Stir', 689.
38 *Economic and Political Weekly*, 9 September 1978, 1542–4.
39 *Times of India*, 28 June 1983, 8.
40 Joshi, 'On Student Unrest'.
41 *University News* (December 1970), 14.
42 V. C. Koshy, 'JNU: A Model University?', *Social Scientist*, 2 (September 1973), 51.
43 S. C. Goel, 'Issues and Options in Higher Education', *University News* (September 1981), 504.
44 'Campus news', ibid. (October 1981), 597.
45 'Administrative Changes over the Years', in *JNU: The Years,* ed. Lochan, 28.
46 The letter was signed by Anand Kumar and ten others: *Times of India*, 18 November 1972, 6.

47 Satish Saberwal, 'Roots of Enervation', *Seminar*, 272 (April 1982), 16.
48 'Report of the Committee to Enquire into the Working of Central Universities', 46.
49 *Times of India*, 2 June 1983, 8.
50 Amiya Rao, 'Jawaharlal Nehru University'.
51 Derived from the *Annual Reports* of the JNU, 1983–2016.
52 'Response to Parliamentary Question Regarding Student Intake in JNU' (August 2013); 'Note on Factual Data on Admissions to Various Programmes of Study for the Year 2015–16', submitted to the Academic Council on 27 May 2016.
53 'Gautum Sharma versus JNU and Others', Judgment of the Delhi High Court, 19 January 2016, www.indiankanoon.org/doc/72050881/. For the plaintiff's account, see Gautum Sharma, 'How I Fought Discrimination at JNU', www.governancenow.com/views/columns/how-i-fought-discrimination-jnu (accessed 1 September 2019).
54 'Response to Parliamentary Question Regarding Faculty Recruitment in JNU' (July 2016).
55 S. Santosh and Joshil K. Abraham, 'Caste Injustice in JNU', *Economic and Political Weekly*, 26 July 2010, 27–9.
56 Arban Kapadia, 'The Problem', *Seminar*, 44 (April 1963), 12.
57 Sadhana Rana, 'Herstory', ibid., 176 (April 1974), 29.
58 Rao, 'A View from the Past', 122.
59 Batabyal, *JNU, the Making of a University*, 186.
60 Pranay Krishna, 'JNUSU: 1993–4', in *30 Years of the JNUSU*, 28.
61 Albeena Shakil, 'JNUSU: 2001–02' in ibid., 38.
62 www.jnu.ac.in/AboutJNU/GSCASH.asp (accessed 6 June 2016).
63 *The Hindu*, 24 February 2016, https://www.thehindu.com/news/cities/Delhi/students-slam-bjp-mla-for-condom-statement/article8273803.ece (accessed 1 September 2019).
64 'Notification on Trans-disciplinary Clusters' (30 October 2013), JNU Co-ordinator Evaluation's Office.
65 According to the Government's Bureau of Statistics, in 2001, SCs constituted 16.2 percent and STs comprised 8.2 percent of the population of India. In 2011, 19.94 and 11.68 per cent of the ST population were enrolled in primary and secondary schools, whereas the comparable percentages for the ST were 19.18 and 9.41 respectively. In 2006–7, for every 100 boys in active education at all levels, there were eighty-eight girls in primary, eighty-two in secondary, seventy-four in colleges and sixty-two in universities: Government of India, 'Socio-Economic Statistics of India, 2011', tables 1.14, 3.11 and 3.7: mospi.nic.in/mospi_new/upload/sel_socio_eco_stats_ind_2001_28oct11.pdf
66 John Rawls, *A Theory of Justice* (Cambridge, MA: Harvard University Press, 1971); cf. Sebastiano Maffettone, 'Sen's Idea of Justice Versus Rawls' Theory of Justice', *Indian Journal of Human Development*, 5 (2011), 119–32.
67 Pramod Ranjan, https://counterview.org/2016/06/30/jnus-deprived-section-students-brought-with-them-experience-thinking-hitherto-alien-to-academia/ (accessed 30 September 2019).

From American dream to nightmare on the left: Student revolts, the 'crèche sauvage' and the slums: The University of Nanterre, 1962–71

by Victor Collet

In 1962 Parisian refinement and university modernization were exported to the city's suburbs. Lacking space in the capital, an autonomous campus based on the North American model carried the support of Gaullist policy makers. Built in record time by the architects Jacques Eugène Marie Chauliat and the building company Moinon, this new annexe of the Sorbonne was located in the middle of a disused military airfield, transplanted into a land of workers and Communists. Surrounded by factories, vacant lots, cafés and Maghreb slums, the Faculty of Letters at Nanterre was born. The year 1962 was also the end of the war in Algeria. A few hundred metres from the airfield independence was celebrated in the immigrant slums of Nanterre, La Folie and Petit Nanterre. Conversely, amongst the communists of the town and in the housing projects, there was a feeling of relief for the local residents who had gone over to fight in the war. At the same time there was an urgent need to remove the slums that damaged the image of municipal communism. With the Algerian war, the divisions among the working population of the city over housing and politics became more entrenched. By 1964 a second city had been planted in the heart of Nanterre.[1] How did the arrival of these renowned intellectuals of the social sciences, bringing with them to the suburbs the daily life of the Faculty of Letters, become integrated into these major upheavals in French politics? The contradictions at work were all in place at Nanterre, for this was precisely where the movement was born that led to the famous days of the student radicalism of May and June 1968. In just a few years, the new campus at Nanterre passed from an American-style dream of modernization to a leftist political nightmare within the Faculty of Letters, reaching its climax in the early part of 1970.

The physical separation between the city of Nanterre and the university influences the folk memory of the era. Two narratives dominate. The first places the slum of Nanterre within its Algerian past. The second inevitably links Nanterre to May 1968 and to the Faculty of Letters, and on to Daniel Cohn-Bendit, the public face of student protest in Paris. This latter narrative is hegemonic. It centres on the notorious student occupation at Nanterre of 22 March 1968 and the shutdown of the campus in May, events which fuelled the wave of protests across Paris. But this focus is flawed. It reduces the excitement of May 1968 to a contrast with the quiescent culture of the subsequent decade. Working-class militancy, as well as the politics of immigration, and the evolution of academic institutions are obscured as the focus is on the lives of those involved in the lifestyles of sexual liberation and other forms of identity politics.[2] This distortion affects our understanding of the events in Nanterre. What remains in the official memory erases what really went on. In particular, memories of 'red Nanterre', mainly focused on May 1968, talk of the shockwave caused by the student uprising and look back to the months and years that preceded it.[3] Loose language turns the memory of May 1968 into something unique. As for the Algerian memory, that is to say, the place of the working class, communist town of Nanterre in the history of this university, this has been, until recently, avoided. Eager to show its independence from the town of Nanterre, the university has always maintained its distance, both in terms of its academic standards and as a social space, from the poor environs which nestle around its perimeter. The real story of 'red Nanterre' is the moment in 1969 and 1970 when that boundary was transcended in the utopian idealism of students, factory workers and North African immigrant families acting in combination.

The third world comes to Nanterre: The student, the intellectual and Paris-Ouest

For over a year, from September 1969 to December 1970, the Faculty of Letters at Nanterre featured in the French national press every week: in *Le Monde*, *Le Figaro*, *L'Humanité*. The newspapers scrutinized the consequences of May 1968. The situation remained explosive in the universities. The media and the political authorities reported regular clashes between increasingly violent splinter groups from the extreme left, activists of the far right and the police. The press examined in particular the role of the new dean of the Faculty of Letters at Nanterre, Pierre Grappin (1964–8) and Paul Ricoeur, the *doyen* (1965–70), then the first president of the new university, René Rémond, and also the various exam boycotts, and the occupation of buildings and halls.[4] The sulphurous reputation of Nanterre endured during the decades that followed, and exists today in part as a memory of the past collectively maintained by the elders of 1968, and, once purified of its political danger, by the new university administration itself, signalled most recently when Cohn-Bendit was invited to join the current Board of Directors.

Yet, when the new university campus was created, nothing was further from the minds of the policy makers than such an 'eruption'.[5] Back in early 1962 the Ministry of Defence agreed to share a portion of its land in the suburbs. This was a military airfield at Nanterre (some 13 km out from central Paris and the Sorbonne), unused since the First World War and located in an area known as La Folie, then simply comprising a small railway station, around which a workers' town had grown up, dominated by the factories, including a car plant, which Citroën took over in 1961. Thus, the countryside had given way to a working-class suburb, and to the building of public housing and infrastructure.[6] The recipient of this handover of land, the Ministry of Education, was trying to cope with the increasing density of student numbers, following the doubling of the number of universities in Paris between 1960 and 1967.[7] Decongestion in the heart of Paris was imperative, so building at Nanterre was expedited from 1963. The Social Sciences building was finished in the summer of 1964 and opened its doors to its first year of students: just under 2000 registered, on what was still a construction site, but which also had a university restaurant and a modern cafeteria. A procession of both younger and less youthful renowned intellectuals quickly followed, from sociology, geography and ethnology (Alain Touraine, Henri Lefebrve, Michel Crozier, Jean Baudrillard, Eric de Dampierre, Marshall Sahlins) and from psychology (Matty Chiva).

On a new campus, with teaching programmes to construct, the new faculty saw the arrival of pioneering teaching teams as creating a special atmosphere: everyone knew virtually everyone else. Students, lecturers, professors and deans crossed paths daily. As reminiscences of former students attest, this mood quickly engendered pedagogical innovation and a relaxation of convention.[8] However, the student numbers at Nanterre swelled too rapidly for the infrastructure. In 1965, a second university residence with capacity for 1,200 permanently changed the appearance of Nanterre. Many students now lived on campus. Recruitment to Nanterre proved more successful than students joining campuses in wealthier parts of Paris such as in the west of the city.[9] Furthermore, Nanterre was fast becoming a militant hotbed. By 1967, the Nanterre residents' association (ARCUN), a very political group, had 800 members, far more than the main student union, UNEF, and it held its meetings in the campus restaurant.

A sports centre and swimming pool were added to the campus at the beginning of 1968, opened by the Minister of Youth Affairs and Sports, François Missoffe. The university residences became fully occupied. There was some negative comment at the time, due to claims of 'mixing' of women and men.[10] In the same year construction began on a large university information centre, and also on a second faculty building, for law and economic sciences. A few other projected restaurants (never fully completed) rounded off the idea of an independent, out-of-town American campus. This new annexe to the Sorbonne was very different from its historic sister in the centre of the city: huge and cut off from the capital, despite significant new transport links, such as the RER 'A' line, started at the end of 1969 (the two Nanterre stations opened in 1972). In the summer of 1964 the new Faculty of Letters was inaugurated with great ceremony by the Prime Minister, Georges Pompidou, and the Minister of National Education, Christian Fouchet.[11] The ceremony began in the Administration Tower, in the Board Room on the seventh floor, a symbol of academic power overlooking a working-class city with its limited public-housing. The neighbouring cities of the French provinces looked on warily at this new development at Nanterre. The visiting ministers and senior officials ended their tour at the university restaurant and in the gigantic halls of the Faculty of Social Science.

As it opened for business, there was a clear division between the city of Nanterre and the new Faculty of Letters. Firstly, the transplantation from the Sorbonne had taken place without the agreement or even consultation with the local municipal authority, the Commune, then controlled by the Parti Communiste Français (PCF). The latter viewed the arrival of this cohort of intellectuals into working-class territory with a suspicious eye. The new site of the university encroached upon land coveted by the PCF mayor, Raymond Barbet, for social housing and cultural facilities, to satisfy an electorate that was 75 per cent working class. No financial compensation was offered.[12] Tension between the campus and the town of Nanterre was there from the beginning. And unrest on the campus quickly made things worse.

Surrounded by the airfield, on an immense construction site, the new Faculty of Letters welcomed the pioneers from Paris. As later recollections attested, they were marked out on their return to Paris by the mud on their shoes. Photographs of the time – those of Élie Kagan and Gérard Aimé – make it look like a no man's land at dawn and not the prestigious Sorbonne whose annexe Nanterre remained until 1970.[13] Politics abhors a vacuum or rather it is nourished by one. Upon this ground, militant radicalism grew quickly. The American dream of the modernizers and the everyday life of Nanterre's neighbourhoods were forced together. In the absence of social and cultural life on the new campus, waves of students, especially those whose rooms faced out onto the slum of Pont de Rouen, crossed over into the city of Nanterre.

For, fuelled by a mixture of anti-colonialism left over from the Algerian conflict and third world anti-imperialism associated with the wave of independence overseas and the Vietnam War, a left-wing mood was in full swing in the French universities in the late 1960s.[14] At Nanterre anarchism gave it an extra twist, focused around the critique of everyday life. Nourished by the lectures of Lefebvre and the Situationist International, this politics grew around the 'Liaison' anarchists and the 'Enragés', leading to the '22 March' movement of 1968. Lefebvre, author of the *Critique of Everyday Life* and a heterodox Marxist, influenced the spread of anti-authoritarian thought, developing Wilhelm Reich's ideas of sexual liberation in the class struggle.[15] Moreover, since 1966 the wave of Maoism sweeping through French universities had also encouraged some students to sing the benefits of leaving the bourgeois university to go into the factories. Nanterre was soon on the radar of the French state's surveillance. SAT, the intelligence service of the Ministry of the Interior, noted in June 1970 that there were five points of contact at Nanterre between the 'left' and the 'immigrants', more than anywhere else in the Ile de France region with the exception of St Denis.[16]

1968 in Nanterre

In order to understand how so many students sought to reach beyond the daily life of the campus to make contact with families and young people in the surrounding slums, we need to appreciate the ideological rearmament of the 1960s. Ever since the war in Algeria (1954–62), anti-colonial fervour and the politics of the third world gave rise to the idea that there might be revolution away from the industrialized countries whilst at the same time the working class was becoming more integrated into the consumer society of the nation. The general ascendancy of anti-imperialism added to the disillusionment of many members of the youth wing of the PCF, then the principal workers' party of France. For many young Maoist activists, the PCF's decision to focus on elections and abandon theoretical debate was proof of embourgoisement at work.[17] Against this backdrop, political debate at Nanterre was transformed. The hold of the PCF over the factories of Nanterre also began to loosen. In 1967 Maoist and Trotskyite militants took over the campaigns against the Vietnam War in the Citroën factories and in the local paper mill, 'Les Papeteries de la Seine'. Trotskyites were strongest in the Vietnam National Committees run by the PCF, whilst the Maoists controlled the 'Comités Vietnam de base'.[18] Although at first there were limited links between students and workers, the strikes during the summer months of 1968 lowered fences. Despite the PCF and the Confédération générale du travail (CGT) retaining control of the Commune of Nanterre, there was an increased belief that a meeting of minds between students and workers was possible. Conditions in the local factories and slums revealed the plight of immigrant workers, families and youth as well as the almost complete absence of PCF and CGT organization.

The low share of students from the neighbourhood at the new campus, at a time when university enrolments were rising rapidly (doubling between 1960 and 1967), added to the tension between the town and the new university. Inevitably, the Nanterre Commune became interested in political activity on the nearby campus. In doing this, it followed the strict guidelines laid down by the PCF in 1967, that is to say, it attempted to stem the tide of Trotskyism and Maoism in the factories.[19] An extra edge was given to the situation in so far as the author of these PCF guidelines was Gaston Plisonnier, himself a PCF official in Nanterre. The PCF now supported an increase in Communist organization on the campus. Other moves were less effective. In September 1968, Michel Dufour, a young councillor, was put in charge of monitoring the teaching unit of the Union des Étudiants Communistes (UEC) and of overseeing the anti-leftist struggle in the Faculty of Letters.[20]

At the same time as militancy was emerging on campus, the PCF was losing its grip in the workers' neighbourhood. In 1967 a Nanterre Palestine Committee with links to the Parti Communiste Marxiste-Léniniste de France (PCMLF) Maoists was set up in the rue des Prés slum. It soon became a general focus for the migrant workers from nearby Sorbiers. The same year several students based themselves at Citroën Nanterre and at the Papeteries de Seine. Exposing the exploitation of immigrant workers quickly became central to the Maoist strategy of undermining the PCF and challenging its control of the working class. From 1968 the Maoists added the Vive La Revolution (VLR) group, the issue of adult literacy and the crèche sauvage (all of which are discussed below) to its activities, and in the summer of 1970 a new group, 'Oser lutter' was set up in Petit Nanterre. Several young protesters radicalized by the extreme left moved into a small pavilion linked to the Cimade, a support association for foreign residents, and, renamed the 'Painted House', it became the headquarters of pro-immigrant activities. A second Palestine Committee was also set up in the university residence building.[21] Only the slum area of the Pont de Rouen separated the two committees.

Unrest on the new campus of Nanterre followed the pattern of student protest elsewhere across the Paris region during 1968, but with a local flavour. The first student uprising was the work of Daniel Cohn-Bendit's '22 March' group, dominated by anarchist and libertarian organizations. Over the following months, this group expanded its work among students and organizations of all stripes through its academic and political critique, focused on the problems of daily life and consumerism.[22] In May several members of the '22 March' published a Maoist pamphlet in the factories, its populism evident in its slogan 'Serve the people'.[23] They enlisted young people from the slums in the movement. After the '22 March' group dissolved on 12 June, the door opened to other organizations at Nanterre, characterized by a Marxist anti-imperialism, the trademark of the far left since the war in Algeria.[24] There was the Trotskyite Jeunesse Communiste Révolutionnaire (JCR), which denounced the '22 March', as did the Maoist Union Jeunesse Communistes, Marxistes-Léninistes (UJC-ML), seeking to reclaim the lessons of the month of May, for example, in the newspapers *Red* and *VLC*, in so doing pushing other leftist groups to defend their legacy of the missing revolution.[25] For all those involved, the strikes of May and June 1968 seemed to show that the contradictions of capitalism were at their height. Of the strikes every Marxist said to all who would listen that 'the Marxist theory is very powerful because it is true'.[26]

At Nanterre, the dissolution of the '22 March' group gave way to two movements better equipped to face the despair which followed the National Assembly elections in June, in which the Gaullist party, the Union pour la Défense de la République (UDR), was victorious.[27] There was now a sharp rise in the membership of the PCF and the CGT, both nationally and in the city of Nanterre. On the campus, there were different developments. The dissolution of many organizations, such as '22 March', UJC-ML, JCR, Fédération des Étudiants Révolutionnaires (FER) and the Comité de Liaison des Étudiants Révolutionnaires (CLER), caused a wide scattering of militants in Nanterre. At the same time many students excluded by the university authorities at Nanterre after the events of May had been deported to the Sorbonne. One activist summed up the mood: 'When there was no leadership that was really the great decline ... a lot of people including me were a little left to themselves, not wanting things to resume "Professor-book, nicely, nicely" as if nothing had happened. They were willing to listen to people who wanted to keep trying to do some activism and give something to identify with.'[28]

Vive la Revolution

By July 1968 the largest strikes of the twentieth century in France were over. Despite the length of the strike at the Citroën plant in Nanterre, the situation there normalized quickly. On the campus things were different. During the summer the UJC-ML broke up into several smaller factions. One of these was 'Vive le communisme' (VLC), established in September 1968. The VLC, attempting to forget the denunciation of the '22 March' group by the Maoists – described as 'petty bourgeois', 'corporatist' and 'student' – quickly lost all the sympathy that had accumulated for students within the 'Comités Vietnam de base'. From April onwards several Parisian members of the UJC-ML had been trying to redress the balance. Roland Castro and Jacques Barda from the Paris School of Architecture came to Nanterre and were severely critical of the situation. Castro became the undisputed figurehead of the new VLR at the start of the new academic year in September 1968. And Barda led other activity at Nanterre, transforming the avant-garde magazine *Tout!* after 1970.[29] Castro and Barda were also sent to Nanterre because of their supposed ability to deal with anarchist groups, for their humour and irony, far removed from the dogmatism of their predecessors.[30] At Nanterre they made some concessions. They understood the anti-authoritarian and anti-institutional mood prevailing on the campus, as well as the strength of the anarchist and Trotskyist groups.

By September 1968, the ingredients for an alliance between the working class, students and immigrants were in place at Nanterre. At least in the minds of the leftist groups, Castro and Barda had broken with most of the Parisian leaders of the Maoist Gauche prolétarienne (GP). Daniel Bensaid, co-founder with Cohn-Bendit, of the '22 March' group, was sent to Nanterre to join the JCR. A handful of students from the '22 March', including Annette Lévy-Villard, quickly joined the Vive le Communisme (VLC), followed by other students: 'Comités Vietnam de base' members (from the Censier site of the Sorbonne and from high schools such as Buffon in Paris) and from other action committees.[31] Workers in the factories of Citroën-Nanterre, Hispano-Suiza in La Garenne-Colombes and Renault-Flins (led by the Maoist Tiennot Grumbach) followed too.[32] Once temporary restrictions were lifted, workers began eating in the university restaurant.[33] Opprobrium against the Maoist groups was quickly forgotten. Loyalty to the principles of May 1968 fast became the common denominator of this heterogeneous mix of activists. The theoretical rigour of Marxism-Leninism faded. Except for a few occasions the VLC stayed away from the epicentre of protest in the 'Mao' wave that was being led by young Communists and by some students at the École Normale Supérieure in Paris, who were eager 'to return to the fundamentals of Leninism' and who adopted the slogans of 'activism towards the people' and supported 'the lessons of Chairman Mao and the cultural Revolution'.[34] By contrast, the VLC favoured spontaneity.

During 1969, the VLC took root in Nanterre, becoming its militant base. The group was able to adapt quickly to the Nanterre environment, accepting temporary anti-fascist alliances with other groups,[35] even supporting the candidacy of the Trotskyist, Alain Krivine, in the presidential election of 1969.[36] Significantly, the VLC refused to give up activity on campus, focus solely on the factories and observe strict secrecy. Consequently, the doors of many factories were closed to the VLC. As the local UEC branch stated, 'We condemn the actions of leftist elements in the campus. These elements, often foreign to Nanterre (as with Geismar), have decided to transform the campus into a "red base" to carry out their "ultra-revolutionary activities" in the direction of factories and slums and prevent the return to college. Such actions can only [...] isolate students from the rest of the population.'[37] Hiring in the factories now became rigorously monitored. Workers at Citroën-Nanterre, for the most part Maoist activists, and many members of the CGT were fired after June 1968.[38] Even with some basic committees established, the VLC was not centralized. Unlike other left-wing groups, it lacked a training school. This reinforced a sense of a hybrid activism: initiatives that were often innovative, but the VLC remained an ephemeral organization.

And then from the autumn of 1969, the VLC developed a more independent register: anti-racist, joining in with the youth and gay liberation front, and creating a women's group.[39] There was a change of name too: from 'Vive le Communisme' to 'Vive la Révolution' (VLR). After May 1968 many activist groups in France redefined the struggle against colonialism in the third world, relocating it in the here and now in the defence of immigrant workers. The committees for Palestine made this change with the creation of the movement of the Arab workers in the early 1970s. The new nomenclature, VLR, reflected these shifts. And there was fresh iconography. Black-and-white newspapers, unillustrated except for the mark of the hammer and sickle, with their incessant citations of Marxist-Leninist thinkers, made way in the summer of 1970 for a new colourful and satirical press.[40] The VLC had already begun to move towards supporting the immigrant struggles earlier in the year. In May 1969 an article entitled 'Football in the Fight against Revisionism' announced the need to expose the contradictions of the PCF in its treatment of immigrants. Actions as seemingly innocuous as the construction of a football field for the youth of the housing estates and slums could help, the article suggested.[41] Later that year, in November, the VLR displayed a photomontage of the new modern Parisian headquarters of the PCF together with a stack of squalid immigrant shacks. The VLR stated its new priority to 'extend the fight to all aspects of capitalist social life'. The group now committed itself to work in the factories, neighbourhoods and shantytowns, considered to be politically neglected. 'If we work on Nanterre, this is because, first, there is particularly strong oppression (slums, squalid public-housing, no jobs for young people) [...] liaising directly with our work in some of the factories.'[42] Schools were also included. For example, in March 1970, in response to accusations of violence and unrest in the schools, the VLR made an announcement: 'What the leftists want ... change life.'[43]

The VLR was thus changing direction at the same time as many other organizations, turning to those who were seen as 'non-voting' and 'unproductive', to 'young people and immigrants, the elderly and women'.[44] Competition with the PCF was never far away. As the formative experience of numerous Maoist leaders in France, the PCF remained for the VLR a centre of attraction as well as conflict. The VLR was obsessed with the challenge of gradually replacing the PCF, with its focus on elections and on the labour aristocracy.[45] Several VLR activists stood for elected positions in municipal government, the emergency services and public housing projects in Nanterre (namely, Marguerites and the Provinces Françaises) in order to take the fight to the immigrant workers. There was another strategy too. In early 1970, a crèche was established on the Nanterre campus. This nursery would seal the alliance between students on the one side and immigrants and workers on the other, subverting the bourgeois order of the university, whilst creating a 'student movement at the service of the revolutionary proletariat' in the neighbouring slums.[46] Here lay the new agenda of the VLR. Let us turn now to describing how the 'crèche sauvage' ('wild nursery') fulfilled this mission.

The crèche sauvage at Nanterre

In February 1970 a nursery, similar to those already operating in Paris, at the Censier and Fine Arts sites of the university, was opened at Nanterre, in an unused room of building D in the Faculty of Letters. VLR quickly renamed and popularized the crèche as the 'crèche sauvage' (wild nursery). The VLR's objectives were clear: to accommodate the children of staff, students, teachers and neighbourhood families. They wanted to fight against the shortage of nursery places as well as their high price and to refuse 'dumbed down' principles. For the VLR the economic struggle over the provision of a crèche was subordinate to the fight against bourgeois and pro-family ideology. They wanted liberation of women's time too. Their crèche campaign followed certain principles: it was 'animated by men as well as women'; it sought 'the participation of all involved in the education of children'; and they called for a 'free nursery, open every day of the week, any time, to children from three months to six years or more, and without separation by age'.[47] The crèche at Nanterre also followed the example of the Fine Arts crèche in Paris, in allowing parents to demonstrate whilst their children were being looked after.[48] The atmosphere was summed up like this: 'The kids in Hall D of the Faculty of Letters: balloons fly, cubes pile up, a race from one end of the room to the other, a fall, screaming, crying, but not for long: the good mood is restored. Here kids can express themselves freely: the crèche sauvage.'[49]

'Catherine C' quickly became one of the main activists in the crèche. Born in Paris in 1947, her father was a railway foreman who died in 1956 of tuberculosis, leaving Catherine's mother with four children.[50] Her mother came from a conservative bourgeois family. She was very religious, she related to the young priest workers at Colombes and she worked alongside many Algerian and Moroccan families after the Liberation. Catherine grew up in the small family home, before going to a religious boarding school, then she was moved abruptly to a public high school in the suburbs. She spent a year in the United States supported by a grant. Enchanted by that trip she found a violent contrast when she arrived in the concrete and slums of Nanterre in September 1964. She studied sociology alongside other members of the '22 March' group, taught by Alain Touraine, and she learned of the theories of Wilhelm Reich from Henri Lefebvre. She missed the events of May 1968, but returned to Nanterre in the autumn, joining the VLC through Annette Lévy-Willard, and earned a living from giving literacy classes in the Nanterre slums. By February 1970, Catherine C was at the forefront of the movement.

From the middle of February, anticipating a national childcare centres meeting, the VLR made links between the crèche and the class struggle in the context of Nanterre in a leaflet entitled 'Crèches Sauvages towards the Revolution'.[51] As Catherine C recalled:

There were at least two ideas: to open the university to the outside, a big motivation since 1968; coming out of the ghetto to go to the larger ghetto that was the social configuration of Nanterre at the time: the immediate area was full of immigrants; and the second goal was also intramural: to take the college out of its confinement to purely student affairs – Maoist students to the field and vice versa.[52]

The factories of the Nanterre neighbourhood became involved. From April the committee at Citröen Nanterre required workers to send their children to the nursery, without much result. But the pressure grew. The crèche symbolized the rejection of privilege, and an awareness around the slums of a 'great power'. The crèche was seen as the means to support 'immigrant women who no longer have time to breathe' and 'their kids ... [who] ... have to go play in the mud or piss in bourgeois nurseries'.[53] Under the emblem of the sickle and the bottle (as a hammer), the nursery was adorned with new qualities: anti-racist, 'open to all children whether they are French or immigrants'; teaching 'because of a revolutionary care for them and not as nannies specialised in spanking and tickles'; responsible because 'all parents can come and see how it goes'; and against economic exploitation and the bourgeois order 'because it is free and because bourgeois citizens cannot accept seeing the students in the slums and the slum dwellers in the universities'.[54]

In fact, the nursery was just one part of the reorientation of the VLR in 1970. But it was a catalyst, an initiative 'that encouraged the workers and residents of Nanterre to identify with the political project of the leftists on campus, as it was a useful service, acting on specific issues affecting their lives and their work'.[55] The nursery softened the image of 'spoil-sport' splinter groups. But it also drew attention to relations between the university and the slums, because it was open to the babies of immigrants at a time when the police were preventing local youth from going onto the campus.[56] The stigma of brown babies and criminal immigrants used by the faculty authorities, and by other opponents, pushed the VLR into making the crèche its main weapon against the bourgeois order.[57] The open-access stance of the VLR over the nursery distinguished it both from other political organizations on the Nanterre campus and from other nurseries in Paris.[58] As the VLR's newspaper observed, 'This crèche sauvage is the Tower of Babel. The students' sons and daughters play with slum children more than do their mothers, Portuguese and Algerian arriving every morning and being picked up every evening. "French. Immigrants. Same bottle!"'[59]

However, the crèche soon became a victim of its own success. Within a couple of months, the average age of the children in the nursery grew rapidly. By May 1970, children and adolescents from the slums comprised the majority in the nursery. The operation became more perilous. Catherine C picks up the story: 'Gradually things turned and it was hard to keep control as more and older children came to the nursery. Some also took advantage of the situation – instead of being in school, they were here without their families' knowledge. That prompted those who had children there deliberately to remove them.'[60] At the same time, clashes on campus between revolutionary groups, conservative activists and the police repeatedly endangered the nursery. In April there were several rampages attributed to the extreme right that led to the partial destruction of the furniture in the nursery.[61] The crèche was caught in the crossfire. The university authorities threatened parents from the neighbourhood with expulsion from France if they returned their children to the crèche. Right-wing opinion denounced the crèche, and so did some communist students and Trotskyites who accused the VLR of perverting its original objectives.[62] Communist teachers and other employees of the university denounced the rising violence, linking it to an alliance between foreign criminals and 'Maoist thugs manipulated by those in power'.[63] Politicized, and involved in controversy that went beyond its doors, as well as being the focal point for marginal revolutionary groups, young people and immigrants, the nursery at Nanterre came to symbolize a nightmare faced by successive deans of the university.

Towards the end of February 1970, clashes broke out on the campus between leftist groups. Across France student protests had increased in the new year, following the introduction of the new law on higher education, devised by Minister of Education, Edgar Faure. For example, at Alsace several faculties were severely disrupted. At Nanterre on 28 February a Maoist commando attacked and left for dead two communist students. Realizing that Nanterre was out of control, Paul Ricoeur, the university dean, suspended the academic privilege preventing police from entering the campus. The establishment of a permanent police presence on campus ignited the fuse, strengthening the resolve of the Maoist groups. On 2 March, at the request of the Dean, the police intervened, following an exams boycott by right-wing protesters at the Faculty of Letters that drew in hundreds of leftist activists.[64] Some were wearing helmets and were armed with iron bars. A pitched battle with the police ensued, lasting two days, as the police tried to end the conflict between right-wing and left-wing students. Unprepared, the police fled on the first day, coming back better prepared on the second, destroying cars parked on campus, encircling hundreds of students in the restaurant and firing tear gas. Only the intervention of the *gendarmerie mobile* ended the battle of Nanterre. Then on 4 March, an unusually cold and snowy day, Maurice Grimaud, the Prefect of the Police of Paris, visited the devastated campus, stirring more reaction.[65] Although the immediate violence ended, incidents on campus continued: protests in the university restaurant and on the faculty. The presence of immigrant youths was seen, even by extreme left groups, as the product of an irresponsible and populist initiative by the VLR.[66] Jean Capelle, former rector of the University of Dakar (in Senegal), now deputy for the Dordogne, urged the government to do something: 'To remove the slum from the doors of the University, should not the teenagers of the slum, North Africans for the most part who are the people behind the disruptive elements, be sent home in the hope of a return to normalcy?'[67]

Initially seen by the university authorities as a small minority, these left groups were now regarded as more significant. The publicity given to the crèche sauvage, together with the upsurge of militancy in the Nanterre neighbourhood, brought matters to a head. So the crèche sauvage blended with a new strategy towards migrant workers on the part of leftist splinter groups. It shook the academic authorities. This microcosm of militant turmoil reinforced the strange identity of a campus located in a workers' town, where a communist municipal authority stood aside.

The restoration of order and the legacy of 1968

Precisely when the nursery actually closed in the summer of 1970 remains unclear. Catherine C remembers leaving Nanterre abruptly in June. During the spring of 1970, the nursery had been supplanted by larger and more permanent crèches in other faculties. VLR remained active in the provision of childcare through the spring of 1970. In the middle of May, following the closure of the Renault-Flins factory and the accompanying street protests that culminated in an attack on the Meulan town hall, VLR activists joined with Portuguese residents in the Nanterre neighbourhood to set up a new centre at Villeneuve-la-Garenne, with a nursery and literacy classes, and doctors and lawyers too.[68] Later, in September, a film was made documenting these events by Marc Hatzfeld, entitled 'On vous parle de Flins'. In July 1970, there was a memorable weekend when thirty children from the Nanterre slums were taken to the beaches of Normandy.[69] However, despite the wishes of some groups, such as the local Women's Liberation Movement (MLF), the nursery did not survive.

Having been so undermined for two years, the university authorities responded during 1970. René Rémond, the new president of Nanterre, quickly put in place controls. These came through card control at the campus entry gates, more discreet security services, a new modus operandi with nearby residents and better relations with the Prefecture. At the end of 1970 leftists were barred from the university cafeteria, the staff there having become tired of the permanent state of hostility.[70] Led by Raymond Barbet, the Commune of Nanterre also clamped down on the influence of militant leftists particularly in the factories.[71] The PCF took up the cause of migrant worker exploitation. Inconspicuous before 1968, the PCF newsletter now focused on the Nanterre factories and on the problems of local youth, devoting fifteen articles to these issues between September 1968 and October 1970.

As for the immigrant workers of Nanterre, residual unemployment had stimulated their political commitment, but not necessarily their conversion to the ideas of the militant left. The language of the different factions of the left remained incomprehensible to the newcomer. The VLR modified its rhetoric when it came into contact with struggling immigrants, a position that was attacked as too spontaneous by the orthodox left of Trotskyites, Maoists and PCF.[72] However, even the VLR had alienated immigrant workers on 22 March 1968 when it decided to go for an all-out strike, with no pause and no economic compensation. This move widened the gap between the students on one side and the youths and workers in the slums on the other. Many departed from the movement at this point. Mohammed Kenzi recalled how this decision was misunderstood in the Nanterre neighbourhood, how it reflected an ever-present cultural distance between the students and the immigrant workers. Social and cultural boundaries still enforced the divisions between the academic world and outside, in spite of the cracks that had been made by the events of 1968. As Kenzi said:

> One morning Said took me on one side reproaching me for my behaviour in running away from my compatriots. He claimed that I was too crushed by European concepts, that leftism had pushed me into another dimension, that I detached myself from a certain reality. He wanted me to come back down to earth to put myself on the same level as my brothers.[73]

Conclusion

Away from the capital, the students of Nanterre had transformed the dream of an American campus into an institutional nightmare. This no man's land became an incubator for the challenges faced by French universities in the late 1960s and early 1970s. What seemed an ideal location, isolated away from normal life – where the transformation of the French university could take place – instead intersected with the rise of militant groups focused on antifascist, anti-colonial and anti-imperialist campaigns.

The infants of Nanterre shifted the gaze of French society. Social inequality became visible from the windows of the new university. Poverty, poor housing and the conditions of the workers affected many students on the campus.[74] There were other factors too, not least the burning memory of the Algerian War, together with the conflict in Vietnam and Palestine. The threshold of political sensitivity had changed, just as France was boasting about economic growth and its 'trente glorieuses' (as the period of rapid economic growth, 1945–75, is sometimes known). The rhetoric of the revolutionary groups at Nanterre might give the impression of an excess of Marxist and Maoist theory over reality. However, amongst the student activists at Nanterre were those with memories of families involved in the resistance during the Second World War, and of deported Jews.[75] Some of them, for example, Tiennot Grumbach, had been out to Algeria to help build socialism in the newly independent country. The plight of the immigrant workers reminded this younger generation of the battles of yesterday and reawakened the revelations of torture in Algeria. These experiences and memories also distanced Nanterre students from the PCF and the Parti Socialiste, the latter in particular tarnished by the war in Algeria. In their place, the Vietnam committees gained ascendancy. Maoist buzzwords of the cultural revolution were everywhere. Intellectuals idealized the immigrants of the working-class suburbs, comparing them to the peasants of China and the displaced Palestinians. Thus 1970 was a brief moment when immigrant workers came together with the far left to destabilize not just higher education but society too.

After 1970 Parisian academics continued to debate Marx, Engels and Mao, deciding how to include the immigrant worker as the paragon of this new phase of the class struggle. At Nanterre, the prefecture of the town and the new university presidential team orchestrated a return to serenity on the campus. Several VLR members were arrested or imprisoned later in the year.[76] No longer at the heart of the 'red' suburbs, Nanterre now became a laboratory of heterodoxy in the return to order. The VLR and other militant groups fell away, and the rhetoric of seizing power from the university authorities disappeared. The weakness of VLR reflected a profound change in the activism on the left in the universities in the France of the late 1960s. These small organizations represented a distinct social movement separate from the traditional political parties and trade unions. Today in the Faculty of Letters at Nanterre the symbolism of the years 1968–70 is forgotten: it is either repressed or mythologized. Fifty years later, Nanterre is known as a suburban campus and not a revolutionary university forged in the climactic aftermath of May 1968.

Notes

1   Victor Collet, *Nanterre, du Bidonville à la Cité* (Marseille: Agone, 2019).
2   Kristin Ross, *Mai 68 et ses Vies Ultérieures* (Marseille: Agone, 2010).
3   Jean-Pierre Duteuil, *Nanterre 65-66-67-68: Vers le Mouvement du 22 mars* (Mauléon: Acratie, 1988).
4   Charles Mercier, *René Rémond et Nanterre. Les Enfants de 68. Contribution à l'Histoire d'un Universitaire et d'une Université Iconiques (1968–1976)* (Lormont: Le Bord de l'eau, 2016), 179.
5   Henri Lefebvre, *L'Irruption de Nanterre au Sommet* (Paris: Anthropos, 1968).
6   Gilbert Wasserman, *Nanterre: une Histoire* (Paris: Temps Actuels, 1982).

7   Charles Soulié (ed.), *Un Mythe à Détruire? Origines et Destin du Centre Universitaire Expérimental de Vincennes* (Saint-Denis: Presses Universitaires de Vincennes, 2012).
8   'Interview with Georges Augustins' (5 March 2014), 'Faire et arpenter l'histoire de l'université de Nanterre', Labex, Nanterre Université (hereafter Labex).
9   Marie-Chantal Combecave-Gavet, 'De Nanterre la Folie à Nanterre Universit (1964–1972): Histoire d'une Institution Universitaire, (unpublished PhD thesis, Université Paris-X-Nanterre, 1999).
10 Rachel Mazuy and Danièle Le Cornu, 'Chronologie des Événements à Nanterre en 1967–1968', *Matériaux Pour l'Histoire de Notre Temps*, 11–13 (1988), 133–5.
11 Ibid.
12 Collet, *Nanterre, du Bidonville à la Cité*, 170–1.
13 Ibid., 165–8.
14 Christoph Kalter, 'Les Damnés de "Nanterre." Extrême Gauche, Tiers-Monde et Années 68 en France', in *Les Années 68. Un Monde en Movement*, ed. Geneviève Dreyfus-Armand (Paris: Editions Syllepse, 2008), 62–81.
15 'Interview with Catherine C.' (13 November 2008), Labex.
16 Daniel Gordon, 'A Nanterre, ça Bouge: Immigrés et Gauchistes en Banlieue, 1968 à 1971', *Historiens et Géographes*, 385 (2004), 77–9.
17 Daniel Gordon, 'Immigrants and the New Left in France, 1968–1971, (unpublished PhD thesis, University of Sussex, 2001), 254.
18 Christoph Kalter, 'Les Damnés de "Nanterre"' ; Duteuil, *Nanterre 65-66-67-68*, 76–77.
19 'Carton Citroën', Société d'Histoire de Nanterre.
20 'Interview with Michel Dufour' (4 January 2012), Labex.
21 'Interview with Chérif Cherfi' (29 October 2009), Labex.
22 Jean-Pierre Duteuil, 'Les Groupes Politiques d'Extrême Gauche à Nanterre', *Matériaux Pour l'Histoire de Notre Temps*, 11–13 (1988), 113–14.
23 'Tribune du 22 mars' (8 June 1968), Fonds Mouvement Étudiants, 1968–70, F Delta 813/7, La Contemporaine, Nanterre Université.
24 Kalter, 'Les Damnés de "Nanterre"', 62–80.
25 Fonds Mouvement Étudiants, 1968–70, F delta 813/7-12, La Contemporaine.
26 Erik Neveu, 'Rétablir les Établis', *Savoir/Agir*, 6 (2008), 49–58.
27 Boris Gobille, *Mai 68* (Paris: La Découverte, 2008), ch. 4.
28 'Interview with Catherine C.'
29 Manus Christian McGrogan, '*Tout !* in Context 1968–1973: French Radical Press at the Crossroads of Far Left, New Movements and Counterculture' (unpublished PhD thesis, University of Portsmouth, 2010).
30 Manus McGrogan, *Tout ! Gauchisme, Contre-Culture et Presse Alternative dans l'Après-Mai 68* (Paris: L'échappée, 2018), 21.
31 Christophe Bourseiller, *Les Maoïstes. La Folle Histoire des Gardes Rouges Français* (Paris: Plon, 2008), 187–9.
32 Ibid., 187–9.
33 Fonds Mouvements Étudiants 1968–1973, F Delta 813/13, La Contemporaine.
34 Prochinois et Maoïstes en France, *Dissidences*, Special Issue, 8 (2010), 9–11.
35 *Vive le Communisme*, 2 (December 1968), 4.
36 Ibid., 4 (May 1969), 1, 6; *Vive le Communisme*, 5 (June 1969), 8–9.
37 Communiqué de l'UEC-Nanterre, *L'Eveil*, 20 September 1968, 2. Alain Geismar was the leader of the national union of university teachers (SNESUP).
38 'Interview with Nadja Ringart', https://around1968.history.ox.ac.uk (accessed 1 September 2019)

39 Bourseiller, *Les Maoïstes*, 187-9.
40 McGrogan, '*Tout !* in Context 1968-1973'.
41 *Vive le Communisme*, 4 (Mai 1969), 4.
42 *Vive la Révolution: journal marxiste-léniniste maoïste*, 1 (November 1969), 8-9.
43 Ibid., Supplement, 6 (June 1970).
44 Ibid., 1 (November 1969), 8-9.
45 Frédérique Matonti and Bernard Pudal, 'L'UEC ou l'Autonomie Confisquée (1956-1968)', in *Mai-Juin 68. Connaissez Vous Vraimment mai 68?* ed. Dominique Damamme et al. (Paris: L'Atelier, 2008), 130-43; Romain Vila, 'De la Divergence à la Dissidence: Études des Rapports Entre les Maoïstes Français et le Couple PCF/CGT (Mai 1966-Juin 1970)', *Dissidences*, 8 (2010), 81-96.
46 *Vive le Communisme*, 3 (March 1969), 13.
47 'Pour une Crèche Sauvage à Nanterre', Fonds Mouvements Étudiants, 1968-1973, F Delta 813/12, La Contemporaine.
48 Fonds Françoise Picq, F Delta Res 0612/13, 17, La Contemporaine; 'Comité d'Action Censier', ibid.; 'Interview with Catherine C.'
49 *Vive la Révolution*, 3-4 (March 1970), 8.
50 'Interview with François C' (younger brother of Catherine C), March 2013, Labex.
51 Fonds France Mouvements Étudiants 1968-1973, F Delta 813/12, La Contemporaine.
52 'Interview with Catherine C.'
53 Fonds Mouvements Étudiants 1968-1973, F Delta 813/12, La Contemporaine.
54 *Base Ouvrière-Obrera*, 21 April 1970, 2, Fonds Mouvements Étudiants 1968-1973, F Delta Res 813/13, La Contemporaine.
55 *Vive la Révolution*, 3-4 (March 1970), 2.
56 Ibid., 6 (June 1970), 5.
57 'Fermeture, Ségrégation, Provocation and Crèche Crucifiée', Fonds France Femme, La Contemporaine, F Delta 150/81; *Vive la Révolution*, 5 (April 1970), 6.
58 Daniel, 'A Nanterre, ça Bouge'.
59 *Vive la Révolution*, 3-4 (March 1970), 8.
60 'Interview with Catherine C.'
61 *Vive la Révolution*, 6 (June 1970), 5.
62 Fonds Mouvements Étudiants 1968-1973, F delta 813-9-13, La Contemporaine; 'France Femme', F Delta 150, ibid.
63 *L'Eveil*, 20 February 1970, 15.
64 Fonds Mouvement Étudiants 1968-1973, F Delta 813/10, La Contemporaine; *Le Figaro*, 27 February 1970, 4.
65 'Situation à Nanterre - Monsieur Grimaud JT 20H du 4 mars 1970', https://www.youtube.com/watch?v=wy0chY4r9Hk (accessed 12 July 2019).
66 'Résident Rouge' (tract), Fonds Mouvement Étudiants 1968-1973, F Delta 813/9-13, La Contemporaine.
67 'Incidents Notables Survenus à la Fac des Lettres de Nanterre' (2 June 1970), CAC 19810075, Article 37, Archives Nationales.
68 *Vive la Révolution*, 6 (1970), Supplement, 1; ibid., 7 (September 1970), 9.
69 'Une Crèche à Nanterre', Fonds France Femme, La Contemporaine, F Delta 150.
70 Charles Mercier, *Autonomie, Autonomies: René Rémond et la Politique Universitaire aux Lendemains de Mai 68* (Paris: Publications de la Sorbonne, 2015), 189-90.
71 'Demande de Réponse aux Justes Interrogations d'une Camarade', Fonds Gaston Plissonnier, 264 J/32, District Archives of Bobigny.
72 Mohammed Kenzi, *La Menthe Sauvage* (Paris: Jean Marie Bouchain, 1984), 73.

73 Ibid., 110.
74 'Interview with Annette Lévy-Willard', https://around1968.history.ox.ac.uk (accessed 1 September 2019).
75 'Nous Vengerons nos Pères. Un Film Documentaire de Florence Johsua et Bernard Boespflug' (2017), http://www.histoire-immigration.fr/agenda/2018-07/nous-vengerons-nos-peres (accessed 12 July 2019).
76 Jean-Claude Vimont, 'Les Emprisonnements des Maoïstes et la Détention Politique en France (1970–1971)', *Criminocorpus*, 2015, http://journals.openedition.org/criminocorpus/3044 (accessed 9 August 2019).

20

The reform universities of West Germany: Bochum, Konstanz and Bielefeld

by Stefan Paulus

Continuity and change

Higher education reform in the Federal Republic of Germany (FRG) was not only discussed in the 1960s and 1970s, but earlier. The occupation of Germany by Allied troops in 1945 provided the anti-Hitler coalition powers with the chance to shape the future development of the German higher education system according to their own ideas. The Allies were aware that the German universities would play a key role in the education and training of future political, economic, scientific and cultural elites.[1]

But the impact on the German higher education system of the Americans, the British, the French and the Soviets was very different. The French and Soviet side had a much more rigorous policy. Soviet ideological penetration of the German universities developed rapidly,[2] while the French were active in the foundations of universities at Mainz (May 1946) and Saarbrücken (April 1948), as well as the 'Verwaltungshochschule' (Science Administration) Speyer (January 1947), designed on the model of the French 'Ecole Supérieure d' Administration'.[3] American and British academic experts laid emphasis on the self-reform of German universities, without undue coercion from outside.[4] The United States had been initially sceptical about the founding of new universities in their zone of occupation.[5] This attitude changed on 4th of December 1948 with the establishment of the Free University of Berlin (FU) as a Western counterpart to the older Sovietized University of Berlin in the eastern part of the city.[6] In the British occupation zone a 'Commission of Enquiry to Examine the Need to Reform in Universities and Colleges of University Status' submitted in autumn 1948 a widely regarded report.[7] A prominent member of the commission was the Oxford philosopher Alexander Dunlop Lindsay, who was engaged at that time in founding a new British university at Keele.[8] The report recommended the opening up of the German universities, the launching of a Studium Generale curriculum, plus an improvement of studentships.[9]

But there was nearly no practical implementation of these recommendations during the period of occupation. The Germans believed that a new start for the universities should not be imposed from outside, but by building on the German university tradition established before 1933.[10] Nonetheless, during that time important elements of the American and British university and science system were introduced into West German reform discourse, which later went on to dominate the reform process of the 1960s and 1970s.[11]

Momentum behind higher education reform was strengthened by the recommendations for the development of scientific institutions published by the influential 'Wissenschaftsrat' (German Council of Science and Humanities, hereafter WR) in 1960.[12] The situation at the beginning of the 1960s was exacerbated by steadily increasing student numbers that could not be absorbed by the nineteen existing universities and seven technical universities in the Federal Republic. According to the WR the number of students had almost doubled between the 1950/1 (c. 111,000) and 1960/1 (205,000). In the following five years the numbers rose even further by 45,000 to about 250,000.[13] There was no broad agreement in 1960 about whether the expansion of existing universities alone could cope with this development. The only way that would simultaneously allow for a quantitative growth of universities and a profound reform of higher education seemed to lie in the establishment of new universities. This shift towards a quasi-progressive and expansive university policy was part of the wider planning boom that 'unfolded fully from the middle of the decade, but declined again in the 1970s'.[14]

Finally in 1961 the education expert Hans Werner Rothe produced a memorandum for the establishment of a university at Bremen with the American campus university as the model.[15] Rothe's concept applied not only to Bremen, but in its principles to all the new foundations. It marked the beginning of a series of similar memoranda.[16] In addition to the publications of the WR, Rothe's memorandum, although never implemented in its pure form, can be considered as one of the most influential documents of the era.[17] Recent studies even suggest that the Bremen memorandum even led to a pan-European model.[18]

Concerning the architectural face of the new university, Rothe imagined a central square or 'forum' composed around the core of a large lecture hall, the student union, the Rector's building and other administrative buildings, as well as the university library. In his plan Rothe grouped different areas concentrically around this forum: first the individual departments and research institutes, a university church, then several halls of residence embedded in a generous park-like environment with sports facilities, a botanical garden and unreserved land for future expansion of the various faculties and institutes, together with a larger area for the individual university hospitals.[19] This was the first time that the idea of the forum was introduced and it influenced the memoranda and architectural designs of the other first campus universities at Bochum, Bielefeld and Konstanz.[20]

New universities were initially rejected in the Federal Republic because of the anticipated costs.[21] Unlike in the United States, where the construction and expansion of universities had always included private (non-governmental) endowment or forms of public–private partnership, in Germany the development of universities and of science was traditionally considered as the primary task of the state. Such a massive multi-billion Deutschmark-devouring expansion of higher education had to be legitimized by the state to taxpayers. Therefore the chancelleries and the ministries of education argued that the re-establishment of science and of the universities would not only relieve the old universities but also implement long discussed reforms. So the financial expenses of the new universities would be considered as a soft alternative investment in the modernization of German higher education as a whole.[22]

Campus architecture

In the Federal Republic at the beginning of the 1960s, the prevalent idealized image of a self-contained university system was mainly influenced by the American campus universities. The American model had itself emerged in the wake of the numerous new universities set up in the second half of the nineteenth century, but the tradition went back to the University of Virginia, founded in 1819.[23] A century later, in the 1950s and 1960s, the special American experience in the construction and design of the entire university complexes demonstrates the transatlantic orientation of the discussion of new universities in Germany.[24] One influence that can hardly be underestimated in the discussion about the shape of the campus university in the Federal Republic was the experiences of German students and scholars who had experienced academic exchanges after 1945.[25]

The campus concept, which meant the construction of a completely new university, offered the opportunity to disregard existing structures and to implement reforms that had been discussed in German higher education circles for several years.[26] The Rothe memorandum, together with the recommendations of the WR, conceived an interplay of internal structure and external configuration in the new university. These ideas were particularly evident in the report drawn up under the chairmanship of Hamburg education scientist, Hans Wenke, in his recommendations for establishing the University of Bochum, published in December 1962.[27] Two years before, in May 1960, the state government of North Rhine-Westphalia had decided to build a new university in the Ruhr area. On 18 July 1961, it was agreed to give the university to Bochum.[28] Wenke's founding memorandum stated that Bochum would relieve the universities of Köln and Münster, as well as the RWTH at Aachen.[29] An additional task of the new university would be the creation of new relationships between the different disciplines. Inspired by the American departmental system, the memorandum proposed smaller departments ('Fachbereiche') instead of the traditional large faculties. A reorganization of the internal structure would be combined with a corresponding reorganization of the overall architectural design of the university. Moreover, the Bochum memorandum made a distinct schematic representation of the special relationship between the various departments on the future campus in the form of an organizational chart that reflected Rothe's influence. Similar to the Bremen plan, departments would be concentrated around a central forum. Also closely connected student dormitories and apartments at Bochum referred back to Rothe's 1960 plan as well as to the suggestions of the WR.[30]

The development of the Ruhr University at Bochum, the first and largest of the new West German universities, took place at a phenomenal rate. From the groundbreaking in the spring of 1964 through to the completion of the first building complexes and the start of lectures in November 1965 and onto the final the completion of the main construction work in the early 1970s, it was undoubtedly a remarkable achievement, considering its size.[31] This was accomplished by using the most modern methods, of using standardized prefabricated reinforced concrete materials, a process partly borrowed from the United States.[32] The architectural competition for the design of the campus was initiated in parallel with the drafting of the foundation memorandum. Renowned architects such as Alvar Aalto, Walter Gropius, Mies van der Rohe and the Stuttgart University designer Horst Linde all entered the competition. There were a total of eighty-five designs submitted on 14 February 1963.[33] The competition jury finally decided to award the commission to the architectural firm Hentrich and Petschnigg. Their plan comprised thirteen largely uniform low-rise buildings, lying in two parallel rows from west to east in which the individual departments were housed. At the centre of the whole system was the forum complex, surrounded from two rows of blocks. Around the forum the architects placed a pearl necklace shape comprising the cafeteria, the auditorium, an art museum, the university library, the student union, the musical centre and the university administration.[34] The student dormitories were not placed on the campus but just to the north in a separate quarter, connected by a footbridge to the university forum.[35]

The establishment of a university on the south shore of Lake Konstanz was an initiative that went back to the former Baden-Württemberg Minister-President Kurt Georg Kiesinger in 1959.[36] Initially there was little local enthusiasm for a new university and instead a demand to focus on the expansion of existing universities. Konstanz was rejected as a location because of its peripheral position.[37] Nonetheless, a memorandum supporting the establishment of research universities in Baden-Württemberg was presented, which borrowed – as well as Bochum – significantly from Rothe's memorandum.[38]

On 27 February 1964, based on this memorandum, the state parliament decided to establish a university at Konstanz. A short time later the founding committee was appointed to draw up a structural plan for the new university, meeting for the first time on 21 March 1964.[39] The report of the founding committee was presented in June 1965. It was an impressive document. Similarly to the new universities at Bochum and Bielefeld, the committee aimed to establish a comprehensive university relieving pressure elsewhere, but they also wanted to establish a dedicated research university with a reduced range of subjects, a student population limited to 3,000, with around 100 professorships. This goal of building Konstanz as a 'new character of university' demonstrated the committee's opinion that the ideal of the German university in existence since Wilhelm von Humboldt founded the University of Berlin in 1810 – that is, the idea of the unity of research and education – was no longer justified. In order to strengthen its research mission, the university limited itself to three faculties subdivided into different subject areas. Traditional faculties such as theology, law and medicine were avoided. Instead, in addition to the faculties of natural sciences and of philosophy, Konstanz established a faculty of social sciences, a novelty in the Federal Republic.[40]

The founding committee at Konstanz deliberately excluded the issue of future growth in its report. Unlike other new universities in the FRG, Konstanz sought to maintain the limit of 3,000 students. Because of this relatively low number of students, living conditions on the future university campus were given particular emphasis. Accommodation was provided for about one-third of the students in dormitories on-campus, and in the immediate neighbourhood apartments for professors and lecturers would be built.[41]

Planning for the campus at Konstanz began in the autumn of 1964. A five-member construction management board was formed. In the first place it decided on where the future university would be placed within the landscape around the town: in a highly exposed location 3 km away from the town centre, opposite the island of Mainau on the Bodensee, sloping towards the Gießberg ridge.[42] The choice of location was not only motivated by concerns for scenic surroundings, but also because of the mission of Konstanz as a reform university dedicated to research. In other words, the peripheral location corresponded to the elitist character of the university.[43]

The construction of temporary buildings at the University of Konstanz began in the summer of 1966 with an official groundbreaking ceremony on 21 June. During 1967 the planning finally began.[44] The recommendations of the report of the founding committee were reflected in the architectural design of the University.[45] The centralization of all facilities on campus presented a new challenge not only from an architectural point of view, but also socially, scientifically and in terms of organization. In fact, the Konstanz plan moved even more towards centralization than either Bochum or Bielefeld. The various buildings were so compact and closely interwoven that Stefan Muthesius described the system as a 'one-building institution'.[46] Like Rothe's campus model for Bremen, and similar to the solutions adopted in Bochum and Regensburg, there was a forum-like space at the centre of the entire system comprising the library, the lecture hall with auditorium, the Rector's building, the canteen and cafeteria, the data-centre and various shops. To the north of the central zone lay the natural sciences with its departments and laboratories and in the south were philosophy and social sciences. The forum at Konstanz was different from other new campuses because of its dense structure. However, the basic principle – to provide a university organism with a pulsating 'heart' along the lines of the American campus university – was apparent in Konstanz as elsewhere. And it can also be seen in the efforts to involve the surrounding countryside in the overall planning.[47]

Construction work had just begun in Bochum when another new university project was started in the state of North Rhine-Westphalia. Chaired by the famous sociologist Helmut Schelsky, the founding committee for a university in eastern Westphalia began work at the end of 1965. Unlike in Bochum the issue of relieving existing universities played less of a part. The University of Bielefeld was established in 1969 as a dedicated reform university, with interdisciplinary work as its focus. Schelsky demanded a more research-orientated university, with teaching and research strengthened under its umbrella. Like Konstanz, a limited range of subjects would be offered and the student numbers kept to 3,600. In order to achieve this goal there would be no medical or engineering faculties at Bielefeld. In November 1969 teaching began with only three faculties: mathematics, law and sociology.[48]

Perhaps more than in any other German university of the 1960s and 1970s, Schelsky's ideas about interdisciplinarity were reflected in the overall architectural design at Bielefeld. There was the closest possible networking of all subjects and academic institutions. The founding concept at Bielefeld was in the architectural form of unified living.[49] The architectural competition for the new university was more modest compared with that of Bochum, but it was characterized by significantly more pragmatic ideas. In the summer of 1970, the decision was made to award the design to the young Berlin architects Helmut Herzog, Klaus Köpke, Peter Kulka, Wolf Siepmann and Katte Töpper. In the eyes of the university planners, these architects realized best the conceptual specifications for the new campus: interdisciplinarity, functionality and flexibility.[50] The Bochum campus had been criticized as too giant and spacious. This error would not be repeated in Bielefeld, although here too there were large block buildings arranged into parallel rows. However, these blocks were brought together more closely in Bielefeld and connected by intermediate buildings.[51] At the centre was a two-story glass-covered reception hall almost 300 m long. On its ground floor, there were social facilities. The hall provided weatherproof access to the cafeteria, the auditorium, the university swimming pool and other departmental buildings via an upstairs gallery. The university library could be reached as the gallery extended across the circumference of the entire second-floor creating links between the departments.[52] This design feature for promoting community won for the reception hall at Bielefeld the North Rhine-Westphalia architecture prize of the Association of German Architects in the category 'social communication' awarded in 1979.[53]

Like in Bochum and Konstanz concrete, glass and metal were the preferred materials at Bielefeld. As the architecture critic Gerhard Ullmann wrote in 1978, one year after the completion of the Bielefeld campus, 'the architects have employed a cool rationality and good sense of aesthetic effect'.[54] These materials kept costs low and enabled relatively quick construction. The building process also allowed this: huge manufacturing plants were established onsite, so that concrete parts could be cast for the new buildings on the spot. In this way, new university construction established new benchmarks in building systems and planning rules.[55]

Curriculum design

The new West German universities not only introduced modern and efficient management and administrative structures, they also established new training and teaching methods. There was a focus on closer cooperation between adjacent departments (interdisciplinarity); there were new small-group teaching methods, the modification of teaching content and also greater interaction between theory (study) and practice (working life). In 1965 at the opening of the University of Bochum, the then Minister of Culture for North Rhine-Westphalia, Paul Mikat, noted: 'The urgent need to relieve the existing universities through new ones should not mean shunning new ways, then the new university would simply be "copies" and the impulse to the overall reform of the German university system could run out.'[56] This statement shows that the new universities were expected to be accompanied by a new approach. The three new universities focused on this in quite different ways. In Bochum, in line with C. P. Snow's 1959 emphasis on two cultures, the university tried to integrate the different academic cultures of the humanities and natural sciences and engineering in one campus. Specifically, theology, humanities and the social sciences stood alongside medicine, mathematics, physics, chemistry, biology and electrical engineering. There was also a department of mechanical and structural engineering, a subject otherwise only provided in the technical universities.[57]

A different approach was taken at the University of Konstanz. Student numbers were restricted; there were only three faculties (natural sciences, social sciences and philosophy), in contrast to the comprehensive programme of subjects at Bochum. At Konstanz the concept of a research-orientated educational focus defined the new university. Interdisciplinarity took a unique form at Konstanz. As the founding committee stated in its 1965 report, 'Biology would establish a centre of gravity in the faculty of science.'[58] Konstanz also gave a new prominence to the behavioural sciences. For the first time in Germany, they came under the umbrella of a faculty. There was not only sociology, but also statistics, psychology, political science, economics, geography and law. The latter usually formed a separate faculty of its own. In its regulations on study, Konstanz also broke new ground by efficiently coordinating lectures, tutorials, laboratories, working groups and colloquia with each other. Finally at Konstanz, there was a new degree. The Anglo-Saxon model of the bachelor of arts was introduced in the natural sciences replacing the traditional pre-diploma. Thus the BA was integrated into the German university system some time before the so-called 1999 'Bologna process' of harmonizing higher education across Europe.[59]

At the University of Bielefeld there were initially also only three faculties structured into various departments: the faculty of law and social sciences, the faculty of philosophy, and the faculty of mathematics and natural sciences. As at Konstanz there was a cap on the total number of students at 3,600, although that was quickly abandoned. In 1968 the 'Centre for Interdisciplinary Research' (ZIF) opened to coordinate cooperation between scientists from different fields.[60] The Bielefeld founding committee also proposed innovation in the curriculum design, stating that the new university must make a contribution to the reform of course design as well as study and examination regulations. A major element here was the intensification of supervision study by professors in small groups with a maximum of thirty students.[61]

There is no doubt that this combination of reform and new priorities in research and teaching in these universities had an impact on professional practice. Especially in the initial phase the new universities set particular emphasis on creating a reformed scholarly culture, particularly by appointing younger teaching staff with international experience.[62] It was envisaged that the future professors and other academic staff should fit with the principles of reform, also to overcome the decrepit and partly authoritarian structures of the traditional 'Ordinarienuniversität'.[63] For example, the Konstanz founding committee stated that the appointment should not be determined by the demands of the individual professor, but should 'carry through structural features […] and take into account the overall interest of the university'.[64] Likewise, the founding fathers of the University of Bochum considered it an absolute necessity that the envisaged administrative, structural, teaching and research reforms could only be successfully implemented through appointing adequate teaching staff.[65] For example in the case of Bielefeld, young university teachers such as the sociologist Niklas Luhmann and the historian Hans-Ulrich Wehler were brought to the new university. Both proved not only to be masters of their craft, but also loyal to their first and only universities. Not for nothing Wehler, the founder of historical social science in Germany, was called 'the mastermind of the Bielefeld school'.[66]

State, locality and philanthropy

The establishment of universities in Germany has always been considered the domain of the locality or the individual states (Länder). The historical evolution of federalism ensured that higher education policy fell primarily under the responsibility of the individual state governments. Only with the Higher Education Act of 1976 were the federal government's powers and role defined in relation to the university sector.[67] However, the establishment of the WR by both the federal and state governments in September 1957 meant that the search for a common co-ordinated higher education and science policy had begun. The WR was composed mainly of members of the Max Planck Society and the West German Rectors' conference. Its task was to work towards an overall plan for the advancement of science.[68]

In 1960 the WR published the *Empfehlungen des Wissenschaftsrates zum Ausbau der wissenschaftlichen Einrichtungen, Teil I: Wissenschaftliche Hochschulen*. Here the focus was still on the expansion of the existing twenty-six universities and technical universities in West Germany. But for the first time the establishment of new scientific universities was also identified as an objective. It was clear that existing institutions would not be capable of coping with the growth of science students. This view was confirmed by the experience of other European countries such as UK, France and the Benelux, where governments were responding to rapidly rising student numbers with plans for university reform and expansion. However, in Germany the WR not only sought new universities to relieve the old, but also to take up the mission of reform.[69]

In addition to its recommendations of 1960, two years later the WR published a further report. The main focus here was not on the outer shape of the new universities but on their internal structure. This report recommended the centralization of administrative structures through the introduction of the presidential system and strengthening of the Rector's office. It also proposed a new organization of the modern university in the shape of division into departments or more focused faculties. The traditional great faculties, controlled by full professors, were valid as too cumbersome and not oriented towards interdisciplinarity.[70]

For the state governments, when it came to locating the new universities, their primary concern was to create campuses far away from the traditionally favoured regions. It was obvious that this would lead to rivalry between local authorities. For example, in North Rhine-Westphalia it was not decided at first whether the new university would come to Bochum or Dortmund. Since the late 1950s both cities had suffered from the increasingly serious crisis in the mining industry in the Ruhr, and therefore both cities sought the new contract. After all, much was expected from the establishment of the university and its anticipated impact on the demographic, economic, cultural and educational development of the city and the region.[71] In order to influence the final decision of the state Parliament, local regional and national political parties launched campaigns.[72] As philosophy professor (first at Bochum and later at Bielefeld) Herman Lübbe noted, 'The local authorities were in hard competition as candidates and outdid each other in making attractive offers of plots of land and other forms of university infrastructure.'[73]

Philanthropists and entrepreneurs played no significant role in the first reform universities of West Germany. This was due on the one hand to the central role of the federal and local state, and on the other to the fact that in Germany especially in the higher education sector there was no comparable tradition of private funding to the United States or the UK. Nevertheless, the new universities had substantial social and economic support at the local and regional level. So-called friendship societies comprising representatives of politics, culture, church and business campaigned for the establishment of each new university and continue to contribute financially even to this day. Similarly, the local chambers of commerce (IHK) advanced the process of establishment.[74]

The student and staff experience

The new universities of the 1960s were not only associated with the reform of higher education, but also with identifying students from educationally disadvantaged backgrounds and regions. Shortly after its teaching operations began in the winter semester of 1965/6, the University of Bochum investigated student expectations of study and living conditions. A representative survey of 2,200 out of the then total of 4,000 students gave some insights. From the survey quite significant differences can be seen between the first reform universities in West Germany and the traditional universities. In Bochum, the proportion of students from working-class families (7.5%) as well as the children of employees (40%) was above the national averages of 5 per cent and 28 per cent respectively. The majority of children of working-class parents matriculated in subjects in the field of social sciences: philosophy, pedagogy, psychology and history of science. Notably in Bochum the proportion of female students at 24 per cent was slightly below the national average of 29 per cent. The number of local students – 58 per cent from the immediate Ruhr area and 85 per cent from North Rhine-Westphalia – clearly shows, as the founders intended, the regional importance and impact of the new university. Furthermore, 7 per cent of the students were already married or had children, a figure twice as high as in the old universities (3%). In this way the University of Bochum benefitted disproportionately from the emerging new forms of adult education and increasingly accepted rights for all the students in the higher education sector.[75]

In terms of residence it is striking that at the University of Bochum, compared to the national average, twice as many students lived in areas adjacent to the campus: 26 per cent versus 12 per cent, but also twice as many students commuted between their home and the university 44 per cent versus 19 per cent. The possibility of commuting led to the increased motorization of the student body which in turn removed the cost of renting a room or apartment, resulting in a significant reduction in the total cost of study. Ultimately this factor helped to fulfil the intention of the founders of the University of Bochum to democratize education. At the same time the university became for many students purely a place of learning and not one of residence.[76]

Convenience of location and proximity to home stands out in the findings of the survey as the reasons given by students for studying at Bochum (51%). In second place 26 per cent stated that the positive working and living conditions at Bochum were important. The fact that Bochum was a new reform university played hardly any significant role in the decision. Whilst many students praised Bochum, in its initial phase, with its tight-knit working and living conditions, for being much more manageable compared to the old universities, an equally significant number of respondents criticized the lack of cultural offerings in the city. Such responses did not undermine the evident euphoria concerning the founding of the university. It was just that Bochum was not a university city with comparable cultural institutions to be found in Berlin, Munich and Hamburg. This was true for the two other new universities in Konstanz and Bielefeld.[77]

In Bochum and in Konstanz, more than in Bielefeld, the design concept focused on settling a comparatively high proportion of students in dormitories in the immediate vicinity of the new campus. This approach was an innovation in the history of the German university. There was hardly any tradition in Germany of a spatially close connection between study and living in comparison to Anglo-Saxon universities.[78] In the following years Bochum and Konstanz established a high rate of student residence in their dormitories. In 1986 only about 10 per cent of students nationwide had a dormitory room. In Konstanz it was 23.4 per cent, and in Bochum in 1990, 18.2 per cent of the students lived in dormitories.[79]

How attractive were the new universities for the teaching staff in these first years? Unlike for the students there is hardly comparable statistical material for professors; hence, one needs to resort to more or less representative testimonies from individuals. There were a range of factors attracting young scholars to the three reform universities. Many appointments meant a full professorship for the first time and a move into a secure university career. Furthermore, the new universities offered the chance to implement a reform agenda and to pioneer the growth of new disciplines. This was accompanied by what was usually a very collegial and non-bureaucratic style, especially in relation to younger colleagues within their division or department. The jurist and later CDU politician Kurt Biedenkopf, appointed to the University of Bochum to a chair of public law in 1964, and Rector from 1967–9, recalled the pioneering mood of the first professors: 'For the majority of us we were the first appointed to the chair at the Ruhr University. For many it was the first encounter with the Ruhr. What fascinated us especially was the new beginning, the chance to participate in the reform of the German university.' In addition, Biedenkopf pointed out the fact that existing professorial chairs showed little interest in going to Bochum. An invitation from Bochum meant accepting a new establishment.[80] The founding fathers of reform such as Schelsky came to realize that his younger colleagues at the University of Bielefeld in the environment of 1968 wanted to implement even more extensive ideas of reform. Niklas Luhmann, who joined Bielefeld in 1968 at the instigation of Schelsky, pointed out in a later interview that 'those who owed their position to Schelsky began to rebel and identify with the liberation movements of the 60s'.[81]

Despite or because of being reform friendly, the new universities of West Germany were not immune to the turbulence of student protest in 1968. Although the main centres of the protest movement were in older university cities such as Berlin, Göttingen, Hamburg and Frankfurt, there was also fierce controversy between students, professors and the state ministries in Bochum, Konstanz and Bielefeld. For example, in late May 1968, the German Bundestag's decision to introduce emergency laws led in Bochum to two demonstrations on the campus under the slogan of 'third parity', that is, equal participation of professors, assistants and students in the university decision-making bodies.[82] Academic self-government came under discussion elsewhere, and these battlefronts did not always stay within the university. In 1972 there was in Konstanz a dispute between the university and the Baden-Wurttemberg Ministry of Culture that created a particularly curious position. The university refused to accept the Ministry's design for its constitutional structure. The Ministry's proposal contained a significant tightening of control over the Senate compared to the constitutional order of 1966. When the Ministry signalled in June 1972 that it was unwilling to back down, the university refused to cooperate. All the professors at Konstanz were in solidarity and no successor could be found to the Rector Gerhard Hess. The Ministry was forced to use a 'state commissioner' as provisional head of the institution. Only at the end of November 1973 did the University Senate again choose a Rector from its own ranks.[83]

Reflections

In terms of their structure and architectural design, these three universities played a pioneering role. Between 1965 and 1975, there were twenty-six new or re-established universities. By the early 1980s the total number of universities in the Federal Republic had more than doubled, from twenty-six to fifty-five, and in many the forum-like space idea was taken up.[84] Despite their differences these new universities demonstrate how the previously unknown concept of a campus university came to be implemented in Germany. The Anglo-Saxon idea of the campus university, especially in its American expression, served as the model for the design and the shape of the new West German universities. The countless descriptions of American academic environments and their effects can hardly be overestimated. In the 1960 Bremen memorandum, Rothe described how the American campus university in its Germanized form as an appropriate model, an umbrella under which new forms of scientific and no less important social interaction should be established.[85]

What has become of all the expectations associated with the new universities? Did the higher educational utopias of the 1960s become concrete and did the campus concept as intended help reform higher education in Germany? In retrospect the answers to these questions must be ambivalent. Looking at the initial memoranda and the associated architectural plans, it becomes apparent that in the case of the first campus universities the novel interaction between architecture and structural reform undoubtedly created a new atmosphere of life and work.[86] The whole university system in Germany, particularly at the level of research and teaching, was modernized on a high level; the aspect of interdisciplinarity was strengthened and the access to higher education was indeed democratized.[87]

But despite these efforts it is also clear from today's perspective that the typical American or Anglo-Saxon campus atmosphere could only be realized on a very limited extent in the German context. The defining motif of Rothe's plan, to create with the help of campus conditioning a new type of academic community, proved ultimately unworkable. This is due primarily to different mentalities and traditions. Whilst in the United States and UK living on campus is a matter of course for a significant proportion of students, German campus universities offer no comparable social and cultural campus life. German students traditionally prefer to live in town centres and not in the unattractive urban borders, where the new universities were settled.[88]

Looking back, one further development came along, which counteracted the expected reform impetus. After the period of expansion, higher education soon became less of a political priority. The reform agenda at the Ministries of Education lost momentum following the experience of 1968.[89] At the same time the funding of the new universities could not keep up with the increasing numbers of students. Within a few years the new universities had become mass universities, despite their original plans to admit a manageable number of students.[90] This also applied to the relatively elitist institution of the University of Konstanz. Four decades after its establishment, one of its most prominent founders, Ralf Dahrendorf, conceded that 'the whole thing was created for 3000 students, being a graduate of there was special. "A small Harvard on the Bodensee" some called the planned institution. It became as you know totally different'.[91] Also from the perspective of 1990, Kurt Biedenkopf was critical of the University of Bochum. He pointed out that the new universities of the 1960s and 1970s became a fixture in the German landscape, but the original intentions of reform remained stuck in many areas. Like Dahrendorf he identified the increase in student numbers as one of the main barriers, with negative consequences for research and teaching.[92]

Notes

1   Manfred Heinemann, *Umerziehung und Wiederaufbau. Die Bildungspolitik der Besatzungsmächte in Deutschland und Österreich* (Stuttgart: Klett-Cotta 1981); Corine Defrance, *Les Alliés occidentaux et les universités allemandes 1945–1949* (Paris: CNRS, 2000); idem., 'Die Westalliierten als Hochschulreformatoren (1945–1949): ein Vergleich', in *Zwischen Idee und Zweckorientierung. Vorbilder und Motive von Hochschulreformen seit 1945*, ed. Andreas Franzmann, Barbara Wolbring (Berlin: Akademie Verlag, 2007), 35–46; Stefan Paulus, *Vorbild USA? Amerikanisierung von Universität und Wissenschaft in Westdeutschland* (München: Oldenbourg Verlag, 2010), 95–146; Moritz Mälzer, *Auf der Suche nach der Neuen Universität. Die Entstehung der Reformuniversitäten Konstanz und Bielefeld in den 1960er Jahren* (Göttingen: Vandenhoek & Rupprecht, 2016), 23–92.

2   *Hochschuloffiziere und Wiederaufbau des Hochschulwesens in Deutschland, Teil 4: Die sowjetische Besatzungszone*, ed. Manfred Heinemann (Hildesheim: Klett-Cotta, 1997); *Amerikanisierung und Sowjetisierung in Deutschland 1945–1970*, ed. Konrad H. Jarausch and Hannes Siegrist (Frankfurt am Main: Campus Verlag, 1997).
3   *Hochschuloffiziere und Wiederaufbau des Hochschulwesens in Westdeutschland*, Teil 3: *Die Französische Zone*, ed. Manfred Heinemann (Hildesheim: Klett-Cotta, 1991); Corine Defrance, 'Die Franzosen und die Wiedereröffnung der Mainzer Universität, 1945–1949', in *Kulturpolitik im besetzen Deutschland*, ed. Gabriele Clemens (Stuttgart: Franz Steiner Verlag, 1994), 117–30; Mälzer, *Auf der Suche nach der Neuen Universität*, 48–92.
4   Karl-Ernst Bungenstab, *Umerziehung zur Demokratie? Re-education-Politik im Bildungswesen der US-Zone 1945–1949* (Düsseldorf: Bertelsmann, 1970); James F. Tent, *Mission on the Rhine. Re-Education and Denazification in American-Occupied Germany* (Chicago: University of Chicago Press, 1982); *Hochschuloffiziere und Wiederaufbau des Hochschulwesens in Westdeutschland 1945–1951, Teil 1: Die Britische Zone*, ed. Manfred Heinemann (Hildesheim: Klett-Cotta, 1990); *Hochschuloffiziere und Wiederaufbau des Hochschulwesens in Westdeutschland 1945–1951, Teil 2: Die US-Zone*, ed. Manfred Heinemann (Hildesheim: Klett-Cotta, 1990);
5   James F. Tent, 'Der amerikanische Einfluss auf das deutsche Bildungswesen', in *Die USA und Deutschland im Zeitalter des Kalten Krieges, vol. 1: 1945–1968*, ed. Detlef Junker (Stuttgart, München: Deutsche Verlagsanstalt, 2001), 601–11; Stefan Paulus, 'The Americanization of Europe after 1945?: The Case of the German Universities', *European Review of History/Revue Européene d'Histoire*, 9 (2002), 241–53; Paulus, *Vorbild USA?*, 151–2.
6   James F. Tent, 'The Free University of Berlin. A German Experiment in Higher Education, 1948–1961', in *American Policy and the Reconstruction of Germany, 1945–1955*, ed. Jeffry Diefendorf and Axel Frohn (Cambridge: Cambridge University Press, 1993), 237–56; Paulus, *Vorbild USA?*, 171–203; Mälzer, *Auf der Suche nach der Neuen Universität*, 43–4.
7   Studienausschuss für Hochschulreform, *Gutachten zur Hochschulreform* (Hamburg: Control Commission for Germany, 1948); David Phillips, 'The Rekindling of Cultural and Intellectual Life in the Universities of Occupied Germany with Particular Reference to the British Zone', in *Kulturpolitik im besetzen Deutschland*, ed. Clemens, 102–16.
8   David Phillips, 'Lindsay and the German Universities: An Oxford Contribution to the Postwar Debate', in Philipps, *Investigating education in Germany. Historical Studies from a British Perspective* (London: Routledge, 2016), 105–21.
9   *Gutachten zur Hochschulreform*, 65–84.
10 'Marburger Hochschulgespräche 1946/1947/1948', in *Dokumente zur Hochschulreform 1945–1959*, ed. Rolf Neuhaus (Wiesbaden: Steiner, 1961), 260–2; 'Schwalbacher Richtlinien (1947)' in *ibid.*, 262–88; 'Gutachten zur Hochschulreform vom Studienausschuss für Hochschulreform 1948 (Blaues Gutachten)' in *ibid.*, 289–368.
11 Paulus, *Vorbild USA?*, 373–448.
12 *Wissenschaftsrat, Empfehlungen des Wissenschaftsrates zum Ausbau der wissenschaftlichen Einrichtungen, Teil I: Wissenschaftliche Hochschulen* (Tübingen: Steiner, 1960).

13 *Wissenschaftsrat, Abiturienten und Studenten. Entwicklung und Vorschätzung der Zahlen 1950 bis 1980* (Bad Godesberg: Steiner, 1964), 19.
14 Gabriele Metzler, *Konzeptionen politischen Handelns von Adenauer bis Brandt. Politische Planungen in der pluralistischen Gesellschaft* (Paderborn: Verlag, 2005), 419.
15 Hans Werner Rothe, 'Über die Gründung einer Universität zu Bremen. Denkschrift vorgelegt der Universitätskommission des Senats der Freien Hansestadt Bremen (1961)', in *Dokumente zur Gründung neuer Hochschulen. Anregungen des Wissenschaftsrats, Empfehlungen und Denkschriften auf Veranlassung von Ländern in der Bundesrepublik Deutschland in den Jahren 1960-1966*, ed. Rolf Neuhaus (Wiesbaden: Franz Steiner Verlag, 1968), 265-482.
16 Klaus Kafka, 'Amerikanische Universitäten', *Bauwelt*, 5 (1972), 171-5; Rudolf Pörtner and Klaus Thinius, 'Zwei Wochen in Amerika', *ibid.*, 176-77; Stefan Paulus, 'Zwischen konzentrierter Stille und Weltoffenheit: Zur Idee der Campus-Universität im Kontext westdeutscher Universitätsneugründungen der 1960er Jahre', in *Ein Campus für Regensburg* (Regensburg: Regensburger Universitätsverlag, 2007), 37-55; Mälzer, *Auf der Suche nach der Neuen Universität*, 101-6.
17 Paulus, *Vorbild USA?*, 483-501.
18 Stefan Muthesius, *The Postwar University. Utopianist Campus and College* (London and New Haven: Yale University Press, 2000), 223.
19 Ibid., 365-6.
20 Muthesius, *Postwar University*, 220-46; Paulus, 'Zwischen Konzentrierter Stille und Weltoffenheit'; idem., *Vorbild USA?*, 502-19; Mälzer, *Auf der Suche nach der Neuen Universität*, 96-106.
21 Ernst Anrich, *Die Idee der deutschen Universität und die Reform der deutschen Universitäten* (Darmstadt: Wisenschaftliche Buchgesellschaft, 1956), 101-2; Stefan Grüner, 'Vom Ort einer Universitätsgründung. Bildungsboom und Hochschulpolitik in Westdeutschland während der "Langen" 1960er Jahre', in *Stätte des Wissens. Die Universität Augsburg 1970-2010. Traditionen, Entwicklungen, Perspektiven*, ed. Werner Lengger et al. (Regensburg: Schnell & Steiner, 2010), 74-6.
22 Sekretariat der Kultusministerkonferenz and Sekretariat der Westdeutschen Rektorenkonferenz, *Das Hochschulwesen in der Bundesrepublik Deutschland. Probleme und Tendenzen. Stand Juli 1965* (Bonn: Deutschen Akademischen Austauschdienst, 1966), 26.
23 Richard P. Dober, *Campus Planning* (Washington: Reinhold, 1963), 85; Hanno-Walter Kruft, *Geschichte der Architekturtheorie*, 4th edn. (München: C. H. Beck, 1995), 398.
24 Paulus, 'Zwischen Konzentrierter Stille und Weltoffenheit'.
25 Paulus, *Vorbild USA?*, 483-519.
26 Mälzer, *Auf der Suche nach der Neuen Universität*, 101-4.
27 'Empfehlungen zum Ausbau der Universität Bochum. Denkschrift des Gründungsausschusses veröffentlicht vom Kultusministerium des Landes Nordrhein-Westfalen (Dezember 1962)' in Neuhaus (ed.), *Dokumente zur Gründung neuer Hochschulen…1960-1966*, 209-13.
28 Josef Hofmann, 'Bochum oder Dortmund? Auseinandersetzung um den Universitätsstandort', in *Materialien zur Geschichte der Ruhr-Universität Bochum. Die Entscheidung für Bochum*, ed. Gesellschaft der Freunde der Ruhr-Universität Bochum (Bochum: Lippenmühlen & Dierichs, 1971), 29-32.
29 Ludwig Adenauer, 'Wie es Dazu Kam. Bochum – Neue Leitbilder für die Gründung einer Universität' in ibid., 13-28; Mälzer, *Auf der Suche nach der Neuen Universität*, 225-42.

30 'Empfehlungen zum Ausbau der Universität Bochum', 209–13, 248–9.
31 Hans-Günther Bierwirt and Hans Thol, 'Ruhr-Universität Bochum – Zehn Jahre Planen und Bauen', in *Materialien zur Geschichte der Ruhr-Universität Bochum*, 149–86; Muthesius, *Postwar university*, 224–31; Cornelia Jöchner, 'Wo Wege Sich Kreuzen. Die Räumliche Logik des Bochumer Campus', in *Ruhr-Universität Bochum: Architekturvision der Nachkriegsmoderne*, ed. Richard Hoppe-Sailer (Berlin: Gebrüder Mann Verlag, 2016), 31–46.
32 *Hochschulplanung. Beiträge zur Struktur- und Bauplanung*, vol. 3, ed. Horst Linde (Düsseldorf: Werner Verlag, 1970), 121–231; Klaus Jan Philipp, 'Beton im Hochschulbau' in *Ruhr-Universität Bochum*, ed. Hoppe-Sailer, 157–64.
33 'Ideenwettbewerb für die Ruhr-Universität Bochum in den Jahren 1962/1963' in *Materialien zur Geschichte der Ruhr-Universität* Bochum, 47–54; Alexandra Apfelbaum and Frank Schmitz, 'Universitas durch Dichte. Der Ideenwettbewerb zur Ruhr-Universität 1962/63' in *Ruhr-Universität Bochum: Architekturvision der Nachkriegsmoderne*, ed. Hoppe-Sailer, 59–78.
34 Maximilian Thurn, 'Die Bauidee der Ruhr-Universität Bochum. Eine Kritische Würdigung' in *Materialien zur Geschichte der Ruhr-Universität Bochum*, 63–108; Fridel Hallauer, 'Baugeschichte der Ruhr-Universität Bochum. Fakten und Wertungen nach 25 Jahren', in *Universität und Politik. Festschrift zum 25jährigen Bestehen der Ruhr-Universität Bochum*, vol. 1, ed. Burkhard Dietz et al. (Bochum: Laumanns, 1990), 201–42; Muthesius, *Postwar university*, 224–31; Paulus, *Vorbild USA?*, 502–7.
35 Bierwirt and Thol, 'Ruhr-Universität Bochum. Zehn Jahre Planen und Bauen', 178–85; Hallauer, 'Baugeschichte der Ruhr-Universität Bochum. Fakten und Wertungen nach 25 Jahren', 212, 226–8.
36 Klaus Oettinger and Helmut Weidhase, *Eine Feste Burg der Wissenschaft. Neue Universität in einer alten Stadt Konstanz am Bodensee* (Konstanz: Stadler Verlag, 1986), 10.
37 Paulus, *Vorbild USA?*, 513.
38 'Denkschrift der Regierung über die Errichtung von Wissenschaftlichen Hochschulen in Baden-Württemberg, Einleitung und Teil I', *Konstanzer Blätter für Hochschulfragen* 2 (1963), 9–21.
39 Oettinger and Weidhase, *Eine feste Burg der Wissenschaft*, 14–17.
40 'Die Universität Konstanz. Bericht des Gründungsausschusses' (1965) in *Dokumente zur Gründung neuer Hochschulen … 1960–1966*, ed. Neuhaus 571–2, 573–91.
41 Ibid., 576, 617–18.
42 Wenzel Ritter von Mann, 'Die Städtebauliche Eingliederung der Universität Konstanz', *Konstanzer Blätter für Hochschulfragen* 14 (1967), 15–16.
43 Muthesius, *Postwar University*, 236; Universität Konstanz, *Gebaute Reform. Architektur und Kunst am Bau der Universität Konstanz* (München: Hirmer Verlag, 2016).
44 Oettinger and Weidhase, *Eine Feste Burg der Wissenschaft*, 25–6; Mälzer, *Auf der Suche nach der Neuen Universität*, 355–74.
45 'Die Universität Konstanz. Bericht des Gründungsausschusses' (1965)', 622–6.
46 Muthesius, *Postwar University*, 236.
47 Oettinger and Weidhase, *Eine Feste Burg der Wissenschaft*, 40–59.
48 'Gründungsausschuss der Universität im Ostwestfälischen Raum' (1966) in *Dokumente zur Gründung neuer Hochschulen … 1960–1966*, ed. Neuhaus, 203–6; Peter Lundgreen, 'Strukturmerkmale einer Reformuniversität', in *Reformuniversität*

Bielefeld 1969-1994. Zwischen Defensive und Innovation, ed. Peter Lundgreen (Bielefeld: Verlag für Regionalgeschichte, 1994), 25–8; Martin Löning and Gerhard Trott, *Die Universität Bielefeld. Eine Geschichte in Bildern* (Erfurt: Sutton Verlag, 2003), 9–34; Mälzer, *Auf der Suche nach der Neuen Universität*, 243–83.

49 Muthesius, *Postwar University*, 355–74.
50 Klaus Köpke et al., *Universität Bielefeld, Bauen in der industriellen Welt. Eine Dokumentation zur Architektur der Universität Bielefeld. Kunsthalle Bielefeld vom 19. Oktober – 16. November 1975* (Bielefeld: Kunsthalle, 1975).
51 Erwin Heinle and Thomas Heinle (ed.), *Bauen für Lehre und Forschung* (Stuttgart & München: Deutsche Verlags-Anstalt, 2001), 133.
52 Muthesius, *Postwar University*, 243.
53 'Erbaulich', 31 March 2019, https://www.uni-bielefeld.de/kultur/baukultur/?__xsl=/unitemplate_2009_print.xsl.
54 Ibid.
55 Linde (ed.), *Hochschulplanung*, 121–231.
56 Paul Mikat, 'Universitäts-Gründungsprobleme in Nordrhein-Westfalen', in *Festschrift zur Eröffnung der Universität Bochum*, ed. Wenke and Joachim H. Knoll (Bochum: Verlag Ferdinand Kamp, 1965), 12–13.
57 'Empfehlungen zum Ausbau der Universität Bochum', 210–42.
58 'Die Universität Konstanz. Der Gründungsausschuss' (1965) in *Dokumente zur Gründung neuer Hochschulen, 1960-1966*, ed. Neuhaus 577–8.
59 Ibid., 578, 591–8.
60 'Gründungsausschuss der Universität im Ostwestfälischen Raum (1966)', 203, 205; Gerhard Sprenger and Peter Weingart, 'Zentrum für Interdisziplinäre Forschung' in *Reformuniversität Bielefeld*, 397–410; Mälzer, *Auf der Suche nach der Neuen Universität*, ed. Lundgreen, 255–73, 447–50.
61 'Gründungsausschuss der Universität im Ostwestfälischen Raum', 205.
62 Mälzer, *Auf der Suche nach der neuen Universität*, 220, 330, 340.
63 Anne Rohstock, *Von der 'Ordinarienuniversität' zur 'Revolutionszentrale'? Hochschulreform und Hochschulrevolte in Bayern und Hessen 1957-1976* (München: Oldenbourg Verlag, 2010), 262–72.
64 'Die Universität Konstanz. Der Gründungsausschuss', 615.
65 'Empfehlungen zum Ausbau der Universität Bochum', 253.
66 Detlef Horster, *Niklas Luhmann* (München: C.H. Beck, 1997), 37–47; Paul Nolte, *Hans-Ulrich Wehler. Historiker und Zeitgenosse* (München C.H.Beck, 2016), 60–104.
67 Westdeutsche Rektorenkonferenz, *Hochschulrahmengesetz – Hochschulgesetze der Länder der Bundesrepublik* (Bad Godesberg: Westdeutsche Rektorenkonferenz, 1976).
68 Olaf Bartz, *Der Wissenschaftsrat: Entwicklungslinien der Wissenschaftspolitik in der Bundesrepublik Deutschland 1957-2007* (Wiesbaden: Steiner Verlag, 2007).
69 *Wissenschaftsrat, Empfehlungen des Wissenschaftsrates zum Ausbau der wissenschaftlichen Einrichtungen, Teil I: Wissenschaftliche Hochschulen* (Tübingen: Steiner, 1960), 52.
70 'Anregungen des Wissenschaftsrates zur Gestalt neuer Hochschulen (1962)', in *Dokumente zur Gründung neuer Hochschulen*, ed. Neuhaus, 1–72.
71 Klaus Tenfelde, 'Die Ruhr-Universität und das Milieu des Reviers', in *Schöne Neue Hochschulwelt. Idee und Wirklichkeit der Ruhr-Universität Bochum*, ed. Wilhelm Bleek and Wolfhard Weber (Esse: Klartext, 2003), 43–56; Mälzer, *Auf der Suche nach der Neuen Universität*, 235–42.

72 Fritz Heinemann and Gerhard Petschelt, 'Die Stadt Bochum und die Ruhr-Universität' in *Festschrift zur Eröffnung der Universität Bochum*, ed. Wenke and Knoll, 57–62; Oettinger and Weidhase, *Eine Feste Burg der Wissenschaft*, 11–13.
73 Hermann Lübbe, 'Aufbau Nach dem Wiederaufbau. Ein Rückblick auf die Gründung der Ruhr-Universität Bochum' in *Universität und Politik*, ed. Dietz et al., 316–17.
74 Heinrich Kost, 'Die Gesellschaft der Freunde der Ruhr-Universität Bochum e. V.' in *Festschrift zur Eröffnung der Universität Bochum*, Wenke and Knoll, ed. 64.
75 Karin Hartewig, 'Die Stadt Bochum und die erste Studentengeneration an der Ruhr-Universität' in *Universität und Politik*, ed. Dietz et al., 177–99.
76 Ibid., 191–2.
77 Ibid., 192–8.
78 Hallauer, 'Baugeschichte der Ruhr-Universität Bochum', 214–16, 226–8; Mälzer, *Auf der Suche nach der Neuen Universität*, 137–44.
79 Oettinger and Weidhase, *Eine Feste Burg der Wissenschaft*, 90; Hallauer, 'Baugeschichte der Ruhr-Universität Bochum', 212.
80 Kurt H. Biedenkopf, '25 Jahre Ruhr-Universität: Erfüllte und Nichterfüllte Reformerwartungen' in ibid., 295.
81 Horster, *Niklas Luhmann*, 38.
82 Rasch and Wilke, 'Von Einer Abteilung ohne Gebäude und Studenten', 283–5.
83 Oettinger and Weidhase, *Eine Feste Burg der Wissenschaft*, 33.
84 Rainer A. Müller, *Geschichte der Universität. Von der Mittelalterlichen Universitas zur deutschen Hochschule* (Hamburg: Callway, 1996), 106.
85 Paulus, *Vorbild USA?*, 373–524.
86 Hallauer, 'Baugeschichte der Ruhr-Universität Bochum', 226–8.
87 Mälzer, *Auf der Suche nach der Neuen Universität*, 402–40.
88 Arnd Morkel, *Erinnerung an die Universität. Ein Bericht* (Vierow bei Greifswald: SH-Verlag, 1995), 75–6.
89 Mälzer, *Auf der Suche nach der Neuen Universität*, 402–40.
90 Peter Glotz, *Im Kern verrottet? Fünf für zwölf für Deutschlands Universitäten* (Stuttgart: Deutsche Verlags Anstalt, 1996); Biedenkopf, '25 Jahre Ruhr-Universität', 296–8; Stefan Paulus, *Vorbild USA?*, 520–50.
91 Dahrendorf, *Gründungsideen und Entwicklungserfolge der Universität*, 9.
92 Biedenkopf, '25 Jahre Ruhr-Universität', 296–7.

Afterword: The utopian universities in historical perspective

by Peter Mandler

Like many of the contributors to this volume, I grew up with and in a utopian university. My parents were among the first to join the University of California at San Diego when it opened in the early 1960s, and I spent my formative years there. Halfway across the world, my wife spent her early years in the late 1950s and early 1960s at another, very different kind of utopian university – Makerere University, in Kampala, discussed by Miles Taylor in his chapter – where *her* parents also came to work in its early post-war years. Perhaps we baby-boomers are not the best people to assess objectively the distinctive features of the utopian universities of the 1960s; the owl of Minerva will undoubtedly fly higher as dusk deepens to night; and yet, after fifty years, with our historians' hats on we ought to be able to make a provisional assessment. To do so requires looking back to the early phases of the development of higher education, to see what was more and what was less radically changed by the experiences of the 1960s, and also forward to the age of mass higher education that most developed countries have since entered, to see what legacies the utopian universities bequeathed. Where did the expanding opportunities, the redrawing of the map of knowledge, the pedagogical innovations, the new experiments in living come from and how much has survived?

First, audience. In Martin Trow's influential formulation, the new universities of the 1960s represented the passage from elite to mass higher education.[1] Numerically this captures well the period in the UK when participation in higher education rose from around 2–3 per cent of eighteen- and nineteen-year-olds, the level sustained before the Second World War, to a temporary peak of 14 per cent around 1970. Both 'elite' and 'mass' however do not tell the full or perhaps even the true story. Apart from Oxford and Cambridge, the British higher education system before the 1960s was not the training ground for elites. Medicine and the higher reaches of the civil service aside, most of the professions that we consider elite today did not require or even value higher education – law, banking, accountancy, engineering and similar professions all relied mostly on apprenticeship systems. Even for the social elite – the national elite in the case of Oxbridge, regional elites in the case of 'redbricks' – university was not necessarily the place one acquired one's social skills or connections or first steps on the ladder to more power and wealth. Churchill didn't go to university; neither did Joseph or Neville Chamberlain; neither did many millionaires. Instead, most higher education before the 1960s was aimed at a small number of unusual professions which had come to require more formal training – medicine for an elite, school-teaching for the rest. As late as 1960, over 40 per cent of all university graduates went into teaching as their first destination.[2]

Most of these 'elite' university students were middle class, a lot of them lower middle class and about a quarter working class, school teaching being one of the principal means of social mobility in the first two-thirds of the twentieth century. The impact of the new universities, therefore, was mostly to *diversify* the middle-class nature of higher education, and to make higher education a more normal part of middle-class life, rather than an optional social experience, or a training for a small handful of specific jobs. In this role, they formed part of a broader re-working of middle-class identity, which the sociologist Mike Savage has identified as more meritocratic, distanced from hereditary social elites, modern, technical, 'strategically mobile'.[3] Partially in imitation of the American model, they made 'student life' integral to middle-class socialization and tied higher education more closely to a much wider range of professional and managerial careers.

The redrawing of the 'map of learning', to which so many of the chapters in this book attest and which was undoubtedly a key feature of nearly all the utopian universities, looks somewhat different in the light of this social change. As many of the individual chapters show, the new British universities were founded in the shadow of the 'two cultures' controversy. They were expected to fill up with earnest young men making the technological revolution through study of and research in the sciences, and the new map of learning was drawn to ensure that they also imbibed enough of 'Western civilization' both to carry on the traditional universities' civilizational mission and to ensure a humane application of potentially dangerous or hubristic new knowledges. Thus foundation years that required both arts and sciences, combined honours degrees, and interdisciplinary programmes that crossed the 'two cultures' boundary: environmental studies, development studies, management, behavioural sciences.

In fact, however, the expansion of higher education in the 1960s led to a sustained 'swing away from science'.[4] Why did this happen? Perhaps, as an anxious political establishment speculated, there was an emerging anti-science bias in the culture, evidenced in growing public scepticism about nuclear power and 'Frankenstein' medical breakthroughs such as heart transplantation.[5] Or there may have been a more benign turn not away from science but towards self-expression and self-discovery among baby boomers, favouring some traditional humanities subjects but mostly subjects that lay between arts and sciences – the 'social studies', including psychology, sociology, communications and even law and business. Above all, however, new entrants were less likely to enter with the scientific qualifications required to pursue science at degree level and were more likely to gravitate towards new subjects without rigorous prerequisites. And, from the 1970s onwards, they were also more likely to be women. As the OECD observed, these were trends not specific to the UK but common to most developed countries, part of the international phenomenon that we call 'the Sixties' (though much of it was situated in the early 1970s) as this volume seeks to show.[6] The lasting curricular legacy of the utopian universities lay therefore not in combinations of arts and sciences but in new social studies subjects, also an international phenomenon. Interdisciplinary programmes that drew on these subjects proved longest lasting – environmental and development studies, American studies and other area specialisms – and even more enduring in the UK were single-honours programmes in psychology and sociology, which grew rapidly in popularity, especially after 1990 when women began to form a majority of all undergraduates in Britain.[7]

The sciences did flourish in the new universities, but not at undergraduate level – rather, they thrived on the Cold War boom in state- and corporate-sponsored scientific and technological research. This growing divide between teaching-oriented arts and social sciences and research-oriented sciences surely exacerbated the generational conflicts that emerged in the utopian universities in the late 1960s, as Jon Agar's conclusion suggests. Student culture drifted to the left and to the counterculture at the same time as institutional interests, at least financial ones, gravitated towards the state and industrial establishments, the divide which set the stage for E. P. Thompson's dramatic political interventions at Warwick. This divide would continue to widen in the 1970s when the polytechnic sector absorbed a growing share of the higher education market. Institutions that some establishment interests had imagined would become bastions of applied, vocational studies closely linked to industry attracted instead a younger generation more attuned to self-expression and social service – and to the creative arts, which were drawn into higher education, thanks not to the utopian universities but to the polytechnics. North-East London Polytechnic became a hotbed of student activism in the early 1970s to rival Essex and the LSE, not because polytechnics recruited a lot of aircraft-design students but because they recruited a lot of art students.

The utopian universities did, however, make a more distinctive contribution to this embroilment of universities in the culture wars by developing and mainstreaming new forms of 'student life', self-consciously aloof from established interests, that generated an idea of a distinctive 'student estate'. The modernist, often brutalist architectural forms with which they experimented set the tone, as did the greenfield sites deliberately distanced from town centres, unlike most redbricks. Partly for this reason, and partly because again they drew serendipitously on a cultural shift towards generational change and self-expression, the utopian universities did not – again unlike the redbricks – consider themselves local or regional institutions, even though their origins often lay in just such a rooted identity. They drew on national catchments and mixed up together young people from different class backgrounds and all parts of the country into a student estate with its own distinctive values and behaviours.[8] In this respect, the new universities of the post-1992 period, which had originated as colleges and polytechnics, followed their example. Though they too had originally been rooted in local communities, and deliberately sited in town centres to lower barriers to widening participation, they ended up participating in the same cultivation of 'student life', opening halls of residence, converging with the utopian universities in curriculum and internal culture, and even sometimes making their own utopian experiments, suitably adapted to the changed circumstances of the late twentieth and early twenty-first centuries. What began as a collection of urban colleges in Hull, badged as the University of Humberside in 1992, has since metamorphosed into a campus university on a new waterside site near Lincoln, rebadged as the University of Lincoln in 2001, looking remarkably like a 1960s utopian university in postmodern clothing.

What remains of the utopian universities, at least in the UK, is a new model of a university largely driven by the new types of students who were drawn into this first phase of expansion and by the unusual circumstances of generational change in the 1960s that happened to coincide with that expansion. This model retains some of its striking utopian features – the bubble of 'student life' set apart from society, a vehicle for self-discovery and self-expression, a new map of learning more oriented to the relation of self and society – but has adapted to new, larger waves of expansion that succeeded from the late 1980s onwards. For some time this adaptation was made easier by the fact that the country as a whole was coming to look more like the more diverse middle classes who entered the new universities in the 1960s: that is, deindustrialization and social, cultural and occupational change were making more of the country liberal, white-collar, public-sector and 'middle class'. But it may be the case that the present moment marks a point where the most enduring features of the utopian universities are finally being fully eclipsed, as social, cultural and economic changes move in other directions from those established in the 1960s. Economic imperatives and a new managerial culture have tended to reconcile the tug of war between teaching and research in favour of the latter. The impact of labour-market precarity since 2008 and a new tuition fee regime may have aligned students' own needs and desires with those economic and managerial imperatives. The swing away from science finally ended, and went into reverse, after 2008. It is unclear whether students want any longer to live in a bubble set apart from society, and in any case as participation rates rise to 50 per cent it becomes harder to distinguish students as an estate of their own. Education is now a more important determinant of class and culture than it once was, but young people and their educated parents will be more likely to share that class and culture than to find in it an aggravating source of division. But who knows? Further into the twenty-first century we may discover other, more enduring legacies of the utopian universities of the 1960s.

Notes

1   Martin Trow, *Problems in the Transition from Elite to Mass Higher Education* (Berkeley: Carnegie Commission on Higher Education, 1973).
2   R. K. Kelsall et al., *Six Years After: First Report on a National Follow-Up Survey of Ten Thousand Graduates of British Universities in 1960* (Sheffield: University of Sheffield, 1970), 33.
3   Mike Savage, *Identities and Social Change in Britain since 1940: The Politics of Method* (Oxford: Oxford University Press, 2010), 19, 67, 76, 83–5, 221–35.
4   Peter Mandler, 'The Two Cultures Revisited: The Humanities in British Universities since 1945', *Twentieth-Century British History*, 26 (2015), 400–23.
5   Ayesha Nathoo, *Hearts Exposed: Transplants and the Media in 1960s Britain* (Basingstoke: Palgrave Macmillan, 2009).

6 *Development of Higher Education 1950–1967* (Paris: OECD, 1971), 129–30.
7 Peter Mandler, 'The Rise of the Social Sciences in British Education, 1960–2016', in *The History of Sociology in Britain: New Research and Revaluation*, ed. Plamena Panayotova (London: Palgrave Macmillan, 2019), 281–99.
8 Keith Vernon, 'Engagement, Estrangement or Divorce? The New Universities and their Communities in the 1960s', *Contemporary British History*, 31 (2017), 501–23.

Index

22 March (political group) 345

Aachen 359
Aalto, Alvar 360
Abercrombie, Jane 311
Abercrombie, Patrick 269
Aberdeen, University of 180, 193
Academy of Science 311 (Australia)
accommodation, *see* student residences
Accra 272, 275
Adams, Walter 280–1
Adelaide 1, 305–6
Adelaide, University of 308–9
Administrative Staff College 175
African studies 61, 63, 127, 279
agriculture 129, 212–13, 229
agronomy 96
Aimé, Gérard 344
Airthrey Castle 197, 231
Alagh, Yoginder K. 335
Aldeburgh 95
Aldeburgh Festival 95
Aldershot 60
Alexander, Norman 271, 279
Alexander, Sir Kenneth 191
Algeria 111, 353
Algerian War 341–2, 344, 346, 353
Algerians 349–50
All India Students Association 331
All Souls College, Oxford, *see* Oxford, University of
Allcorn, Derek 144
Allen West and Co (radar manufacturer) 60
Allen, David 80
Allen, H. C. ('Harry') 63
Allen, Sir George 273
Alsace 350
Ambedkar, B. R. 338
Ambleside 183
American Council of Learned Societies (ACLS) 165

American studies 8, 38, 58, 61, 63, 91, 93–5, 97, 99, 101, 127, 180, 377
American University of Rome 153
Amis, Kingsley 10
anatomy, comparative 127, 131
Anderson, Albert 215, 247
Anderson, Sir Robert 9–10, 228
Angers, University of 3
Anglin, Douglas 279
Anglo-American Commission (on the West Indies) 274
Angola 270
Angola Estudos Gerais Universitários de 270
animal physiology 130–1
Annales school 160
Annan, Lord Noel 6, 24, 84, 94, 105–9, 116, 128, 226
Anselm, St 145
anthropology 96, 148–50, 292, 311
Arab studies 180
archaeology 133
Archard, Peter 114
Architects Co-Partnership 107
Architectural Association 7
architecture (subject) 75, 182
architecture (university) 7, 11, 21–3, 29–31, 73, 79, 84, 90, 92, 98, 105, 107–8, 110, 112, 117, 129, 146, 162, 177, 215, 270, 272–3, 278, 287, 292, 293–5, 298, 325, 328, 341, 358–62, 367, 377
Argentina 2
Armagh 209, 211, 213
Arnold, Thomas 78
art, history of 147
Arthurs, Harry 294
artificial intelligence (AI) 132
Arts Council 43, 196, 233
arts, *see* humanities
arts, applied 75
arts, fine 75, 93, 95, 233, 298

Aryamehr University of Industry 7
Ashby, Eric 270–1, 276, 279, 306–7
Ashford 141
Ashton family 224, 230
Ashton, Robert ('Bob') 94
Ashworth, John 133
Asian studies 61, 63, 127, 311, 329
Asquith, Cyril 270
Associated Electrical Industries Ltd 60
Association of German Architects 362
Association of University Teachers 193
Aston University 23, 29
Athens University of Economics and Business 153
Atomic Energy Commission (Australia) 311
atomic energy research 125
Atomic Energy Research Establishment 41
Attlee, Clement 40, 89
Auden, W. H. 152
Augustine, St 142
Australia 1, 3, 5–7, 9, 47, 76, 145, 180, 215, 231, 269, 305–17
Australian National University (ANU) 1, 59–60, 306, 310, 325
Aylmer, Gerald 82–3
Ayr 193–4
Azikiwe, Nnamde ('Zik') 275

Bacon, Francis 169
Bacon, Sir Edmund 236
Baden-Württemberg 360, 367
Baghdad, University of 7
Baker, Ron 291
Balliol College, Oxford, *see* Oxford, University of
Ballymena Academy 209
Balme, David 272
Baltimore 43
Banda, Hastings 277
Bandaranaike, S. W. R. D. 274
Banham, Reyner xix, 31
Banks, Tony 83
Bannerjee, D. 326
Bannerjee, Sumanta 329–30
Banton, Michael 44
Barbados, University of West Indies, *see* West Indies, University of
Barbet, Raymond 344, 352

Barcelona 1, 4
Barclays Bank 234
Barda, Jacques 346
Barlow, Sir Alan 6, 122, 193
Barraclough, Geoffrey 94
Barratt, Sir Charles 225
Basutoland, Bechuanaland and Swaziland, University of 42, 278–9, 281
Batabyal, Rakesh 323, 336
Bath, University of 29
Baudrillard, Jean 343
Bauhaus 38, 294
Beaton, Cecil 92
Beaver, Stanley 41
Bechuanaland (Botswana) 278
Becket, Thomas 142, 145
Bedford College, *see* London, University of
Belfast 211, 218
Belfast, Queen's University 176, 208, 210, 212, 215–18, 270, 315
Belgium 1, 269
Bell Telephone Laboratories 232
Beloff, Michael 6–7, 27
Ben-Gurion University of the Negev 4
Benelux countries 364
Bennet-Clark, Thomas 96–7, 129, 134
Bennett, W. A. C. 291, 300 n. 8
Bensaid, Daniel 347
Bentwich, Norman 41
Berghahn, Volker 59
Berkeley, University of California, *see* University of California
Berlin 1, 366–7
Berlin, University of 360
Bernal, J. D. 78
Bernstein, Henry 150
Bevington, John 178
Bhattacharya, K. R. 329
Biedenkopf, Kurt 366, 368
Bielefeld, University of 358, 360–7
Bilbao, University of 1
Biobaku, Saburi 279–80
biochemistry 129–32, 147, 179
biology 75, 81, 93, 96–7, 127–32, 179, 182, 185, 191, 196, 212, 215–16, 272, 363
biophysics 96, 129
Birds Eye 133

Birkbeck College, *see* University of London
Birks, Tony 26
Birley, Robert 2, 40, 76, 276, 280
Birley, Sir Derek 217–18
Birmingham 162
Birmingham Post 28
Birmingham, University of 23, 25, 142–3, 178, 180, 224, 229
Birrell, James 241
Black and White 280
Blackburn Cathedral, Canon Theologian of 180
Blackpool 183
Blake, John 42, 50 n. 31, 279
Blantyre 277
Blin-Stoyle, Roger 127–8
Blinkhorn, Martin 179, 187 n. 19
Bochum 364
Bochum, University of 358–68
Boden, Margaret 132
Bologna process 363
Bombay, University of 324
Bomont, R. G. 195
Bonn 3
Bootham School 90
Borthwick Institute for Historical Research 74
Boston (USA) 161
botany 81, 127, 129, 133
Bottisham 79
Bowden, B. V. 125
Boyle, Edward 37
Boyson, Rhodes 116
Bracknell 133
Bradbury, Malcolm 6, 23, 95, 97–8
Bradley, Denise 305
Brasenose College, *see* Oxford, University of
Brasilia University 1, 5, 7
Brawijaya, University of 3
Brazil 1–2, 5-6, 9
Bremen, University of 240, 358–9, 361, 367
Brenchley, Lord 55
Brent North (parliamentary constituency) 116
Brentwood 232
Bridge of Allan 198
Bridges, Lord 64

Bridgewater, Shepheard and Epstein (architects) 7, 38, 177
Brigate Rosse 4
Briggs, Lord Asa 3–4, 25, 46, 60–1, 64–5, 82, 90, 127–8, 132, 232–4
Brighton 60, 125, 210, 223, 225
Brisbane 306, 309
Bristol, University of 29, 81, 224, 230
British Association of American Studies (BAAS) 63, 66
British Broadcasting Corporation (BBC) 41, 108, 114
British Columbia 287, 291
British Columbia, University of (UBC) 291–2, 297
British Council 196
British Film Institute 152
British Journal of Sociology 44
British Nylon Spinners Ltd 60
British Overseas Airways Corporation (BOAC) 133
British Petroleum (BP) 133
British Sociological Association 44
Britten, Benjamin 95
Broadbent, Malcolm 97
Brockbank, Philip 80–1
Brown, Edmund G. 'Pat' 251
Brown, Ian 31
Brussels 1
Brussels School of International Studies 153
Buck, Antony 113
Buckingham, University of 10, 88 n. 35
Burges, Dr N. A. 215–17
Burgess, Anthony 152
Burmah Oil 133
Burnaby 291
business studies 4, 8, 64, 130, 176, 179, 181, 183, 376
Butler, 'Rab' 48, 78, 105–6
Butterworth, Lord Jack 8, 17 n. 45, 60, 77, 162, 164–5, 176, 231, 271
Buxted Ltd 60

Cadbury Foundation 166
Cahill, Rowan 312
Caine, Sir Sidney 273, 281
Calcutta, University of 324
California 93, 251, 253
California Master Plan for Higher Education 252–3

California State University 252
California, University of 2–3, 5–6, 8, 10, 58, 94, 251–67, 290
  UC Berkeley 37, 43, 66, 93–4, 106–7, 111–13, 251–3, 255, 257–60, 264–5, 295–7
  UC Irvine 246
  UC San Diego 7, 258, 375
  UC Santa Barbara 251, 258
  UC Santa Cruz 258
Cambodia 329
Cambridge 79, 106
Cambridge, University of 5, 21–3, 27, 65, 73, 75–6, 79–85, 89–94, 96–9, 142–5, 147, 150, 160, 175, 182, 209, 271–2, 274, 295, 375
  Churchill College 5, 96, 107, 124, 136
  Clare College 270
  Gonville & Caius College 125
  Jesus College 272
  King's College 74, 97, 106, 128
  Magdalene College 143
  New Hall 124, 209
  St John's College 90, 100
Cambridgeshire 79
Cameron, Ewen 201
Campaign for Nuclear Disarmament (CND) 82–3
Canada 1, 3, 5, 7, 43, 231, 269–70, 277, 279, 287–99
Canadian Association of University Teachers 297
Canadian studies 303 n. 70
Canberra 1, 59, 306
Canterbury 125, 141–2, 146, 152, 154, 194, 224, 227
Canterbury Christchurch University 153
Canterbury, Archbishop of 153
Canterbury, University of 142
Capelle, Jean 351
Capon, Kenneth 107–8, 110
Carey, John 152
Carleton University 279
Carnegie Foundation 276
Carolynne 178
Carr-Saunders, Alexander 271–3
Carter, Sir Charles 77, 176, 178–9, 182–3, 185, 235
Cassidy and Ashton (architects) 177

Castro, Roland 346
Catalonia, University of 1
Catherine 'C' (Nanterre) 349–51
Central Electricity Generating Board (CEGB) 133
Central Florida, University of 1
Central Intelligence Agency (CIA) 329
Centre for South African Studies 82
ceramics 39, 42
Ceylon (Sri Lanka) 269, 272
Ceylon, University of 269–70, 274
Chablo, Diane 25
Chagula, Wilbert 276
Chamberlain, Joseph 376
Chamberlain, Neville 376
Chamberlin, Powell and Bon (architects) 25
Chandra, Bipin 331, 335
chapels 8, 16 n. 36, 38, 79, 85, 272
chaplaincy 177
Charlton, John 41
Charterhouse School 40
Chatham 155 n. 24
Chaucer, William 142, 145
Chaudhuri, Nag 328
Chauliat, Jacques Eugène Marie 341
Checkland, Peter 181, 186
Chelmsford 125, 227
Chelsea College 151
chemistry 75, 81, 83, 93, 96–7, 107, 128–31, 135, 178–9, 210, 215–16, 363
Chicago 4
Chicago, University of 8, 66, 289
Chilver, Guy 143
Chilver, Sir Henry 217–18
China 1–2, 41, 55, 63, 270, 273
Chinese studies 186, 278
Chitty, Anthony 241, 278
Chiva, Matty 343
Chomsky, Noam 56
Chorley 183
Christchurch (New Zealand) 142
Christopherson, Sir Derman 8, 143, 147, 151
Church of Scotland 193
Churchill College, Cambridge, *see* Cambridge, University of
Churchill, Winston 105–6, 108, 118 n. 17, 124, 305, 376

Citroën (car manufacturer) 342, 345–7
Civil Rights Act (1964) 251
Clare College, Cambridge, see University of Cambridge
classics 41, 75, 144, 148, 151, 153, 180, 186, 272
Clayton, Keith 96, 128, 133
Clunies Ross, Ian 307, 316
Coates, Rick 112
Cobbett, William 158
Cockcroft, John 130
Cockcroft, Sir Wilfred H. 217
cognitive sciences 131
Cohen, Sir Andrew 76
Cohn-Bendit, Daniel 341–2, 345
Colchester 106–7, 110, 113, 125, 194, 227
Cold War 8, 55–9, 62–6, 106, 135, 253, 270, 288, 307–8, 316, 377
Cole, G. D. H. 161
Coleraine 207, 209, 211–16, 218
Coleridge, Samuel Taylor 78
colleges of advanced technology (Britain) 5, 10, 84, 121–4, 126, 130, 151, 311
Colman, Timothy 230
Colonial Office 271, 277
Columbia University 57, 289
Colville, John 124
Comberton 79
Comité de Liaison des Étudiants Révolutionnaires (CLER) 346
Comités Vietnam de base 345–7
Commission on Higher Education in the Colonies 270–1, see also Asquith, Cyril
Committee of Vice-Chancellors and Principals (CVCP) 17 n. 35, 176, 195
Committee on Higher Education in Northern Ireland 207, 209, 213–15, 218, see also Lockwood, Sir John
Commonwealth (British) 5–6, 63, 74, 142, 196, 269–81, 288, 307
Commonwealth Tertiary Education Commission (CTEC) 305, 312, 316
communications 38, 298–9, 376
Communist Party (USA) 262
Communist Party of Great Britain 116, 167

Communist Party of India 327, 329, 331
community colleges (California) 252
comparative studies 107
Complutense University of Madrid 4
computer science 4, 132, 179, 298
computers 121, 132
Conant, James 8, 40, 43, 47, 289
Condon, Julia 21
Confédération générale du travail (CGT) 345–7
Confederation of British Industry (CBI) 9, 17 n. 45
Congo and Ruanda-Urundi, University of 269
Congo, Republic of the (Democratic Republic of the Congo) 269
Congress Party (India) 3, 327, 329–30
Connell, R. W. 311–12
conservation 128
Conservative party 10, 37, 55, 63, 65–6, 105–6, 113, 116, 151, 201
continuing education, see extra-mural studies
Cook, Sir James 215
Coombs, H. C. ('Nugget') 305–6
Cordingly, Alan 25
Cornell University 180
Cornwall 106
Cornwallis, Lord 141, 231, 254 n. 1
Cottrell, Tom 195–6, 198
Courtaulds 60
Coventry 27–9, 125, 159, 162, 169, 225, 227, 231
Coventry University 35 n. 74
Coventry, Bishop of 227
Cox, Christopher 271, 273–4, 277–8, 280
Crabtree, Derek 152
Craig, David 185
Craigavon 209, 211, 213
Cranbrook, Lord 234
Cranfield Institute of Technology 218
creative writing 95, 180
crèche sauvage, see Nanterre, University of (Paris X)
criminology 298
Crosland, Anthony 10, 35 n. 74, 112
Crowther Report on Education (1959) 77
Crozier, Michel 343
Cubitt, James 7, 275, 280

Cum Grano 40
Cumbernauld 193–4
Cumbria 183–4
Cumbria, University of 186
Cunliffe, Marcus 63
curriculum design 2, 4–5, 7–8, 37–8, 40–8, 57, 61, 64, 73, 75, 78, 85, 90, 94, 105–7, 109–10, 117, 131, 143, 147–8, 162–3, 165, 175, 178–82, 184, 186, 269, 271–4, 277–9, 289, 291–2, 298–9, 362–3, 377, *see also* individual subjects
Cygnet 47

Dahrendorf, Ralf 368
Daiches, David 6, 278
Daily Mail 122
Dainton, Lord Frederick 17 n. 45, 126, 199
Dakar, University of 270, 351
Dalton, Hugh 41
Danquah, Joseph 275
Dar es Salaam, University College of 81, 276, 278–9
Darlow, Stephen 143
Dartmouth College 233
Darwin, Charles 145
Datta, Asis 336
Davies, Zelda 209, 211
Davin, Anna 169
Dawbarn, Graham 273, 278
de Dampierre, Eric 343
De Gaulle, Charles 111
De Montfort University 32
De, Barun 329
Deakin University 306, 310, 315
Dean, Sir Maurice 200
Defence, Ministry of (Britain) 90
Defence, Ministry of (France) 342
Delft Institute of Technology 137 n. 22
Delhi, University of 324
demography 96
Denmark 1, 5
Derby, Edward Stanley, 18th Earl of 175, 230
Derbyshire, Andrew 79
Derry (Londonderry) 207–16, 219, 247, *see also* Londonderry
Desai, Morarji 330
development studies 64, 93, 101, 329, 376

Devon 114
Dick, Marcus 95
Disney, Walt 1
Divine, Father 254
Djenné 275
Dodd, Edward 166
Donnelly, Harry 195
Doraiswamy, Rashmi 336
Dortmund 364
Dover 141
Drayton, Harold 277
Drew, Jane 7, 272
Driver, Christopher 176, 184
Dublin 208
Duff, Sir James 271
Dufour, Michel 345
Dumfries 193–4
Duncan, Alexander 194
Dundonald, James (pseud.) 6, *see also* Lawlor, John
Dunn, Tommy 196
Durham, University of 21, 143, 271

Eaborn, Colin 130
Eagan, Ronald 247
Earlham Hall 92
Earp, Alan 277
East Africa, University of 215, 271, 277–8, 281
East Anglia, University of (UEA) 1, 5–8, 22, 30–1, 45–6, 55, 60, 78, 81, 89–101, 121, 123, 128–30, 132–3, 135, 149, 175, 180, 195, 197, 226–7, 229–34, 243, 278
École Normale Supérieure 347
ecology 96, 131, 179
Economic and Political Weekly (*EPW*) 328–9
economic history xix, 95
economics xix-xx, 8, 42, 75, 80–1, 90, 95–6, 101, 107, 109, 147, 163, 191, 176, 181, 185, 215–16, 329, 363
Edge Hill College (later University) 183, 186
Edinburgh 192
Edinburgh, University of, 80–1, 193, 195, 274
education 42–4, 61, 75, 81, 180, 183–4, 196, 215–16

Education Act 1944 (Britain) 76, 105–6, 159
Education Act 1962 (Britain) 24
Education and Science, Department of 8, 183, 199
Educational Institute of Scotland (EIS) 193
Edwards, David 143, 146
Egypt 273
Eidgenössische Technische Hochschule (ETH) Zürich 137 n. 22
Electricity Council 133
electron microscopy 131
Eliot, George 159
Eliot, T. S. 145, 152
Eliot, Valerie 152
Elizabeth II, Queen 198, 234, 248
Elizabeth, Queen, the Queen Mother 232
Elizabethville (Lubumbashi) 269
Elliott, Julian 241, 278
Elliott, Walter 271
Ely, Isle of 228
embryology 131, 134
Emerson, Ralph Waldo 253
Engels, Friedrich 353
Engerman, David 57
engineering 75, 123–4, 129, 162, 179–80, 186, 196, 363
English xx, 61, 63, 75, 81, 90–1, 93, 95, 97, 99, 101, 127, 132, 147–8, 151, 179–80, 182, 185, 196, 216
environmental science 4, 93, 96–7, 101, 128–9, 133, 178–9, 185, 215–16, 303 n. 70, 376–7
Epping 106
Epstein, Gabriel 177
Erickson, Arthur 242, 294–5
Eros, John 43
Essex 105–6, 225, 227
Essex County Council 106, 227–8
Essex, University of 1, 4, 7, 9–10, 22–3, 25–7, 29–30, 45, 64, 78, 83, 105–17, 121, 133, 135, 141, 148, 175, 178, 184, 195, 197, 225, 227–30, 232–4, 244, 278, 377
Esso 133
ethnology 343
Eton College 76, 276
European languages 81, 95, 151, 180–1, 186

European studies 61, 93–4, 101, 127, 132
European Union 153
Ewing, W. T. 209, 216
Exeter, University of 21, 28, 183, 215
extra-mural studies 39, 43, 163, 234, 239 n. 65, 183, 185, 217, 272, 274, 291

Faber and Faber 152
Fairhurst, Harry 80
Falkirk 193–4
Falmer 142
Family Reform Act 235
Farquharson, Alexander 44
Farquharson, Dorothea 44
Faure, Edgar 350
Federal Bureau of Investigation (FBI) 253
Fédération des Étudiants Révolutionnaires (FER) 346
Feldman, David 160
Feltrinelli, Istituto 163
Fiji 94, 278–9
film studies 147
Financial Times 29, 231
Finer, Samuel H. 41, 44
First World War 224
Fisher, Norman 76, 79
Fitzgerald, Will 169
Fletcher, Ronald 81
Flinders University 1, 305, 310–11, 313, 316
Florence 162
Floud, Jean 43
Flowers, Lord Brian 132
Foakes, Reginald 143
Folkestone 141
Food Research Institute 129
Ford Foundation 58, 232, 279
Ford Motor Company 64, 232
Foreign Affairs 65
Fortune 254
Foster, Norman 233
Foucault, Michel 260
Fouchet, Christian 343
Fourah Bay College 278
Fox, Eric 143
France 1, 3–5, 8, 10–11, 47, 95, 111, 160–1, 163, 167, 234, 270, 341–53, 357, 364
Frankenberg, Ronnie 43
Frankfurt 367

Franks, Lord 9, 78
Fraser, Malcolm 315
free university movement 3, 47, 114, 312–13, 316
Freeman, Christopher 132
Freetown 278
Freie Unversität Berlin 1, 357
French studies 180, 184, 298
Froelicher, Hans 43
Fry, Maxwell 7, 272
Fulton, Sir John 38, 42, 60, 62–3, 77, 246, 271, 278

G. I. Bill (Servicemen's Readjustment Act) 3, 288
Gajendragadkar, Dr P. B 326
Gandhi, Indira 248, 323–4, 327, 331
Gannochy Trust 231
Gardeners' Question Time 41
Gardner, Dr T. Lyddon 233
Gash, Norman 203 n. 32
Gauche Prolétarienne (GP) 347
Geismar, Alain 347, 354 n. 37
Gemmell, Alan 41
General Electric Company (GEC) 232, 290
genetics 81, 129–31
geography 8, 75, 96, 132, 185, 216, 343, 363
geology 96, 128
Georgetown 277–8
Germany, West 1, 3, 5–6, 8, 11, 39–40, 46, 58–9, 234, 357–68
Ghana 272, 275
Gibraltar 273
Gifford, Anthony 115
Glasgow 44, 124, 178, 193
Glasgow Herald 230
Glasgow, University of 39, 41, 193, 195
Glen, Major J. A. 209
Glover, Major Gerald 214
Glover, Mary 41
Gold Coast 275
Gold Coast, University of the 180, 272, 274
Goldwater, Barry 251
Gonville and Caius College, Cambridge, *see* Cambridge, University of
Gonzalez, Mike 112–13
Gordon Memorial College 273

Göttingen 367
governance 2, 4, 9–10, 46–7, 75, 84, 98, 109–10, 112, 114–15, 117, 144, 149, 158, 175–7, 182, 186, 192, 200, 259, 262, 264, 269, 276, 287, 296–7, 306, 314, 326, 367
government 107, 117, *see also* political science
Gower, Lawrence 279–80
Granada Television 38, 47
Grant, Michael 274
Grappin, Pierre 342
Gray, Fred 128
Gray, Hanna Holborn 2
Green, J. R. 160
Green, T. H. 90
Greenland 43
Greenwich, University of 153
Greer, Germaine 313
Gribbin, John 128
Griffith University 306, 309, 315
Griffiths, Allen 130
Grimaud, Maurice 351
Gropius, Walter 7, 79, 292, 294, 360
Grumbach, Tiennot 347, 353
Guardian 47, 114
Gulbenkian Foundation 233
Gurney family 92
Guyana, University of 277–8

Hagenbuch, Walter 143
Hailsham, Lord 55, 63, 65
Halberstadt, Raphael 114
Hale, John 162–6
Hale, Sir Edward 74, *see also* Report of the Committee on University Teaching Methods
Hall, Sir Noel 175–6
Halmos, Paul 44–5
Halsey, A. H. 44
Hamburg 161, 366–7
Harland and Wolff (shipyard) 209
Harrison, Austen 272
Harrison, Royden 160
Harrod, Roy 152
Harvard University 8, 40, 43, 47, 252, 259, 265, 289, 368
Harwood, Elain 25, 28
Hatzfeld, Marc 351

Haynes, Dr John 225
Headmasters' Conference 76
Heaney, Seamus 152
Hebrew University of Jerusalem 269
Henderson, A. B. ('Brum') 209, 211
Henley 175
Henstock, Ralph 216
Hentrich and Petschnigg (architects) 360
Herklots, H. G. G. 26
Hertford College, *see* Oxford, University of
Hertfordshire 79
Herzog, Helmut 362
Heslington Hall 74
Hess, Gerhard 367
Hewison, Robert 66
Hewitt, Halford 116
Heywood, John 184
Higher Education Act 1976 (West Germany) 364
Higher Education Facilities Act (USA) 3
Higher Education Research and Development Society of Australasia (HERDSA) 314
Higher Technological Education, Committee on 6, 122, *see also* Percy, Lord Eustace
Hindi 81
Hispano-Suiza (car manufacturer) 347
histology 131
Historic England 80
history 42, 63, 75, 82–3, 90–1, 101, 133, 148, 153, 162–3, 165, 180, 185, 196, 215–16
history of science 180, 183, 365
Hobart 306
Hobsbawm, Eric 172 n. 58
Hogben, Lancelot 78, 277
Hoggart, Simon 47
Holford, Lord William 7, 25, 27, 146, 155 n. 14
Holroyd, Dr Ronald 59–60
Holst, Imogen 95
Hong Kong Universities Grants Committee 278
Hong Kong, Chinese University of 42, 271, 246, 278
Hong Kong, University of 269

Hooker Craigmyle (fundraising consultancy) 230
Hooker, Michael 230
Horne, Donald 309, 317
Horwood, Thomas 39
Howard University 274–5
Howard, Maurice 62
Hughes, Robert 313
Hull 228, 272, 378
Hull, University of 21, 25, 144, 217
humanities 61, 93, 96, 106, 126–9, 135, 144, 148, 151, 153, 179, 182, 196, 210, 215, 265, 288, 292, 299, 325, 363
Humberside Polytechnic 32
Humberside, University of 378
Hume, John 214–15, 247
Hunter, Ian 45
Huntingdonshire 228
Huntington, Samuel P. 65
Hussey, Stanley 180
Hutchins, Robert 2, 8, 126, 289
Hyderabad, University of (India) 81

Ibadan, University of 7, 272, 274–5
Iceland 43
Iliffe, Alan 42
Illinois-Chicago, University of 4
Imperial Chemical Industries (ICI) 59–60, 181, 195
Imperial College, *see* London, University of
Impington 79
Inch, Dr Thomas 114, 135–6
India 1, 3, 5–6, 9–11, 180, 269, 323–38
Indian Commission on Historical Research 329
Indian Education Commission 1964–6 324
Indian History Congress 329
indigenous studies 303 n. 70
Indonesia 3–4
Ingold, Sir Christopher 96, 128, 130
Ingra, David 151
Institute for Advanced Architectural Studies 74, 78–9
Institute for Development Studies (IDS) 63–4
Institute of Food Research 96
Institute of Marketing 181
Institute of Physics 41

Institute of Sociology 44
institutes of technology (Australia) 326
institutes of technology (India) 325
Inter-University Council for Higher Education Overseas (IUC) 271–2, 274–5, 277, 280
International Marxist Group 116
international relations 180
International Socialists 116
International Sociological Association 90
Inverness 193–4
Ipswich 227
Iran 6
Irvine, James 271, 274
Irvine, University of California, *see* California, University of
Irving, Terry 312
Islam 80
Israel 4, 38, 269
Italy 4, 9, 90, 163, 277, 279

Jackson, Lord Willis 209
Jagan, Cheddi 277
Jamaica 273
James Cook University 241, 306
James, Clive 313
James, Lady Cordelia 80, 83
James, Lord Eric xix, 3, 6, 73–4, 76–85, 125–6, 131, 231
James, Louis 148
Janata Party (JNP) 327, 329–30
Japan 269
Japanese studies 186
Jarman, Derek 147
Jawaharlal Nehru University Student Union (JNUSU) 248, 326–7, 330–1, 336
Jawarhalal Nehru University (JNU) 1, 10, 248, 308, 323–38
Jenderal Soedirman University of 3
Jenkins, Gwilym 181
Jennings, Ivor 269–70, 272, 274
Jerusalem 142
Jesus College, *see* University of Cambridge
Jeunesse Communiste Révolutionnaire (JCR) 346–7
Jews 177, 353
Jha, Akhileshwar 327
John O'Gauntlet 178

Johnson, Lyndon 251
Johnson, Samuel 288
Johnson-Marshall, Stirrat 78–80
Jordanstown 218
Joseph, Sir Keith 22
Joshi, Amar Chand 326, 330

Kagan, Élie 344
Kaldor, Nicholas 152
Kampala 272, 375
Karat, Prakash 327, 329
Karmel, Peter 305–6, 310, 312–13, 316
Kashmir 337
Katritzky, Alan 96–7, 102 n. 31, 130
Kaunda, Kenneth 276, 279, 281
Keele University 1–3, 5, 9, 22, 25, 28, 30–1, 37–53, 62, 90, 144, 149, 179, 184, 224, 269, 278–9, 315, 357
Keith-Lucas, Bryan 143–6
Kent (county) 141, 146
Kent County Council 141, 225, 231
Kent, University of 1, 7–9, 16 n. 36, 27, 30, 121, 130, 141–54, 176, 227–8, 231–4
Kenya 277
Kenzi, Mohammed 352
Kerby, Captain Henry 65
Kerr, Clark 2–3, 6, 58, 66, 68 n. 34, 107, 226, 246, 251–67, 290
Keynes, John Maynard 145
Khartoum, University of 80, 180, 273–4
Kidd, Alan 226
Kidd, Sir Robert 209, 211
Kieser, Alfred 58
Kiesinger, Georg 360
kinesiology 298
King Edward VII Medical College (Singapore) 273
King's College, Cambridge, *see* Cambridge, University of
King's College, London, *see* London, University of
King's Lynn Technical College 130
Kirk, Herbert 213
Klein, Michael 65
Kneale, William 8
Knights, L. C. 80
Koc, Wacek 184
Köln, University of 359

Konstanz, University of 358, 360–3, 366–8
Köpke, Klaus 362
Koshy, V. C. 326–7, 329
Kothari, Professor D. S. 324
Krishna, Pranay 336
Kristeva, Julia 152
Krivine, Alain 347
Kukreja, C. P. 325
Kulka, Peter 362
Kumar, Girja 329
Kumasi 272
Kuwait, University of 270

L'Humanité 342
La Trobe University 306, 309, 314
labour history 167
Labour party (Australia) 312, 314
Labour party (Britain) 10, 28, 39, 41, 111, 125, 194, 201, 228
Lagos 272
Lagos, University of 279–80
Lamb, Hubert 96, 132–3
Lancashire 178, 183–4
Lancashire County Council 175–6
Lancaster 125, 175, 178, 183
Lancaster City Council 175
Lancaster Town and Gown Club 234
Lancaster, University of 1, 8, 25, 29–31, 121, 129–30, 150, 175–86, 195–6, 225, 227–31, 233–5, 281
languages 81
Laos 329
Lasdun, Denys 7, 92, 129
law 75, 148, 180, 360–1, 363, 376
Law Society 182
Lawlor, John 41
Lawrence, Jon 160
Lawrenson, Tom 180
Le Corbusier (Charles-Édouard Jeanneret) 38, 293–4
Le Figaro 342
Le Monde 342
Le Page, Bob 81
Le Play House 44
Leamington Spa 158, 163
Leary, Timothy 98
Leatherland, Charles 225
Leavis, F. R. 80–1
Leavis, Q. D. 80
Lebanon 2

Ledger, Philip 95
Leech, Geoffrey 186
Leech, Rachel 131
Leeds, University of 22–7, 29–31, 132, 163, 271
Leeson, Spencer 77
Lefebvre, Henri 343–4, 349
Leff, Gordon 82
Leicester Polytechnic 32
Leicester, University of xix, 21, 24, 45
Lenin, V. I. 338
Lennard-Jones, John 41
Leopardville (Kinshasa) 269
Lethbridge, University of 242, 294
Leverhulme Trust 184, 217
Lévy-Villard, Annette 347, 349
Lewin, Kurt 257
Lewis, Gwynne 166
Lewis, Kenneth 113
Leyland Motors 229
Leys, Colin 280–1
Libya 7
Lincoln 378
Lincoln, University of 32, 378
Linde, Horst 360
Lindsay, Lord Alec 2–3, 39–43, 45–8, 357
Linehan, Sean 65
Linfoot, Dennis 143
Ling, Arthur 28
linguistics 81, 132, 180, 186
Listener 59, 61
literature 107, *see also* American studies; English
literature, African 148, 196
literature, American 148
literature, Caribbean 148
Liverpool 161, 224
Liverpool, University of 22, 24–5, 27, 31, 66, 106, 183, 186, 215, 224, 230
Lloyd Wright, Frank 293, 295
Lockhart, Sir John Bruce 165
Lockwood, Sir John 209–12, 215–17, 279
Lodge, Oliver 23
logic 132
loi Faure 8, 350
London 59, 63, 66, 143, 151, 153, 192, 217, 227, 272, 275
   Royal Festival Hall 272
London County Council 78–9

London School of Economics (LSE), *see* University of London
London, University of 24, 38, 45, 81, 130, 159, 269–75, 280
  Bedford College 129, 178
  Birkbeck College xix, 209, 279
  Imperial College 124, 151, 153, 181
  King's College 96, 129
  LSE 40, 43–4, 64, 76, 80–1, 83, 96, 109, 113, 116, 144, 197, 234, 271, 273, 281, 377
  Queen Mary College 180
  SOAS 150
  UCL 90, 96, 181, 275, 311
Londonderry (city), *see* Derry
Londonderry (county) 209
Londonderry Corporation 214
Londonderry County Borough 209, 211–12
Longshoremen's Union 254
Look and Learn 160
Loughborough University 23, 29
Lovanium, University of 269
Low, Anthony 281
Löwe, Adolf 40
Lowen, Rebecca 56–7
Lowestoft Fisheries Laboratory 129
Lübbe, Herman 365
Lucas, Eric 272
Luhmann, Niklas 363, 366
Lusaka 277–8
Lutman, Stephen 148
Lycée Buffon 347
Lyman, Richard 296

Macdonald, John B. 291
MacKinnon, Archie 292
Mackintosh, Athole 95
Mackintosh, John and Sons 231, *see also* Mackintosh of Halifax, Lord 231
Mackintosh of Halifax, Lord 55, 231
Maclean's Magazine 288
Macmillan, Harold 62, 106, 108, 122
Macmurray, John 39
Macquarie University 306, 309–12
MacRobert Trust 231
Madras, University of 324
Magdalene College, Cambridge, *see* Cambridge, University of
Magee University College 208–14, 216–18

Magee, Martha 208
Mainz, University of 357
Makerere University of 271–2, 274, 277, 375
Malawi 273
Malawi, University of 271, 277
Malaya 81, 270, 274, 278
Malaya, University of 81, 270–1, 273–4, 277–8
Malta 269
management studies 8, 58–9, 129, 179–81, 186, 329, 376
Manchester College of Science and Technology 124–5
Manchester Grammar School xix, 74, 76–7, 83
Manchester xix, 30, 178, 224, 272
Manchester, University of 3, 22–3, 42, 64, 82, 175–6, 178–9, 183–4, 196, 224, 230, 315
Mandela, Nelson 83
Manley, Gordon 129, 178
Mannheim, Karl 40, 43
Mao Zedong 338, 353
Maoism 65, 313, 344–53
Marconi 29, 125
Margaret, Princess 47
Margate 141
Marginson, Simon 314
Marina, Princess of Kent 232
marine biology 129, 212
Marlborough College 230
Marlowe, Christopher 145
Marsden, Lorna 301 n. 20
Marshall Plan 58–9
Marshall, Robert Matthew-Johnson (architects) 7, 78–9, 215
Martin, Graham, 143
Martin, Helen 233
Martin, Sir Leslie 311, 315
Marx, Karl 338, 353
Marxism 296, 329, 331, 344, 346–8, 353
Maseru 278
Massachusetts Institute of Technology (MIT) 6, 27, 40, 56, 106, 124, 132, 134, 293
Mataram, University of 4
mathematics 75, 93, 97, 107, 121, 126–9, 134, 179–80, 216, 361, 363
Matheson, Louis 313–14

Max Planck Society 364
Mazzariol, Giuseppe 294
McAteer, Eddie 215, 247
McCarthy, Joseph 43, 56
McConnell, Jack 201
McConnell, Thomas 37
McFarland, Sir Basil 214
McFarlane, James ('Mac') 94
McGill University 125
McTaggart-Cowan, Patrick 291, 297
medicine 75, 129, 179, 186, 196, 311, 360, 363, 375–6
medieval studies 180
Medway 153, 155 n. 24
Melbourne 306, 313
Melbourne, University of 308, 313
Mellers, Wilfrid 81
Melly, George 82
Menzies, Robert 307, 311–12
meteorology 96, 128, 133
Meterological Office 133
Michigan State University 275
microbiology 96, 131
Mid-Kent College 151
Middle East Technical University 4
Mikat, Paul 362
Milan 163
Miller, Jonathan 152
Milner-White, Eric 74–5
Milton, Keynes 1
Milton, John 97, 288
Minerva 6, 324
Missoffe, François 343
Moberly, Sir Walter 6, 22, 26, 224
Moçambique, Estudos Gerais Universitários de 270
Moinon (builders) 241
Mol, W. H. 209
molecular sciences 128
Mona, University of the West Indies, *see* University of the West Indies
Monash University 306, 309–10, 313–14
Moncton, Université de 1, 300 n. 8
Montgomery, David 165–7
Montgomery, G. R. 230
Montreal 125
Moodie, Graeme 10, 84
More, Thomas 169
Morgan, David 144, 148

Morrell, J. B. 73–4, 227, 231
Morris, Charles 271, 279, 280, 286 n. 56
Morris, Henry 79–80
Mott, Nevill 125
Munich 366
Münster, University of 359
Murdoch University 306, 315
Murray, Bill 179
Murray, Dame Rosemary 209
Murray, Sir Keith 3–4, 8, 75–6, 84, 195, 203 n. 26, 223, 225–6, 232, 234, 236, 242, 305, 307
music 8, 75, 81–2, 93, 95, 152, 181, 186, 233
Muthesius, Stefan 61, 121, 361

Nairobi 277
Nalanda, University of 11
Namier, Sir Lewis 82, 160
Nanterre (town) 341–5, 347–9, 352–3
Nanterre, University of (Paris X) 1, 4, 10, 247, 341–53
  crèche sauvage 348–51
Nanyang University 273
Nasser, President 273
National and Local Government Officers' Association 193
National Coal Board (NCB) 74, 76, 181
National Council for Academic Awards (NCAA) 10, 217
National Health Service 159
National Training Laboratories Institute for Applied Behavioral Science 257
National Union of Australian University Students 313–14
National Union of Students (NUS) 10, 23, 30, 116, 193
National University of Ireland 208
nationalism 6, 193, 200–1, 215, 270, 273, 275–6, 279
natural sciences 130, 144, 179–80, 197, 360, 363
Nayak, Pradip 83, 245
Nehru, Jawaharlal 323–4, 330
Netherlands 5, 7
Neuberg, Victor 160
neurobiology 132
New Brunswick 288
New College, *see* Oxford, University of

New Delhi 1, 3, 328
New Hall, *see* Cambridge, University of
New Hampshire 233
New Society 158
New South Wales (state of Australia) 306, 309–10
New York 43, 94, 165, 289
New York Review of Books 56
New York University 56, 279
New Zealand 6, 242, 269, 278
Newcastle, University of (Australia) 306
Newcastle, University of (Britain) 183
Newman, John Henry 89, 91, 292
Newsom, Sir John 79
Newsome, Audrey 45
Niblett, Roy 4
Niemeyer, Oscar 1, 7
Nigeria 7, 272, 275, 280–1
Nis, University of 1
Niven, Alastair 196
Njoku, Eni 279
Nkrumah, Kwame 275
Norfolk 90, 92, 96, 225, 234
Norfolk County Council 225
Norman, Richard 81
North Atlantic Treaty Organization (NATO) 133
North Korea 329
North Rhine-Westphalia 359, 361–2, 364
North-East London Polytechnic 377
Northern Ireland 83, 207–19
Norway 1
Norwich 73, 76, 89, 92, 99, 125, 129, 193, 223–5, 227, 230, 234
Norwich City Council 92
Norwich Technical College 130
Norwich Union (insurance company) 231
Nottingham, University of 21, 23, 62
Nouse 81
Nowell-Smith, Patrick 144
Nssuku, University of 275
Nuffield College, *see* Oxford, University of
Nuffield Foundation 42–3, 133, 180, 232
Nuffield, Lord 74
Nuneaton 27
nursing 217
Nuttgens, Patrick 78
Nyerere, Julius 275

O'Neill, Terence 207–8, 211, 213–14
Observer 178
oceanography 96, 128
Oddie, G. B. 25
Odense, University of 1, 5
Ogilvie, Lady Mary 80, 131, 195
Ogilvie, R. 330
Ontario 287, 289–91, 295
Oomen, T. K. 335
Open University 1, 10, 184
Oriel College, *see* Oxford, University of
Orlando 1
Orleans, University of 3, 272
Ormskirk 183
Ortolano, Guy 124
Owens College 30
Oxford Delegacy for Extra-Mural Studies, *see* Oxford, University of
Oxford, University of 5, 7–8, 21–3, 27, 41, 45, 61, 73, 75–6, 78, 80–5, 89–92, 96–7, 99, 106, 142–5, 147, 150, 160, 175, 230, 271, 274, 290, 295, 357, 375
  All Souls College 79
  Balliol College 22, 39, 41, 82, 271, 273
  Brasenose College 175
  Delegacy for Extra-Mural Studies 39
  Hertford College 271
  New College 271
  Nuffield College 74, 144, 271–2
  Oriel College 290
  Queen's College 77, 79
  Rhodes House 90
  St Anne's College 80, 195
  St Catherine's College 5
  St John's College 107, 176

Pacific studies 279
Pakenham, Thomas 125
palaeontology 131
Palestine 345, 348, 353
Palmer, Bryan 168
Papeteries de la Seine, Les (factory) 345
Papua New Guinea, University of 6, 307–8
Paris 135, 149, 153, 341–2, 344–5, 347–51
Paris School of Architecture 346
Paris, University of 1, 3, 341–4, 346–9

Park, Dr A. T. 209, 211
Parker, Alastair 82
Parkin, Frank 144
Parthasarathy, Gopalaswami 325–7, 330
Parti Communiste Français (PCF) 343–6, 348, 352–3
Parti Communiste Marxiste-Léniniste de France (PCMLF) 345
Parti Socialiste 353
particle physics 125
Paulus, Stefan 58
peace studies 180
Peacock, Alan 80–1, 87 n. 26, 195
Pears, Peter 95
pedagogy 365
Peel, Arthur, 2nd Earl 175
Peers, Edgar Allison, see Truscot, Bruce (pseud.)
Pei, I. M. 278
Penguin Books 158
Penson, Lillian 271
Percy, Lord Eustace 6
performing arts 298
Perham, Margery 271–2
Perkin, Harold 65, 82, 135, 179–80
Perth (Australia) 306
Perth (Scotland) 193–4
Peterborough (Canada) 290
Pevsner, Nikolaus 78
philanthropy 5, 9, 223–4, 228–36
Philippines 2, 6
Phillips, G. C. 230
philosophy 61, 75, 77–8, 95, 126–7, 130, 132, 134, 144, 147–8, 180, 271, 288, 313, 363, 365
Phule, Jyotirao 338
physical sciences 61, 96–7, 106–7, 127–8, 130, 151, 215, 337
physics 75, 93, 97, 107, 129, 133, 147, 179, 181–3, 185, 216, 363
Piening, Chris 110–11
plant physiology 96
Plato 22, 39
Plessey (electronics and communications company) 232
Plisonnier, Gaston 345
Plumb, J. H. 94
political science xix, 75, 82, 95, 109, 147–8, 150, 180, 292, 363

politics, see political science
Polwarth, Lord 231
polytechnics (Britain) 5, 10, 32, 377
Pompidou, Georges 343
Port Elizabeth, University of 37
Portland (USA) 47
Porton Down (laboratory) 114–15, 135
Portugal 270, 350–1
Post Office (Britain) 29, 74
postgraduate students, see students, postgraduate
Powell, Enoch 83, 113–14
Pratt, Cranford 279
Preston 177
Price, Cedric 31
Price de Solla, Derek 125
Princeton University 57, 135
Pristina, University of 1
Proctor, Harvey 83
Progressive People's Party (Guyana) 277
psychology 42–4, 48, 75, 132, 148, 179, 181, 257–8, 343, 363, 365, 376–7
Pullan, Brian 82

Quebec 288
Québec, Université du 1, 300 n. 8
Queen Mary College, see London, University of
Queen's College, Oxford, see Oxford, University of
Queensland (state of Australia) 306, 310
Quirk, Roger 123

Radhakrishnan, Dr S. 324
Radical Student Alliance 112
radio astronomy 125
Radio Research Station 182
Raffles College 273
Rai, Usha, 327
Rajasthan 336
Ralphs, Sir Lincoln 225
Ramsey, Archbishop Michael 55, 75
Ramsgate 141
Rand Afrikaans University 4
Ranjan, Pramod 338
Rao, Amiya 328, 332
Rao, Nina 335–6
Rao, V. K. R. V. 331
Ratcliffe, Jack 182
Rathbone family 224, 230

Rawls, John 337
Reading, University of 21, 26, 38, 230
Reagan, Ronald 149, 251, 261
Rebbeck, Dr Denis 209
Red 346
Rée, Harry 79, 81
Reed College 47
Rees, Dr Henry 225
Regensburg, University of 361
Reggio, University of 9
Reich, Wilhelm 344, 349
Reims, University of 3
religious studies, *see* theology
Rémond, René 342, 352
Renault-Flins (car manufacturer) 347, 351
Report of a Joint Working Group on Computers for Research (1966) 132
Report of the Committee on Australian Universities (1957), 305, 307, 315, *see also* Murray, Sir Keith
Report of the Committee on the Future of Tertiary Education in Australia (1964) 315, *see* Martin, Sir Leslie
Report of the Committee on University Teaching Methods (1964) 184, 196
Representation of the People Act (1969) 159
Republican Party 251
Reynolds, Philip 183
Rheinisch-Westfälische Technische Hochschule (RWTH) 359
Rhodes House, *see* Oxford, University of
Rhodesia 82–3, 273, 280–1
Rhodesia and Nyasaland, University College of (UCRN) 80, 271, 273, 279–81
Ribeiro, Darcy 3
Richardson, Alan 62
Ricoeur, Paul 342, 351
Riddell, Anne 80
Rivett, Patrick 181, 186, 234
Robbins Report on Higher Education (1963) 5–6, 8–11, 48, 108–9, 113, 121, 126, 159, 193–5, 200, 208, 225, 229, 311–12
Robbins, Lord Lionel 5, 8, 11, 76–7, 80–1, 124, 127, 131, 159, 175, 191, 195, 198, 209–10, 226, 311
Robinson, Joan 152

Robinson, Robert 130
Rockefeller Foundation 44, 57–8, 63, 277
Rogers, Colin 112–13
Rome 142
Rootes, Lord William 29, 162, 231
Roscoe family 224, 230
Rose, Hannan 149
Ross, Alec 183
Ross, C. R. ('Dick') 95
Ross, Murray 3, 289–90, 292, 296, 298–9
Rothe, Hans Werner 3, 240, 358–61, 367–8
Rotherham 228
Rousseau, H. Jac 271
Rowntree Trust 33 n. 34, 75, 231
Rowntrees 73–4
Royal Institute of Chemistry 182
Royal College of Science and Technology 124
Royal Festival Hall, *see* London
Royal Institute of British Architects (RIBA) 7, 62, 177
Royal Institution 134
Royal Military College (Australia) 311
Royal Society 81, 96, 97, 127, 129
Rugby (school) 78, 176
Ruggles-Brise, Sir John 227
Russell, Bertrand 114, 145–6
Russell, Lionel 183
Russian (language) 180, 186, 216
Rutherford, Ernest 125, 145, 271
Rutland and Stamford (parliamentary constituency) 113
Rutter and Draper (architects) 278

St Martin's College 183
Saarbrücken, University of 357
Saatchi, Roy 111
Saberwal, Satish 329, 331
Saffron Walden 105
Sahni, Balraj 327, 331
Sahlins, Marshall 343
Said, Edward 152
Sainsbury, David 233
Sainsbury, Lisa 233
Sainsbury, Robert 233
Salamanca 294
Salerno, University of 9
Salford, University of 29
Salisbury (Harare) 273, 280
Salisbury, Marquess of 123

Samara Yuvajan Sabha 328
San Diego, University of California, *see* California, University of
San Francisco 251, 254
Sanderson, Michael 234
Sanford, Nevitt 43
Sankore 275
Santa Barbara, University of California, *see* California, University of
Santa Cruz, University of California, *see* California, University of
Santiago de Compostela 143
Sargent, Dick 163
Sartre, Jean-Paul 114
Savage, Mike 376
Savio, Mario 112, 253
Sawston 79
Schelsky, Helmut 361, 366–7
School of Oriental and African Studies (SOAS), *see* London, University of
science 1, 4–5, 8, 38, 40–2, 46, 48, 58, 61–2, 74–5, 77–8, 80, 82, 90, 93, 95–7, 106–10, 121–35, 145, 147, 151, 160, 162, 181–4, 191, 193, 196–7, 205 n. 74, 212, 232, 261, 265, 288, 292, 306, 309, 325, 337, 358–9, 363–4, 376–8
Scientific Man-power, Committee on (Britain) 6, 193
Scotland 37, 41, 80, 191–201, 225, 231
Scott, Tom 39
Scottish Education Department 193
Scottish National Party (SNP) 194
Scottish Office 192, 194–5
Scrutiny 81
Second World War 1, 3, 6, 8, 27, 41, 57–9, 89, 95, 98, 106, 143, 175–6, 193, 197, 209, 215, 235, 269, 274, 306, 317, 375
Seers, Dudley 63
Senegal 270
Shakespeare, William 288
Shannon, John 74
Shannon, Richard 94
Shattock, Michael 27, 169
Shaw, George Bernard 288
Shaw, Roy 43
Sheffield, University of 25

Sheldon, Oliver 73
Shell Canada 296
Shell UK 133
Shils, Edward 14 n. 14, 44, 276
Shimmin, Sylvia 181
Shine, W. B. 330
Shrum, Gordon 291–2, 298, 300 n. 26
Sibley, Frank 180
Sicily 124
Siepmann, Wolf 362
Sierra Leone 278
Simmons, James 216
Simon Fraser University 1, 287, 291–3, 294, 296–8
Simon, Lord 3
Singapore 273
Singapore, University of 278
Singh, Amrik 330, 332
Singh, Yogendra 326
Situationist International 344
Sloman, Albert 4, 6, 27, 66, 105–17, 178, 229, 232
Slough 182
Smart, Ninian 180
Smirnoff (vodka brand) 233
Smith, Brian 128
Smith, Ian 82, 280
Smith, John Maynard 127–8, 131
Smithson, Alison 25
Smithson, Peter 25
Snow, C. P. 8, 61, 89–90, 95, 106, 123, 125, 182, 362, *see also* 'two cultures'
social history 8, 157, 160–1, 164–8, 180
social organization (subject) 216
social policy 109, 148, 180
Social Science Research Council (USA) 57
social sciences 8, 48, 57–8, 60–1, 75, 80–1, 83, 93, 96, 106–7, 109, 126, 132, 143–5, 147–8, 151, 153, 179, 186, 196, 199, 200, 210, 212, 261, 265, 293, 299, 306, 325, 329, 337, 341, 343, 360–1, 363, 377
social studies 39, 42, 61, 93, 95, 101, 126–7, 215, 272, 180, 376
Society for Research into Higher Education 6
Sociological Review 38, 43–4

sociology xix, 4, 8, 39, 43–4, 48, 75, 81, 95, 107, 109, 117, 147–51, 163, 191, 292, 311, 343, 361, 363, 376–7
Solidarity (JNU) 331, 336
Sorbonne, see Paris, University of
South Africa 1, 4, 10, 37, 65, 76, 83–4, 116, 234, 273, 276, 280–1
South America 6
South Australia (state) 309
South Kent College 153
South Korea 2–3, 330
South Pacific, University of the 278–9
Southampton 77
Southampton, University of 21
Southeast Asian Command (SEAC) 273
Soviet studies 69 n. 69
Soviet Union 1, 41, 55, 57–8, 62, 66, 89, 106, 270, 277, 288, 357
Spain 1, 275, 294
Special Institutions of Scientific and Technological Education and Research (SISTERs) 124
Spence, Basil 7, 62, 78, 240
Spence, Robert 152
Spratt, Brian 151
Sproul, Gordon 258
Sputnik satellite 28, 57, 124, 135, 288
Sri Lanka Freedom Party 274
St Andrews, University of 193, 203 n. 32, 271
St Anne's College, Oxford, see Oxford, University of
St Augustine, University of the West Indies, see University of the West Indies
St Catherine's College, Oxford, see Oxford, University of
St John's College, Cambridge, see Cambridge, University of
St John's College, Oxford, see Oxford, University of
St Paul's Cathedral School 107
St. Etienne, University of 3
Staffordshire 31, 39–40
Staffordshire County Council 47
Stanford University 56–7, 135, 296
statistics xix 363
Steele, Tom 39–40

Steiner, George 152
Sternberg, Saul 40
Stewart, W. A. C. (Campbell) 39, 43, 45–8
Stillman, John 38
Stirling 191, 193
Stirling Town Council 198
Stirling, J. F. 195
Stirling, James 25
Stirling, Paul 144
Stirling, University of 1, 5, 10, 37, 191–201, 225, 228, 230–4, 239 n. 65, 248
Stirlingshire, East 193–5
Stoke-on-Trent 39
Stone, Lawrence 80
Stone, W. G. ('Bill') 225
Stormont 208, 214–15, 247
Strathclyde, University of 195
Streeten, Paul 64
Stresa 90
Student Federation of India 327
student protest 2, 4, 46–7, 56, 64–5, 83–4, 98, 111–17, 134–5, 149, 157, 197–8, 234, 274, 280–1, 295–6, 313, 328, 330, 335, 341, 345, 347, 350–1, 367
student residences 4, 7, 11, 26–7, 30–2, 37, 45–6, 48, 60, 125, 145–6, 158, 165, 176, 185, 197–200, 210–12, 226, 229, 231–2, 273, 278–80, 290, 296–7, 315, 343, 345, 358, 365–6, 377
Students for a Democratic Society (SDS) 111
students, African-American 265
students, Latino 265
students, native American 265
students, postgraduate 1, 9, 11, 80, 153, 176, 181, 183, 210, 212, 216
students, women 9, 46, 288, 299, 323, 327, 331, 335–7, 343, 347, 376–7
Stuttgart, University of 360
Sudan 273
Suez crisis 274
Suffolk 29, 95
Sukarno, President 3
Sun 37, 114
Sunday Telegraph 29
Supply, Ministry of (Britain) 41

Sussex, University of 1, 3–9, 15 n. 26, 22–3, 29, 31, 37–8, 42, 45–6, 55, 59–67, 74, 76, 78, 81–2, 90, 97, 121–2, 127–31, 135, 141, 143–4, 151, 159, 175, 180–1, 195, 197, 210, 225, 228, 232–4, 240, 271, 278, 281, 310, 325
Sutherland, Stuart 132
Suva 278
Swarthmore College 40, 254, 259
Swaziland (Eswatini) 278
Swinnerton-Dyer, Sir Peter 218
Sydney 306
Sydney, University of 42, 306, 308, 310, 312–13, 315
Syiah Kuala University 3
Sykes, Pete 47
Symons, Tom 290, 296, 298, 303 n. 70
Szanton, David 57

Talbot Tool Company 60
Tanzania 275, 277, 279
Tavistock Institute of Human Relations 257
Tawney, R. H. 39, 43
Taylor, A. J. P. 123
Taylor, Dr Harold 37, 45
Taylor, Miles 160
Taylor, Thomas 273
teacher training 23, 31, 38, 42–5, 48, 62, 183, 196, 210, 212–13, 217, 276, 279, 292, 312
technology 1, 8–10, 39, 60, 62, 106, 121–6, 132, 135–6, 193, 196, 210, 266, 306, 311, 325
Technology, Ministry of 200
Templeman, Geoffrey 142–5, 149, 231
Thanet, Isle of 141
Thatcher, Margaret 100, 116, 149, 151, 192, 200–1
theatre studies 180–1
theology 75, 143–4, 147–8, 180, 299, 360, 362
Thistlethwaite, Frank 8, 59–61, 66, 77, 90–101, 129, 233
Thom, René 134
Thom, Ron 294–5, 302 n. 50
Thomas, Brinley 63
Thomas, Nick 65

Thompson, Dorothy 169
Thompson, E. P. 134, 157–61, 163–8, 377
Thompson, Paul 109
Tilsley, Gordon 225
Timbuktu 275
Times Educational Supplement 149
Times Higher xx, 191
Times, The 55, 59–60, 62, 160, 280
Todd, Lord Alexander 130
Tonbridge 153
Töpper, Katte 362
Toronto 288–9, 293
Toronto, University of 288–90, 294, 297, 302 n. 50
Toulmin, Stephen 132
Touraine, Alain 343, 349
Tours, University of 3
Tout! 346
Townsend, Peter 44, 108–9
Toyne, Herbert 225
Trades Union Congress (TUC) 165
Trainer, James 199
Treasury (Britain) 3, 74, 99, 122–3, 192, 194, 213, 223, 225–7, 229, 234
Trent University 287, 290–3, 294–5, 299, 303 n. 70
Trento, University of 4
Trevelyan, G. M. 160
Triesman, David 111–14
Trinidad and Tobago 277
Trinity College Dublin 81, 106, 184, 208, 210
Tromso, University of 1
Trotskyites 344–5, 347, 350, 352
Trow, Martin 375
Truscot, Bruce (pseud.) 26–7, 30
Tsvetkova, Natalia 58
Turing, Alan 146
Turkey 4, 330
Twente, University of 5, 7
'two cultures' 8, 61, 106, 123, 182, 362, 376

Uckfield 60
Uganda 272, 277
Ullmann, Gerhard 362
Ulster Polytechnic 207, 217–18
Ulster Television 209
Ulster Unionist Party 208

Index

Ulster, New University of 1, 5, 10, 50 n. 31, 207–19, 225, 231, 279
Unicorn 280
Unilever 133
Union des Étudiants Communistes (UEC) 345, 347
Union Jeunesse Communistes, Marxistes-Léninistes (UJC-ML) 346
Union pour la Défense de la République (UDR) 341, 346
United Nations 76
United States Agency for International Development (USAID) 63, 279
United States of America 1, 3, 8–9, 11, 27, 37, 40, 43, 45–8, 56, 58–60, 63–6, 76, 83, 89, 91, 106–7, 111, 114, 135, 142, 149, 161–3, 165, 167, 180, 217, 229–31, 233–4, 270, 274–5, 277, 287, 289, 291–3, 295, 298, 307, 311, 341, 343–4, 349, 352, 357–8, 361, 365, 367–8, 376
Universities Grants Committee (UGC), Britain 3–4, 9–10, 22, 25, 28, 39–40, 42, 47–8, 59, 62, 73–6, 83, 99–100, 106–7, 110, 116, 122, 129–31, 142, 162, 168, 175–81, 192–5, 197–200, 209, 211–13, 217, 223–6, 228–9, 232–3, 235–6, 307
Universities Poetry 38
Universities Review 82
University Challenge 45
University College London (UCL), *see* University of London
University College of North Staffordshire, *see* Keele University
University Education Commission 1948 (India) 324
University Grants Commission (India) 325–6, 332
utopia, concept of 2–3, 39, 159, 168–9, 269

van der Rohe, Mies 292–3, 360
Vancouver 1, 294
Venables, Sir Peter 209
Vendler, Helen 152
Verwaltungshochschule 357
Veysey, Laurence 261–2
Vick, Arthur 41
Victoria (state of Australia) 306, 309, 314

Victoria, University of (Canada) 291
Vietnam War 10, 56, 64–5, 84, 111–12, 114, 116, 135, 295, 314, 316, 329–30, 344–5, 353
Vile, Maurice 144, 148
Villeneuve-la-Garenne 351
Vincennes, University of (Paris VIII) 4, 344
Vine, Fred 96
Virginia, University of 359
visual arts 180
Vive la Révolution (VLR) 345, 347–53
Vive le Communisme (VLC) 346–7, 349
VLC (newspaper) 346–7
von Humboldt, Wilhelm 360
Vrije University 1

Wachuku, Jaja 285 n. 53
Wade, Anthony 146
Waikato, University of 6, 15 n. 29
Walata (Oulata) 275
Wales 201
Wales, University College of (Cardiff) 43, 45
Wales, University College of (Swansea) 38
Wall, Pat 83
Ward, Basil 182
Warwick, University of 1, 4, 9, 21–2, 27–31, 46, 60, 64, 78, 83, 97, 121, 130, 134–5, 157–69, 176, 197, 225–8, 231–2, 234–6, 271, 377
Warwickshire 158, 162, 227
Warwickshire County Council 27, 162
Washington DC 274–5
Washington, University of (Seattle) 293
Waterfront Employers Association 254
Waterhouse, Alfred 30
Watford Grammar School 79
Watt, Ian 93–5
Webb, Beatrice 161
Webb, Sidney 161
Weber, Max 149
Wehler, Hans-Ulrich 364
Wenke, Hans 359
West German Rectors' conference 364
West Indies, University of 271, 273–4, 277

Barbados campus 278
Mona campus 81, 273, 278
St Augustine campus 277
West Kent College 153
Western Australia (state) 306, 310
Western Australia, University of 242
Western Cape, University of the 1
Western Sydney, University of 310
West-Taylor, Catherine 79
West-Taylor, John 74-5, 78-9, 225
Whitehouse, Alec 143
Whitlam, Gough 312, 314-15
Whyte, William 197
Wilcher, Lewis 273
Wilford, Hugh 66
Williams, Bruce Rodda 42, 44, 315
Williams, Eric 274-5, 277
Williams, Gwyn xx
Williams, Penry 82
Williams, Raymond 93
Williamson, Mark 81, 131
Wills family 230
Wilson, Angus 95
Wilson, Charles 8, 195
Wilson, Harold 10, 64, 125
Winchester College 76, 78, 82, 85
Winstanley, Gerrard 169
Wiseman, Jack 81
Wissenschaftsrat 358-9, 364
Witwatersrand, University of the 276, 280
Wivenhoe 107, 142
Wolfenden, Lord John 195
Wolff, Robert 2
Wollongong, University of 306, 310
Women students, *see* students, women
Women's Liberation Movement (MLF) 351
women's studies 303 n. 70
Woodford 306
Woolf, Virginia 146
Work in America 258

Workers Educational Association (Australia) 311
Workers Educational Association (Britain) 183
Wye College 150
Wyvern 110-11, 113

Yale University 233, 293
Yardley Cosmetics 233
Yechury, Sitaram 248
York (Britain) 73, 75, 90, 97, 125, 193, 223-4
York Academic Trust (later York Academic Promotion Committee) 74-5, 77, 225
York City Council 74
York Civic Trust 73-4, 227
York Conservation Trust 73
York Georgian Society 73
York University (Canada) 142, 287, 289-94, 296-9
York, County Borough of 228
York, University of xix-xx, 1, 3, 5-6, 8, 27, 29-30, 33 n. 34, 45, 73-84, 121, 125-7, 130-1, 142, 144, 175, 178, 180, 195, 197, 227, 231-4, 245, 278, 281
Yorke, Rosenberg and Mardall (architects) 28
Yorkshire 227
Youell, George 38
Young, Michael 44
Yugoslavia 1

Zambia 273, 276-7, 281
Zambia, University of 241, 278-9
Zeeman, Chris 134
Zinn, Howard 56
Zomba 277
zoology 81, 127, 129
Zuckerman, Sir Solly 6, 95-7, 123, 128-9, 133, 143

Lightning Source UK Ltd.
Milton Keynes UK
UKHW021551061220
374652UK00003B/165